p_h	hysteresis loss density	W/m^3
p_M	magnetic moment	$A \cdot m^2$
P_c	core loss	W
P_g	air gap power	W
P_w	winding loss	W
q, Q	electric charge	C
q	subscript for quadrature axis	
Q	reactive power	VA
r	radius	m
R	resistance	Ω
$\mathcal{R}$	reluctance	A/Wb
s	distance	m
s	Laplace operator	
s	slip	
S	complex or apparent power	VA
t	time	s
T	time period	s
T	temperature	°C or K
v, V	voltage	V
V	volume	m^3
w	energy density	J/m^3
w	width	m
W	energy	J
x	linear displacement	m
X	reactance	Ω
y	linear displacement	m
Y	admittance	℧
z	linear displacement	m
z	wavelength	m
Z	impedance	Ω
α	temperature coefficient of resistance	/°C
α	delay angle	deg, rad
β	rotor angle	deg, rad
γ	operator $= 1\angle 120° = \epsilon^{j2\pi/2}$	
γ	half angular width of magnet	rad
γ	transformation ratio	
γ	angular width of phase band	rad
δ	angle	deg, rad
δ	lamination thickness	m
∂	partial differential	
ϵ	natural logarithm base = 2.718	
ζ	damping ratio	/s
η	efficiency	
θ	stator angle	rad
λ, Λ	flux linkage	Wb
μ_0	permeability of free space $= 4\pi(10^7)$	
μ_r	relative permeability	
ν	linear velocity	m/s
π	3.1416	
ρ	resistivity	$\Omega \cdot m$
τ	time constant	s
ϕ	phase angle of current	deg, rad
Φ	magnetic flux	Wb
ω	angular frequency	rad/s
ω_s	stator frequency	rad/s
ω_r	rotor frequency	rad/s
ω_{mech}	mechanical angular velocity	rad/s or r/min

Electric Machines and Drives

Electric Machines and Drives

Gordon R. Slemon

University of Toronto

ADDISON-WESLEY PUBLISHING COMPANY, INC.

Reading, Massachusetts • Menlo Park, California • New York
Don Mills, Ontario • Wokingham, England • Amsterdam • Bonn
Sydney • Singapore • Tokyo • Madrid • San Juan • Milan • Paris

This book is in the
ADDISON-WESLEY SERIES IN ELECTRICAL ENGINEERING

Library of Congress Cataloging-in-Publication Data

Slemon, Gordon R.
Electric machines and drives / Gordon R. Slemon.
p. cm.
Includes index.
ISBN 0-201-57885-9
1. Electric driving. 2. Electric machines. I. Title.
TK4058.S539 1992
621.31'042—dc20 91-12449
CIP

4 5 6 7 8 9 10-MA-9594

Preface

In writing this book, I have had two basic objectives in mind: to provide the sort of introduction to electric machines that I feel every student of electrical engineering should have, and to provide a more in-depth treatment of electric machines and drives for those who may wish to know more about the subject.

Chapters 1 through 6 address the first objective by providing a selection of topics considered suitable for a one-term course. During this introductory course the student should encounter a number of basic concepts:

- How ferromagnetic materials behave and why they are basic components of most useful electric machines.
- How a transformer works, first ideally, and then taking into account some of its imperfections.
- What are the basic useful mechanisms for producing force and torque from electric energy sources and how these mechanisms can be exploited to produce useful types of electric machines such as commutator, induction, and synchronous.
- How all types of electric machines are inherently capable of converting energy from a mechanical source to electric form as generators and also of converting electric energy to mechanical form in a motor or drive.

- How an adequate analytical model, usually in the form of an equivalent electric circuit, can be developed for a device.
- That engineers, recognizing that all models are approximate representations of reality, choose a model which is no more complex than necessary for the application at hand.

I have assumed that students taking an introductory electric machines course will have some prior knowledge of electric and magnetic field concepts, but without much depth in ferromagnetics, and also some knowledge of elementary electric circuit analysis.

It is my conviction that the essence of engineering is design. Accordingly, the end objective of each phase of preparatory study should be to increase the student's capability to design useful devices and systems to meet the needs of society. The educational route to this objective follows the sequence of aspects: an understanding of the physical processes, the derivation of approximate models, the use of analytical techniques and, finally, design.

Throughout the first six chapters I have attempted to develop a sound physical understanding of the energy conversion processes utilized in magnetic devices, transformers, and machines. To emphasize the importance of this approach, consider the force tending to close the air gap in a ferromagnetic core. This force may be calculated by use of the principles of energy conservation and virtual displacement. While this method is both analytically simple and powerful, it provides little appreciation of the origin of the force or its area of action. The insight arising from physical visualization is an important ingredient in arriving at an appropriate model of the device. Use of a purely mathematical approach without adequate attention to the physical model can frequently lead to serious error.

Modeling is an art which will develop as knowledge and experience grow. Considerable emphasis has been given in the book to the freedom to choose a model which is just adequate to meet the present need for performance prediction. This choice cannot be made in the abstract, but requires an assessment of the actual numerical parameter values. Insofar as possible, without incurring undue complexity, the governing parameters of each device have been related directly to its dimensions and to the properties of its materials to assist in developing the engineering judgment basic to modeling.

Chapter 1 begins with a review of magnetic field concepts. It then introduces the basic modeling ideas of equivalent magnetic and electric circuits. An understanding of ferromagnetic materials is based on the visualization of magnetic domains and how the orientation boundaries of the domain can be moved to produce intense fields around closed paths including those with air gaps. A treatment of permanent-magnet materials is integrated into this introduction in view of the increasing importance of permanent-magnet devices and machines.

Chapter 2 deals with the understanding and basic modeling of a two-winding transformer, followed by a few of its most important operating properties. The important parameter of leakage inductance is treated from a physical rather than a coupled-circuit point of view.

Basic principles of electromechanical energy conversion machines are introduced in Chapter 3. Again, an attempt has been made to develop a physical understanding of the various machine types before launching into the more detailed discussion of these machines in the following chapters.

Commutator or direct-current machines have been considered first in Chapter 4, not because of their importance relative to the induction and synchronous machines of Chapters 5 and 6, but rather because they are so widely used in electric-machine laboratories in experiments on all types of rotating machines. Where this laboratory use is not of significance, the study of the dominant categories of induction and synchronous machines can be undertaken immediately after Chapter 3.

Since a majority of all electric machines are of the induction type, particular emphasis has been given to the concept of the rotating magnetic field and its interaction with currents induced in a squirrel-cage rotor. The elements of space-vector notation are introduced as a compact and convenient means of representing such rotating fields. The features of a double-cage or deep-bar rotor are included in the introductory phase of study because they are characteristic of essentially all induction motors encountered in practice.

The treatment of synchronous machines in Chapter 6 emphasizes the basic nature of such machines as a source of alternating current, in contrast with the more conventional modeling as a voltage source. This approach builds naturally on the previous modeling of induction machines. A discussion of permanent-magnet machines and their use in electronically switched drives is included.

Chapters 7 through 11 may form the basis for several selections of material for a second course for those students who choose to pursue further the fascinating and important specialty of electric machines and drive systems. Alternatively, these chapters may be useful for later reference when the practicing engineer discovers a need to know somewhat more than was covered in the brief introductory course.

Magnetic systems and transformers are revisited in Chapter 7, which presents more concepts on the modeling of ferromagnetic devices, some design concepts to aid users in appreciating operating limitations, and a number of magnetically based devices and transformers frequently encountered in practice. Instructors may wish to incorporate selected parts of this chapter into the introductory course syllabus if time permits.

A rapidly increasing number of electric drives require sources of variable voltage, current, and/or frequency. Most of such sources are produced through the use of semiconductor switches. Chapter 8 presents a brief introduction to the idealized structure and behavior of the basic power-semiconductor systems in common use as converters.

Chapter 9 presents a treatment of electric drives using commutator machines. While the trend in variable-speed drives is now away from these d-c motors, they are still practically significant. Also, they provide a simple and useful appreciation of the transient behavior of electromechanical systems for speed and position control.

Induction machines are revisited in Chapter 10 but this time with a view to developing an understanding of transient as well as steady-state performance. While induction machines will continue to dominate constant-speed drive applications, they will also be used extensively in variable-speed drive systems incorporating variable-frequency electric supply from power-semiconductor converters. In such applications, the transient behavior is of increased relevance. The transient analysis is based on space-vector concepts, which are particularly convenient for induction machines because of their cylindrical symmetry.

The discussion of synchronous machines in Chapter 11 includes both the important class of synchronous generators used in electric supply systems, and synchronous motors which are increasingly being used in controlled-speed electric drives. Again, analytical emphasis is on transient modeling and performance. Various types of permanent-magnet machines are given special attention because of their major and increasing application in drives.

Selected material from Chapters 7 through 11 has been used as a basis for a senior-undergraduate elective course on controlled electric drives at the University of Toronto. Also, a graduate course on modeling of electric machines has been based on a selection from the same chapters.

Most of the models developed in this book are in the form of equivalent circuits. A major reason for this choice is that nonlinear parameters can be represented directly in circuit form in relation to the controlling variables. Very little space has been given to methods of analyzing these models since it can be assumed that parallel courses in electric circuits, differential equations, computer programming, and systems analysis will have provided an adequate range of analytical techniques. Where appropriate, some of these techniques have been used to derive typical operating characteristics.

Most of the models derived in this book are appropriate for the solution of both dynamic and steady-state performance. They can therefore be integrated into system representations. Transient solutions for some simple situations have been included in the book. For more complex situations, the analytical and simulation methods developed in companion courses on control systems may be employed.

A substantial number of problems have been included at the end of each chapter. In most instances, the pertinent section of the book is indicated at the end of the problem statement. These problems draw on all the concepts of understanding, modeling, analysis, and design. Answers for most of the problems have been included in a separate section to reassure the reader of progress. To the instructor who wishes to use any of these problems as test or quiz assignments, I would suggest that a new set of parameter values be used.

A significant proportion of the problems relate to design. For these, some engineering judgment may have to be exercised in arriving at an appropriate approximation and in choosing materials, configurations, and dimensions. Where answers are given to such problems, they should be regarded as typical rather than definitive. Many engineering applications have been introduced in the problem sections only, partly to demonstrate that, with a good grounding in basic concepts, a very wide range of engineering systems can be understood, analyzed, and devised.

Lists of the principal symbols used in the book have been printed on the endpapers. Each vector and phasor quantity has been identified through the use of a normal letter with an arrow above the symbol rather than the common use of boldface type. I consider the latter practice unfortunate since instructors and students cannot write in boldface. In expressing data, the International System (SI) units and notation have been employed. Conversion factors to other unit systems are given in an appendix.

Substantial sections of the book have been derived in revised form from the book *Electric Machines*, by Alan Straughen and myself, published in 1980 by Addison-Wesley Publishing Company. As I acknowledge this, I would add that it has been a pleasure for me to work with this excellent publisher again. Also, a number of sections and problems have been incorporated into the text in revised form from one of my earlier books, *Magnetoelectric Devices—Transducers, Transformers, and Machines*, published in 1966 and now out of print. I am grateful to the publishers of that book, John Wiley and Sons, for permitting me to use this material in the present work.

I am grateful to my colleagues and graduate students in the Power Devices and Systems Group at the University of Toronto for their many helpful suggestions and criticisms. Finally, I wish to acknowledge my deep gratitude to my wife, Margaret, not only for typing the manuscript, but mainly for her continued support during the gestation period of this book.

Toronto G.R.S.

Contents

CHAPTER 7 241

Magnetic Systems and Transformers — Revisited

CHAPTER 8 307

Power Semiconductor Converters

CHAPTER 11 463

Synchronous Generators, Motors, and Drives

Appendixes

Answers to Problems 545

Index 549

CHAPTER 1

Magnetic Systems

This book is concerned with a wide range of electromagnetic devices that are used for conversion of electric energy to or from mechanical energy or, in some instances, to transform electric energy to a more readily usable form. The machine systems considered include (1) generators for the production of electric power from hydraulic, steam, or gas turbines, (2) transformers to transform electric energy from its most convenient generation voltage to a voltage level suitable for transmission and then to transform it to voltages appropriate for distribution and use, (3) motors to convert electric energy to mechanical form, (4) electronic converters that change the available electric power to other desired forms with controllable voltage, current, or frequency, and (5) drive systems consisting of motors and converters capable of controlling the speed or position of mechanical loads.

Most electric machines convert energy by use of a magnetic field as an intermediary. Therefore, it is appropriate to begin with a discussion of the concepts, materials, and structures that are involved in producing magnetic fields in the required form, location, and intensity. In particular, the means of directing and controlling magnetic fields, in much the same way as electric currents are directed by electrical conductors, will be examined.

1.1

Magnetic Field Concepts

Magnetic fields are produced by (1) electric currents, (2) permanent magnets, or (3) electric fields that are changing with time. The first two are the subject of this chapter; the last can be ignored in the context of this book because it is significant only for high-frequency systems that radiate electromagnetic energy.

Let us start with Ampere's Circuital Law, which relates the intensity of a magnetic field to the current that produces it. The law may be expressed in the form

$$\oint \vec{H} \cdot d\vec{\ell} = \int_A \vec{J} \cdot d\vec{A} \quad \text{A} \tag{1.1}$$

It is important that expressions such as this be read with a view to visualizing the physical field. Therefore, let us first consider a system that produces a relatively uniform magnetic field within a confined space.

Figure 1.1 shows cross sections of a coil of N turns of conductor uniformly wound around a torus made of any nonferromagnetic material. When a current i is passed through the coil, a magnetic field with an intensity denoted by the vector quantity $\vec{H}$ is produced within the torus. Consider the closed path shown within the torus at radius r. Because of the circular symmetry, the magnitude of the field intensity must be the same at all points along the path. According to Eq. (1.1), the product of the magnetic field intensity directed along this path and the length of the closed path is equal to the sum of all the electric current passing through the area enclosed by the path. Each of the N turns

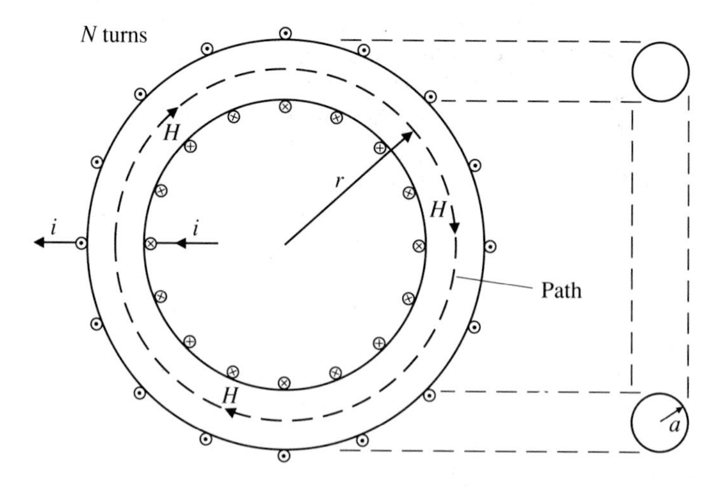

Figure 1.1 Coil wound on toroidal core showing magnetic field intensity $\vec{H}$.

of the coil penetrates this enclosed surface. Thus, the right-hand side of Eq. (1.1) is equal to Ni or, more generally, to the product of the conductor current density J and the total cross-sectional area of the N conductors. Thus, at radius r,

$$H_r(2\pi r) = Ni \qquad \text{A} \tag{1.2}$$

or

$$H_r = \frac{Ni}{2\pi r} \qquad \text{A/m} \tag{1.3}$$

Application of the circuital law for a circular path at any radius that is not within the torus shows no net enclosed current. Therefore, the magnetic field is effectively confined to the volume within the torus.

Examination of Eq. (1.3) shows that the field intensity will be greatest along a path at the inside edge of the torus and least at the outer side of the torus. If the cross-sectional radius a of the torus is small relative to the radius of the torus, the value of H at the average value of path radius r may be used as a good approximation to the field over the whole cross section.

For our purposes, the most important property of a magnetic field is described by its magnetic flux density, denoted by a vector quantity $\vec{B}$. For free space, and practically for all nonferromagnetic materials, this quantity is related to the magnetic field intensity vector by a constant μ_0 having a value of

$$\mu_0 = 4\pi \times 10^{-7} \qquad \text{T·m/A} \tag{1.4}$$

so that

$$\vec{B} = \mu_0 \vec{H} \qquad \text{T} \tag{1.5}$$

the unit of flux density being the tesla (T).

Just as current density $\vec{J}$ integrated over the cross-sectional area of a conductor gives the current i, the integration of the flux density over the cross-sectional area of the magnetic field gives the magnetic flux Φ measured in webers (Wb). Thus, for the torus of Fig. 1.1,

$$\Phi = \bar{B}A \qquad \text{Wb} \tag{1.6}$$

where, for the torus of Fig. 1.1,

$$A = \pi a^2 \qquad \text{m}^2 \tag{1.7}$$

and $\bar{B}$ is the average magnitude of the flux density directed perpendicular to the area A. Magnetic flux is a scalar quantity and is continuous around a closed path, similar to the way electric current is continuous in a conducting path.

The next concept that is required is that an electric current that is changing with time sets up in the space both within and around the conductor an electric field of intensity $\vec{\mathcal{E}}$ which opposes the change in current. Figure 1.2 shows this electric field around a cross section of the toroidal coil of Fig. 1.1. Usually we are interested in the voltage that is induced either in the turns of the toroi-

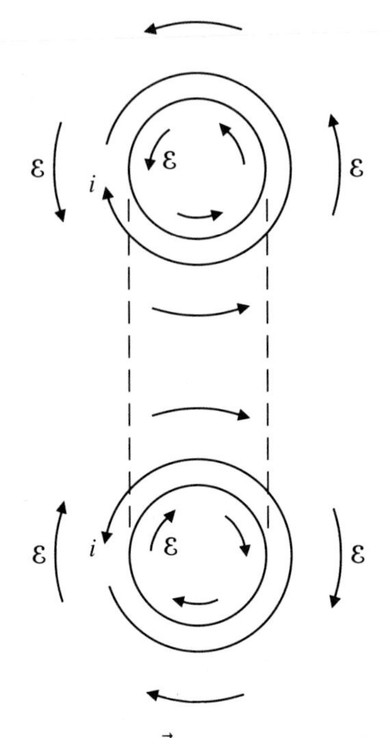

Figure 1.2 Electric field intensity $\vec{\mathcal{E}}$ caused by increasing current i in toroidal coil.

dal coil or in any other coil that encircles the toroidal core. In this text, the induced voltage is given the symbol e and is the negative of the electromotive force (emf) of Faraday's Law. The induced voltage may be visualized as the voltage that would occur across the ends of a slightly opened loop. By adaptation of Faraday's Law, the induced voltage around an essentially closed path such as a turn of the toroidal coil is related to the flux Φ enclosed by the expression

$$e_{path} = \oint \vec{\mathcal{E}} \cdot \vec{d\ell} = \frac{d\Phi}{dt} \qquad \text{V} \tag{1.8}$$

If the path is taken along the total length of the conductor in the toroidal coil, each turn will be a nearly closed path enclosing flux Φ. For the whole coil, the induced voltage will be

$$e = \oint_{coil} \vec{\mathcal{E}} \cdot \vec{d\ell} = \frac{d(N\Phi)}{dt} \qquad \text{V} \tag{1.9}$$

The quantity $N\Phi$ is the total flux linked by the coil as it wraps N times around the flux Φ and is denoted as the flux linkage λ measured in webers. For the toroidal coil,

$$\lambda = N\Phi \qquad \text{Wb} \tag{1.10}$$

The voltage induced in a coil may now be expressed as

$$e = \frac{d\lambda}{dt} \qquad \text{V} \tag{1.11}$$

The direction or polarity of the induced voltage is always such as to oppose a change in the flux linkage.

1.2

Equivalent Magnetic and Electric Circuits

In situations where the magnetic field is confined to a definite volume of space, such as in the torus of Fig. 1.1, it is convenient to introduce the concept of a magnetic circuit, just as an electric circuit is used to denote constrained electrical conducting paths. This involves introduction of the concept of a magnetomotive force (mmf) $\mathfrak{F}$. For a closed path, it is a scalar quantity equal to both sides of the circuital law of Eq. (1.1). Thus,

$$\mathfrak{F} = \int_A \vec{J} \cdot d\vec{A} = \oint \vec{H} \cdot d\vec{\ell} \tag{1.12}$$

For the torus, the mmf is

$$\mathfrak{F} = Ni \qquad \text{A} \tag{1.13}$$

While the unit of $\mathfrak{F}$ is often called an "ampere turn," the proper unit is the ampere because the number of turns is dimensionless.

The mmf may be related to the magnetic flux in essentially the same way that potential difference or voltage is related to current in an electrical conductor. For the torus, with a path length

$$\ell = 2\pi r \qquad \text{m} \tag{1.14}$$

and a cross-sectional area A, the relation is

$$\mathfrak{F} = H\ell = \frac{B}{\mu_0}\,\ell = \Phi\,\frac{\ell}{\mu_0 A} \qquad \text{A} \tag{1.15}$$

The constant of proportionality between $\mathfrak{F}$ and Φ is the reluctance $\mathcal{R}$ of the magnetic path and is defined by the relation

$$\mathcal{R} = \frac{\mathfrak{F}}{\Phi} = \frac{\ell}{\mu_0 A} \qquad \text{A/Wb} \tag{1.16}$$

The relation between the variables may now be modeled by the equivalent magnetic circuit of Fig. 1.3. Reluctance $\mathcal{R}$ in a magnetic circuit is analogous to re-

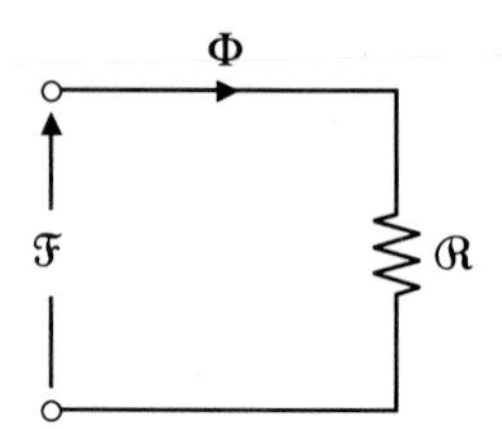

Figure 1.3 Equivalent magnetic circuit.

sistance R in an electric circuit, both being proportional to the path length and inversely proportional to the cross-sectional area.

The flux linkage λ of the toroidal coil of Fig. 1.1 can be related to the coil current i by combining Eqs. (1.3, 1.5, 1.6), and (1.10):

$$\lambda = N\Phi = NBA = N\mu_0 HA = N^2 \frac{\mu_0 A}{\ell} i \qquad \text{Wb} \tag{1.17}$$

The constant of proportionality between flux linkage and current is denoted as the inductance L of the coil, expressed in henries (H):

$$L = \frac{\lambda}{i} \qquad \text{H} \tag{1.18}$$

The induced voltage in the coil can now be related to the time rate of change of current. From Eqs. (1.11) and (1.18),

$$e = \frac{d\lambda}{dt} = L\frac{di}{dt} \qquad \text{V} \tag{1.19}$$

If a voltage v is applied to the terminals of a coil of resistance R and inductance L, the current i is governed by the expression

$$v = Ri + e = Ri + L\frac{di}{dt} \qquad \text{V} \tag{1.20}$$

Therefore, the coil may be modeled by the electric equivalent circuit of Fig. 1.4.

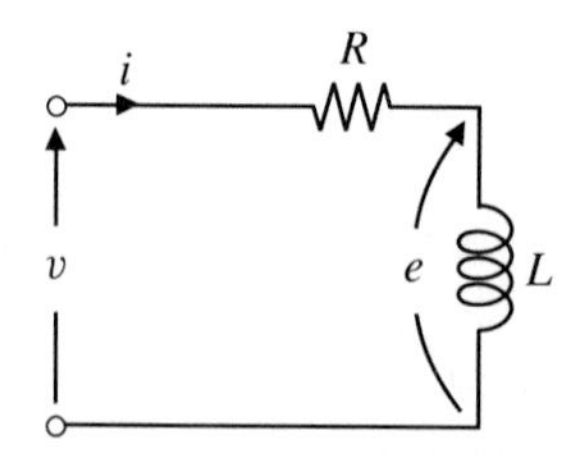

Figure 1.4 Equivalent electric circuit.

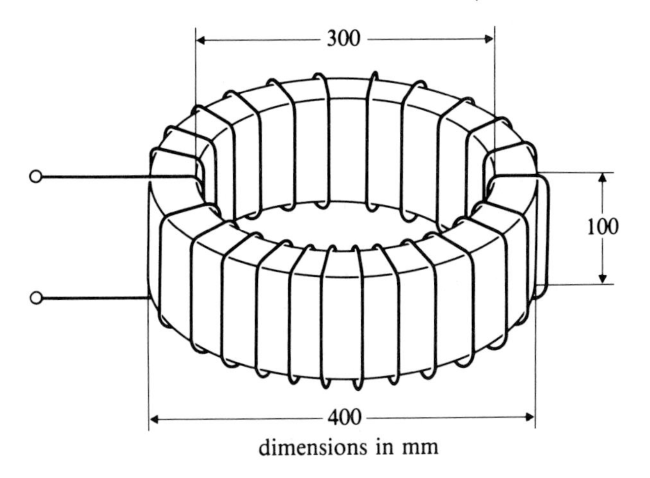

Figure 1.5 Diagram for Example 1.1.

EXAMPLE 1.1 Figure 1.5 shows a toroidal coil wound on a plastic ring of rectangular cross section. The coil has 200 turns of round copper wire 3 mm in diameter.

(a) For a coil current of 50 A, find the magnetic flux density at the mean diameter of the ring.

(b) Find the inductance of the coil, assuming the field within the coil is uniform and equal to that at the mean diameter.

(c) Determine the percentage error in assuming uniform flux density in the ring.

(d) Given the resistivity ρ of copper as $1.72 \times 10^{-8}\ \Omega \cdot \text{m}$, determine the parameters of the electric equivalent circuit of Fig. 1.4.

Solution

(a) At the mean diameter

$$H = \frac{Ni}{2\pi r} = \frac{200 \times 50}{0.35\pi} = 9095 \qquad \text{A/m}$$

$$B = \mu_0 H = 4\pi \times 10^{-7} \times 9095 = 0.0114 \qquad \text{T}$$

(b) Assuming $B = 0.0114$ T,

$$\Phi = BA = 0.0114 \times 0.1 \times 0.05 = 57.15 \qquad \mu\text{Wb}$$

$$\lambda = N\Phi = 200 \times 57.15 \times 10^{-6} = 11.43 \qquad \text{mWb}$$

$$L = \frac{\lambda}{i} = \frac{11.43 \times 10^{-3}}{50} = 0.229 \times 10^{-3} \qquad \text{H}$$

Alternatively,

$$\mathcal{R} = \frac{\ell}{\mu_0 A} = \frac{0.35\pi}{4\pi \times 10^{-7} \times 0.1 \times 0.05} = 175 \times 10^6 \quad \text{A/Wb}$$

$$L = \frac{N^2}{\mathcal{R}} = \frac{200^2}{175 \times 10^6} = 0.2286 \quad \text{mH}$$

(c) For $0.15 < r < 0.2$ m,

$$B = \frac{\mu_0 Ni}{2\pi r} = \frac{4\pi \times 10^{-7} \times 200 \times i}{2\pi r} = 4 \times 10^{-5} \frac{i}{r} \quad \text{A/Wb}$$

$$\Phi = \int_{0.15}^{0.2} 0.1 B \, dr \quad \text{Wb}$$

$$\lambda = N\Phi = 200 \times 0.1 \times 4 \times 10^{-5} i \int_{0.15}^{0.2} \frac{1}{r} \, dr \quad \text{Wb}$$

$$L = \frac{\lambda}{i} = 8 \times 10^{-4} i \ln\left(\frac{0.2}{0.15}\right) = 0.2301 \quad \text{mH}$$

$$Error = \frac{0.2301 - 0.2286}{0.2301} \times 100 = 0.65\%$$

(d)

$$R = \rho \frac{\ell}{A} = \frac{1.72 \times 10^{-8} \times 200 \times 0.3}{\pi \times 0.0015^2} = 0.146 \quad \Omega$$

$$L = 0.229 \quad \text{H}$$

1.3

Energy in a Magnetic Field

The process of setting up a magnetic field in a coil involves supplying energy to the coil. Suppose a voltage v is applied to the N-turn toroidal coil of Fig. 1.1. Let us assume that the field inside the torus is reasonably uniform in magnitude. From Eq. (1.20), the power entering the coil at any instant in time is

$$p = vi = Ri^2 + i \frac{d\lambda}{dt} \quad \text{W} \tag{1.21}$$

The first term in Eq. (1.21) is the power being dissipated as heat in the coil conductor, and the remaining part is the power p_B flowing into energy storage in the magnetic field. Thus,

$$p_B = i\frac{d\lambda}{dt} = NAi\frac{dB}{dt} = A\ell H\frac{dB}{dt} \qquad \text{W} \tag{1.22}$$

When the flux density B is zero, the energy W stored in the magnetic field will also be zero. Thus, the stored energy may be expressed as

$$W = \int p_B\,dt = \int_0^B A\ell H\,dB = \int_0^B \frac{A\ell}{\mu_0} B\,dB = (A\ell)\frac{B^2}{2\mu_0} \qquad \text{J} \tag{1.23}$$

Because the product $A\ell$ is the volume of the space enclosed by the coil, the density of energy storage in the magnetic field is

$$w = \frac{B^2}{2\mu_0} \qquad \text{J/m}^3 \tag{1.24}$$

The stored energy can also be expressed in terms of the overall parameters and variables of the coil. From Eqs. (1.18) and (1.22),

$$p_B = i\frac{d\lambda}{dt} = Li\frac{di}{dt} \qquad \text{W} \tag{1.25}$$

The energy stored is then given by

$$W = \int p_B\,dt = \int_0^i Li\,di = \frac{Li^2}{2} \tag{1.26}$$

Alternate forms of this energy expression are

$$W = \frac{\lambda i}{2} = \frac{\lambda^2}{2L} \qquad \text{J} \tag{1.27}$$

The stored energy may also be expressed in terms of the magnetic circuit variables as

$$W = \frac{\mathcal{R}\Phi^2}{2} = \frac{\mathcal{F}^2}{2\mathcal{R}} = \frac{\mathcal{F}\Phi}{2} \qquad \text{J} \tag{1.28}$$

1.4

Ferromagnetism

The relation of Eq. (1.5) that the magnetic flux density B is equal to the magnetic field intensity H multiplied by the constant μ_0 is strictly true only for a vacuum. However, it is true to within less than 0.001% for most materials. Of the basic elements, only a few such as iron, nickel, and cobalt display a markedly different relationship at normal operating temperature. Using core materials containing these elements and their alloys, the flux or flux density produced by a given field intensity or coil current can be enormously increased. This phenomenon is known as *ferromagnetism*.

A basic understanding of ferromagnetism can be developed through use of a relatively simple model of the atom as a positively charged nucleus surrounded by an ordered cloud of negatively charged electrons. Each electron may be considered to be in an orbit around the nucleus, as shown in Fig. 1.6(a). The electric charge on each electron is -1.6×10^{-19} C. A negative charge revolving clockwise around a circular path can be considered as equivalent to an electric current i directed counterclockwise around the path. Such a loop of current will produce a magnetic field proportional to the current i and to the area A enclosed by the loop. This source of magnetic field is denoted as the magnetic moment

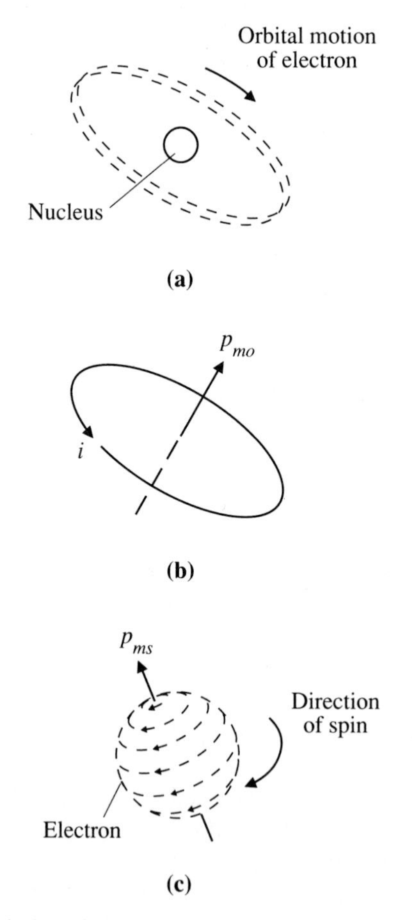

Figure 1.6 (a) Orbital motion of electron, (b) equivalent current loop and magnetic moment, and (c) electron spin and its magnetic moment.

$\vec{p}_{mo}$, a vector having a magnitude iA and a direction normal to the plane of the loop in the positive direction of the field intensity that it would produce, as shown in Fig. 1.6(b).

In addition to its orbital field, each electron produces a magnetic field that can be visualized by considering the electron as a charge cloud spinning on its own axis, as shown in Fig. 1.6(c). The result is a spin magnetic moment, $\vec{p}_{ms}$. Magnetic moments of electrons occur only as multiples of a basic quantity having a magnitude of 9.27×10^{-24} A·m^2. This quantum property is analogous to the discrete charge on each electron of -1.6×10^{-19} C. The spin magnetic moment $\vec{p}_{ms}$ has one unit of this basic value, and the orbital magnetic moment $\vec{p}_{mo}$ is either zero or an integral multiple of this value.

With each electron acting as a source of magnetic field, one might expect all atoms and materials to produce a continuous magnetic field. However, in the atoms of many elements the electrons are arranged either symmetrically or at random about the nucleus so that the magnetic moments due to their spin and orbital motions cancel out, leaving the atom with no net magnetic moment. Nevertheless, the atoms of more than one-third of the known elements lack this symmetry, so that they do possess a net moment.

Even when a material consists of atoms that each have a net magnetic moment, the arrangement of atoms in the material frequently is such that the magnetic moment of one atom is canceled out by that of an oppositely directed near neighbor. Of the basic elements, only iron, nickel, and cobalt have the magnetic moments of adjacent atoms aligned when at room temperature. Two other elements in the rare-earth series, dysprosium and gadolinium, have the same property, but only at lower temperature.

Most of the useful magnetic materials are based on iron, which is both plentiful and well aligned magnetically. It is usually alloyed with other elements, depending on the desired properties. These other elements are not restricted to those three or five that are naturally aligned but may include others whose atoms have a net magnetic moment. This includes such elements as vanadium, chromium, and manganese and many of the rare-earth elements.

Most of the interesting magnetic materials are crystalline, and the direction of alignment of the magnetic moments is normally along one of the crystal axes. One might then expect each crystal to act as a small permanent magnet; this is known to be the case with very small iron filings, which tend to stick to any iron object. However, it was noted in Section 1.3 that considerable energy is required to produce a magnetic field in free space (Eq. [1.24]). If there is an alignment of magnetic moments within the crystal that requires lower energy than the complete alignment along one crystal axis with a field external to the crystal, it will be preferred. Figure 1.7 shows a typical alignment in an iron crystal which has three mutually perpendicular easy axes of magnetic alignment. The magnetic moments have arranged themselves into four magnetic domains, each directed along a crystal axis of easy alignment. The magnetic flux produced by the aligned moments finds a closed path within the crystal, and only a negligible amount can be observed near the domain boundaries outside the crystal.

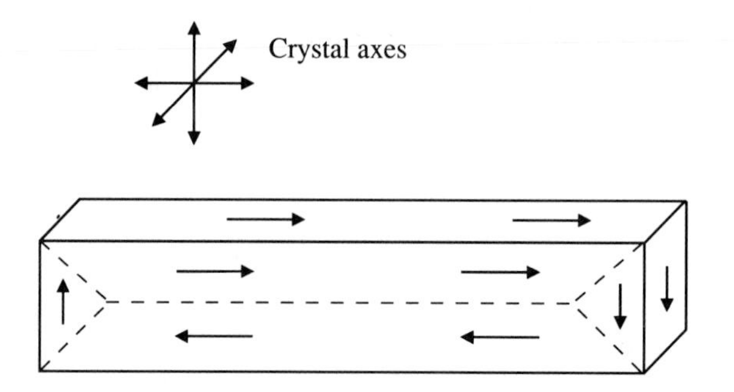

Figure 1.7 Orientation of magnetic domains in a crystal of iron.

In its natural state, a block of iron consists of many crystals that normally are randomly oriented. Each crystal may contain many of these magnetic domains, each oriented along one of the three mutually perpendicular crystal axes. Each domain can be considered as a small permanent magnet. The domains are normally shaped and oriented so that flux closure paths are set up within the crystal, similar to that shown in Fig. 1.7. In most ferromagnetic materials, the domains are relatively small, having typical dimensions of the order of 1 to 100 μm.

1.5 Magnetization

Let us now examine the process whereby the orientation of domains in a ferromagnetic material may be changed by application of an externally produced magnetic field. Suppose a magnetic moment $\vec{p}_m$ is placed in a magnetic field of flux density $\vec{B}$, as shown in Fig. 1.8. The moment will experience a torque

$$\vec{T} = \vec{p}_m \times \vec{B} \qquad \text{N}\cdot\text{m} \tag{1.29}$$

tending to rotate the moment toward alignment with the field.

Next, consider the cross section of a crystal shown in Fig. 1.9(a) consisting of a number of parallel oriented domains, each of such a small width that there is no significant magnetic flux out of the crystal ends. In the microscopic cross section y between two oppositely oriented domains, the orientation of atomic magnetic moments does not change abruptly from one direction to the other but does so gradually over a domain wall, typically with a width of a few hundred atomic diameters or over a distance of 10^{-8} to 10^{-7} m, as shown in highly expanded form in Fig. 1.9(b). In this figure, the transition in alignment can be pic-

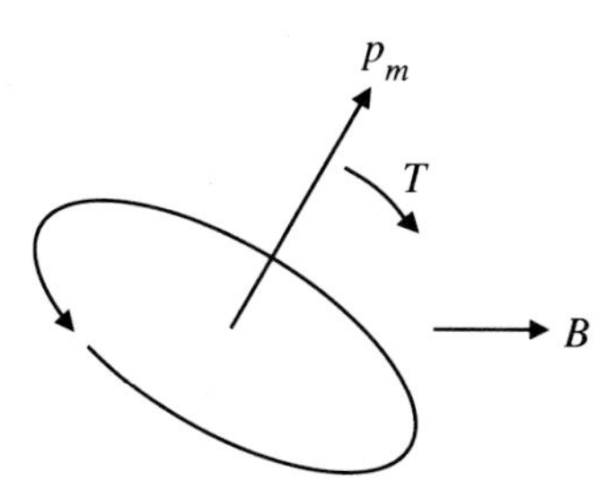

Figure 1.8 Torque on a magnetic moment in a magnetic field.

tured as similar to a twist in a narrow paper strip, with the direction of the moment being always across the width of the strip.

Suppose an external magnetic field $\vec{H}$ is now applied to this crystal in a direction parallel to the orientation of the domains, as shown in Fig. 1.9(b). Magnetic moments already aligned with the field $\vec{H}$ will experience no torque, as will those that are oppositely aligned. It is only those moments in the transitional domain wall that experience a torque, tending to rotate them into line with the applied field $\vec{H}$. In Fig. 1.9(b), the effect will be to move the wall downward, increasing the width of the upper domain and reducing the width of the lower domain. The overall effect in the crystal is to change the original domain pattern of Fig. 1.9(a) toward that of Fig. 1.9(c), with larger domains directed to the right and smaller domains to the left. There will now be a net magnetic flux out of the right-hand face of the crystal. This flux may now pass into an adjacent crystal that is similarly oriented and so continue around a closed path in the magnetic core.

Suppose we could arrange a toroidal core of iron with all of its crystal axes aligned along the circumferential axis of the torus and with all of its magnetic moments similarly aligned along the same axis. A cross section of the core then might be visualized as minute current loops, as shown in Fig. 1.10(a). Inside the material, each loop current is adjacent to an oppositely directed loop current, and the effects thus can be considered to cancel. It is only at the outer surface that there is no cancelation. Therefore, the effect is equivalent to a fictitious surface current, as shown in Fig. 1.10(b).

In iron, each atom has a magnetic moment of 2.2 times the basic quantum of 9.27×10^{-24} A·m². The spacing between iron atoms is about $d = 2.27 \times 10^{-10}$ m. Thus, the area occupied by a single atom in the core cross section will be d^2. Each atomic magnetic moment may be visualized as a current i_M flowing in a loop of area d^2. Thus, for one atom

$$p_M = i_M d^2 = 2.2 \times 9.27 \times 10^{-24} \quad \text{A} \cdot \text{m}^2 \tag{1.30}$$

and

$$i_M = 0.394 \times 10^{-3} \quad \text{A} \tag{1.31}$$

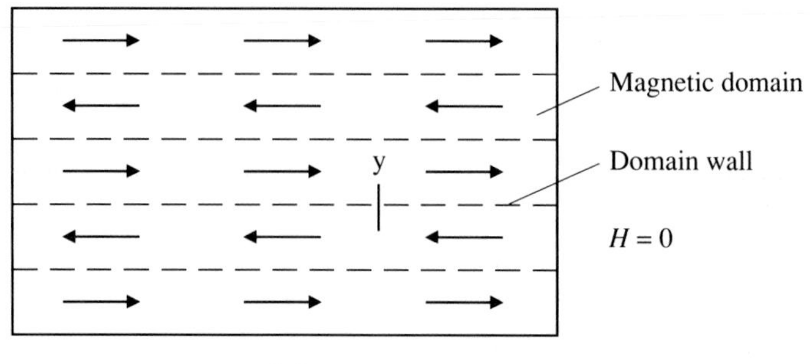

(a)

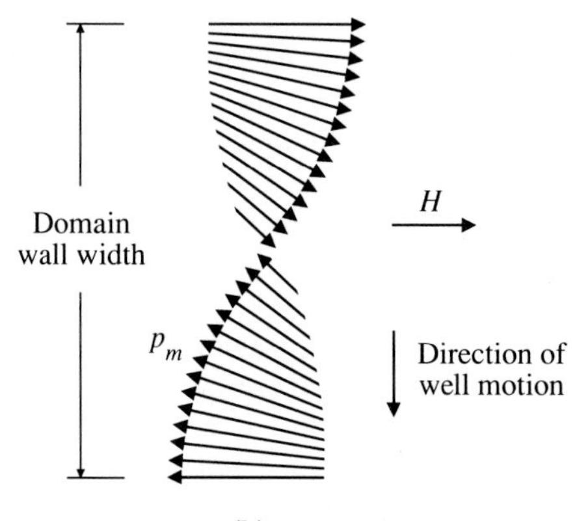

(b)

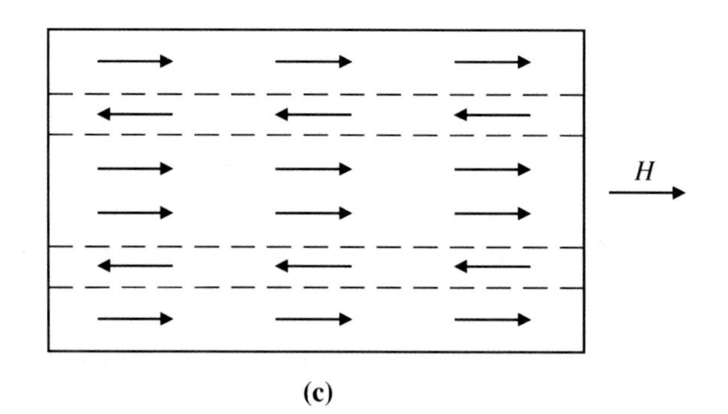

(c)

Figure 1.9 (a) Cross section of crystal with no net magnetic flux, (b) domain wall at section "a" in an applied field $\vec{H}$, and (c) crystal domains following domain wall motion.

The effect of a totally aligned single layer of magnetic moments is seen to be equivalent to a current i_M encircling the torus. There will be one such current for each atomic spacing d. The equivalent magnetic field intensity within the torus then will be

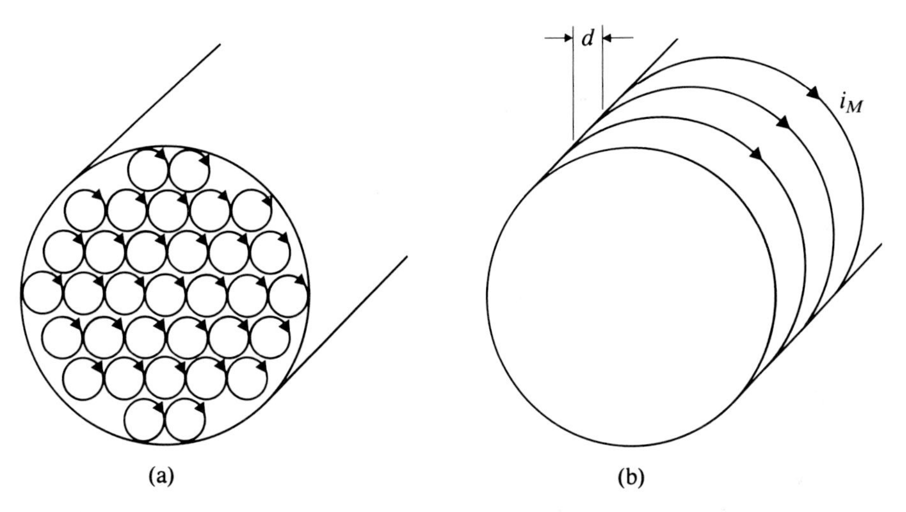

Figure 1.10 Magnetization of a ferromagnetic torus.

$$H_{equiv} = \frac{i_M}{d} = \frac{i_M d^2}{d^3} = \frac{p_M}{d^3} \qquad \text{A/m} \tag{1.32}$$

For cubic structures, d^3 is the volume occupied by one atom. Therefore, Eq. (1.32) expresses the maximum possible aligned moment per unit of volume.

This aligned magnetic moment per unit volume is denoted as the magnetization $\vec{M}$. It represents the effect of the magnetic moments in the material just as the vector $\vec{H}$ represents the magnetic field intensity due to external currents. For iron, the maximum possible magnetization is

$$\hat{M} = \frac{i_M}{d} = \frac{0.394 \times 10^{-3}}{2.27 \times 10^{-10}} = 1.73 \times 10^6 \qquad \text{A/m} \tag{1.33}$$

Thus, the effect of perfectly aligned magnetic moments in a toroidal iron core is equivalent to the effect of a coil around a nonmagnetic torus with a current of 1730 A per millimeter of length around its circumference. The flux density produced in the material by its ideally aligned magnetic moments is, by analogy with Eq. (1.5),

$$\hat{B} = \mu_0 \hat{M} = 4\pi \times 10^{-7} \times 1.73 \times 10^6 = 2.18 \qquad \text{T} \tag{1.34}$$

In practice, this is a very intense field.

As the temperature is increased, the random motion of the atoms increasingly disturbs the alignment of the net moments of adjacent atoms, thus reducing the net aligned magnetization M, as shown in Fig. 1.11. The Curie temperature, T_C, is the value at which alignment can be considered to be completely random, producing no net field in the material. For iron, this temperature is 770 °C, for nickel 358 °C, and for cobalt 1115 °C.

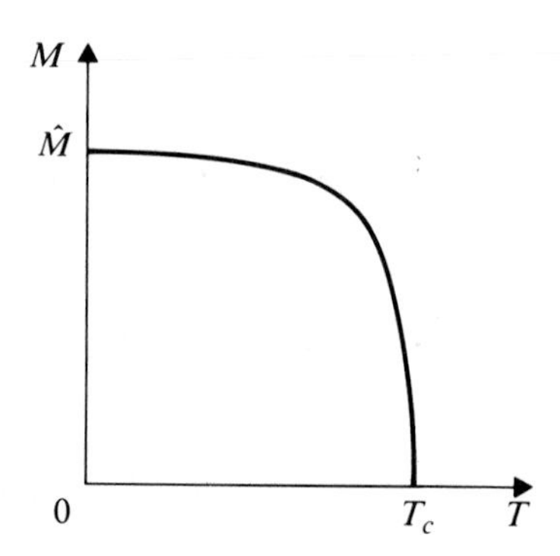

Figure 1.11 Decrease of magnetization with temperature.

In practice, we are interested in two general classes of ferromagnetic materials, one in which the magnetic alignment can be easily changed by use of a small external magnetic field and another in which the alignment strongly resists any effect of an external field. The most frequent example of the first class, denoted as *magnetically soft materials*, is iron alloyed with small amounts of other elements such as silicon. When it is rolled into thin sheets, the crystals are elongated and partially aligned in the direction of rolling. Annealing tends to relieve internal stresses and makes it easy for domain walls to be moved by an external field. Suppose a toroidal core is made of this soft iron and a magnetic field intensity H is applied to it using a uniformly wound coil. The total magnetic flux density will be that due to the coil plus the effect of the imperfectly aligned magnetic moments:

$$\vec{B} = \mu_0(\vec{H} + \vec{M}) \qquad \text{T} \tag{1.35}$$

A typical result is the measured magnetization curve for sheet steel, as shown in Fig. 1.12. With an applied magnetic field intensity of 1000 A/m, the value of flux density in a nonferromagnetic core would be $4\pi \times 10^{-7} \times 10^3 = 0.00125$ T. This is seen to be relatively insignificant in comparison with the contribution made by the alignment of magnetic moments in the sheet steel producing a flux density of 1.8 T, i.e., some 1440 times as great.

Figure 1.12 also shows magnetization curves for cast steel and cast iron. In these materials, the crystal alignment is more random; the crystals contain many impurities, defects, and strains, all of which make it more difficult for domain walls to be moved and which also reduce the maximum flux density that can be achieved.

The second class of magnetic materials is denoted as *hard* or *permanent magnet materials*. These are constituted and processed in such a way as to lock in a near-permanent alignment of their magnetic moments. This is accomplished by appropriate selection of the alloying elements, by the production of small crystals that naturally have discontinuity at their boundaries, and by the production of internal stresses, typically by quenching the material when hot.

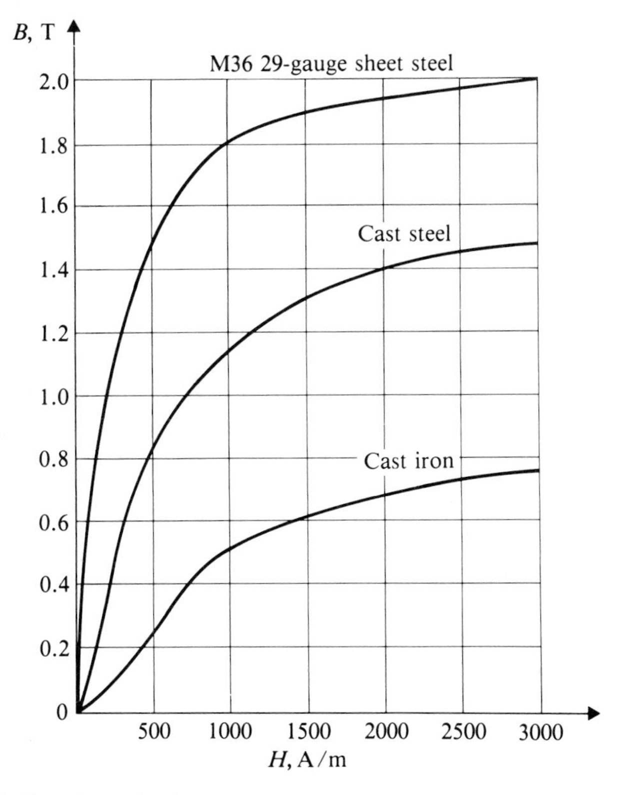

Figure 1.12 Magnetization curves.

1.6

Hysteresis

Suppose the coil in Fig. 1.1 is wound on an iron torus in which the domains are initially oriented such that the net flux density across a section of the core is zero. As the coil current is increased from zero, applying the field intensity H waveform shown in Fig. 1.13(a), the flux density B, averaged over the cross section of the core, increases, as shown in Fig. 1.13(b). At first, little increase is observed in B until H reaches a value sufficient to shift the most readily moved domain walls. The flux density B then increases by a substantial amount due to domain wall motion, with only a minor increase in H. When most of the domains are aligned along the crystal axes nearest to the direction of the applied field H, further application of a larger field has the effect of rotating these domains more nearly into line with H, in spite of their tendency to be directed along the crystal axes. Such rotation requires much greater values of H than are required for domain wall motion. The additional flux density achieved by ad-

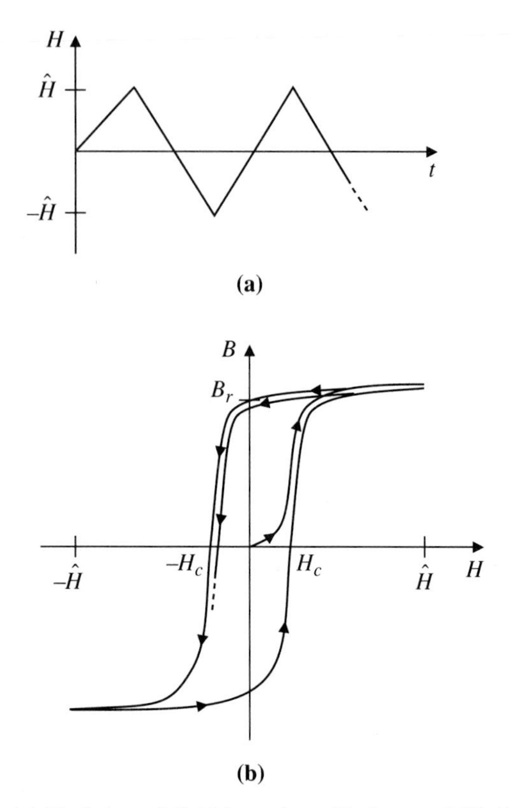

Figure 1.13 (a) Variation of field intensity with time and (b) flux density-field intensity locus leading to closed *B-H* loop.

dition of more field H gradually decreases as the iron approaches saturation.

Now suppose that the coil current is gradually reduced. First, the domains that were rotated due to the applied H will relax back toward their axes of crystal orientation. However, most of the domain walls will remain in position as the applied field H is reduced to zero, leaving a residual flux density B_r in the core. It is only when a reversed field H, sufficient to move the domain walls, is applied that the flux density will reduce toward zero and reverse. Figure 1.13(b) shows that the effect of repeated cycling between fields of maximum values $\hat{H}$ and $-\hat{H}$ eventually results in a *B-H* loop which closes on itself each cycle of reversal. Such a loop is known as a *hysteresis loop*.

Figure 1.14 shows a typical family of hysteresis loops for several values of H, each after the loop has reached a steady repetitive state. For loops involving small values of overall maximum flux density, the change in flux density is produced mainly by domain wall motion. It is only as the flux density approaches saturation that its increase is due mainly to domain rotation.

After a large value of positive H has been applied to the core and then removed, the flux density drops to its residual value B_r, sometimes referred to as

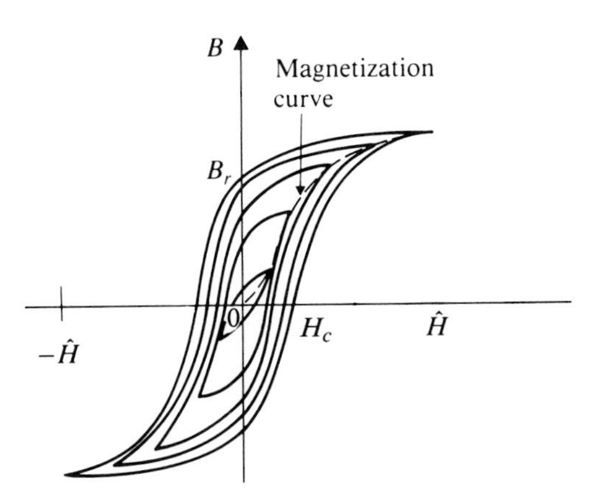

Figure 1.14 Family of steady-state hysteresis loops.

its *remanence*. The negative value of magnetic field intensity ($-H_c$) required to reduce the flux density to zero is known as the *coercive force* or *coercivity* of the material.

If the tips of the first quadrant of a family of loops are joined, the result is the dashed line shown in Fig. 1.14. This is known as the *magnetization curve* of the material. Examples of such curves have been presented in Fig. 1.12. The limited amount of information contained in the magnetization curve is adequate for many applications.

In the analysis of many magnetic systems, a further simplifying approximation is permissible. A typical magnetization curve for sheet steel is shown in Fig. 1.15. While this curve is nonlinear, it can be approximated by a straight line over a limited range of flux density as shown. This line can then be modeled by the relation

$$B = \mu_r \mu_0 H \qquad \text{T} \tag{1.36}$$

The quantity μ_r is known as the *relative permeability* of the material. It is a dimensionless number indicating the flux density in the material relative to that which would occur with an air core for the same magnetic field intensity.

The hysteresis loops of Fig. 1.14 show the *B-H* relationship when the applied field is increased and decreased consistently, as shown in Fig. 1.13(a), i.e., without intermediate reversals. Figure 1.16(a) shows a more complex variation of H involving several reversals of the rate of change H with time. Each intermediate change of increase to decrease of H creates a minor *B-H* loop, as shown in Fig. 1.16(b).

Ferromagnetic alloys have been developed to produce a variety of *B-H* relationships; an example is shown in Fig. 1.17. In this nickel-iron material, the domains can be almost perfectly aligned in the direction of the desired magnetic field. The freedom from impurities and stresses in the material is such that a

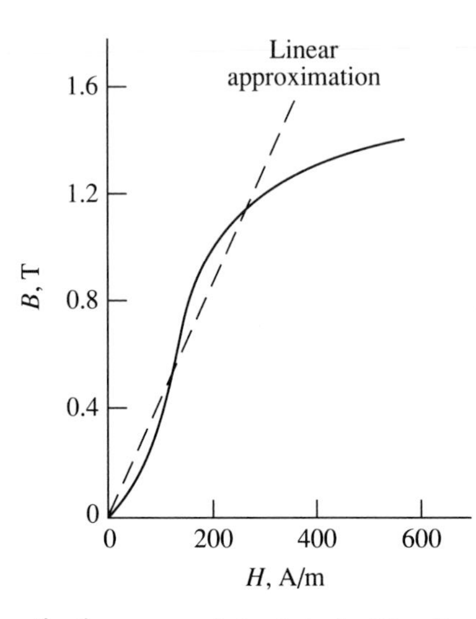

Figure 1.15 Magnetization curve of sheet steel with a linear approximation.

field intensity of less than 10 A/m or 10 mA/mm of magnetic path length is required to move the domain walls. When the wall motion is complete, the material is effectively saturated.

Permanent magnet materials are designed to maintain their saturated value of magnetization M in spite of a large value of reversed magnetic field H. Examples of B-H loops for two typical permanent magnet materials are shown in Fig. 1.18. The first of these is a mixture of the rare-earth element neodymium with iron and boron (Nd-Fe-B). The second is a ferrite material made of the oxides of iron and other magnetic elements such as barium or strontium. These materials may be oriented or magnetized by applying a large positive magnetic field intensity H, which in turn sets up the saturation value of magnetization $\hat{M}$ in the material. As the field H is reduced to zero, the flux density reduces to the residual value B_r. As a reversed field H is applied, the flux density decreases along a line having a slope only slightly greater than μ_0, indicating, from Eq. (1.35), that the value of magnetization is relatively unchanged at $\hat{M}$. This straight line relationship continues until sufficient field intensity is applied to cause the magnetization M to reverse as indicated by the abrupt change in slope at the reversed field intensity H_D. For these materials, any reversed field H lesser in magnitude than H_D may be applied to the magnet without substantially changing its magnetization. Thus, excursions along the straight line portion can occur in both directions without any appreciable minor loop. However, if a reversed field greater than H_D is applied, the magnetization will be partially reversed, and the new B-H relation will be an essentially straight line of slope near μ_0 parallel to the original B-H line.

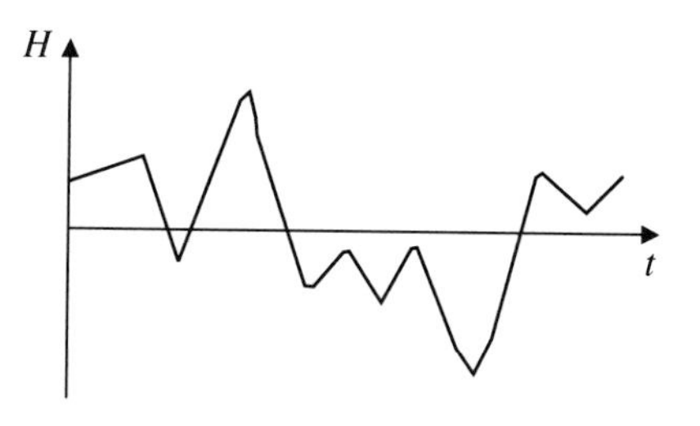

(a)

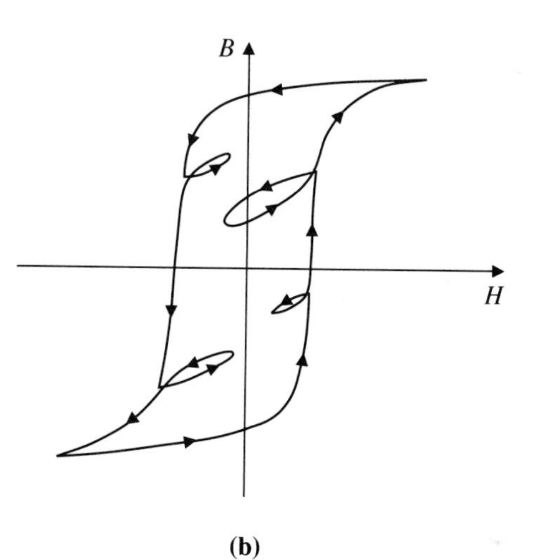

(b)

Figure 1.16 (a) Variation of magnetic field intensity with time and (b) minor *B-H* loops due to reversals in direction of increase or decrease in *H*.

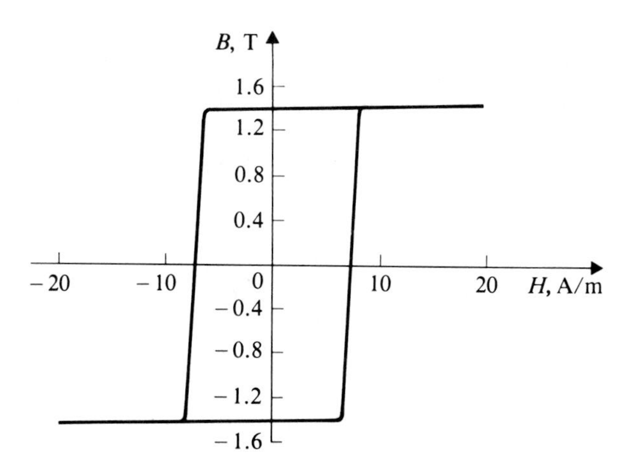

Figure 1.17 *B-H* loop for deltamax (50% Ni : 50% Fe).

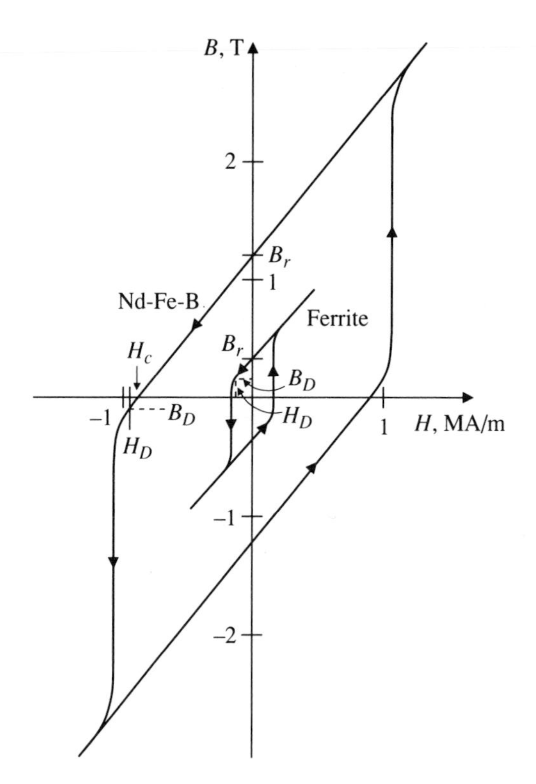

Figure 1.18 *B-H* loops for neodymium and ferrite permanent magnet materials.

The straight line relation for the useful *B-H* range of the permanent magnet material may be modeled by the equation

$$B = B_r + \mu_r \mu_0 H \qquad \text{T} \tag{1.37}$$

In some permanent magnet materials such as samarium cobalt, the value of flux density B_D at which magnetization begins to reverse may be highly negative, while in others it is still positive.

EXAMPLE 1.2 A toroidally wound coil has 1000 turns. Its core is made of the sheet steel for which a *B-H* magnetizing curve is given in Fig. 1.15. The torus has a mean diameter of 250 mm and a square cross section of 25 × 25 mm.

(a) Using the linear approximation to the *B-H* curve, determine the relative permeability of the core material.

(b) Determine the reluctance of the core using this linear approximation.

(c) Determine the corresponding value of coil inductance.

Solution

(a) Using a point on the linear approximation:

$$B = 1.15 \text{ T}, \quad H = 260 \text{ A/m}$$

$$\mu_r = \frac{B}{\mu_0 H} = \frac{1.15}{4\pi \times 10^{-7} \times 260} = 3520$$

(b) $$\mathcal{R} = \frac{\ell}{\mu_r \mu_0 A} = \frac{\pi \times 0.25}{3520 \times 4\pi \times 10^{-7} \times 0.025^2} = 284\,100 \quad \text{A/Wb}$$

(c) $$L = \frac{N^2}{\mathcal{R}} = \frac{1000^2}{284\,100} = 3.52 \quad \text{H}$$

1.7

Hysteresis Loss

If the magnetic flux in a ferromagnetic material is periodically reversed, energy will be dissipated as heat in the process of moving the domain walls past the impurities and strains in the crystal structure. Consider a toroidal ferromagnetic core with a coil connected to an electrical source. Suppose the magnetic field intensity in the magnetic material is being cycled around the hysteresis loop of Fig. 1.19 between the maximum limits $\hat{H}$ and $-\hat{H}$ by varying the coil current

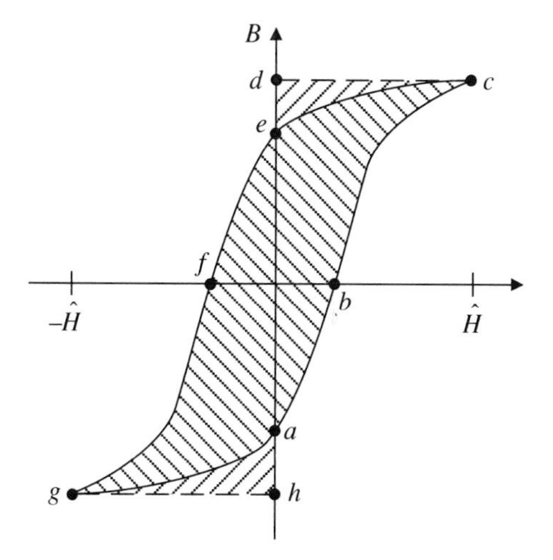

Figure 1.19 *B-H* characteristic showing hysteresis loss as the area of the loop.

direction periodically. From Eq. (1.23), the energy input per unit volume to the magnetic field in changing the flux density from point a to point c is

$$\Delta w = \int_{B_a}^{B_c} H\,dB \qquad \text{J/m}^3 \tag{1.38}$$

This energy input is represented by the area enclosed between the B-H curve and the B axis of Fig. 1.19, i.e., the area *abcda*. When the field intensity is now reduced from $\hat{H}$ to zero, the flux density decreases to B_e, releasing the amount of energy per unit volume represented by the area *cdec*. This energy is returned to the coil and to the electrical source. If H is now increased in a negative direction to $-\hat{H}$, the energy per unit volume supplied to the field from the source is given by the area *efghe*. If H is again reduced to zero, the energy per unit volume returned to the source is given by the area *ghag*.

The cycle of H has now been returned to its starting point, and the stored energy at the end of the cycle must be the same as it was at the beginning. Thus, there is a loss of energy per unit volume during this cycle, represented by the areas

$$abcda - cdec + efghe - ghag = abcefga = \text{area of loop} \tag{1.39}$$

Therefore, the energy dissipated per unit volume of material and per complete cycle is the area enclosed by the B-H loop for the particular limiting values of maximum H or B. If this is designated as w_h, the power dissipated in a core of volume V, cycled at a frequency f, due to this hysteresis effect will be

$$P_h = w_h f V \qquad \text{W} \tag{1.40}$$

This is known as the *hysteresis loss*.

The volume V of the torus can be expressed as the product of its cross-sectional area A and its flux path length ℓ. Suppose, for a core, we rescale its B-H characteristic, such as the one shown in Fig. 1.17, to represent the relation between the magnetic flux $\Phi = BA$ and the mmf $\mathfrak{F} = H\ell$. The area of the resultant Φ-$\mathfrak{F}$ loop then represents the hysteresis loss per cycle in the toroidal core.

The energy loss per cycle is a nonlinear function of the maximum value of flux density $\hat{B}$, as can be seen from the typical family of loops in Fig. 1.14. The hysteresis loss relationship can be modeled approximately for a given magnetic material by

$$p_h = k_h f \hat{B}^n \qquad \text{W/m}^3 \tag{1.41}$$

where k_h and n are empirically determined constants. The value of the exponent n is usually in the range $1.5 < n < 2.5$.

EXAMPLE 1.3 A toroidal core of rectangular cross section 10×20 mm with a mean diameter of 80 mm is fabricated using deltamax material for which a B-H loop is shown in Fig. 1.17. If the flux density is cycled between ± 1.4 T,

(a) Determine the hysteresis energy loss per cycle.

(b) Determine the hysteresis power loss at a frequency of 400 Hz.

Solution The volume of the torus is

$$V = 0.01 \times 0.02 \times 0.08\pi = 50.3 \times 10^{-6} \qquad \text{m}^3$$

The energy loss per unit volume is approximately

$$w_h = 4 \times 1.4 \times 7 = 39.2 \qquad \text{J/m}^3$$

(a) The energy loss per cycle is

$$W_h = w_h \times V = 39.2 \times 50.3 \times 10^{-6} = 1.97 \qquad \text{mJ}$$

(b) The hysteresis loss is

$$P_h = W_h f = 1.97 \times 10^{-3} \times 400 = 0.79 \qquad \text{W}$$

1.8

Eddy Current Loss

An additional power loss known as *eddy current loss* occurs in a conducting ferromagnetic material as a result of the rate of change of flux density with time. In Section 1.1, it was noted that a magnetic field that is being changed with time sets up around itself an electric field that tends to oppose the change in magnetic field.

Consider a toroidal core and coil of the form shown in Fig. 1.1. A current i in the N-turn coil produces a magnetic field intensity H in the core. With a ferromagnetic core material, a flux density B will be established in the core. Let us begin by assuming that this flux density is uniform over the cross section of the torus, as shown in Fig. 1.20.

Consider now a circular path within the core with a radius r. If the flux density B is increasing with time, a voltage e_r will be induced along this path. From Eq. (1.11),

$$e_r = \frac{d\Phi_r}{dt} \qquad \text{V} \tag{1.42}$$

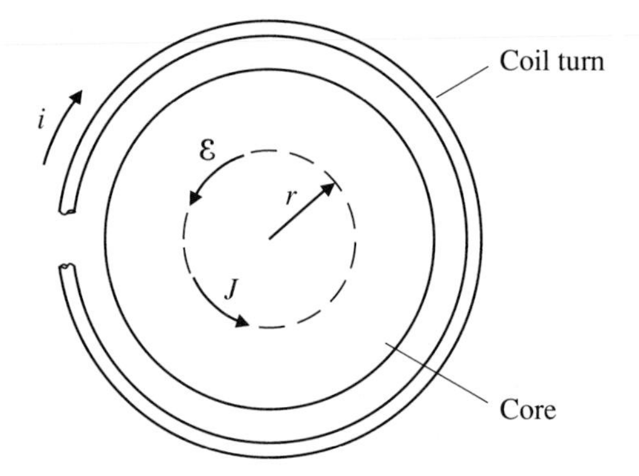

Figure 1.20 Cross section of a conducting toroidal core showing eddy current density J.

Because e_r is the summation of the electric field intensity $\mathcal{E}_r$ around the path and Φ_r is the magnetic flux enclosed by the path,

$$\mathcal{E}_r(2\pi r) = \pi r^2 \frac{dB}{dt} \qquad \text{V} \tag{1.43}$$

and

$$\mathcal{E}_r = \frac{r\,dB}{2\,dt} \qquad \text{V/m} \tag{1.44}$$

If the ferromagnetic material in the core is an electrical conductor, this electric field intensity will produce a current density J in the material

$$J = \frac{\mathcal{E}}{\rho} \qquad \text{A/m}^2 \tag{1.45}$$

where ρ is the resistivity of the material. This current circulating within the core will have two effects. First, a power loss will occur wherever there is such a current density, the loss density being

$$p_e = \rho J^2 \qquad \text{W/m}^3 \tag{1.46}$$

This is the eddy current loss. Second, the applied magnetic field H in the core and the resultant flux density B will not be uniform over the cross section. According to the circuital law of Eq. (1.1), the mmf around a toroidal path in the core at a radius r from the center of its cross section will be due to the current enclosed within the path, i.e., the current Ni of the coil plus the effect of the current density J between the radius r and the outside of the core. When the current i is increasing, the circulating current density J will be in the direction opposite to i, and the effect will be to set up less flux density than would have occurred in a nonconducting core. It is only at the core surface that the flux density will be due to the coil current i alone. If the coil current i is alternating at high frequency, the induced currents in the core are such as to inhibit change in core flux density, particularly near the center of the core cross section. The net effect is that both the flux and the circulating currents tend to be concentrated near the outer surface of the core. This is known as the *magnetic skin effect*.

If the magnetic skin effect was allowed to be significant, the utilization of the core would be inefficient because only its outside skin would be effective as a flux conductor. Therefore, means are required to ensure that the flux density remains substantially uniform over the core cross section. For the majority of ferromagnetic materials that are electrically conducting, the principal means is to construct the core of thin sheets or laminations that are electrically insulated from each other. This breaks up the conducting path. As an additional measure, the resistivity of the ferromagnetic material is increased above that of iron, typically by the addition of a few percent of silicon.

Figure 1.21(a) shows a half section of a toroidal core made of a thin continuous strip of thickness δ and width w. An enlarged section of one lamination is shown in Fig. 1.21(b). Eddy current of density J tends to flow along one side of the lamination and back along the other side. This current density is greatest at the outside surface and tends toward zero along the center plane of the lamination. Assuming the flux density B to be uniform, the total flux contained within the dotted path of thickness $2x$ is

$$\Phi = Bw2x \qquad \text{Wb} \tag{1.47}$$

The induced voltage around the periphery of the path is therefore

$$e = \frac{d\Phi}{dt} = w2x\frac{dB}{dt} \qquad \text{V} \tag{1.48}$$

By symmetry, this can be visualized as an electric field intensity $\mathcal{E}$ along the path of length approximately $2w$:

$$\mathcal{E} = \frac{e}{2w} = x\frac{dB}{dt} \qquad \text{V} \tag{1.49}$$

The current density at the lamination surfaces is then given by

$$J = \frac{\mathcal{E}}{\rho} \qquad \text{A/m}^2 \tag{1.50}$$

The eddy current power loss per unit of core volume can now be found by integration over a unit length of a lamination:

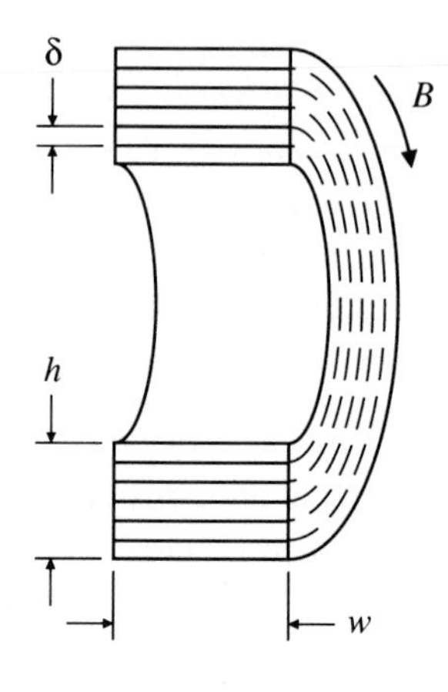

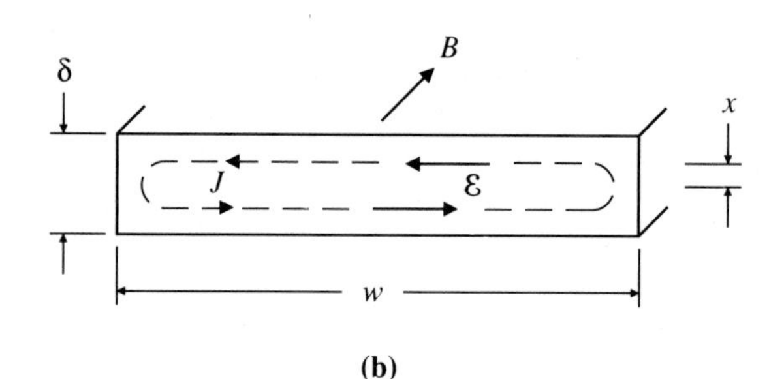

Figure 1.21 (a) Section through toroidal core made of a thin strip and (b) eddy current density J within one lamination.

$$
\begin{aligned}
p_e &= \frac{1}{\delta w} \int \rho J^2 \, dV \\
&= \frac{1}{\delta w} \int_{-\delta/2}^{\delta/2} \rho \left(\frac{x}{\rho} \frac{dB}{dt} \right)^2 wx \, dx \qquad (1.51) \\
&= \frac{\delta^2}{12\rho} \left(\frac{dB}{dt} \right)^2
\end{aligned}
$$

Examination of Eq. (1.51) suggests that the eddy current loss is proportional to the square of the rate of change of flux density. For the frequently encountered case of sinusoidally varying flux density,

$$B = \hat{B} \sin \omega t \qquad \text{T} \qquad (1.52)$$

and the eddy current loss density averaged over the strip cross section is of the form

$$p_e = k_e \omega^2 \hat{B}^2 \qquad \text{W/m}^3 \tag{1.53}$$

where k_e is a constant for the material with thickness δ. From Eq. (1.51), this constant would be expected to be inversely proportional to the material resistivity.

Equation (1.51) would further suggest that the eddy current loss would be proportional to the square of the lamination thickness δ. This is not found to be the case with most of the lamination thicknesses used in practice. The process of flux density distribution and change is somewhat more complex than this simple approach predicts. Further, the process of rolling a thin strip produces some strains near the surfaces which affect the magnetic properties in very thin laminations. As a result, the eddy current loss does reduce as lamination thickness is reduced, but not as greatly as the square law would predict.

Typical lamination thicknesses are in the range 0.3 to 1 mm for 50 to 60 Hz operation. They may be as thin as 0.02 mm for high-frequency devices.

An alternative approach to the control of eddy current losses is to use cores made of ferrite material. These ferrites consist of a mixture of oxides of iron and other elements such as manganese, nickel, or zinc. As oxides, they are poor electrical conductors with high resistivity. Thus, cores can be made in solid block form while still maintaining low eddy current losses for frequencies into the megahertz range. The maximum flux density of ferrites is much lower than that of iron because a large fraction of the volume is taken up by oxygen atoms, which are nonmagnetic. A *B-H* loop for a typical ferrite material is shown in Fig. 1.22.

1.9 Core Losses

The total of the hysteresis loss and the eddy current loss is known as the *core loss*. Manufacturers of magnetic materials provide measurements of the core loss for various grades of material, usually in per unit of mass, for ranges of values of lamination thickness, frequency, and peak value of sinusoidally alternating flux density.

Figure 1.23 shows curves of peak flux density B versus core loss p_c in W/kg for three values of lamination thickness and at a frequency of 60 Hz. The core material is of a grade used in many small and medium motors and generators. Most transformers use a somewhat higher quality grade of steel with somewhat lower core loss.

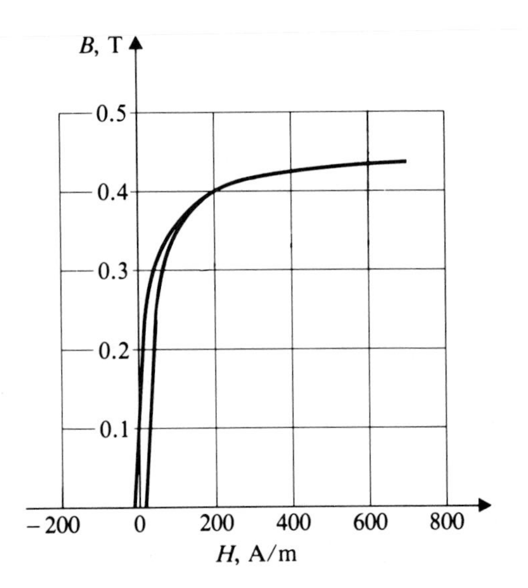

Figure 1.22 *B-H* loop for a typical ferrite material.

Figure 1.24 shows curves of core losses as a function of frequency for the same grade of steel as in Fig. 1.23. Because this is a log-log plot, it can be noted that the core losses tend to be nearly proportional to frequency at low values

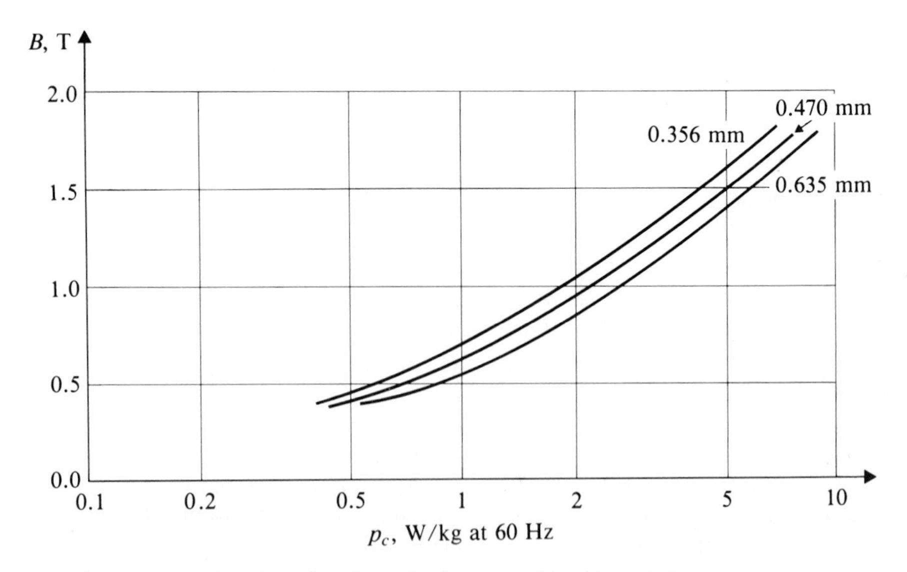

Figure 1.23 Core loss for three thicknesses of M-36 steel sheet.

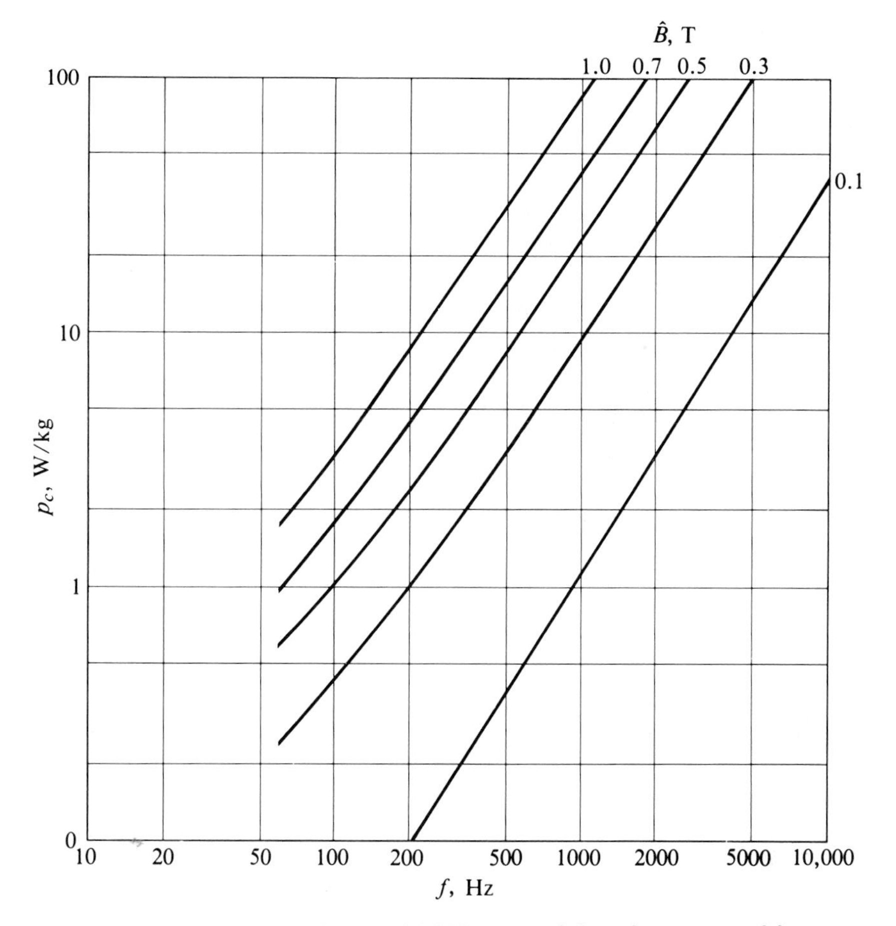

Figure 1.24 Core losses for M-36 0.356-mm steel sheet for a range of frequency.

of frequency because of the predominance of hysteresis loss. At higher frequencies, the eddy current losses that are proportional to frequency squared predominate.

1.10 Alternating Excitation

Magnetic cores normally operate on alternating excitation because it is only by change in flux that voltage can be induced and because change in any direction is limited by saturation. A variety of excitation waveforms is encountered in practice. However, the most common excitation is with sinusoidal voltage because of its ready availability from commercial power supplies. Suppose such a sinusoidal voltage v is applied to the coil of resistance R wound on a toroidal ferromagnetic core where

$$v = \hat{v} \cos \omega t \qquad \text{V} \tag{1.54}$$

From Eqs. (1.19) and (1.20), this voltage is related to the current and flux linkage of the coil by

$$v = Ri + \frac{d\lambda}{dt} \qquad \text{V} \tag{1.55}$$

The flux linkage λ is related to the coil current by the core dimensions and by the appropriate B-H loop for the core material.

In many situations involving highly efficient devices, the voltage drop across the resistance will be negligible in comparison with the rate of change of the flux linkage term in Eq. (1.55). Thus, in the steady state, with sinusoidal applied voltage, the flux linkage is also forced to be essentially sinusoidal:

$$\lambda \approx \frac{\hat{v}}{\omega} \sin \omega t \qquad \text{Wb} \tag{1.56}$$

In turn, the flux density in the core also must be sinusoidal. The coil current waveform is then whatever must flow to maintain the required sinusoidal flux density and induced voltage waveforms. This current is usually referred to as the *exciting current.*

The waveform of the exciting current can be determined graphically, as shown in Fig. 1.25. At very low frequencies, the relation between λ and the exciting current i_e is of the same shape as the static hysteresis loop, rescaled to provide

$$\lambda = NAB \qquad \text{Wb} \tag{1.57}$$

$$i_e = H\ell \qquad \text{A} \tag{1.58}$$

for a core of area A and path length ℓ. At higher frequency, the $\lambda - i_e$ loop is somewhat broadened due to the effect of eddy currents and is sometimes referred to as the *dynamic loop.*

It can be noted from Fig. 1.25 that, for sinusoidal flux linkage, the exciting current is nonsinusoidal. It has alternating symmetry in the steady state and thus is made up of a component of fundamental frequency plus odd-order harmonics. The fundamental component of the exciting current can be represented by a phasor $\vec{I}_{1e}$ consisting of two components, as shown in the phasor diagram of Fig. 1.26(a). The component $\vec{I}_c$ is in phase with the applied voltage phasor $\vec{V}$ and, because v is sinusoidal, accounts for all the power dissipated in the core. The component $\vec{I}_{1m}$ is in phase with the flux linkage phasor $\vec{\Lambda}$ and lags the applied voltage $\vec{V}$ by 90°.

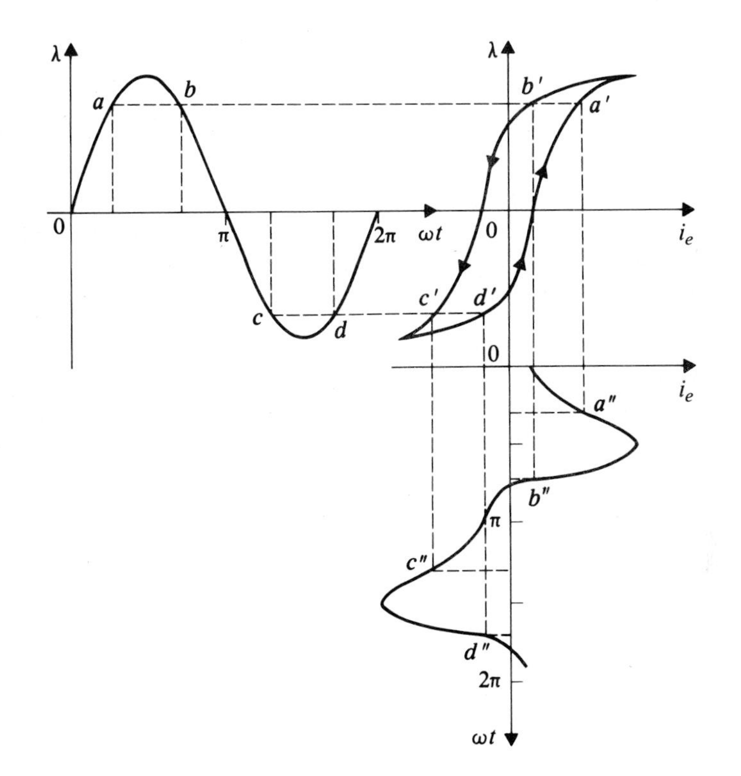

Figure 1.25 Exciting-current waveform for an inductor.

With sinusoidal excitation, the inductor can be represented approximately by the fundamental frequency equivalent circuit of Fig. 1.26(b). The harmonic currents have been ignored. The resistance R_c provides for the core loss while the inductive reactance $X_m = \omega L_m$ carries the fundamental magnetizing current

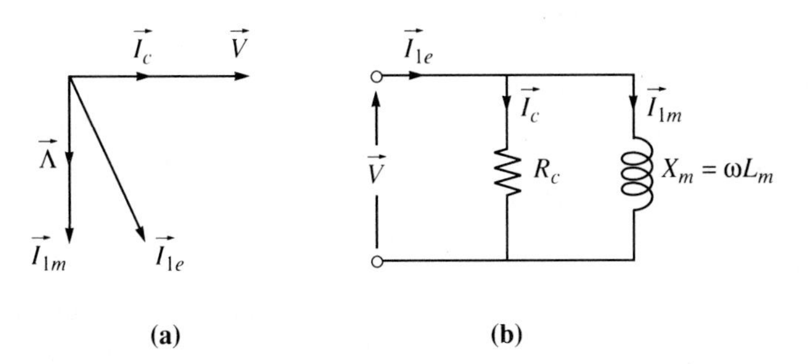

Figure 1.26 (a) Phasor diagram of exciting-current components and (b) fundamental equivalent circuit for an inductor.

I_{1m}. The values of both R_c and L_m are nonlinear functions of the peak value of flux density and of the frequency, although constant values of resistance and inductance may sometimes be used over a restricted range of the variables.

EXAMPLE 1.4 A toroidal core has a cross section of 30 × 50 mm and a mean diameter of 150 mm. A coil is to be wound on the core and connected to a 115-V (root mean square, rms), sinusoidal, 60-Hz supply. The maximum flux density in the core is to be 1.5 T.

(a) Find the required number of turns on the coil. Coil resistance may be ignored.

(b) The core material is M-36 sheet steel with a a density of 7760 kg/m^3. Find the core los

(c) Determine the appropriate value for the c

Solution

(a) Given that

$$\hat{v} = \omega N A \ell \hat{B},$$

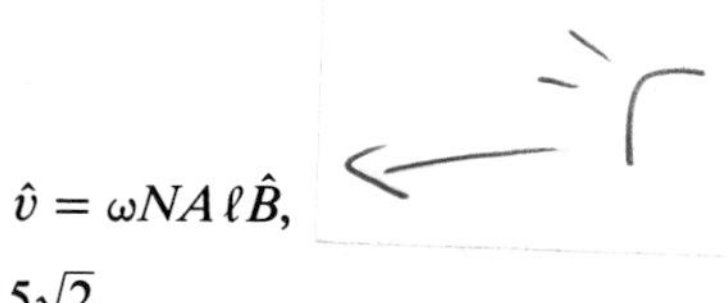

$$N = \frac{115\sqrt{2}}{2\pi \times 60 \times 0.03 \times 0.05 \times 0.15 \times \pi \times 1.5} = 407 \quad \text{turns}$$

(b) From Fig. 1.23, for $\hat{B} = 1.5$ T and $\delta = 0.47$ mm, $p_c = 5$ W/kg:

$$P_c = 0.03 \times 0.05 \times 0.15\pi \times 7760 \times 5 = 27.4 \quad \text{W}$$

(c)
$$R_c = \frac{V^2}{P_c} = \frac{115^2}{27.4} = 483 \quad \Omega$$

1.11

Production of Magnetic Flux in Air Gap

In some magnetic devices such as transformers, the main magnetic flux is confined to a closed magnetic path as in the torus. However, in machines it is generally necessary to establish flux between a stationary and a moving part. This involves causing flux to cross an air gap. The creation of flux in an air gap is also necessary for other ferromagnetic devices such as electromagnets.

We have seen that establishing a flux density up to about 1.5 T is relatively easy in iron, requiring an mmf that is usually less than 0.5 A/mm length of iron path, the reason being that the iron domains are permanently magnetized and the applied field merely aligns them. In the air gap, however, a flux density of 1.5 T requires a field intensity of about 1200 A/mm. The following discussion

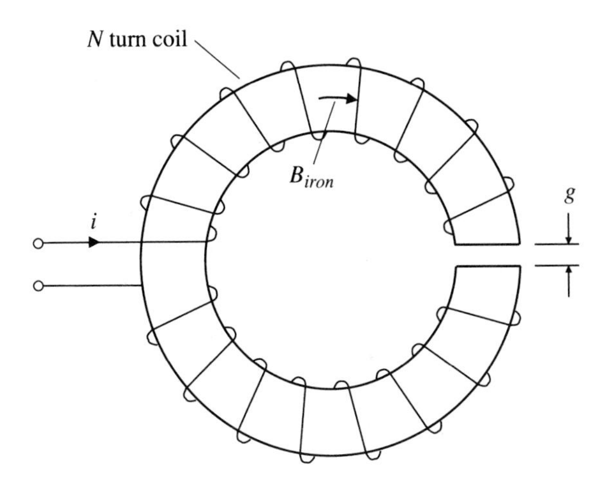

Figure 1.27 Toroidal core with air gap.

will explain how an iron structure can be used to concentrate the effect of a coil to create a magnetic field in an air gap.

Figure 1.27 shows an iron torus with an air gap of length g cut across it. There is a closed path for flux around the torus and across the gap. If the gap is short, it may be assumed that the flux will be mainly confined to a gap area A equal to the core cross-sectional area. (In fact, the flux will fringe outward from the gap edges by a small amount.) Because flux must be continuous around the path,

$$\Phi_{iron} = \Phi_{air} \qquad \text{Wb} \tag{1.59}$$

Because both iron and air paths are assumed to have the same area A,

$$B_{iron} = B_{air} \qquad \text{T} \tag{1.60}$$

For the iron path, the value of field intensity H_{iron} corresponding to B_{iron} can be found from a B-H characteristic such as the one shown in Fig. 1.12. For air,

$$H_{air} = \frac{B_{air}}{\mu_0} \qquad \text{A/m} \tag{1.61}$$

From the circuital law, Eq. (1.1), the mmf of the coil required to establish the flux density can be obtained:

$$\begin{aligned} \mathfrak{F} = Ni &= \oint \vec{H} \cdot d\vec{\ell} \\ &= H_{iron}\ell_{iron} + H_{air}g \qquad \text{A} \end{aligned} \tag{1.62}$$

Suppose an air gap 2 mm long is cut in the core of Example 1.4. To establish a flux density of 0.8 T in the sheet steel of Fig. 1.12, $H_{iron} = 120$ A/m. For the air gap,

$$H_{air} = \frac{0.8}{4\pi \times 10^{-7}} = 637 \qquad \text{kA/m} \tag{1.63}$$

Thus,

$$Ni = 120 \times 0.15\pi + 637\,000 \times 0.002 = 57 + 1274 = 1331 \qquad \text{A} \tag{1.64}$$

Note that only 4.3% of the mmf is required for the iron path. Most of the mmf has been concentrated on the air gap. Only a small error would have occurred if the mmf required by the iron path had been ignored.

If the above calculation is repeated for a number of values of flux density, a curve of coil flux linkage versus coil current can be produced, as shown in Fig. 1.28. This curve is nearly linear until the flux linkage reaches about 0.8 Wb, i.e., when the flux density in the core reaches the region of 1.3 T. At higher flux density, the effect of iron saturation becomes increasingly evident. Thus, an air-gapped core can be used to produce a nearly constant inductance. For a given torus shape and coil, the value of this inductance will be much higher than with an air core.

In the analysis of magnetic systems with air gaps, the concept of a magnetic equivalent circuit is frequently useful. In Section 1.2, a path of magnetic flux Φ requiring an mmf $\mathfrak{F}$ was represented by a reluctance $\mathcal{R}$. In the torus of Fig. 1.27, the air gap can be represented from Eq. (1.16) by the reluctance

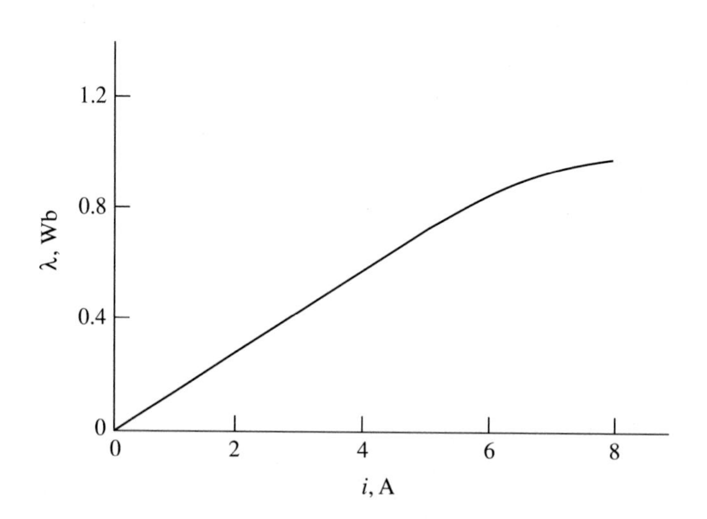

Figure 1.28 Flux linkage-current relation for the air-gapped core of Fig. 1.27.

$$\mathcal{R}_g = \frac{g}{\mu_0 A} \qquad \text{A/Wb} \tag{1.65}$$

Because the iron path requires much less mmf than the air path, considerable approximation is permitted in evaluating its reluctance. As long as the flux density in the core is not too high, the *B-H* relation of the core material may be approximated from Eq. (1.36), as

$$B_{iron} = \mu_r \mu_0 H_{iron} \qquad \text{T} \tag{1.66}$$

where μ_r is the average slope of the magnetization characteristic over the range of B and H to be used. The reluctance of the iron core then can be expressed as

$$\mathcal{R}_{iron} = \frac{\mathcal{F}_{iron}}{\Phi_{iron}} = \frac{\ell_{iron}}{\mu_r \mu_0 A_{iron}} \qquad \text{A/Wb} \tag{1.67}$$

The toroidal magnetic system of Fig. 1.27 can now be modeled by the magnetic equivalent circuit of Fig. 1.29.

The previous analysis shows that most of the mmf of the uniformly distributed toroidal coil of Fig. 1.27 is used to cause the flux to cross the air gap. If the *N*-turn coil is now concentrated around a limited arc of the toroidal core, as shown in Fig. 1.30, the path flux will be relatively unchanged. Physically, the region inside the coil will be subjected to a higher value of field intensity *H* than with the uniformly distributed coil. This will cause a higher torque on the magnetic moments in the adjacent domain walls, tending to cause them to move. However, to maintain flux continuity around the toroidal path, there must be simultaneous motion of domain walls all around the path at the same time. Torque from domains within the region of the coil thus is transferred to other domains along the path, much as torque is transferred along a shaft from its local point of application.

In situations where a constant magnetic field is required in an air gap, a magnetic system using a permanent magnet may be used. Figure 1.31 shows such an arrangement, consisting of a block of neodymium permanent magnet material, two blocks of iron, and an air gap. For simplicity, let us assume that

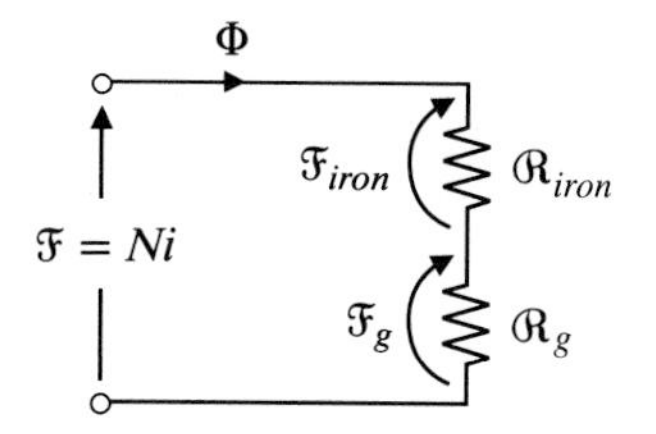

Figure 1.29 Equivalent magnetic circuit for the air-gapped core of Fig. 1.27.

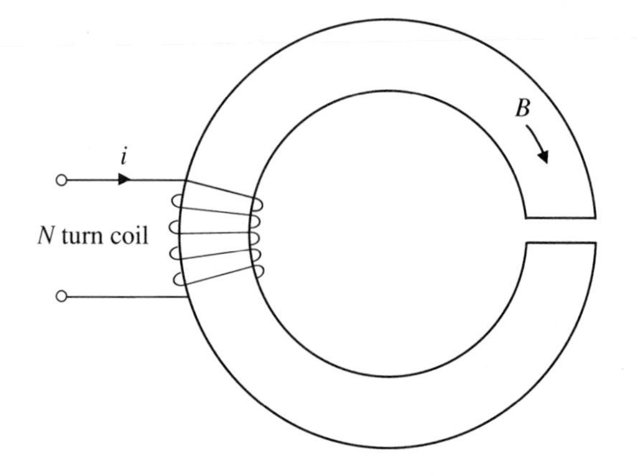

Figure 1.30 Toroidal core with concentrated coil.

the magnet flux is confined to the path shown. (In practice, there will be some leakage flux around the magnet and between the upper and lower iron plates.)

From the circuital law (Eq. [1.1]) around the flux path,

$$\oint \vec{H} \cdot d\vec{\ell} = H_m \ell_m + H_{iron} \ell_{iron} + H_g g = 0 \qquad \text{A} \tag{1.68}$$

The integral is equal to zero because no current is enclosed. If the flux density in the iron is less than about 1 T, its mmf and its reluctance may safely be ig-

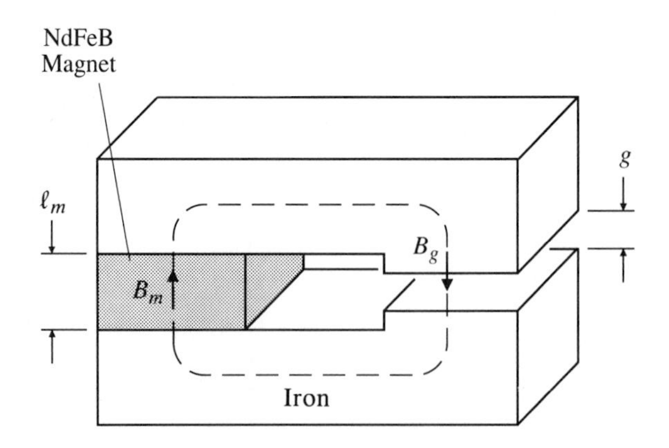

Figure 1.31 Permanent magnet producing flux density in air gap.

nored. As shown in Eq. (1.37), the *B-H* characteristic of the permanent magnet material can be modeled by the relation

$$B_m = B_r + \mu_r \mu_0 H_m \qquad \text{T} \tag{1.69}$$

For Nd-Fe-B magnet material, the residual flux density (Fig. 1.18) is about 1.1 T and the relative permeability μ_r is about 1.05. For continuity of flux,

$$\Phi = B_m A_m = B_g A_g \qquad \text{Wb} \tag{1.70}$$

Insertion of Eq. (1.69) into Eq. (1.68) gives

$$\left(\frac{B_m - B_r}{\mu_r \mu_0}\right)\ell_m + \frac{B_g}{\mu_0} g = 0 \qquad \text{A} \tag{1.71}$$

Elimination of B_m from Eqs. (1.70) and (1.71) gives the following relation for the useful air-gap flux density:

$$B_g = \frac{B_r}{\left(\dfrac{A_g}{A_m} + \mu_r \dfrac{g}{\ell_m}\right)} \qquad \text{T} \tag{1.72}$$

EXAMPLE 1.5 In the permanent magnet device shown in Fig. 1.31, the air gap has a length of 1.5 mm and a cross section of 60 × 40 mm. The magnet is of neodymium-iron material for which a *B-H* characteristic is shown in Fig. 1.18. The magnet has a cross section of 60 mm^2 and a length of 2.5 mm. The permeability of the iron sections is very large, and fringing of flux may be ignored. Determine the flux density in the air gap and in the magnet material.

Solution For the air gap, $A_g = 2400$ mm^2 and $g = 1.5$ mm. For the magnet, $A_m = 3600$ mm^2 and $\ell_m = 2.5$ mm. From Fig. 1.18, for Nd-Fe-B, $B_r = 1.2$ T and $H_c = 910$ kA/m:

$$\mu_r = \frac{B_r}{\mu_0 H_c} = \frac{1.2}{4\pi \times 10^{-7} \times 910\,000} = 1.05$$

From Eq. (1.72),

$$B_g = \frac{B_r}{\dfrac{A_g}{A_m} + \mu_r \dfrac{g}{\ell_m}} = \frac{1.2}{\dfrac{2400}{3600} + 1.05 \times \dfrac{1.5}{2.5}} = 0.93 \qquad \text{T}$$

From Eq. (1.70),

$$B_m = \frac{\Phi}{A_m} = \frac{B_g A_g}{A_m} = 0.93 \times \frac{2400}{3600} = 0.62 \qquad \text{T}$$

1.12

Modeling of Magnetic Systems

Application of the reluctance concepts of the previous section frequently allows us to model complex magnetic systems with adequate accuracy using a magnetic circuit approach. For example, consider the magnetic system of a rotating machine shown in cross section in Fig. 1.32(a). Coils on two iron pole structures on the outer or stator structure force flux across the air gaps to the inner rotating cylinder of iron. The outer flux paths are completed by iron yoke sections.

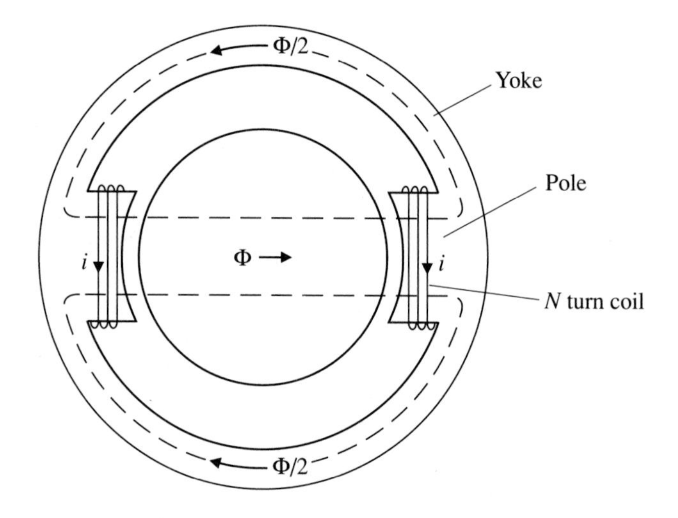

(a)

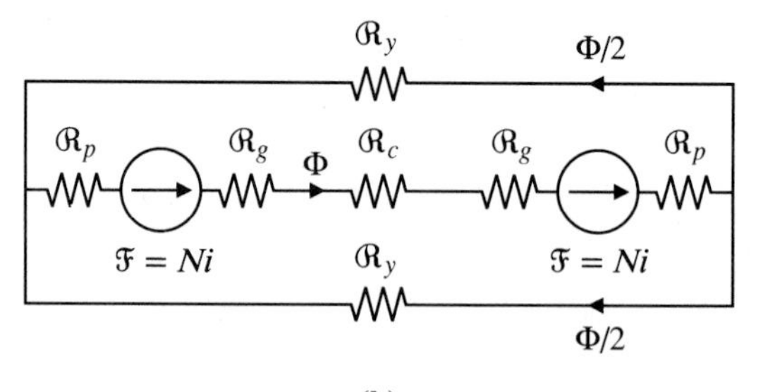

(b)

Figure 1.32 (a) Cross section of magnetic system of a machine and (b) equivalent magnetic circuit.

An equivalent magnetic circuit for this system is shown in Fig. 1.32(b). The identity of the various reluctances is apparent. Each of these reluctances can be evaluated by estimating its length, area, and relative permeability. In a practical system, most of the coil mmf will be used to overcome the air-gap reluctances. Thus, reasonable approximations are permissible in estimating the iron reluctances. Normal methods of circuit analysis then can be used to calculate the path fluxes given the coil mmfs. As iron flux densities are increased, reference should be made to the *B-H* magnetizing curves for the iron sections. Each corresponding reluctance then takes on the form

$$\mathcal{R} = \left(\frac{H_{iron}}{B_{iron}}\right)\frac{\ell}{A} \qquad \text{A/Wb} \qquad (1.73)$$

The concepts of magnetic modeling are extended in Chapter 7.

EXAMPLE 1.6 Figure 1.32(a) shows a cross section of the magnetic system of a two-pole rotating machine. The rotor has a radius of 120 mm and an axial length of 200 mm. Each pole covers an arc of 40° and is of the same axial length as the rotor. The air-gap lengths are each 1.5 mm. All the iron parts may be considered to be made of the material for which the *B-H* magnetizing curve is shown in Fig. 1.15. Each pole has a coil of 360 turns, and these coils are connected in series.

(a) Determine the approximate reluctance of each air gap.

(b) Determine the flux in each pole if the air-gap flux density is to be 0.8 T.

(c) Ignoring the reluctances of all the iron sections, estimate the current required in the coils to produce the flux of (b).

(d) The yoke has an outer radius of 210 mm and a thickness of 25 mm. Determine the flux density in the yoke.

(e) Determine the coil current required for an air gap flux density of 0.8 T if the yoke mmf is included but that in the poles and the rotor core is ignored.

Solution

(a)

$$\text{Pole arc} = r\theta = 0.12 \times \frac{40\pi}{180} = 0.084 \qquad \text{m}$$

$$\text{Gap area} \approx 0.084 \times 0.2 = 0.0168 \qquad \text{m}^2$$

Gap reluctance:

$$\mathcal{R}_g = \frac{g}{\mu_0 A} = \frac{0.0015}{4\pi \times 10^{-7} \times 0.0168} = 71\,200 \qquad \text{A/Wb}$$

(b) Gap flux = pole flux = $0.8 \times 0.0168 = 0.0134$ Wb

(c) Without the iron reluctances, the magnetic circuit of Fig. 1.32(b) consists of two coil mmfs in series with two gap reluctances. Then each coil mmf must be

$$\mathfrak{F}_g = \mathcal{R}_g \Phi = 71\,200 \times 0.0134 = 954 \qquad \text{A}$$

$$i_{coil} = \frac{\mathfrak{F}_g}{N} = \frac{954}{360} = 2.65 \qquad \text{A}$$

(d) The yoke flux is half the gap flux. The flux density is

$$B_{yoke} = \frac{0.0134}{2 \times 0.025 \times 0.2} = 1.34 \qquad \text{T}$$

(e) From Fig. 1.15, the field intensity in the yoke is approximately 450 A/m. The circumferential length of each yoke section (not counting the pole arc) along the mean radius of 197.5 mm is

$$\ell_y = (\pi \times 0.1975) - 0.084 = 0.54 \qquad \text{m}$$

The mmf required by the yoke is

$$\mathfrak{F}_y = H_y \ell_y = 450 \times 0.54 = 241 \qquad \text{A}$$

Taking a closed path around one yoke and through two air gaps, the total mmf is

$$\mathfrak{F}_{coil} = \mathfrak{F}_y + 2\mathfrak{F}_g = 241 + 2(954) = 2149 \qquad \text{A}$$

The coil current is

$$i = \frac{\mathfrak{F}_{coil}}{2N} = \frac{2149}{2 \times 360} = 2.98 \qquad \text{A}$$

PROBLEMS

1.1 A transmission line consists of two parallel cylindrical conductors spaced 2 m apart in air and carrying a current of 800 A.

(a) Determine the magnitude of the magnetic flux density produced by one conductor in the vicinity of the other conductor.

(b) Determine the magnitude and direction of the flux density at the center line between the conductors.

(Section 1.1)

1.2 A coil of 5000 turns is wound in the form of an air-cored torus with a square cross section. The inner diameter of the torus is 60 mm and the outer diameter is 100 mm. The coil current is 0.25 A.

(a) Determine the maximum and minimum values of the magnetic field intensity within the toroidal coil.

(b) Determine the magnetic flux within the torus.

(c) Determine the average flux density across the torus and compare it with the flux density midway between the inner and outer edges of the coil.

(Section 1.1)

1.3 Suppose the current in the coil of Problem 1.2 is increased linearly from zero to 0.25 A in a time of 50 μs. Determine the induced voltage in the coil.

(Section 1.1)

1.4 For the toroidal coil described in Problem 1.2,

(a) Develop a magnetic equivalent circuit giving the value of the reluctance parameter.

(b) If the conductor is circular in cross section with a diameter of 0.5 mm, develop an electric equivalent circuit and give its parameters. The conductor material has a resistivity of $1.9 \times 10^{-8}\ \Omega \cdot$m.

(Section 1.2)

1.5 Refer to the plastic ring shown in Fig. 1.5.

(a) Determine the number of turns required to provide a coil inductance of 15 mH.

(b) Suppose round copper wire with an insulation thickness of 0.2 mm is used and that the coil is wound in five layers. Determine the maximum wire diameter that can be used and find the resistance of the coil at 20 °C.

(Section 1.2)

1.6 For the toroidal coil described in Problem 1.2, carrying a current of 0.25 A,

(a) Find the magnetic field energy per unit volume at the mean diameter of the torus.

(b) Find the total field energy in the torus.

(c) Compare the value of the energy in (b) with that determined from the coil inductance of Problem 1.4 and the coil current.

(Section 1.3)

1.7 A solenoid coil consists of a single layer of 250 circular turns of wire with each turn having a 0.02-m radius. The axial length of the coil is 0.3 m. The coil is self-supporting, containing only air.

(a) Determine the inductance of the coil, assuming that the magnetic field intensity is uniform inside the coil and zero elsewhere.

(b) Find the stored energy in the magnetic field of the coil when the coil current is 18 A.

(Section 1.3)

1.8 On the average, the magnetic moment of an atom of nickel is about 0.6 of the basic quantum of spin magnetic moment (9.27×10^{-24} $A \cdot m^2$). The density of atoms is $9.1 \times 10^8/m^3$.

(a) If all magnetic moments are aligned, what will be the flux density in the nickel?

(b) A useful magnetic material consists of 50% Ni and 50% Fe, the latter having 2.2 Bohr magnetons per atom and a density of 8.5×10^8 atoms/m^3. Assuming all magnetic moments are aligned, determine the flux density in this material.

(Section 1.5)

1.9 An idealized magnetic material has a magnetic moment per unit volume of 1.2 MA/m. All domains are ideally oriented in the direction of magnetization, and all domain walls are fixed in position until a reversed magnetic field intensity of 20 A/m is applied.

(a) Sketch hysteresis loops for this material:

(i) for the case when it is driven cyclically well into saturation

(ii) for the case when the residual flux density is half the maximum residual value.

(b) Determine the magnetic field intensity required to produce a flux density of 1.05 times the maximum residual value.

(Section 1.5)

1.10 A toroidal core of the dimensions of Fig. 1.5 is made of a continuous strip of M-36 sheet steel for which the magnetization curve is shown in Fig. 1.12. The coil has 300 turns.

(a) Determine the coil current to produce an average flux density of 1.4 T in the core.

(b) If the magnetization curve is approximated by a straight line through the origin and the point for $B = 1.2$ T, what is the relative permeability of the core material and the effective inductance of the coil?

(Section 1.6)

1.11 A 350-turn coil is wound on a toroidal steel core having a cross-sectional area of 10^{-4} m^2 and a mean flux path length of 0.15 m. The core material may be considered to have a constant relative permeability of 5000 up to a flux density of 1.4 T.

(a) What rms value of sinusoidal voltage can be induced in the coil at a frequency of 200 Hz if the maximum flux density is to be 1.4 T?

(b) Determine the rms coil current in (a).

(c) If the coil resistance is 15 Ω, what will be the rms terminal voltage for the condition of part (a)?

(Section 1.6)

1.12 A magnetic core has a cross-sectional area of 500 mm^2. It can be assumed to have infinite relative permeability up to a flux density of 1.5 T. Determine the peak amplitude of a voltage of rectangular waveform having a period of 0.15 s that can be applied to the terminals of a 400-turn coil without exceeding the flux density of 1.5 T.

(Section 1.6)

1.13 For the nickel-iron magnetic material for which the limiting *B-H* loop is shown in Fig. 1.17,

(a) Determine the hysteresis energy loss per cycle per unit of volume.

(b) Determine the hysteresis energy loss per cycle in the core shown in Fig. 1.5 if it is made of this material.

(c) If the frequency is 20 Hz, determine the hysteresis loss power in the core.

(d) If a sinusoidal voltage of 120 V (rms) at 20 Hz is applied to the coil and the coil resistance is negligible, how many turns are required to drive the core around the limiting *B-H* loop?

(Section 1.7)

1.14 A magnetic core has a cross-sectional area of 0.015 m^2 and a magnetic path length of 0.9 m. The core material is M-36 sheet steel, 0.356 mm thick, with a density of 7760 kg/m^3. A coil of 40 turns with negligible resistance is wound on the core.

(a) Determine the core loss when a sinusoidal voltage of 220 V (rms) at a frequency of 60 Hz is applied to the coil.

(b) Repeat (a) for the same voltage at 100 Hz.

(Section 1.9)

1.15 A ferrite torus has a cross-sectional area of 100 mm^2 and a mean diameter of 50 mm. A hysteresis loop for the material is shown in Fig. 1.22. The coil on the torus has 200 turns and negligible resistance.

(a) Suppose a rectangular voltage wave having a peak value of 6 V is applied to the coil. What value of the period of the wave will produce a peak flux density of 0.44 T?

(b) Sketch the voltage and current waveforms and determine peak value of the current.

(Section 1.10)

1.16 A core with a cross-sectional area of 500 mm^2 and a magnetic path length of 200 mm has a *B-H* characteristic that may be represented by the piecewise-linear approximation of Fig. 1.33. The coil has 500 turns and negligible resistance.

(a) What rms value of a 300-Hz sinusoidal voltage applied to the coil will produce a flux density of 1.5 T?

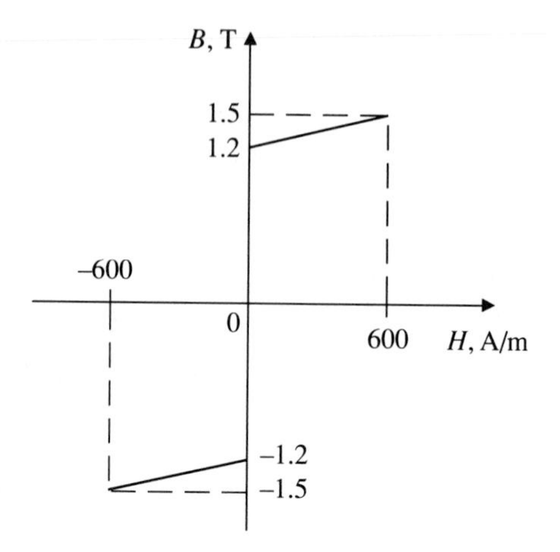

Figure 1.33 Diagram of Problem 1.16.

(b) Sketch the waveform of the magnetizing current under the steady-state condition of (a).

(c) Estimate the peak magnitude of the first three terms in the Fourier series of the current in (b).

(Section 1.10)

1.17 A core of a material for which the *B-H* loop may be approximated by Fig. 1.34 has a magnetic path length of 150 mm. A 180-Hz sinusoidal voltage is applied to the 200-turn coil, the magnitude being adjusted to give a peak

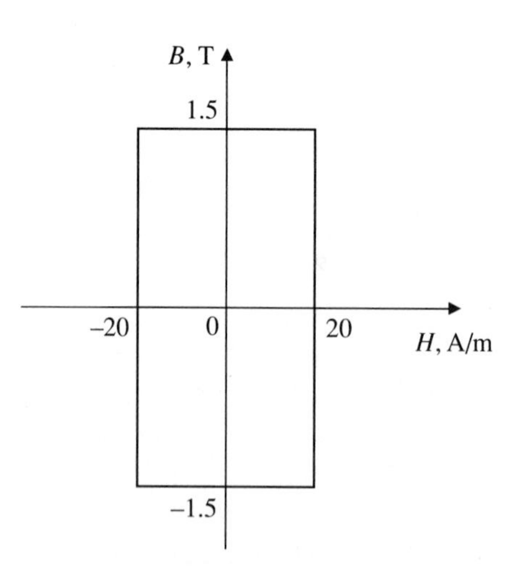

Figure 1.34 Diagram for Problem 1.17.

flux density of 1.5 T. Determine the peak amplitude and frequency of the first three components in the Fourier series of the steady-state coil current.

(Section 1.10)

1.18 When operated at a sinusoidal flux density of peak value 1.4 T at a frequency of 60 Hz, a sheet steel material has a core loss of 3.9 W/kg. The product of its fundamental voltage and its magnetizing current is 9.8 VA/kg. The material density is 7800 kg/m^3. If a core of area 0.015 m^2 and path length 0.9 m is made of this material, determine the equivalent circuit parameters—core loss resistance and magnetizing reactance—and also the terminal voltage in a 40-turn coil when the core is operated at 60 Hz and 1.4 T.

(Section 1.10)

1.19 A toroidal core with a mean diameter of 250 mm and a cross-sectional area of 1000 mm^2 has an air gap of length 1.25 mm cut in it. Fringing of flux around the air gap may be ignored. The core material is cast steel for which a *B-H* characteristic is given in Fig. 1.12. Determine the current required in a coil of 1200 turns to produce a flux density of 1.2 T in the air gap.

(Section 1.11)

1.20 By a process of trial and error, find the flux density and the flux in the air gap of the core of Problem 1.19 when the coil current is 1.2 A.

(Section 1.11)

1.21 The ferrite configuration shown in Fig. 1.35 is commonly known as a potcore. The central cylinder, bottom, and sides are all of one piece. A coil is placed around the central core, and a flat ferrite ring is placed on the top, with a small notch to provide for the coil connections. The coil inductance is adjusted by grinding down the central cylinder to leave an air gap. Suppose the central core has a diameter of 20 mm. Assume that the reluctance of the ferrite is negligible.

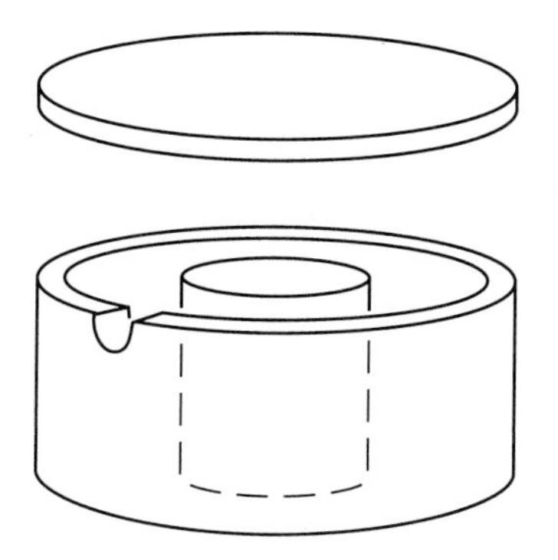

Figure 1.35 Potcore for Problem 1.21.

(a) With a coil of 1200 turns, what should be the length of the air gap to produce a coil inductance of 0.4 H?

(b) If the flux density in the core is not to exceed 0.35 T, what is the peak current allowable in the coil?

(Section 1.11)

1.22 Figure 1.36 shows a cross-sectional view of the permanent magnet assembly of a loudspeaker. A flux density of 1.2 T is required in the cylindrical air gap. The magnet is to be of the neodymium-iron material for which the *B-H* characteristic is shown in Fig. 1.18. This material is to be operated at a flux density of 0.6 T. The permeability of the soft iron may be assumed to be very high. Determine the dimensions of the magnet.

(Section 1.11)

1.23 A toroidal core has a cross-sectional area of 10^{-4} m^2 and a mean flux path length of 0.15 m. The relative permeability of the steel core is approximately 3500 up to a flux density of 1.5 T. The core is cut into two 180° sectors, a coil of 750 turns is wound on one sector, and the core is reassembled leaving two air gaps each of 1.5 mm length.

(a) Determine the values of the reluctances in an equivalent magnetic circuit for this device.

(b) Determine the inductance of the coil.

(c) What error would have occurred in the inductance of (b) if the core reluctance had been ignored?

(Section 1.12)

1.24 The magnetic material in the device shown in Fig. 1.37 can be assumed to have a constant relative permeability of 1500. The coil turns are $N_1 = 800$ and $N_2 = 1600$. Leakage flux may be ignored.

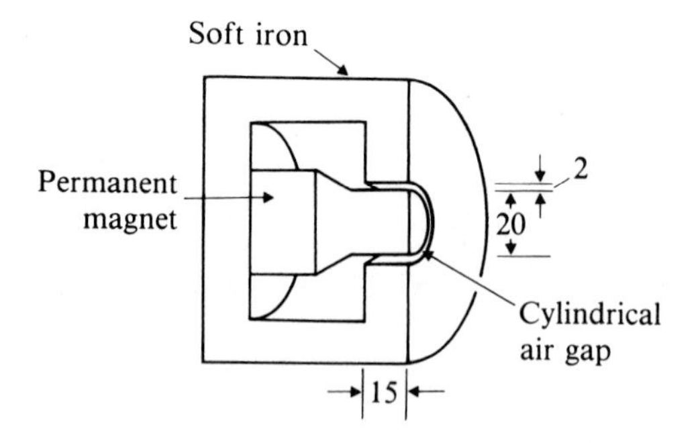

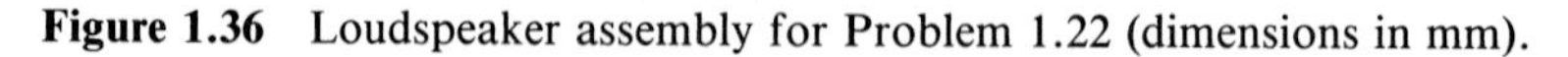

Figure 1.36 Loudspeaker assembly for Problem 1.22 (dimensions in mm).

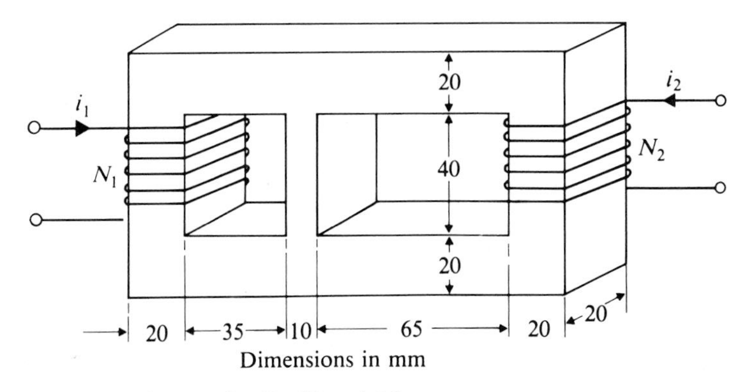

Figure 1.37 Diagram for Problem 1.24.

(a) Draw a magnetic equivalent circuit for this device showing on it the values of the reluctances.

(b) Assuming a current in the N_1-turn coil of 50 mA and that the N_2-turn coil is open circuited, determine the flux linkages of the two coils.

(Section 1.12)

GENERAL REFERENCES

Bobrow, L.S., *Fundamentals of Electrical Engineering*, Holt, Rinehart and Winston, New York, 1985.

Chikazumi, S., *Physics of Magnetism*, John Wiley and Sons, New York, 1964.

Dorf, R.C., *Introduction to Electric Circuits*, John Wiley and Sons, New York, 1990.

Heck, C., *Magnetic Materials and Their Applications*, Butterworth, London, 1974.

Hoyt, W.H., *Engineering Electromagnetics*, 4th ed., McGraw-Hill, New York, 1981.

Kraus, J.D., *Electromechanics*, 3rd ed., McGraw-Hill, New York, 1984.

Morrish, A.H., *The Physical Principles of Magnetism*, John Wiley and Sons, New York, 1965.

Slemon, G.R., *Magnetoelectric Devices—Transducers, Transformers and Machines*, John Wiley and Sons, New York, 1966.

Watson, J.K., *Applications of Magnetism*, John Wiley and Sons, New York, 1980.

CHAPTER 2

Transformers

A transformer is a device to transfer electrical energy from one circuit to another. Usually, it is capable of doing this without any physical electrical connection between the two circuits, a feature that is important for both safety and flexibility. Normally, the voltage at which energy leaves one circuit is different from that at which it enters the other, thus providing a voltage supply appropriate for the particular load. Basic transformer action can be achieved by placing two coils on a closed ferromagnetic core, as shown in Fig. 2.1. The energy transfer then occurs through the medium of a magnetic field linking the two coils. Because induction depends on the rate of change with time of the magnetic field, transformers are used with periodically alternating voltages.

2.1 Ideal Transformers

Most transformers are highly efficient devices in which the performance is close to ideal. In order to understand their operation and to analyze their perfor-

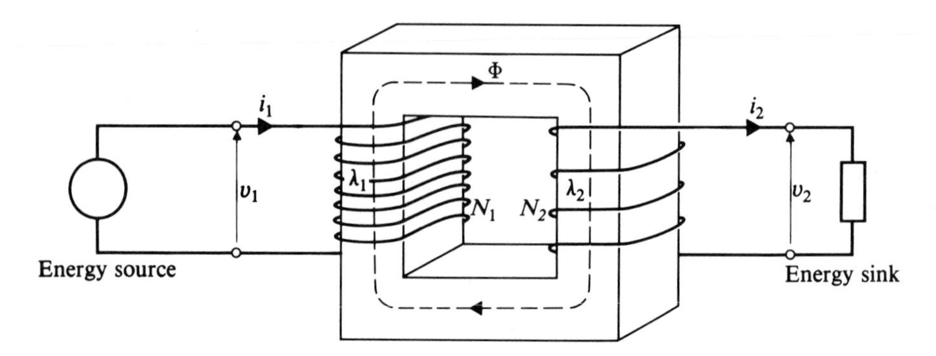

Figure 2.1 Two-winding transformer.

mance, we will first examine the properties of an ideal transformer and then introduce its imperfections and their effects. Figure 2.1 represents a transformer consisting of two coils on a closed ferromagnetic core. Suppose we assume that (a) the winding resistances are negligible, (b) the magnetic field intensity that is required in the core is negligible, i.e., the relative permeability is very high, (c) the core losses are negligible, and (d) the magnetic flux is confined to the ferromagnetic core.

If a voltage v_1 is applied to the terminals of the N_1-turn winding of Fig. 2.1, a magnetizing current will flow, producing a flux Φ and a flux linkage λ_1 in the winding. The rate of change of the flux linkage with time, i.e., the induced voltage e_1, must equal the applied voltage because the voltage drop across the resistance has been assumed to be negligible:

$$e_1 = v_1 = \frac{d\lambda_1}{dt} = N_1 \frac{d\Phi}{dt} \qquad \text{V} \tag{2.1}$$

The flux Φ also links the N_2-turn winding, producing flux linkage λ_2. The rate of change of this flux linkage induces a voltage e_2, which, in the absence of any voltage drop across the winding resistance, is equal to the terminal voltage v_2:

$$e_2 = v_2 = \frac{d\lambda_2}{dt} = N_2 \frac{d\Phi}{dt} \qquad \text{V} \tag{2.2}$$

Therefore, the voltage ratio of this ideal transformer is equal to its turns ratio:

$$\frac{v_2}{v_1} = \frac{N_2}{N_1} \tag{2.3}$$

Suppose a load circuit is now connected to the N_2-turn winding, causing a current i_2 to flow. Considering the flux path around the core, the integral of field intensity H around this closed path is zero from assumption (b) above. Thus, there must be an accompanying current i_1 in the N_1-turn winding so that

$$N_1 i_1 - N_2 i_2 = 0 \qquad \text{A} \tag{2.4}$$

Thus, the current ratio of the ideal transformer is the inverse of the turns ratio:

$$\frac{i_2}{i_1} = \frac{N_1}{N_2} \tag{2.5}$$

Combining Eqs. (2.3) and (2.5), it follows that

$$v_1 i_1 = v_2 i_2 \qquad \text{W} \tag{2.6}$$

i.e., the instantaneous power input is equal to the instantaneous power output.

In the representation of electric circuits containing transformers, it is useful to have a symbol for an ideal transformer. Such a symbol is shown in Fig. 2.2. Because this symbol does not show the direction in which the windings are wound, dots are placed at the ends of the windings which, at any instant, have positive induced voltage simultaneously with respect to the other ends.

For a transformer such as that shown in Fig. 2.1, it is customary to refer to the winding connected to the power source as the *primary winding* while the one connected to the load is referred to as the *secondary winding*. Other terms frequently used to distinguish windings are *high voltage* and *low voltage*. A transformer with a low voltage primary is commonly referred to as a *step-up transformer*, such as would be used to connect a source or generator having a terminal voltage of about 15 kV to a transmission line having a voltage of 230 kV. Similarly, an example of a *step-down transformer* would be one that transforms power from a street distribution line operating at a voltage of up to 35 kV down to the 115- or 230-V levels used in homes and offices.

The choice of voltage level to be used in each part of an electric transmission and distribution network is based largely on economics. Voltages in the range 115 to 765 kV are used for long-distance transmission to minimize the conductor currents and therefore the resistance losses. The voltage level for large generators is chosen to produce the most economical winding and insulation and

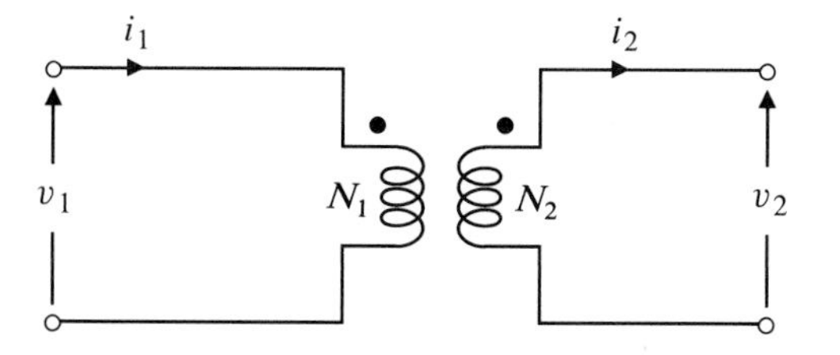

Figure 2.2 Symbol for an ideal transformer.

is usually in the range 10 to 25 kV. Distribution along streets and within large buildings is typically in the range 4 to 35 kV. Domestic electric power is normally provided at a 110- to 240-V level while industrial equipment is frequently supplied at 400 to 600 V. Between the source or generator and the local customer, the electric power may typically be transformed three or more times.

2.1.1 ▪ Transformation of Electric Circuit Elements

Figure 2.3(a) shows a series circuit with resistance, inductance, and capacitance elements connected to the N_2-turn winding terminals of an ideal transformer. The voltage and current in this circuit are related by

$$v_2 = Ri_2 + L\frac{di_2}{dt} + \frac{1}{C}\int i_2\,dt \qquad \text{V} \tag{2.7}$$

Using Eqs. (2.3) and (2.5), the corresponding relation between the voltage and the current in the N_1-turn winding is

$$v_1 = \left(\frac{N_1}{N_2}\right)^2 Ri_1 + \left(\frac{N_1}{N_2}\right)^2 L\frac{di_1}{dt} + \left(\frac{N_1}{N_2}\right)^2 \frac{1}{C}\int i_1\,dt \qquad \text{V} \tag{2.8}$$

Thus, the secondary circuit as seen through the primary winding of the ideal transformer can be represented by the circuit of Fig. 2.3(b) where

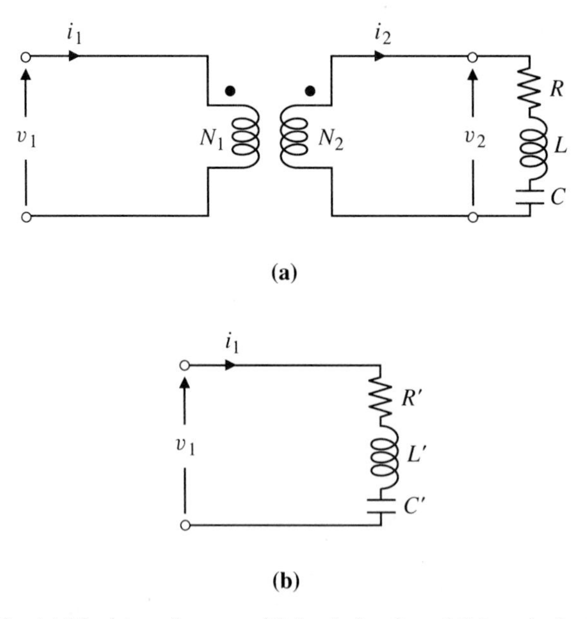

Figure 2.3 (a) Ideal transformer with load circuit and (b) equivalent as seen from primary side.

$$R' = \left(\frac{N_1}{N_2}\right)^2 R \qquad \Omega \tag{2.9}$$

$$L' = \left(\frac{N_1}{N_2}\right)^2 L \qquad \text{H} \tag{2.10}$$

$$C' = \left(\frac{N_2}{N_1}\right)^2 C \qquad \text{F} \tag{2.11}$$

From this example, it follows that any circuit configuration can be transferred across an ideal transformer to allow normal circuit analysis to be performed on that side only. For alternating currents, a load impedance Z connected to the secondary terminals can be transferred across the ideal transformer to become an impedance Z' across the primary terminals where

$$Z' = \left(\frac{N_1}{N_2}\right)^2 Z \qquad \Omega \tag{2.12}$$

In some situations, it is desirable to have the maximum power transferred into a load of resistance R_L from a source that has an internal impedance Z_s. It can be shown that this maximum power transfer occurs when the load resistance as seen through the transformer by the source is the same as the magnitude of the source impedance. From Eq. (2.12), the turns ratio can be so chosen as to provide the appropriate equivalent load resistance as seen by the source.

EXAMPLE 2.1 A power of 10 kW is to be supplied to an electric furnace at a voltage of 50 V. A 230-V, 60-Hz supply is available. Considering the required transformer to be ideal,

(a) What should be its turn ratio?
(b) Determine the primary and secondary current.
(c) Determine the impedance as seen by the 230-V supply.

Solution

(a) Turns ratio: $\dfrac{N_1}{N_2} = \dfrac{V_1}{V_2} = \dfrac{230}{50} = 4.6$

(b) Secondary current: $I_2 = \dfrac{P_2}{V_2} = \dfrac{10\,000}{50} = 200 \qquad \text{A}$

Primary current: $I_1 = \dfrac{N_2}{N_1} I_2 = \dfrac{200}{4.6} = 43.5 \qquad \text{A}$

(c) Load impedance: $Z_2 = \dfrac{\vec{V}_2}{\vec{I}_2} = \dfrac{50}{200} = 0.25 + j0 \qquad \Omega$

The equivalent impedance as seen by the source is

$$Z_1 = \left(\frac{N_1}{N_2}\right)^2 Z_2 = 4.6^2(0.25 + j0) = 5.29 + j0$$

2.2

Equivalent Circuit of a Transformer

The ideal transformer is not a sufficiently accurate model for all purposes. For more accurate prediction of transformer behavior, a more elaborate model including the effects of one or more of the transformer's imperfections is used. In the following sections, the effects of the assumptions stated in Section 2.1 are examined. Each will result in the introduction of one or more elements into an equivalent circuit of the transformer. The choice of which element to retain or ignore will depend on the particular situation being analyzed.

2.2.1 ▪ Winding Resistances

The windings of a transformer are normally made of copper conductor although aluminum is used occasionally. A model that includes the resistances of the two windings of a transformer is shown in Fig. 2.4(a). In the primary winding, the induced voltage differs from the terminal voltage v_1 by the voltage across the winding resistance $R_1 i_1$.

In large transformers, the voltage drop in each of the primary and secondary resistances is usually less than 1% of the terminal voltage and may be as low as 0.1%. Thus, the effect of the winding resistances on voltage relationships often may be ignored. However, in small transformers, the winding resistances may be the most important imperfection.

2.2.2 ▪ Magnetizing Inductance

Suppose the *B-H* relationship of the core material can be characterized by a constant relative permeability μ_r, as discussed in Section 1.11. Also, suppose a voltage v_1 is applied to the primary terminals while the secondary winding is open circuited. Ignoring the winding resistance R_1, a flux linkage λ_1 must be established such that its rate of change matches the applied voltage v_1. This requires a core flux Φ such that $N_1 \Phi = \lambda_1$. For a core of cross-sectional area A, the flux density will be $B = \Phi/A$. The corresponding field intensity H around the core path length ℓ is given by the core model

$$B = \mu_r \mu_0 H \qquad \text{T} \tag{2.13}$$

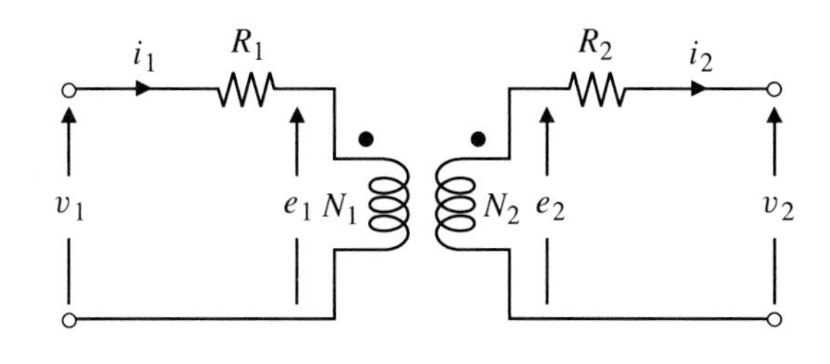

(a)

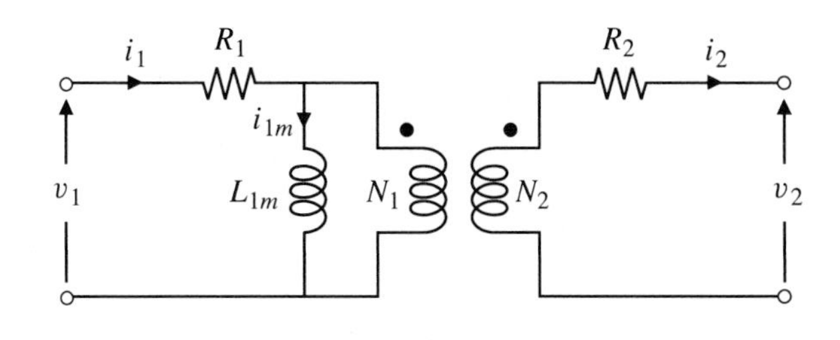

(b)

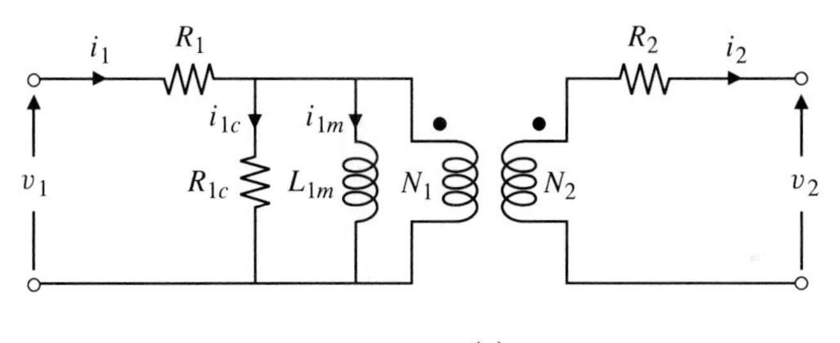

(c)

Figure 2.4 Transformer equivalent circuits including (a) winding resistances, (b) magnetizing inductance, and (c) core loss resistance.

A magnetizing current, denoted as i_{1m}, must now flow in the N_1 turn winding to satisfy the circuital law:

$$N_1 i_{1m} = H\ell \qquad \text{A} \tag{2.14}$$

Combining these relations gives an expression for the magnetizing inductance L_{1m} as seen from the primary winding:

$$\lambda_1 = N_1 \Phi = N_1 BA = N_1^2 \frac{\mu_r \mu_0 A}{\ell} i_{1m} = L_{1m} i_{1m} \qquad \text{Wb} \tag{2.15}$$

An equivalent circuit including this magnetizing inductance is shown in Fig. 2.4(b). This inductance could be transferred over to the secondary side of the ideal transformer, as discussed in Section 2.1.1. Its value would then be

$$L_{2m} = \left(\frac{N_2}{N_1}\right)^2 L_{1m} \qquad \text{H} \tag{2.16}$$

If the magnetizing current flowed from a source connected to the secondary terminals, the magnitude required to produce the same flux conditions would be

$$i_{2m} = \frac{N_1}{N_2} i_{1m} \qquad \text{A} \tag{2.17}$$

In many transformers, the magnetizing current is only a few percent of the allowable or rated winding current. Thus, the effect of the magnetizing inductance can be ignored for many performance predictions. In other situations, a more precise model including the effects of hysteresis and core nonlinearity such as described in Sections 1.7 and 1.10 may be required.

2.2.3 ▪ Core Losses

The hysteresis and eddy current losses that make up the core losses were discussed in Sections 1.7, 1.8, and 1.9. For transformer operation over a limited range of flux density B and frequency ω, the core-loss density may be approximated conveniently by a simple model similar to the eddy-current loss expression of Eq. (1.51.), i.e.,

$$p_c = k_c\left(\frac{dB}{dt}\right)^2 \qquad \text{W/m}^3 \tag{2.18}$$

The induced voltage in an N_1-turn winding is

$$e_1 = \frac{d\lambda_1}{dt} = N_1 A \frac{dB}{dt} \qquad \text{A} \tag{2.19}$$

Thus, for a core volume $A\ell$, the total core loss can be expressed as

$$P_c = A\ell p_c = A\ell k_c\left(\frac{e_1}{N_1 A}\right)^2 = \frac{e_1^2}{R_{1C}} \qquad \text{W} \tag{2.20}$$

where the core loss is represented by a resistance

$$R_{1c} = N_1^2 \frac{A}{k_c \ell} \qquad \Omega \tag{2.21}$$

This resistance can be introduced into the equivalent circuit as shown in Fig. 2.4(c). This circuit model is useful in analyzing the behavior of the transformer on no load.

In most transformers, core losses are only a very small fraction of the power transferred by the transformer. Thus, in most circuit computations, the core loss resistance can be ignored. However, the losses are important in computing efficiency and temperature.

2.2.4 ▪ Leakage Inductance

One of the most important imperfections of a transformer emerges when we remove the assumption that all the magnetic flux is confined to the ferromagnetic core. At first, this assumption might seem to be justified because of the high relative permeability of the core in contrast with that of the alternative air paths. To examine the limitations of this assumption, let us consider first the toroidal transformer shown in Fig. 2.5. A winding of N_1 turns is wound closely around a ferromagnetic torus. A thick layer of insulation is then applied, and a winding of N_2 turns is wound over it. Winding 1 is connected to a source and winding 2 to a load.

Consider a closed path in the core at the mean radius $\bar{r}$. If the core permeability is high, the field intensity in the core can be ignored in the circuital law expression. Thus,

$$N_1 i_1 = N_2 i_2 \qquad \text{A} \tag{2.22}$$

as in the ideal transformer.

The core flux Φ_c will have to be such that the rate of change of the flux linkage λ_1 of winding 1 is equal to the source voltage v_1 (ignoring winding resistance):

$$\lambda_1 = N_1 \Phi_c \qquad \text{Wb} \tag{2.23}$$

Now consider a closed circular path of length $\ell = 2\pi r$ in the space between windings 1 and 2. The net current enclosed by this path is $N_2 i_2$ and, from the circuital law, the field intensity in this region is

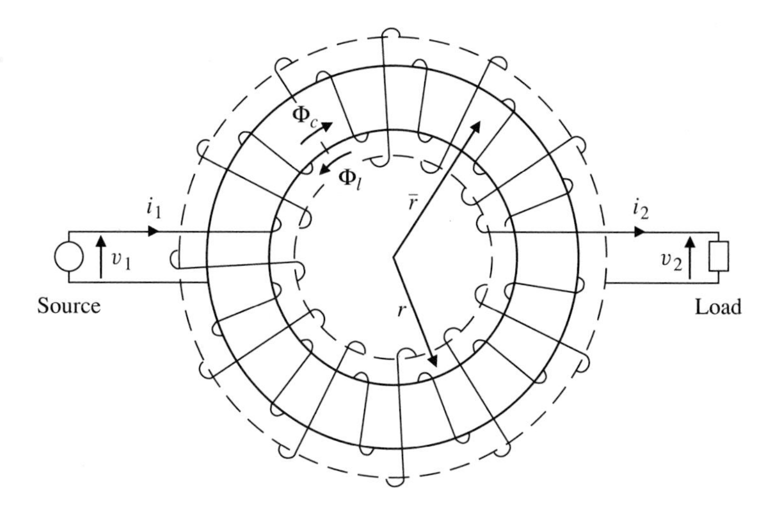

Figure 2.5 Two-winding transformer on toroidal core, showing path for leakage flux.

$$H_L = \frac{N_2 i_2}{\ell} \qquad \text{A/m} \tag{2.24}$$

If the space between the windings has a cross-sectional area A_L and a mean length ℓ, the flux in this area will be

$$\Phi_L = A_L B_L = \frac{A_L \mu_0 N_2 i_2}{\ell} \qquad \text{Wb} \tag{2.25}$$

This flux is directed counterclockwise in contrast with the core flux. The flux linking each turn of winding 2 is then

$$\Phi_2 = \Phi_c - \Phi_L \qquad \text{Wb} \tag{2.26}$$

Because not all of the flux Φ_c of winding 1 links the turns of winding 2, the difference Φ_L is known as the *leakage flux.*

The flux linkage of winding 2 can be expressed as

$$\begin{aligned} \lambda_2 &= N_2 \Phi_2 = N_2(\Phi_c - \Phi_L) \\ &= \frac{N_2}{N_1}\lambda_1 - \frac{N_2^2 A_L \mu_0}{\ell} i_2 \\ &= \frac{N_2}{N_1}\lambda_1 - L_{2L} i_2 \qquad \text{Wb} \end{aligned} \tag{2.27}$$

The difference between the winding 2 flux linkage $\lambda_1 N_2/N_1$ which would have occurred with an ideal transformer and the flux linkage λ_2 of Eq. (2.27) is seen to be directly proportional to the current i_2 and to have the property of an inductance. This leakage inductance L_{2L} as seen by the N_2-turns of winding 2 is

$$L_{2L} = \frac{N_2^2 A_L \mu_0}{\ell} \qquad \text{H} \tag{2.28}$$

Ignoring winding resistances, the voltage at the terminals of winding 2 is

$$v_2 = \frac{d\lambda_2}{dt} = \frac{N_1}{N_2} v_1 - L_{2L}\frac{di_2}{dt} \qquad \text{V} \tag{2.29}$$

The leakage inductance can now be introduced into an equivalent circuit model of the transformer, as shown in Fig. 2.6(a).

The effect of flux leakage can be represented equally well by the equivalent circuit of Fig. 2.6(b). As discussed in Section 2.1.1, the inductance L_{2L} can be transferred across the ideal transformer to become the leakage L_{1L}, as seen by the N_1 turns of winding 1. From Eq. (2.10),

$$L_{1L} = \left(\frac{N_1}{N_2}\right)^2 L_{2L} \qquad \text{H} \tag{2.30}$$

Next, consider a transformer with windings on opposite legs of its core, as shown in Fig. 2.1. On load, the primary and secondary magnetomotive forces will be approximately equal in magnitude. A substantial part of the flux link-

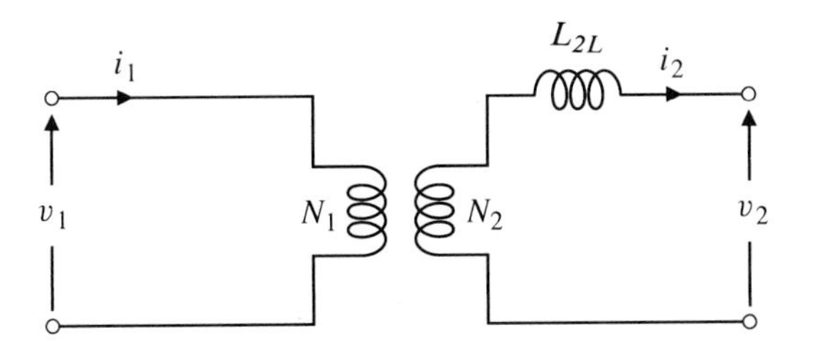

(a)

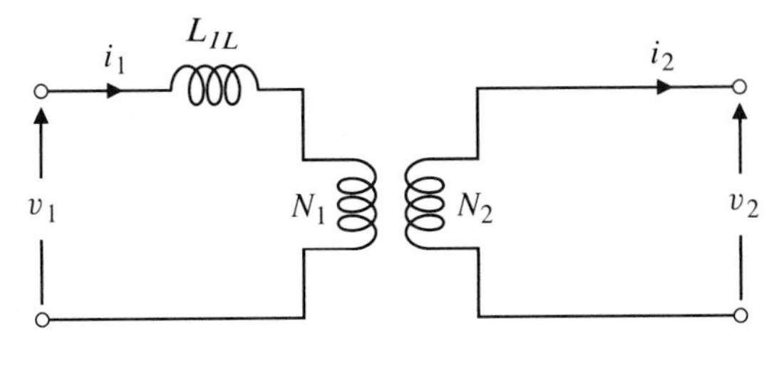

(b)

Figure 2.6 Transformer equivalent circuits including leakage inductance (a) referred to secondary and (b) referred to primary.

ing the primary winding of this transformer may take an air path around the primary coil to avoid the opposing mmf of the secondary coil, i.e., this transformer will have a high leakage inductance. To reduce the leakage, a more usual winding arrangement is that shown in Fig. 2.7(a). Here, the coils have been placed on top of each other on one of the legs. The flux paths are suggested by the dotted lines in Fig. 2.7(b). Assume a loaded transformer with equal and opposite magnetomotive forces produced by the two windings as in Eq. (2.22). The somewhat simplified view of Fig. 2.7(b) shows a flux Φ_1 in the left-hand leg of the core linking winding 1. The major part of this flux continues on as Φ_c around the ferromagnetic core. However, a part of the flux Φ_L passes down the air path between the windings, thus avoiding the opposing mmf of the outer winding 2. The amount of this flux Φ_L will be directly proportional to the mmf of winding 2.

The actual leakage flux pattern is somewhat more complex than shown in Fig. 2.7(b). Some of the leakage flux links only a part of the N_1 turns of winding 1 and some links part of winding 2. However, the major dimensional factors influencing the magnitude of the leakage inductance can be appreciated by examination of Eq. (2.28). To minimize the leakage inductance, the windings should be placed as close as possible to each other, thus minimizing the area of the leakage path. Also, the height h of the leakage flux path between the coils in Fig. 2.7(b) should be made as long as feasible. In some transformers, sections of the two windings are interleaved to achieve close coupling.

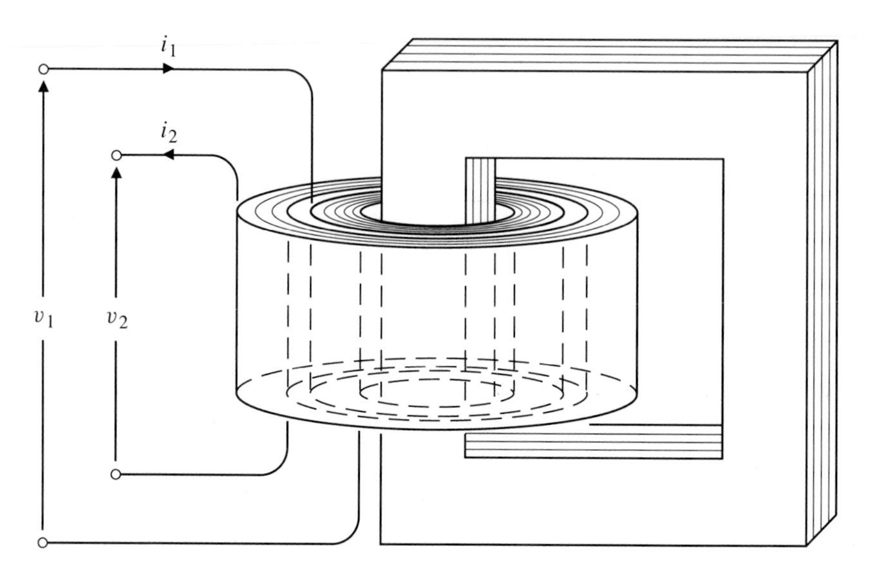

(a)

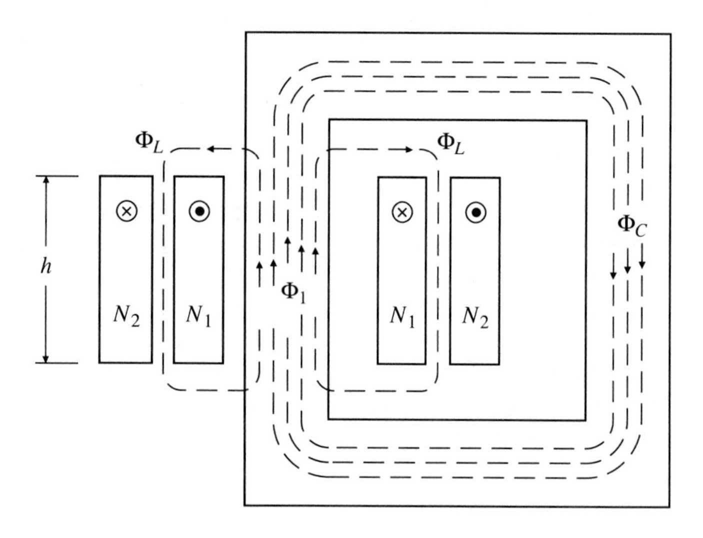

(b)

Figure 2.7 (a) Two-winding transformer with laminated core. (b) Leakage flux pattern.

Some typical winding and core arrangements for power transformers are shown in Fig. 2.8. The shell-type construction of Fig. 2.8(a) has coils similar to those of Fig. 2.7 but the outer core has been divided into two parts, each carrying half the core flux. The core-type construction of Fig. 2.8(b) has a core

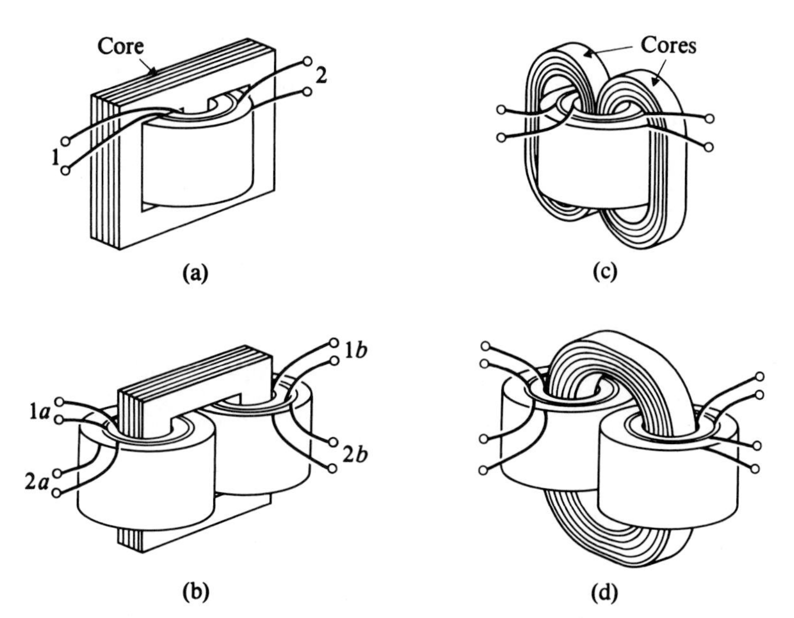

Figure 2.8 Winding and core arrangements. (a) Shell type with core of laminated sheets (1, primary coil; 2, secondary coil). (b) Core type with core of laminated sheets. (c) Shell type with wound core. (d) Core type with wound core.

similar to that of Fig. 2.7, but each leg carries a primary (1) and a secondary (2) coil. The coils *a* and *b* are identical and may be connected either in series or in parallel.

The cores of the transformers in Figs. 2.8(a) and (b) are made of flat laminations of steel, shaped and interleaved so as to leave no effective air gaps at the corners. An alternative construction consists of cores wound from a continuous strip of grain-oriented steel fed through the coil as shown for the shell-type transformer in Fig. 2.8(c) and similarly for the core-type transformer in Fig. 2.8(d).

EXAMPLE 2.2 The core of the transformer shown in Fig. 2.5 has a mean radius r of 75 mm. Its cross section is circular with a diameter of 40 mm. A primary winding of 270 turns of round copper wire 1.1 mm in diameter is wound in a single layer around the core. Insulation 4 mm thick is then applied, and a secondary winding of 700 turns of 0.4-mm-diameter round copper wire is wound over the insulation. The core can be assumed to have a relative permeability of 1500. The resistivity of copper at the expected operating temperature is 200 $\mu\Omega\cdot$m. Derive approximate values for the parameters for an equivalent circuit of this transformer with all values referred to the primary side.

Solution Primary winding resistance:

$$\text{Length of 1 turn: } \ell_1 = \pi \times 0.0411 \quad \text{m}$$

$$\text{Area of wire: } A_1 = \pi \times 0.00055^2 \quad \text{m}^2$$

$$R_1 = N_1 \rho \frac{\ell_1}{A_1} = 270 \times \frac{2 \times 10^{-8} \times \pi \times 0.0411}{\pi \times 0.00055^2} = 0.73 \quad \Omega$$

Secondary winding resistance:

$$\text{Length of 1 turn: } \ell_2 = \pi \times 0.0506 \quad \text{m}$$

$$\text{Area of wire: } A_2 = \pi \times 0.0002^2 \quad \text{m}^2$$

$$R_2 = N_2 \rho \frac{\ell_2}{A_2} = \frac{700 \times 2 \times 10^{-8} \times \pi \times 0.0506}{\pi \times 0.0002^2} = 17.7 \quad \Omega$$

Referred to the primary side:

$$R_2' = \left(\frac{N_1}{N_2}\right)^2 R_2 = \left(\frac{270}{700}\right)^2 17.7 = 2.63 \quad \Omega$$

Magnetizing inductance:

$$\text{Mean length of flux path: } \ell_m = 2\pi \times 0.075 \quad \text{m}$$

$$\text{Area of flux path: } A_m = \pi \times 0.02^2 \quad \text{m}^2$$

$$L_{1m} = N_1^2 \mu_r \mu_0 \frac{A_m}{\ell_m} = 270^2 \times \frac{1500 \times 4\pi \times 10^{-7} \times \pi \times 0.02^2}{2\pi \times 0.075} = 0.366 \quad \text{H}$$

Leakage inductance: Assume that the mean length of leakage flux path is equal to the mean length of core flux path. Area of leakage flux path is

$$A_L = \pi(25.1^2 - 21.1^2)10^{-6} = 0.581 \times 10^{-3} \quad \text{m}^2$$

$$L_{1L} = N_1^2 \mu_0 \frac{A_L}{\ell_m} = \frac{270^2 \times 4\pi \times 10^{-7} \times 0.581 \times 10^{-3}}{2\pi \times 0.075} = 0.113 \quad \text{mH}$$

(Note: the leakage inductance will be slightly higher than this because some leakage flux occurs within the primary and secondary conductors. To verify this, apply the circuital law around a path part way through either of the windings.)

An equivalent circuit for the transformer is shown in Fig. 2.9. The operating frequency will normally be such that the magnetizing reactance $X_{1m} = \omega L_{1m}$ is much larger than the resistance R_1. For calculation purposes, it is usually acceptable to move this magnetizing branch out to the primary terminals. The equivalent resistance of the windings as seen from the primary side is then

$$R_{1e} = R_1 + R_2^1 = 0.73 + 2.63 = 3.36 \quad \Omega$$

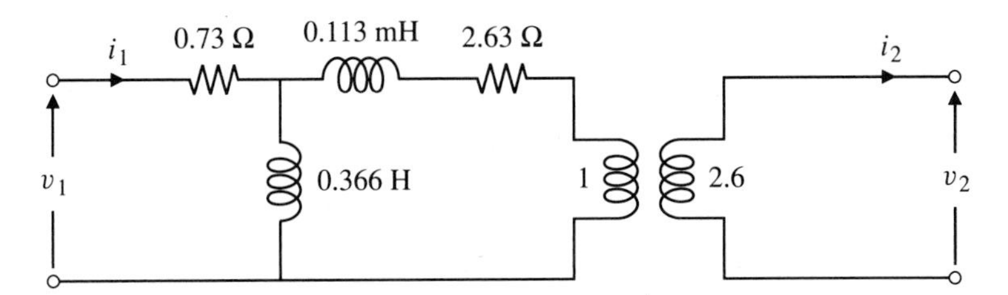

Figure 2.9 Equivalent circuit for the transformer of Example 2.2.

2.3

Transformer Rating

The conditions of frequency, voltage, current, and voltage-current product under which a transformer is designed to operate continuously are collectively termed the *rating* of the transformer. The rated voltage for any winding is restricted by the limit that must be imposed on the maximum flux density $\hat{B}$ in the core. As discussed in Sections 1.7 to 1.10, an excessive value of $\hat{B}$ would produce a large peak value of magnetizing current in the transformer winding and would also produce a large value of the core losses in the iron core. A typical value of peak flux density for steel laminations might be about 1.5 T. For a winding of N_1 turns on a core of effective iron area A, the rated rms voltage for sinusoidal operation at a frequency of ω rad/s would be

$$V_{1\,(rated)} = \frac{\omega N_1 A \hat{B}}{\sqrt{2}} \quad \text{V} \tag{2.31}$$

The rated voltage of a secondary winding of N_2 turns would be in proportion to the turns ratio. Note that the rated voltage depends on the frequency ω of operation. For other than sinusoidal waveform, the same limit of maximum flux density would apply.

The rated current of a winding is the maximum rms value that will not lead to excessive heating of the insulation of the transformer winding. For the oil-impregnated paper insulation used in most power transformers, this maximum temperature is about 100 °C. Other insulation materials may be able to withstand a higher temperature. Transformers are designed to be cooled by air or oil flow over their winding and core surfaces or occasionally by water flow through cooling pipes. For a given difference between the temperature of the surfaces and the temperature of the cooling medium and for a given surface area, the allowable power loss P_L that can be dissipated as heat can be derived. This loss consists of the core loss P_c plus the losses in the winding resistances:

$$P_L = P_c + R_1 I_1^2 + R_2 I_2^2 = P_c + R_{1e} I_1^2 \quad \text{W} \tag{2.32}$$

where

$$R_{1e} = R_1 + \left(\frac{N_1}{N_2}\right)^2 R_2 \quad \Omega \tag{2.33}$$

R_{1e} is the total effective resistance, as seen by the primary winding. If the allowable power loss is P_L and the transformer has a continuous core loss P_c, the remainder is the allowable winding loss. Thus, the rated current for continuous operation is

$$I_{1\,(rated)} = \left(\frac{P_L - P_c}{R_{1e}}\right)^{1/2} \quad \text{A} \tag{2.34}$$

The rated current of the secondary winding would be in inverse proportion to the turns ratio.

If the transformer temperature is considerably below its maximum permissible value, a larger winding current can be allowed to flow for a short time until the insulation temperature approaches its design limit. If this temperature is exceeded, the insulation becomes brittle and its life can be substantially reduced.

Because the voltage and current ratings are essentially independent of each other, the transformer does not have a power rating but instead has a voltampere rating that is the product of the voltage and current ratings for either of its windings:

$$S_{rated} = V_{1\,(rated)} I_{1\,(rated)} = V_{2\,(rated)} I_{2\,(rated)} \quad \text{VA} \tag{2.35}$$

Note that this rating is independent of the power factor of the load. A transformer can be fully loaded when supplying either an inductive or capacitive load that absorbs very little real power. The nameplate of a transformer usually contains data on the rated voltages of the windings, the rated voltampere product, and the rated frequency.

When a transformer is used at other than its rated frequency, its voltage rating must be appropriately adjusted. Similarly, if the cooling conditions are changed, such as might occur with an increase in ambient temperature above the design value, the current rating must be adjusted.

2.3.1 ▪ Per Unit Values

A per unit quantity is defined as

$$\text{Quantity in per unit} = \frac{\text{Actual quantity}}{\text{Base-value quantity}} \tag{2.36}$$

The subscript b denotes a base value. For a transformer, it is usually convenient to choose the rated values of winding voltage and voltampere product as base values.

$$V_{1b} = \text{rated voltage for winding 1 (A)}$$

$$S_b = \text{rated voltamperes for either winding (VA)}$$

Other base values may now be determined from these two. The base current for winding 1 is

$$I_{1b} = \frac{S_b}{V_{1b}} \qquad \text{A} \tag{2.37}$$

The base impedance for winding 1 is

$$Z_{1b} = \frac{V_{1b}}{I_{1b}} \qquad \Omega \tag{2.38}$$

At a base frequency ω_b, there is also a base inductance:

$$L_{1b} = \frac{Z_{1b}}{\omega_b} \qquad \text{H} \tag{2.39}$$

For a transformer with turns ratio N_1/N_2, the base values for the secondary winding are

$$V_{2b} = \frac{N_2}{N_1} V_{1b} \qquad \text{V} \tag{2.40}$$

$$I_{2b} = \frac{N_1}{N_2} I_{1b} \qquad \text{A} \tag{2.41}$$

$$Z_{2b} = \left(\frac{N_2}{N_1}\right)^2 Z_{1b} \qquad \Omega \tag{2.42}$$

$$L_{2b} = \left(\frac{N_2}{N_1}\right)^2 L_{1b} \qquad \text{H} \tag{2.43}$$

The parameters of a transformer are frequently given as per unit (pu) of these base values. For example, the leakage inductance may be stated as the dimensionless quantity

$$L_{1L} = \frac{L_{1L\,\text{(henries)}}}{L_{1b}} \qquad \text{pu} \tag{2.44}$$

From Eqs. (2.30) and (2.43), it is seen that the per unit value is the same if the leakage inductance is transferred to the secondary side of the ideal transformer and divided by the secondary base inductance L_{2b}. Thus, in per unit, the leakage inductance may be simply denoted as L_L. Typical values of leakage inductance for power transformers are in the range 0.015 to 0.2 pu. When the transformer is operated at its rated frequency ω_b, the per unit value of the leakage reactance X_L is equal to L_L pu.

As a further example, the equivalent resistance R_{1e} of the two windings referred to the primary winding (Eq. [2.33]) may be expressed in per unit of the base impedance Z_{1b} as

$$R_{1e} = \frac{R_{1e\,\text{(ohms)}}}{Z_{1b}} \qquad \text{pu} \tag{2.45}$$

The same per unit value would have been obtained if all the resistance had been referred to the secondary side. Typical values of per unit resistance range from as high as 0.1 pu for very small transformers to about 0.002 pu for transformers larger than 100 MVA.

It is frequently convenient to do circuit calculations directly in per unit quantities. For example, the per unit voltage drop across the equivalent resistance for a given per unit current is

$$V_{R\,(\mathrm{pu})} = R_{(\mathrm{pu})} I_{(\mathrm{pu})} \tag{2.46}$$

This approach allows us to dispense with the ideal transformers in our circuit models because the per unit values are the same when all quantities are referred across an ideal transformer.

EXAMPLE 2.3 A 10-kVA, 2300:230-V, 60-Hz distribution transformer has the following equivalent circuit parameters: $R_1 = 5.8\ \Omega$, $R_2 = 0.0605\ \Omega$, $L_{1L} = 85.2$ mH, $L_{1m} = 48$ H, $R_{1c} = 75.6$ kΩ.

(a) Derive the appropriate base values for this transformer.

(b) Express the parameters in per unit.

Solution

(a)

$$S_b = 10\,000 \qquad \mathrm{VA}$$

$$V_{1b} = 2300 \qquad \mathrm{V}$$

$$I_{1b} = 10\,000/2300 = 4.35 \qquad \mathrm{A}$$

$$Z_{1b} = 2300/4.35 = 529 \qquad \Omega$$

$$L_{1b} = 529/(2\pi \times 60) = 1.4 \qquad \mathrm{H}$$

(b) The equivalent winding resistance as seen from the primary side is

$$R_{1e} = R_1 + \left(\frac{N_1}{N_2}\right)^2 R_2 = 5.8 + 10^2 \times 0.0605 = 11.85 \qquad \Omega$$

$$R = \frac{11.85}{529} = 0.0224 \qquad \mathrm{pu}$$

$$L_L = \frac{0.0852}{1.4} = 0.061 \qquad \mathrm{pu}$$

$$L_m = \frac{48}{1.4} = 34.3 \qquad \mathrm{pu}$$

$$R_c = \frac{75\,600}{529} = 143 \qquad \mathrm{pu}$$

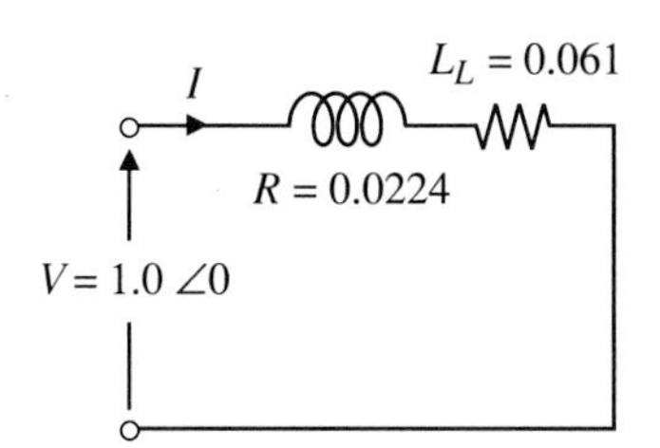

Figure 2.10 Equivalent circuit for Example 2.4.

EXAMPLE 2.4 Suppose the primary winding of the transformer in Example 2.3 is connected to a source of 1 pu voltage and the secondary winding is short circuited. Determine the primary current in per unit and in amperes.

Solution For this example, the transformer may be modeled by a per unit equivalent circuit consisting of its equivalent resistance and leakage inductance in series as shown in Fig. 2.10. The circuit impedance (at rated frequency ω_b) is

$$Z = R + j\omega_b L = 0.0224 + j0.061 = 0.065\angle 69.8 \qquad \text{pu}$$

$$I = \frac{V}{Z} = \frac{1.0\angle 0}{0.065\angle 69.8} = 15.4\angle -69.8 \qquad \text{pu}$$

Thus, the insulation and the mechanical bracing of the windings must be designed to withstand the forces that arise when 15.4 times rated current flows. In the primary winding, the actual current is

$$I_1 = I_{(\text{pu})} I_{1b} = 15.4 \times 4.35 = 67 \qquad \text{A}$$

2.4

Measurement of Transformer Parameters

Designers choose the materials and dimensions of the transformer to produce specified rated values of voltage and voltampere product and to produce specified parameters. When the transformer is manufactured, tests are usually performed to confirm these values. The same tests provide parameters for transformers where no data are available.

The winding resistances may be measured separately using a Wheatstone or Kelvin bridge. The equivalent resistance may then be found using Eq. (2.33). It should be noted that this will be the resistance to direct current. Alternating currents may cause nonuniform distribution of current in the winding conductors and result in a somewhat higher effective resistance at rated frequency.

The magnetizing inductance and the core-loss resistance can be measured by a test in which rated voltage at rated frequency is applied to one winding with the other winding open circuited. The choice of which winding depends on the source available. The magnitude of the voltage V_1, current I_1 and power P_1 on the supply side and the voltage V_2 on the open-circuited side is measured. For this test, an equivalent circuit of the form shown in Fig. 2.4(c), ignoring the winding resistances, usually can be assumed. The input power is then effectively the core loss. The parameters R_{1c} and L_{1m} and the turns ratio $N_2 : N_1$ now can be computed from the measured quantities. Usually, those quantities can be considered constant over the small range of voltage encountered in operation with most transformer installations. If the voltage is to vary widely from its rated value, measurements might be required over the expected operating range.

The leakage inductance and the effective winding resistance can be derived from measurements made on a short-circuited transformer. One of the windings is short circuited, and just enough voltage is applied to the other winding to cause a current of about rated value to flow in both windings. Measurements are made of input voltage, current, and power; from these, the equivalent resistance R_e, the leakage reactance X_L, and the leakage inductance L_L can be computed, as seen from the supply-side winding. The value of R_e then can be compared with that determined from direct-current measurements to assess the effect of frequency on the winding resistance. If desired, the current in the short-circuited winding can be measured to provide a check on the turns ratio.

EXAMPLE 2.5 The following measurements were made on a 20-kVA, 4600:230-V, 60-Hz distribution transformer: Open-circuit test, supplying the low-voltage winding: $V = 230$ V; $I = 3.5$ A; $P = 152$ W; short-circuit test, supplying the high-voltage winding: $V = 225$ V; $I = 4.35$ A; $P = 360$ W. Determine the parameters of the transformer referred to the low-voltage winding.

Solution From the open-circuit test, using the subscript 2 for the low-voltage winding,

$$R_{2c} = \frac{V_2^2}{P_2} = \frac{230^2}{152} = 348 \quad \Omega$$

$$I_{2c} = \frac{V_2}{R_{2c}} = \frac{230}{348} = 0.66 \quad \text{A}$$

$$I_{2m} = (I_2^2 - I_{2c}^2)^{1/2} = (3.5^2 - 0.66^2)^{1/2} = 3.44 \quad \text{A}$$

$$X_{2m} = \frac{V_2}{I_{2m}} = \frac{230}{3.44} = 67 \quad \Omega$$

$$L_{2m} = \frac{X_{2m}}{\omega} = \frac{67}{377} = 0.177 \quad \text{H}$$

From the short-circuit test:

$$R_{1e} = \frac{P_1}{I_1^2} = \frac{360}{4.35^2} = 19.02 \quad \Omega$$

$$R_{2e} = R_{1e}\left(\frac{N_2}{N_1}\right)^2 = 19.02\left(\frac{230}{4600}\right)^2 = 0.048 \quad \Omega$$

$$Z_1 = \frac{V_1}{I_1} = \frac{225}{4.35} = 51.7 \quad \Omega$$

$$X_{1L} = (Z_1^2 - R_{1e}^2)^{1/2} = 48.1 \quad \Omega$$

$$L_{1L} = \frac{X_{1L}}{\omega} = \frac{48.1}{377} = 0.128 \quad \text{H}$$

$$L_{2L} = L_{1L}\left(\frac{N_2}{N_1}\right)^2 = \frac{0.128}{20^2} = 319 \quad \mu\text{H}$$

2.5 Transformer Performance

The parameters that we have developed for the transformer can now be used to predict some significant aspects of its performance. A major concern is the efficiency with which the device transforms its input energy into output energy. Transformers are normally very efficient devices. However, most are connected full time to the electric power system, and their continuous energy losses can represent a substantial operating cost continuing over the long lifetime of the transformer.

Efficiency for an energy conversion device is usually defined as

$$\eta = \frac{\text{output power}}{\text{input power}} \tag{2.47}$$

This may be expressed either in per unit or in percent. In the case of a transformer, active output power is dependent on the power factor of the load. For example, if the load was a capacitor, the output power would be essentially zero, the input power would be equal to the power loss in the transformer, and the efficiency would be zero. Thus, there is no value of efficiency that is a characteristic performance property of the transformer alone. Efficiency can be determined only of a transformer with a specified load. If the power factor of the load connected to the secondary winding is $\cos\phi$, the output power is

$$P_2 = V_2 I_2 \cos\phi \quad \text{W} \tag{2.48}$$

The losses in the transformer consist of the core loss P_c, which will be essentially constant at constant voltage and frequency, and the winding losses, which will depend on the winding current. Thus, the efficiency is

$$\eta = \frac{P_2}{P_2 + P_c + R_{2e}I_2^2} \tag{2.49}$$

If a transformer is to operate near its rated capacity and is to be switched off when not required, it is usual to design the transformer to have its maximum efficiency at or near rated load. On the other hand, a distribution transformer that is connected to the power system for 24 h/day but operates at well below rated output for most of those hours might be designed for maximum efficiency at or near its average output. If the power factor (cos ϕ) is fixed by the nature of the load, then for maximum efficiency

$$\frac{d\eta}{dI_2} = 0 \tag{2.50}$$

from which it can be shown that maximum efficiency occurs when

$$P_c = R_{2e}I_2^2 = R_{1e}I_1^2 \qquad \text{W} \tag{2.51}$$

An alternative figure of merit for a distribution transformer is its "all day" or energy efficiency given by

$$\eta_{AD} = \frac{\text{Energy output during 24 h}}{\text{Energy input during 24 h}} \tag{2.52}$$

This can be determined from the transformer parameters if the load cycle for a typical day is known.

Another important performance characteristic of a transformer is the amount by which the load voltage changes as the load is varied from full to zero. Most electrical apparatus is designed to operate at a specified or rated voltage, allowing for a maximum variation of about 5%. Too high a voltage would overheat a space heater, for example. It would also tend to saturate any magnetic core in a load device. Some loads require constant power. With too low a voltage, the current becomes excessive and overheats the load device and possibly also the transformer.

The voltage regulation of a transformer with constant supply voltage V_1 is defined as

$$\text{Regulation} = \frac{V_{2\,(no\ load)} - V_{2\,(rated\ load)}}{V_{2\,(rated\ load)}} \tag{2.53}$$

where V_2 is the load voltage. Again, as with efficiency, the voltage regulation depends on the power factor of the load and thus is not a unique characteristic of the transformer alone.

Voltage regulation can be readily predicted using an equivalent circuit such as shown in Fig. 2.11(a). The magnetizing branch has negligible effect and can

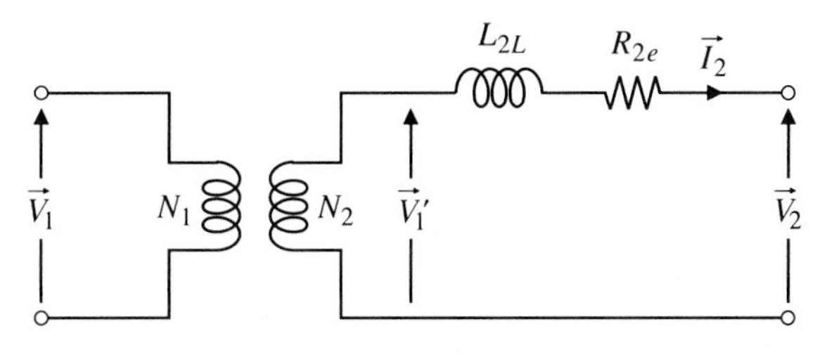

(a)

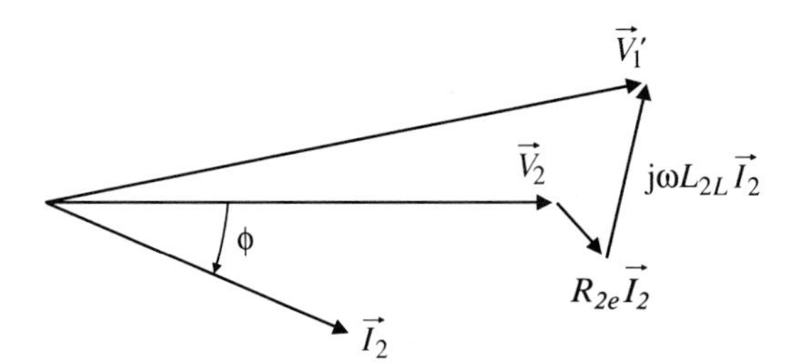

(b)

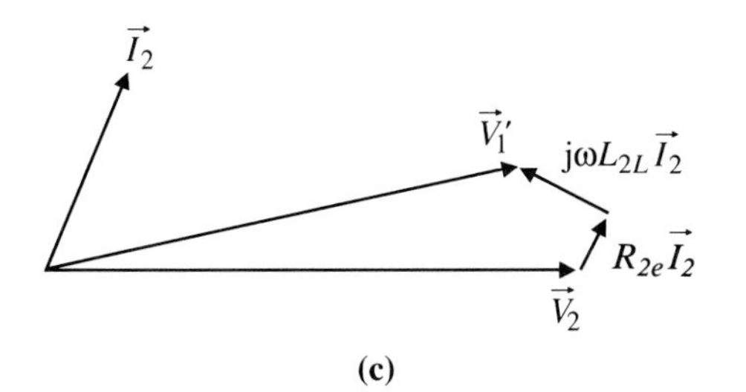

(c)

Figure 2.11 (a) Equivalent circuit for predicting voltage regulation, (b) phasor diagram for lagging power factor load, and (c) phasor diagram for leading power factor load.

be ignored. For convenience, all parameters have been referred to the secondary side. Suppose the source voltage V_1 has been so adjusted as to provide rated load voltage V_2 at rated load of a stated power factor. In practice, this might be accomplished by adjustment of the turns ratio using taps on one of the windings.

Figure 2.11(b) shows a phasor diagram for a somewhat inductive load with the load current lagging the load voltage by angle ϕ. The required supply voltage phasor to supply rated current I_2 at rated voltage V_2 is

$$\vec{V}_1 = \left(\frac{N_1}{N_2}\right)\vec{V}_1' = \frac{N_1}{N_2}\,[\vec{V}_2 + (R_{2e} + j\omega L_{2L})\vec{I}_2] \qquad \text{V} \tag{2.54}$$

When the load is removed, the load voltage magnitude will be

$$V_2 = \frac{N_2}{N_1} V_1 \qquad \text{V} \tag{2.55}$$

using the magnitude of V_1 from Eq. (2.54). These values of V_2 may now be inserted into Eq. (2.53) to obtain the voltage regulation.

For the inductive load of Fig. 2.11(b), the regulation is seen to be positive, i.e., the voltage rises when the load is removed. For a capacitive load as shown in Fig. 2.11(c), the regulation is negative. The most direct way of ensuring the regulation to be within acceptable limits is to use a transformer with a small leakage inductance.

The discussion of transformers is taken up again in Chapter 7. Models for more complex magnetic systems are developed, and connections of transformers for use in three-phase systems are discussed there.

EXAMPLE 2.6 The distribution transformer of Example 2.5 is supplying a load at 230 V and at an inductive power factor of 0.85.

(a) Determine the load at which the transformer efficiency is a maximum and determine the efficiency at that load.

(b) The transformer operates at constant power factor on a daily load cycle that can be approximately represented by the following: 90% of rated load for 6 h, 50% of rated load for 10 h, and no load for 8 h. Determine its energy efficiency under these conditions.

Solution

(a) For maximum efficiency, the winding loss is equal to the core loss. From Example 2.5, $P_c = 152$ W. The equivalent winding resistance referred to the 230-V side is $R_{2e} = 0.048\ \Omega$. Thus, for maximum efficiency,

$$P_c = R_{2e} I_2^2$$

$$I_2 = \left(\frac{152}{0.048}\right)^{1/2} = 56.3 \qquad \text{A}$$

The voltampere product or apparent power at this current is

$$S_2 = 230 \times 56.3 = 12.94 \qquad \text{kVA}$$

i.e., 65% of the rated load of 20 kVA. At this load, the power output is

$$P_2 = 12\,940 \times 0.85 = 11\,000 \qquad \text{W}$$

The efficiency at this load is then

$$\eta = \frac{P_2}{P_2 + P_c + R_{2e} I_2^2} = \frac{11\,000}{11\,000 + 152 + 152} = 0.973$$

(b) The energy output during a 24-h period, expressed in kilowatt-hours, is

$$(0.9 \times 20 \times 0.85 \times 6) + (0.5 \times 20 \times 0.85 \times 10) = 176.8 \qquad \text{kW}\cdot\text{h}$$

Full-load current on the secondary side is

$$I_2 = \frac{20\,000}{230} = 87 \qquad \text{A}$$

The energy loss in the windings during 24 h is

$$0.048[6(0.9 \times 87)^2 + 10(0.5 \times 87)^2] = 2.67 \qquad \text{kW}\cdot\text{h}$$

The energy loss in the core during 24 h is

$$0.152 \times 24 = 3.65 \qquad \text{kW}\cdot\text{h}$$

Thus,

$$\eta_{AD} = \frac{176.8}{176.8 + 3.65 + 2.67} = 0.965$$

EXAMPLE 2.7 The distribution transformer of Example 2.5 is supplying rated load at 0.85 power factor inductive or lagging.

(a) Determine the voltage regulation.

(b) Determine the turns ratio required if the source voltage is fixed at 4600 V.

Solution

(a) The machine is operating at rated load and 0.85 power factor. Let

$$\vec{V}_2 = 230\angle 0$$

$$\vec{I}_2 = \frac{20\,000}{230}\angle\cos^{-1} 0.85 = 87\angle -36.8$$

At 60 Hz,

$$X_{2L} = 2\pi \times 60 \times 319 \times 10^{-6} = 0.12 \qquad \Omega$$

$$\begin{aligned}\vec{V}_1' &= \vec{V}_2 + (R_{2e} + jX_{2L})\vec{I}_2 \\ &= 230\angle 0 + (0.048 + j0.12)87\angle -36.8 \\ &= 239.65\angle 1.4\end{aligned}$$

$$\text{Regulation} = \frac{239.65 - 230}{230} = 0.042$$

(b) To provide 230 V at rated load from the 4600-V supply,

$$\frac{N_1}{N_2} = \left(\frac{4600}{230}\right)(1 - 0.042) = 19.16$$

PROBLEMS

2.1 Consider the transformer shown in Fig. 2.12.

(a) Place polarity dots on windings 3 and 4.

(b) State the number of turns of winding 4.

(Section 2.1)

2.2 A two-winding transformer has 200 turns on its primary winding and 500 turns on its secondary winding. The primary winding is connected to a 230-V sinusoidal supply, and the secondary winding supplies an apparent power 10 kVA to a load. The transformer may be considered to be ideal.

(a) Determine the load voltage and current and the primary current.

(b) Determine the magnitude of the load impedance as seen from the primary side.

(Section 2.1)

2.3 A 5-kVA, 60-Hz, 440:110-V, two-winding transformer is connected to a 460-V supply from which it draws a primary current of 10 A at a lagging power factor of 0.9. The transformer may be considered ideal. Determine the complex impedance of the load circuit connected to the low-voltage terminals.

(Section 2.1)

2.4 Figure 2.13 shows a circuit in which a source that can be modeled as a voltage v_s in series with an inductance L_s is connected to a load consisting of elements L, R, and C by an ideal transformer.

(a) Derive an equivalent circuit in which all the load elements are transferred to the source side of the transformer.

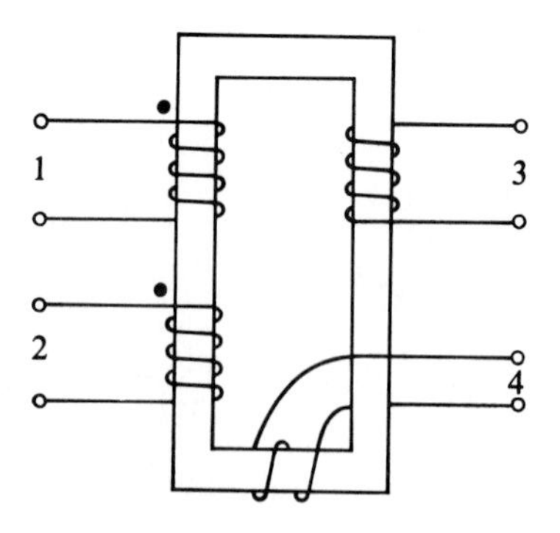

Figure 2.12 Diagram for Problem 2.1.

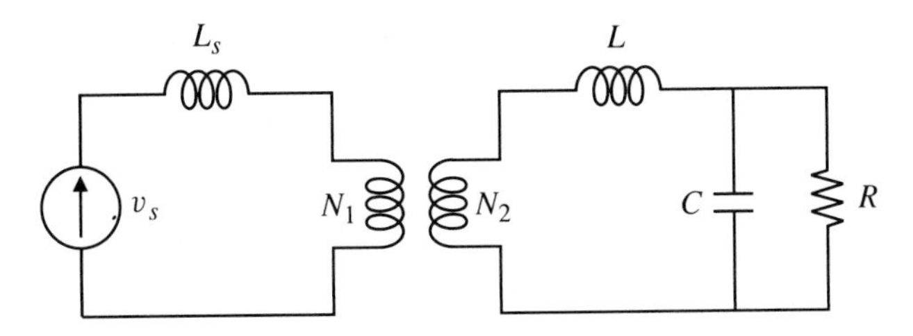

Figure 2.13 Diagram for Problem 2.4.

(b) If $L = 0.1$ H, $R = 15\ \Omega$, $C = 150\ \mu$F, and $N_1/N_2 = 6$, determine the values of the transferred elements.

(Section 2.1)

2.5 A 5-kHz ac generator can be modeled as a source voltage of 250 V (rms) in series with an inductance of 1.2 mH. The generator is to supply power to a resistive load of 0.67 Ω through a transformer that may be considered ideal.

(a) What should be the turns ratio of the transformer to achieve maximum power into the load?

(b) What are the required voltage and current ratings of the two windings of the transformer?

(Section 2.1.1)

2.6 An amplifier is to be connected to a loudspeaker by means of a two-winding transformer. The loudspeaker can be considered as a resistive load of 8 Ω. The amplifier can be modeled as a source of audio frequency voltage in series with a resistance of 1000 Ω. Assuming an ideal transformer, determine the turns ratio that will provide maximum power into the loudspeaker.

(Section 2.1.1)

2.7 A small toroidal ferrite core has a circular cross section with a radius of 7.5 mm and an inner diameter of 30 mm. The primary winding has 1000 turns and the secondary winding has 400 turns. The total thickness of these two windings is 5 mm. Approximately 40% of the total volume occupied by the winding is copper of resistivity $1.72 \times 10^{-8}\ \Omega \cdot$m.

(a) Estimate the resistances of the two windings.

(b) Develop an equivalent circuit of the transformer with all the winding resistances referred to the primary side.

(c) If the core-flux density is not to exceed 0.35 T, determine the maximum value of sinusoidal rms voltage that can be induced in the primary winding at a frequency of 400 Hz.

(d) If the rms current density in the winding conductors is to be 5 A/mm^2, estimate the ratio of the voltage drop across the effective resistance to the induced voltage of (c).

(Section 2.2)

2.8 The core in the transformer described in Problem 2.7 has a relative permeability of 300. Estimate the value of the magnetizing inductance of the transformer as seen from the secondary side.

(Section 2.2.2)

2.9 A transformer has the following parameters: $R_1 = 0.5\ \Omega$, $R_2 = 30\ \Omega$, $L_{1L} = 7.1$ mH, $L_{2m} = 54$ mH, $N_1/N_2 = 0.12$. Draw an equivalent circuit with all parameters referred to:

(a) the high-potential side

(b) the low-potential side.

(Section 2.2)

2.10 A voltage of $v_1 = 200\sqrt{2}\sin(314t)$ is applied to the primary terminals of a transformer. The resulting exciting current can be represented by the Fourier series:

$$i_e = 7.21\sin(314t - 80°) + 2.90\sin(942t - 70°).$$

(a) Determine the rms value of the exciting current.

(b) Calculate the core loss.

(c) Determine the magnetizing reactance, the magnetizing inductance, and the core loss resistance.

(Section 2.2)

2.11 A 20-kVA, 4600:230-V, 60-Hz distribution transformer has a no-load power loss of 280 W. The measured resistance of its high-voltage winding is 10.3 Ω, and the resistance of the low-voltage winding is 0.024 Ω. The transformer is to be immersed in oil in a tank. Convection of air over the tank surface can extract heat at an estimated rate of 25 W/m^2 of surface area and per °C of temperature difference between the tank and the air. If the tank surface temperature is not to exceed 100 °C and the air temperature may be as high as 40 °C, estimate the required surface area of the tank.

(Section 2.3)

2.12 An industrial plant has an electrical load that consists of 150 kW of resistive lighting load plus an induction motor load of 200 kVA at 0.85 lagging power factor. What should be the kVA rating of a transformer to supply this load?

(Section 2.3)

2.13 A 200-kVA, 2300:230-V, 60-Hz single-phase transformer has a primary magnetizing current of 1.8 A. The winding resistances are 0.16 Ω for the primary and 1.5 mΩ for the secondary. The leakage reactance at rated frequency as seen from the primary side is 1.8 Ω.

(a) Using rated nameplate quantities as base values, determine the base voltage, current, impedance, inductance, and complex power for both high- and low-voltage sides of the transformer.

(b) Determine the per unit values of the equivalent winding resistance, the leakage inductance, the magnetizing impedance, and the magnetizing inductance.

(c) Determine the per unit value of the magnetizing current.

(d) If the core loss is 0.8% of the kVA rating, determine the total power loss of the transformer in per unit at rated load.

(Section 2.3.1)

2.14 A 1.5-kVA, 220:110-V, 60-Hz single-phase transformer gave the following test results:

(i) Open-circuit test, low-voltage winding excited:

$$V_2 = 110 \text{ V},\ I_2 = 0.4 \text{ A},\ P_2 = 25 \text{ W},\ V_1 = 220 \text{ V}$$

(ii) Short-circuit test, high-voltage winding excited:

$$V_1 = 16.5 \text{ V},\ I_1 = 6.8 \text{ A},\ P_1 = 40 \text{ W}$$

(iii) Direct-current measurement of winding resistances:

$$R_1 = 0.412\ \Omega,\ R_2 = 0.112\ \Omega$$

Draw an equivalent circuit for this transformer referred to the high-voltage side and calculate the values of its parameters.

(Section 2.4)

2.15 A 10-MVA transformer is required to supply a unity-power-factor industrial process. It is to be operated at full load for 8 h/day and at essentially no load for the remainder of the day. Supplier A offers a transformer with a full-load efficiency of 99.0% and a no-load loss of 0.5%. Supplier B offers a transformer at the same price having a full-load efficiency of 98.8% and a no-load loss of 0.3%.

(a) Which transformer should be chosen?

(b) If energy costs 5.5¢/kW·h, what will be the annual difference in energy cost between the two transformers?

(Section 2.5)

2.16 A 20-kVA, 2200:220-V, 60-Hz, single-phase transformer gave the following test results:

(i) Open-circuit test, low-voltage winding excited:

$$V = 220 \text{ V}, I = 1.52 \text{ A}, P = 161 \text{ W}$$

(ii) Short-circuit test, high-voltage winding excited:

$$V = 205 \text{ V}, I = 9.1 \text{ A}, P = 465 \text{ W}$$

(a) Determine the per unit of rated load at which this transformer has maximum efficiency, and give this maximum efficiency value for a load power factor of 0.8.

(b) The transformer supplies a load that has a daily load cycle that can be approximated by 95% rated load for 8 h at 0.8 power factor lagging, 60% rated load for 8 h at 0.9 power factor lagging, and 5% rated load for 8 h at unity power factor. The load voltage is 220 V. Determine the all-day efficiency of the transformer on this load cycle.

(Section 2.5)

2.17 A 100-kVA, 11 000:2200-V, 60-Hz single-phase transformer has the following equivalent circuit parameters, referred to the low-voltage side:

$$R_{2e} = 0.531 \ \Omega, \ X_{2L} = 2.49 \ \Omega,$$

$$R_{2c} = 4942 \ \Omega, \ X_{2m} = 1441 \ \Omega, \ \frac{N_1}{N_2} = 5.0$$

Determine the voltage regulation of this transformer when supplying its rated load at rated voltage of 2200 V and at a lagging power factor of 0.8.

(Section 2.5)

2.18 A transformer has an equivalent series impedance of $0.01 + j0.05$ pu and a no-load power loss of 0.01 pu, the base for all per unit quantities being the rating of the transformer.

(a) Determine the efficiency of the transformer when supplying rated load at rated voltage and 0.8 lagging power factor.

(b) Determine the voltage regulation for the condition of (a).

(Section 2.5)

2.19 A transformer is required to supply a load that varies from zero to rated value at a power factor of 0.85 lagging. The voltage regulation is not to exceed 5%.

(a) Estimate the maximum permissible value for the per unit leakage inductance of the transformer. The effect of the winding resistances can be ignored.

(b) With rated voltage maintained at the primary side, what will be the steady-state short-circuit current in per unit if the secondary winding is short circuited?

(Section 2.5)

GENERAL REFERENCES

Blume, L.F., *Transformer Engineering*, 2nd ed., John Wiley and Sons, New York, 1951.

Fitzgerald, A.E., Kingsley, C., Umans, S.D., *Electric Machinery*, 5th ed., McGraw-Hill, New York, 1990.

Match, L., *Electromagnetic and Electromechanical Machines*, 2nd ed., Harper and Row, New York, 1977.

Nasar, S.A., *Electric Machines and Transformers*, Macmillan, New York, 1984.

Say, M.G., *Alternating Current Machines*, 5th ed., Pitman, London, 1983.

CHAPTER

3

Basic Principles of Electric Machines

This chapter is intended to develop an understanding of the concepts involved in converting electrical energy to mechanical energy in a motor or converting mechanical energy to electrical form in a generator. As will be shown, motors and generators are similar in their basic principles and differ only in detailed features that are specific to the application.

Almost all electric machines are based on exploiting two basic phenomena: the force exerted on an electric current in a magnetic field and the force produced between ferromagnetic structures carrying a magnetic flux. There are several other energy conversion phenomena for which application has been restricted to very specialized situations. Examples of these are

1. The forces between electric charges, i.e., Coulomb's Law. These electrostatic forces are being exploited to a limited extent in very small micromotors.
2. The piezoelectric effect, which is the tendency of a crystal to change its shape slightly when a voltage is applied to it. Typical applications are in microphones, phonograph pickups, and accelerometers.
3. Magnetostriction, i.e., the tendency of a ferromagnetic material to change shape slightly when magnetized. Major applications are in the

production of ultrasonic waves and in strain gauges for the measurement of very small displacements.

3.1 Forces on Electric Currents

An electric charge of Q coulombs moving at a velocity $\vec{v}$ across a magnetic field of flux density $\vec{B}$ experiences a lateral force of $\vec{F}$ newtons, as shown in Fig. 3.1(a).

$$\vec{F} = Q(\vec{v} \times \vec{B}) \qquad \text{N} \tag{3.1}$$

Note that the result of a cross product of the two vectors $\vec{v}$ and $\vec{B}$ is a vector in the direction that would be followed by a right-hand screw rotating from the first vector to the second.

If the moving charge is in an electrical conductor, its effect may be described by its electric current density vector $\vec{J}$ or by its current i, a scalar. Thus, the force on an incremental volume V of conductor carrying current of density $\vec{J}$ in a field $\vec{B}$ is

$$\frac{\vec{F}}{V} = (\vec{J} \times \vec{B}) \qquad \text{N/m}^3 \tag{3.2}$$

as shown in Fig. 3.1(b).

Most electrical conductors are small enough in cross-sectional area that the current density $\vec{J}$ is essentially uniform across the area and is directed along the conductor length. For this, the most usual situation, the force may be stated as

$$\vec{F} = i(\vec{\ell} \times \vec{B}) \qquad \text{N/m} \tag{3.3}$$

where, as shown in Fig. 3.1(c), $\vec{\ell}$ is a vector along the axis of an increment of the conductor length for which the magnitude of B and its direction relative to the conductor axis can be assumed constant.

The relations of Eqs. (3.1) to (3.3) can be considered as a basis for motoring action, i.e., producing force and thus mechanical energy from an electrical source. The converse relation arises when a conductor is moved with velocity $\vec{v}$ across a magnetic field of density $\vec{B}$. From Eq. (3.1), the charges in the conductor experience a force per unit charge, defined as the electric field intensity $\vec{\mathcal{E}}$ given by

$$\vec{\mathcal{E}} = \frac{\vec{F}}{Q} = \vec{v} \times \vec{B} \qquad \text{N/C or V/m} \tag{3.4}$$

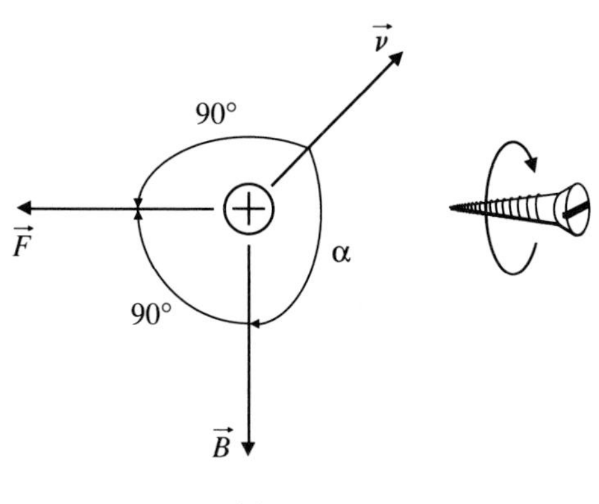

(a)

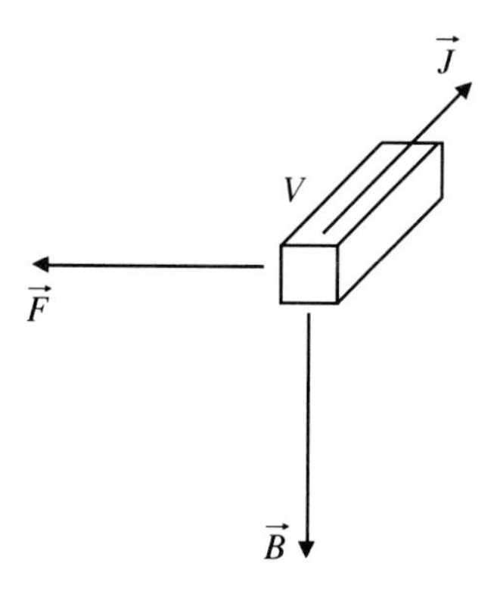

(b)

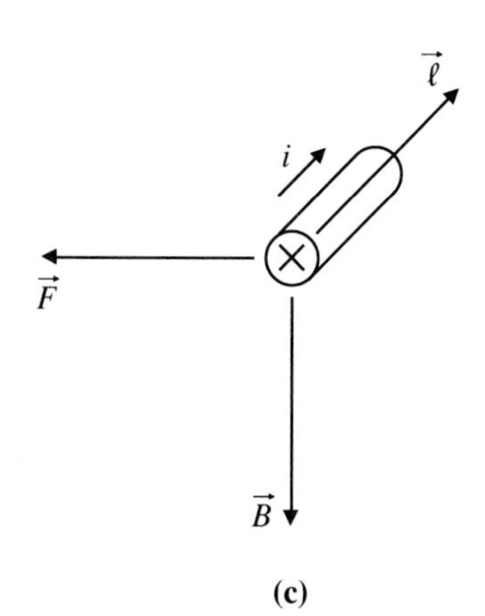

(c)

Figure 3.1 Force on (a) a moving electric charge, (b) a conductor volume element, and (c) a conductor length element in a magnetic field.

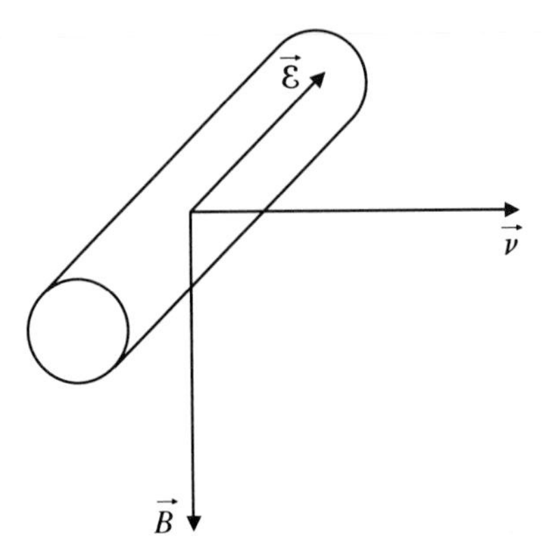

Figure 3.2 Electric field intensity in conductor moving across a magnetic field.

This vector relationship is shown in Fig. 3.2. It is this electric field intensity, integrated along the whole length of a conductor, that produces the induced voltage between its terminals. Therefore, this phenomenon provides a basis for producing a source of electrical energy from mechanical motion.

3.2

Forces on Ferromagnetic Material

Some generators and motors are based directly on the forces on current-carrying conductors in a magnetic field, and the operation of many machines can be adequately predicted on this basis. However, there is another phenomenon that contributes most of the forces and torques produced in common electric machines; this is the force between the aligned magnetic moments of ferromagnetic material. As discussed in Sections 1.4 and 1.5, ferromagnetic material consists of atoms with aligned magnetic moments. A magnetic moment p_m in a flux density B experiences a torque, which tends to align it with the magnetic field, as portrayed in Eq. (1.29) and Fig. 1.8. However, it is another translational force on magnetic moments that is most often exploited in electric machines.

Consider a magnetic moment represented by the loop of current i, as shown in Fig. 3.3. Suppose this loop is placed in a nonuniform magnetic field where the flux density is decreasing in magnitude from left to right. Using the force expression as stated in Eq. (3.2), it can be seen that the major component of force on the loop is radially outward, tending to expand it but not tending to move it. This component is not of use to us. However, there is a smaller component directed perpendicular to the plane of the loop because the flux density is not purely horizontal but has a small component directed radially outward.

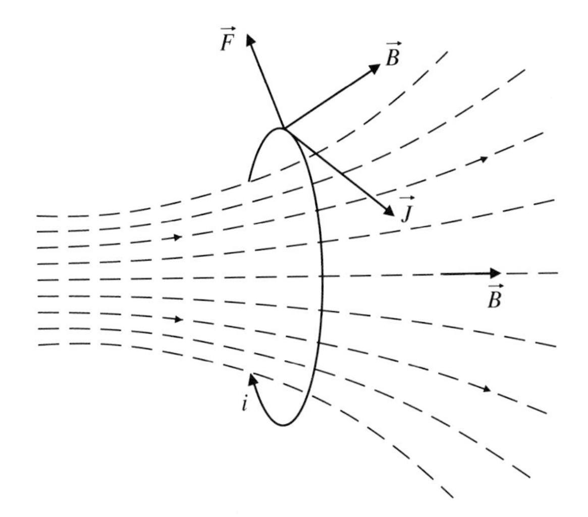

Figure 3.3 Force tending to move a magnetic moment to a region of higher flux density.

Thus, the net force on the loop is such as to move it to the left, i.e., to the region of higher flux density.

As an extension of this concept, consider the two aligned magnetic moments of Fig. 3.4(a). The moment $\vec{p}_{m1}$ on the left produces a magnetic field of reducing flux density along its axis. The second moment $\vec{p}_{m2}$ experiences a translational force $\vec{F}$ attracting it toward the first moment. This is the basis for attraction between magnetized bodies.

Magnetic flux can be made to cross an air gap by aligning the magnetic moments on each side of the gap, as shown in Fig. 3.4(b). The total force between the two blocks of ferromagnetic material then will be the summation of all the forces between moments on the two sides of the gap. As might be expected, this force is dependent on the net aligned magnetic moment per unit volume, i.e., the magnetization M as defined in Section 1.5. Section 3.7 will show that the force per unit of gap area A tending to close the gap is

$$\frac{F}{A} = \frac{\mu_0 M^2}{2} \qquad \text{N/m}^2 \tag{3.5}$$

Essentially all of the flux density in the air gap results from the alignment of the magnetic moments on each side of the gap. From Eq. (1.34),

$$\vec{B} = \mu_0 \vec{M} \qquad \text{T} \tag{3.6}$$

It follows that the force can usually be expressed in terms of the flux density normal to the surface of the air gap as

$$\frac{F}{A} = \frac{B^2}{2\mu_0} \qquad \text{N/m}^2 \tag{3.7}$$

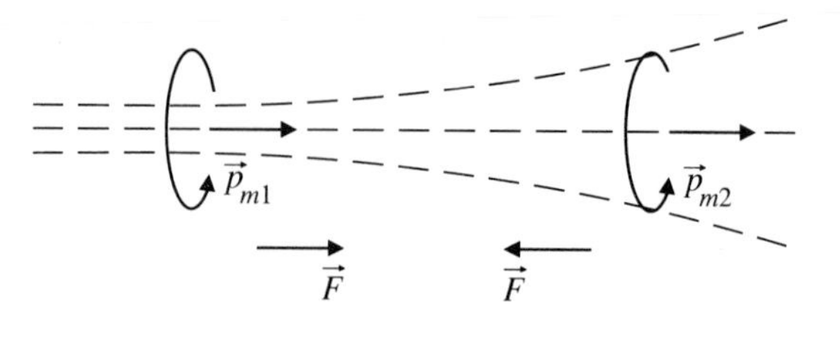

(a)

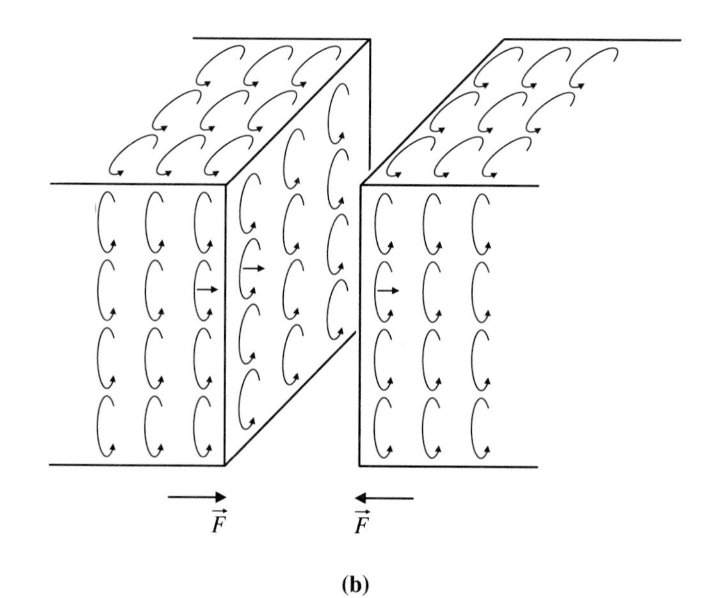

(b)

Figure 3.4 (a) Force of attraction between two aligned moments and (b) force between two faces of an air gap in a ferromagnetic structure.

3.3

Elementary Rotating Machine

The principles of an elementary machine are demonstrated in the arrangement of Fig. 3.5, where a loop of conductor is free to rotate in a magnetic field. The field is created in this instance by permanent magnets on each side, producing a radially directed flux density across the air gaps. A central iron core and an outer iron yoke are used to provide a low reluctance path for the magnetic flux.

Suppose that the two ends of the loop are connected to a pair of conducting slip rings that rotate with the loop. A source of electric current i can now be connected to the loop through a pair of stationary brushes contacting the pair

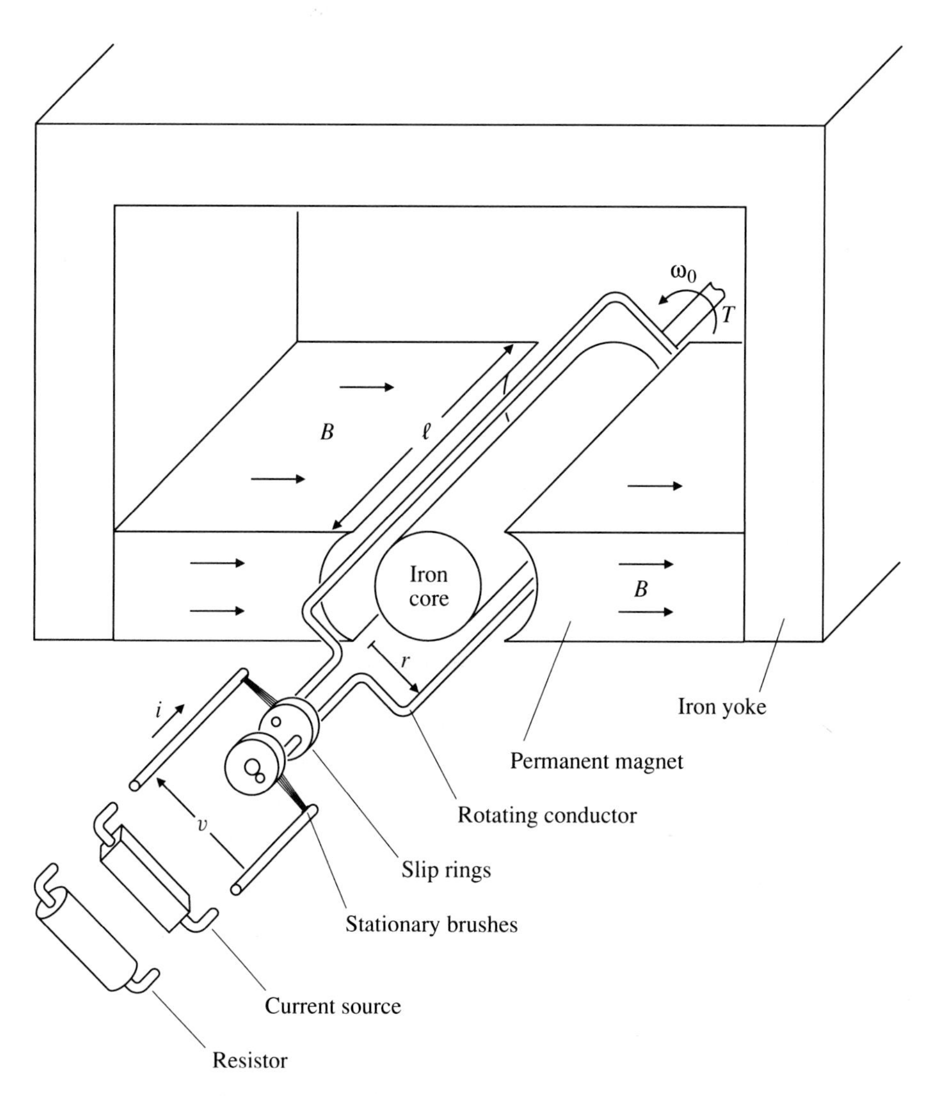

Figure 3.5 Elementary rotating machine.

of slip rings. The direction of the current i in the loop conductors is perpendicular to the flux density B such that there is a downward tangential force F on the left-hand side of the loop, where, from Eq. (3.3),

$$F = B\ell i \qquad \text{N} \tag{3.8}$$

On the other side of the loop, the force is upward. This pair of forces constitutes a torque on the loop of

$$T = 2rF = 2r\ell Bi \qquad \text{N}\cdot\text{m} \tag{3.9}$$

This torque can be used to accelerate the loop or to drive a mechanical load attached to it. If the loop is rotating at angular velocity ω_0 rad/s, the mechanical power being produced will be

$$P_0 = T\omega_0 = 2r\ell Bi\omega_0 \qquad \text{W} \tag{3.10}$$

At the same time, the tangential velocity ν of the conductor will be perpendicular to the flux density B in the air gaps, and this will produce an electric field intensity $\mathcal{E}$ in the conductor in a direction opposite to the direction of the current:

$$\mathcal{E} = \vec{\nu} \times \vec{B} \tag{3.11}$$

where

$$\nu = r\omega_0 \tag{3.12}$$

The voltage induced in the loop is the integral of $\mathcal{E}$ around the loop. Ignoring conductor resistance, the induced voltage will be equal to the voltage v across the conductor terminals. Thus,

$$v = e = 2\ell\mathcal{E} = 2\ell rB\omega_0 \qquad \text{V} \tag{3.13}$$

The electric power input from the current source is therefore

$$P_e = vi = 2\ell rB\omega_0 i = P_0 \qquad \text{W} \tag{3.14}$$

which is seen to be equal to the power P_0 in Eq. (3.10). Thus, as long as the loop is in the magnetic field, and to the extent that power losses are negligible, the machine acts as an ideal motor converting electrical to mechanical power.

Next, suppose that the loop is connected through the slip rings to an electrical load such as a resistor. If the loop is being rotated at angular velocity ω_0 by a mechanical driver, the terminal voltage v of Eq. (3.13) will be applied to the load, causing a load current i out of the left-hand terminal. For a load resistance R,

$$i = -\frac{v}{R} \qquad \text{A} \tag{3.15}$$

With reversed current direction, the direction of the torque T on the loop is also reversed and is now opposite to that of the angular velocity. An input of mechanical power is now required to provide the electrical output power. The machine is now acting as an ideal electrical generator.

For an alternative analysis, consider the view in Fig. 3.6(a), which shows the loop in a plane at an angle of β rad to the horizontal axis. The flux linkage λ of the loop while it is within the magnet arc is

$$\lambda = 2B\ell r\beta \qquad \text{Wb} \tag{3.16}$$

Then, from Eq. (1.11), the induced voltage in the loop will be

$$e = \frac{d\lambda}{dt} = \frac{d\lambda}{d\beta}\frac{d\beta}{dt} = \frac{d\lambda}{d\beta}\omega_0 \qquad \text{V} \tag{3.17}$$

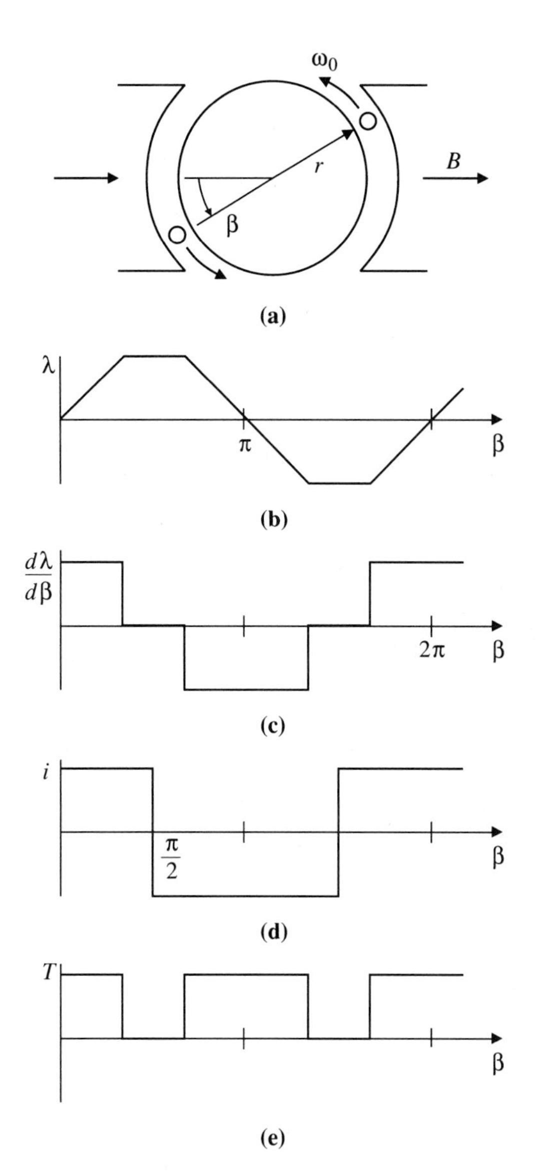

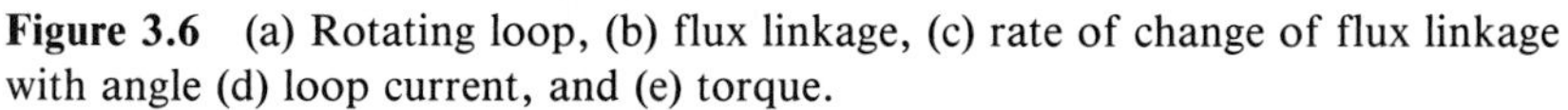
Figure 3.6 (a) Rotating loop, (b) flux linkage, (c) rate of change of flux linkage with angle (d) loop current, and (e) torque.

In this idealized machine, power input equals power output:

$$vi = ei = T\omega_0 \qquad \text{W} \tag{3.18}$$

Thus, from Eqs. (3.17) and (3.18), the torque can be expressed as

$$T = i\,\frac{d\lambda}{d\beta} \qquad \text{N}\cdot\text{m} \tag{3.19}$$

Idealized waveforms of λ and $d\lambda/d\beta$ are shown in Figs. 3.6(b) and (c). It is noted that $d\lambda/d\beta$ reverses each half revolution. Thus, if the current in the conductor were held constant, the torque averaged over a revolution would be zero. One means of producing a motor with unidirectional torque is to reverse the current i from the current source when β is at or near $\pi/2$ and again at $3\pi/2$, as shown in Fig. 3.6(d), giving the pulsating but unidirectional torque of Fig. 3.6(e). This approach is exploited in many motors supplied from electronic sources. A sensor attached to the shaft provides a signal indicating when the switching of the current should occur. It is also the basis of the commutator machines to be discussed in Section 3.5 and Chapter 4.

EXAMPLE 3.1 The essential components of a loudspeaker are shown in Fig. 3.7. The permanent magnet produces a uniform radially directed magnetic flux density of 0.8 T across a cylindrical air gap. A coil of $N = 30$ turns is wound on an insulated cylinder of diameter $d = 20$ mm. When assembled, the coil is inserted in the air gap of the magnet.

(a) Determine the force on the speaker cone as a function of the coil current.

(b) Determine the induced voltage in the coil per unit of horizontal velocity of the coil.

(c) Over most of the audio frequency range, the force produced by the coil current is absorbed by the air being driven by the motion of the cone. Suppose this force is 0.3 N per m/s of cone velocity. Ignoring any resistance in the coil, determine the ratio of coil voltage to coil current.

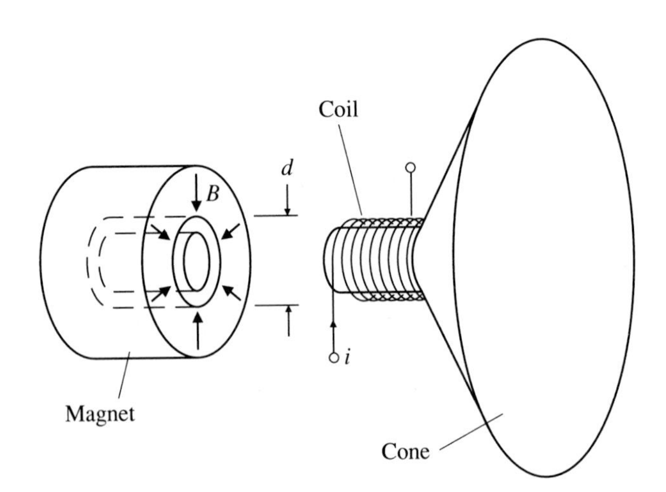

Figure 3.7 Loudspeaker assembly for Example 3.1.

Solution

(a) The flux density and the current are directed perpendicular to each other. Thus, from Eqs. (3.2) or (3.8), the force per unit of coil current can be expressed as

$$\frac{F}{i} = \pi NdB = \pi \times 30 \times 0.02 \times 0.8 = 1.51 \qquad \text{N/A}$$

(b) The velocity ν is perpendicular to the flux density. The induced voltage is the integral of the electric field intensity along the coil length. From Eq. (3.4),

$$\frac{e}{\nu} = \pi NdB = 1.51 \qquad \text{V}\cdot\text{s/m}$$

(c) The load force is

$$F = 0.3\nu \qquad \text{N}$$

With no coil resistance, the coil terminal voltage is

$$v = e = 1.51\nu = 1.51\,\frac{F}{0.3} = 1.51^2\,\frac{i}{0.3} = 7.6i$$

Thus, the speaker acts as a resistive load of 7.6 Ω, as seen by the source supplying the electrical power.

3.4

Placing Conductors in Slots

The elementary machine of the previous section contained a conducting loop rotating about a cylindrical iron core. To provide greater mechanical rigidity, the conductor could be attached to the surface of the iron core and the combined assembly or rotor could be rotated. This arrangement involves a continually changing flux density in the iron core. To prevent excessive eddy current losses, the core may then have to be made of a stack of circular laminations, as discussed in Section 1.8. With this arrangement, the conductor is still in the magnetic field and the relations of Eqs. (3.8) and (3.11) apply.

Let us now consider the possibility of fitting each side of the conductor loop into a slot in the rotor surface, as shown in Fig. 3.8(a). This would allow the air gaps between the rotor and the outer stationary magnetic system or stator to be made just large enough for mechanical clearance. The same flux density could then be produced in the air gaps with a much lower mmf from the stator magnets.

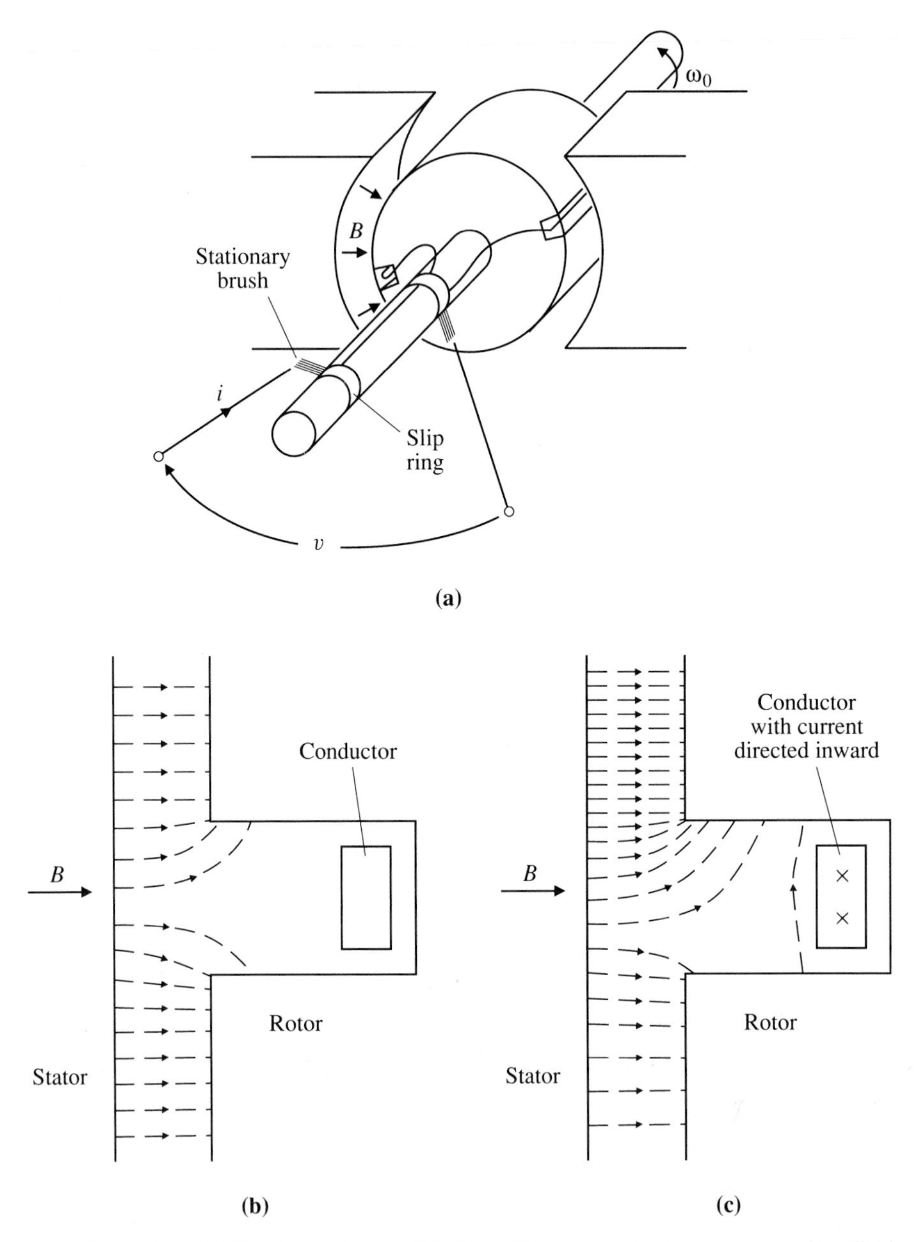

Figure 3.8 (a) Conductor in slot, (b) field with zero conductor current, and (c) field with conductor current.

On first examination, it might appear that the machine would no longer operate as before. The conductor is now in a region of very low flux density because most of the flux takes the short path across the air gap and little penetrates down into the slot, as shown in the field plot of Fig. 3.8(b). The direct torque on the conductor loop carrying current i will still be given by Eq. (3.9) but will be quite small. On the other hand, the flux linkage λ of the loop will still change as the loop is rotated in essentially the same manner as when the conductor was on the surface of the core. Thus, the overall torque as given by Eq. (3.19) should be unchanged.

In explanation, the torque is mainly exerted on the iron core of the rotor, and only a small torque is exerted directly on the coil. As shown in Fig. 3.8(c), when a current is passed through the conductor in the slot, the flux density on one side of the slot near the gap is increased while that on the other side is reduced. From the discussion of Section 3.2, there is a force tending to close an air gap between two sections of ferromagnetic material carrying a magnetic flux. In the small air gap between the stator and rotor of Fig. 3.8(c), this force is directed radially and thus produces no rotational torque. However, in the region of the slot with its current-carrying conductor, the flux pattern is asymmetrical. It is noted from Eq. (3.7) that the force per unit area is proportional to the square of the flux density. Thus, there will be a large downward force on the upper side of the slot and a smaller upward force on the lower side of the slot. The result is a tangentially directed force creating a torque on the rotor core. The magnitude of the torque is correctly given by Eq. (3.19). This is equal to the torque that was predicted for the conductor in the air gap in Eq. (3.9), but the mechanism of force production is different. In many situations, the difference may be ignored for calculation purposes but is conceptually important.

When the conductor is in a slot, the voltage induced in it can no longer be considered to arise from its velocity v across a flux density B as in Eq. (3.12). Rather, the motion of the iron core produces a rate of change of flux linkage of the coil which results in the induced voltage as given by Eq. (3.17). Again, this turns out to be numerically equal to that stated in Eq. (3.13) for the air-gap conductor. The mechanism is different but the overall effect is the same.

3.5

Some Elementary Types of Electric Machines

If the elementary machine of Fig. 3.8(a) is driven at angular velocity ω_0, the voltage at the terminals has the form shown in Fig. 3.9(a). While the slots are under the poles, the rate of change of flux linkage is approximately constant, producing nearly constant induced voltage. When the coil sides are not under the poles, the flux linkage of the coil is essentially constant, and the induced voltage is zero. One cycle of this alternating voltage is produced for each revolution of the rotor. The frequency of this alternation is related to the rotational angular velocity by

$$f = \frac{\omega_0}{2\pi} \qquad \text{Hz} \tag{3.20}$$

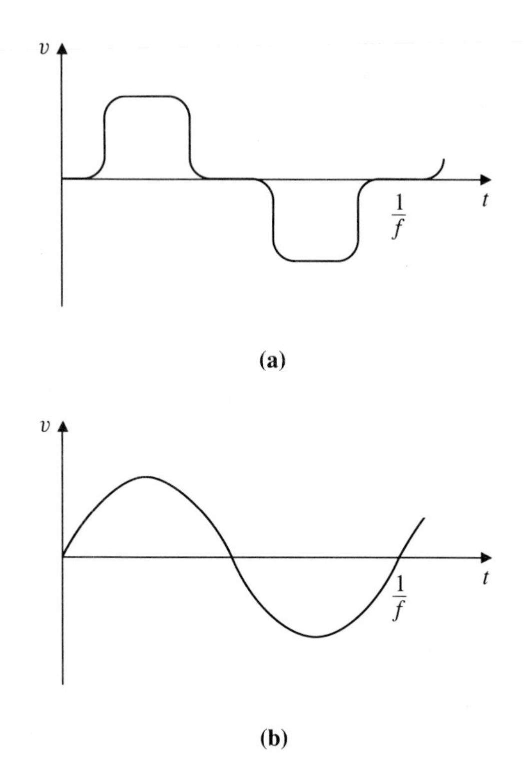

Figure 3.9 (a) Voltage produced by machine of Fig. 3.8(a). (b) Sine wave of voltage.

The most readily usable form of alternating voltage is the sine wave, as shown in Fig. 3.9(b). A major reason for this is that a sine wave retains the same shape when either differentiated or integrated. One way of obtaining such a sinusoidal waveform approximately is to shape the air gaps so that the flux density varies more or less sinusoidally. Other approaches will be discussed in the following sections.

An alternative structure for an elementary machine is shown in Fig. 3.10. In this structure, the magnets are placed on the rotor core and the conductor loop is placed in slots on the inner surface of the stator. The flux in the rotor core remains substantially constant. Thus, the rotor core could be made of solid steel. The stator will experience a time-varying magnetic flux density as the rotor rotates and thus usually must be made of laminations. With the current direction shown in Fig. 3.10, there will be a clockwise torque on the stator and thus a counterclockwise reaction torque on the rotor. The induced voltage in the conductor loop will have the alternating form shown in Fig. 3.9 with the frequency of Eq. (3.20). Because of the synchronism between the electrical frequency and the mechanical speed, machines of the type shown in Figs. 3.8 and 3.10 are known as *synchronous machines.*

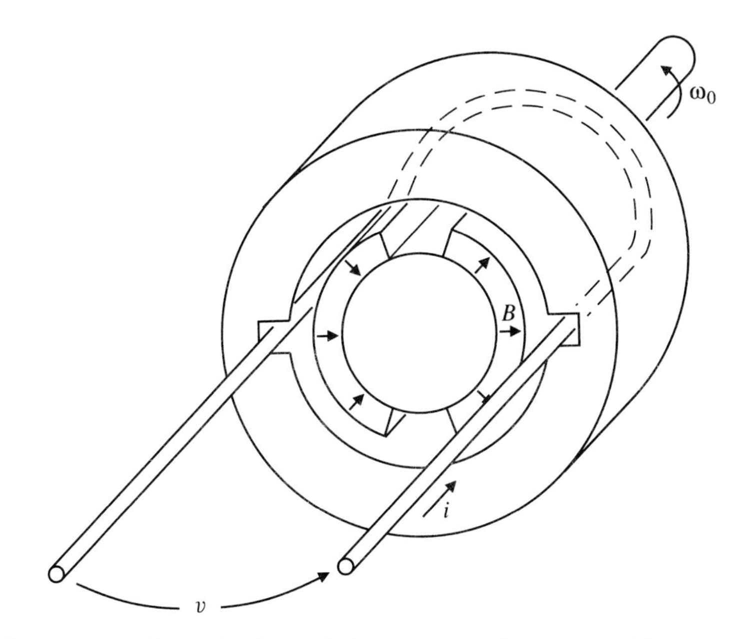

Figure 3.10 Alternative form of elementary synchronous machine.

The magnetic field can be produced by use of a current-carrying coil wound around a ferromagnetic rotor core, as shown in Fig. 3.11. This is known as a *field coil.* The field current can be supplied through insulated slip rings mounted on the rotor shaft and connected to a current supply through stationary brushes. Figure 3.11 also suggests how the air gap can be shaped to produce a desired sinusoidal flux density distribution.

In all the elementary structures introduced thus far, the single-turn loop can be replaced by a multi-turn coil placed in the same slots. The flux linkage then will be increased by a factor equal to the number of turns in the coil.

Better use of the stator surface can be achieved by placing coils in a number of slots over an arc of the stator, as shown in Fig. 3.12. These coils have front and back connections that are shaped to circle around the stator periphery, keeping the rotor space clear. The coils are usually connected in series.

Suppose coils are arranged so that there are N conductors more or less uniformly distributed over an arc of γ rad on each side of the stator. This conductor distribution, with each conductor carrying current i, can be considered as a current sheet of linear density K where, for a radius r,

$$K = \frac{Ni}{r\gamma} \qquad \text{A/m of periphery} \tag{3.21}$$

If the current sheet is in a uniform flux density B directed perpendicular to the current direction, the force per unit of sheet area will be

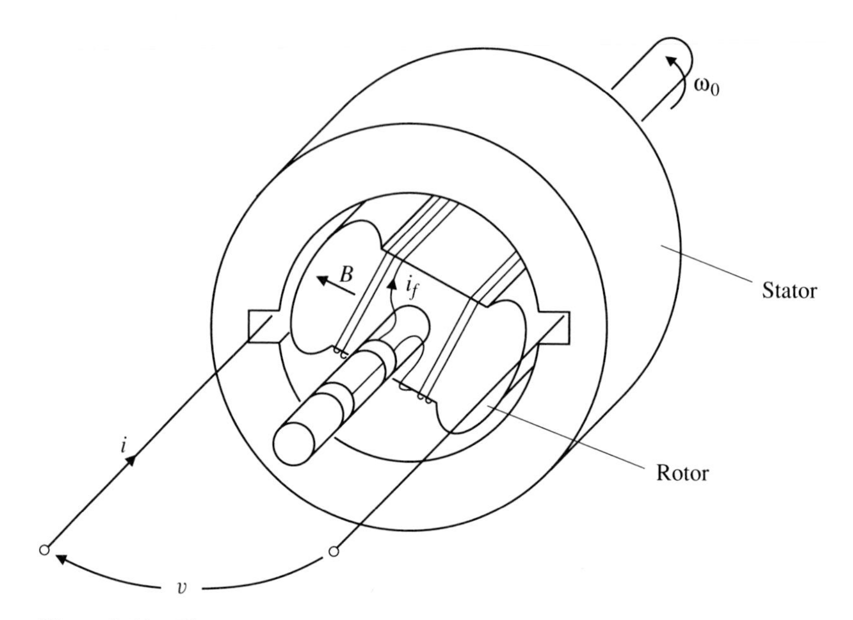

Figure 3.11 Elementary synchronous machine with field coil.

$$\frac{F}{A} = BK \qquad \mathrm{N/m^2} \tag{3.22}$$

The torque due to the conductors in the two arcs will then be

$$\begin{aligned} T &= BK(2\gamma r\ell)r \\ &= 2\gamma r^2 \ell BK \qquad \mathrm{N\cdot m} \end{aligned} \tag{3.23}$$

In addition to making better use of the stator surface, distributing the winding over an arc as shown in Fig. 3.12 has the additional advantage that the induced voltage is more nearly sinusoidal than that induced in any one turn. Figure 3.13 shows the induced voltages in three displaced coils, each voltage being similar in shape to that of Fig. 3.9(a). The sum of these is seen to be approximately sinusoidal.

Suppose the arc occupied by the stator winding in Fig. 3.12 is 60° or $\pi/3$ rad in width. The stator can then accommodate two other similar windings, each of the same width, displaced symmetrically, and $2\pi/3$ and $4\pi/3$ rad ahead of the first winding. For simplicity, Fig. 3.14(a) shows only the center turn of each of the three windings, designated as *abc* on the entry side and $a'b'c'$ on the exit side. When the machine is rotated at angular velocity ω_0, the voltages on these three windings constitute a three-phase set, as shown with idealized waveform in Fig. 3.14(b). Plotted on an abscissa of $\omega_0 t$, the voltage on phase b lags that of phase a by $2\pi/3$ rad or 120°.

Three-phase systems have considerable advantages beyond the ability to make good use of all the inner stator surface of a machine. If the three phase

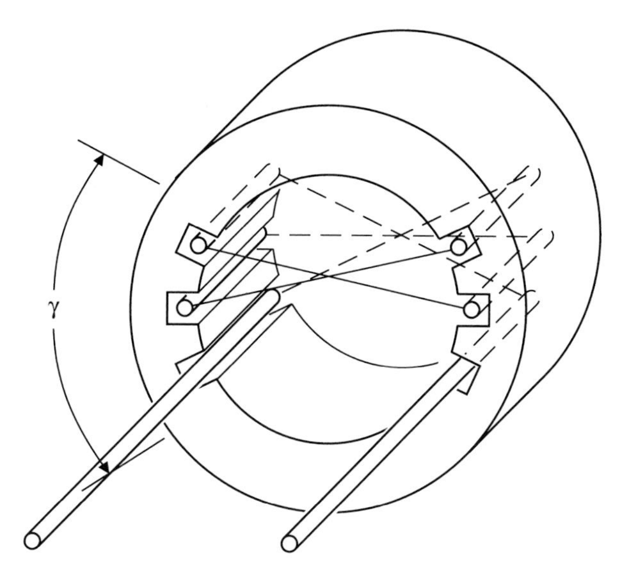

Figure 3.12 Series coils in slots over a stator band of arc γ.

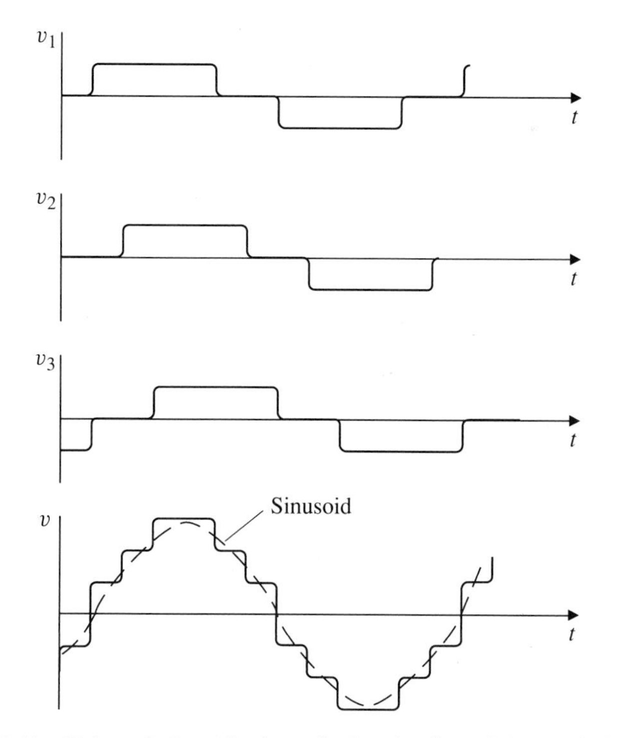

Figure 3.13 Voltage induced in three displaced coils and the total winding voltage $v = v_1 + v_2 + v_3$.

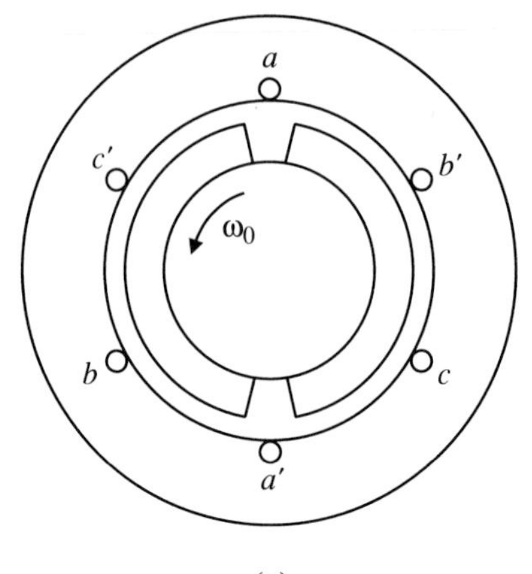

(a)

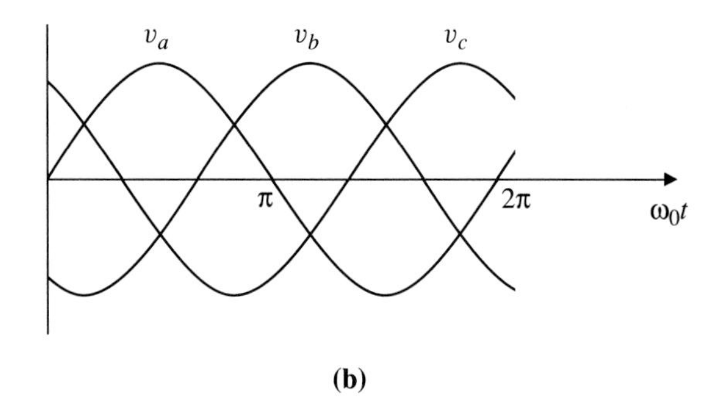

(b)

Figure 3.14 (a) Machine with three-phase windings and (b) three-phase balanced set of voltages.

windings of Fig. 3.14(a) are connected to three identical resistive loads, the load currents will have waveforms similar to those shown for the voltages in Fig. 3.14(b). It may be noted that the sum of the three phase values at any instant is zero. Thus, instead of using six transmission conductors to connect the three machine windings to the three loads, only three conductors are required.

The three-phase machine of Fig. 3.14(a) supplying electric power to three resistive loads can be considered as a synchronous generator. The phase currents will be of such a polarity as to produce a torque on the rotor opposite to the direction of rotation. Ideally, the electrical power output is supplied by an equal mechanical power input from the prime mover driving the generator shaft.

The machine windings can also be connected to a source of three phase voltages. The machine will then produce useful torque only when operated at a speed

$$\omega_0 = \omega_s = 2\pi f \qquad \text{rad/s} \tag{3.24}$$

where f is the frequency of the supply in Hz and ω_s is its angular frequency in rad/s. The machine connected to a source of constant voltage and frequency can be considered as a synchronous motor or as a synchronous generator depending on the direction of power flow. This class of synchronous machines is discussed in Chapter 6.

A serious limitation of a synchronous machine as a motor is its inability to operate at any speed other than its synchronous value. Unless a source of variable frequency is available, its speed cannot be controlled over a range of values. To overcome this limitation, other machine types are required. To introduce these types in elementary form, let us first examine the machine of Fig. 3.15(a), which acts as an electrically controlled brake. A magnetic field of controllable magnitude is produced by use of field coils on the two stator poles. The rotor consists of an iron core with conductors either on the surface or in slots. These conductors are all shorted together at the two ends of the rotor. When the rotor is rotated, a voltage e is induced in each conductor according to Eq. (3.13). Because the conductors are short circuited, a current i will flow such that $i = e/R$, where R is the resistance of each conductor. This current interacts with the flux density to produce a torque that opposes the rotation. The value of this braking torque is dependent on the rotor speed ω_0 and on the flux density, which in turn depends on the current i_f in the field coil, as shown for several values of field current in Fig. 3.15(b).

As a further step in producing variable-speed motor action, consider the machine of Fig. 3.16(a). This is the same as in the brake except that the outer "stator" member is now being rotated at angular velocity ω_s by a prime mover. The pole flux moving past the shorted rotor conductors induces a voltage in them and produces a short-circuited conductor current of the polarity shown. This current is in such a direction as to produce a counterclockwise torque tending to accelerate the rotor to make its angular velocity ω_0 approach that of the revolving stator. The torque will be proportional to the difference between the angular velocities $(\omega_s - \omega_0)$, as shown in Fig. 3.16(b). This machine operates as a clutch for which the torque tends to make the rotor and stator speeds equal. It can accelerate a shaft load up toward stator speed ω_s. If the rotor is driven by its mechanical system such that $\omega_0 > \omega_s$, the torque will reverse and act to bring the speeds toward equality.

The three-phase induction motor achieves a drive characteristic that is basically similar to that of the clutch. Instead of actually rotating the magnetic system of the "stator," a revolving magnetic field is produced by currents in a set of three symmetrically placed windings in a stationary stator, as shown in Fig. 3.17(a). These windings are each distributed over a band, as in Fig. 3.12, but only the center conductor is shown for simplicity. The three windings are supplied with sinusoidally varying three-phase currents of the form shown in Fig. 3.17(b) where the current in phase a is

$$i_a = \hat{i} \sin \omega_s t \qquad \text{A.} \tag{3.25}$$

Consider the instant t_1 in Fig. 3.17(b). The current in phase a is at its maximum positive value while the currents in phases b and c are each half that mag-

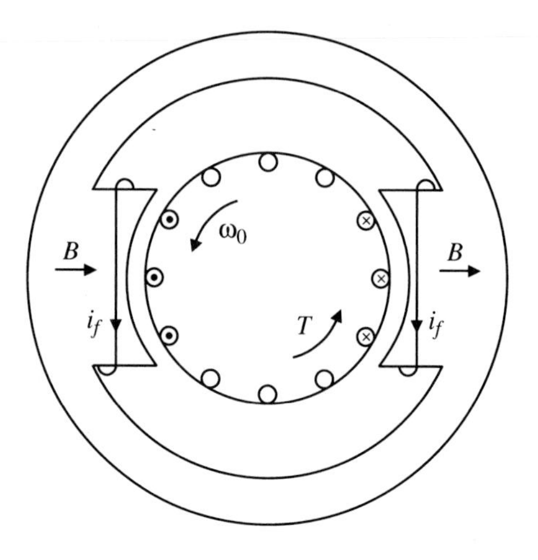

(a)

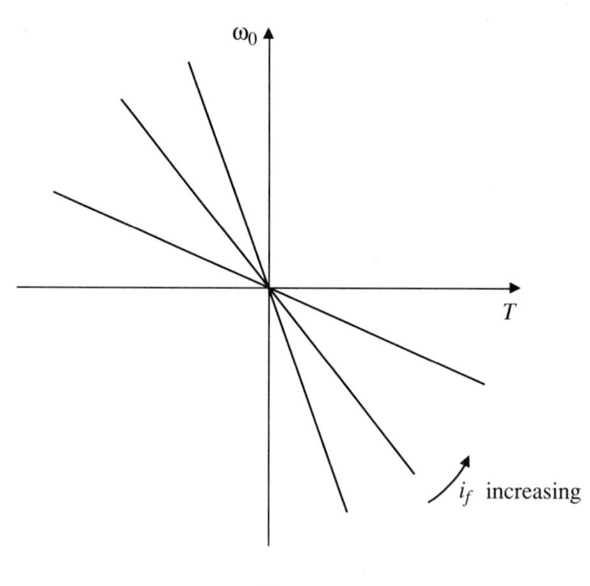

(b)

Figure 3.15 (a) Electrically controlled brake and (b) speed-torque relation.

nitude and negative. The corresponding current pattern in the stator conductors is suggested by the t_1 portion of Fig. 3.17(c). The currents are such as to produce a horizontal flux density across the air gap to the rotor. At time t_2, which is 60° later in the current cycle, i_c is maximum negative while i_b and i_a are half the maximum value positive. The magnetic field has now been rotated by 60°

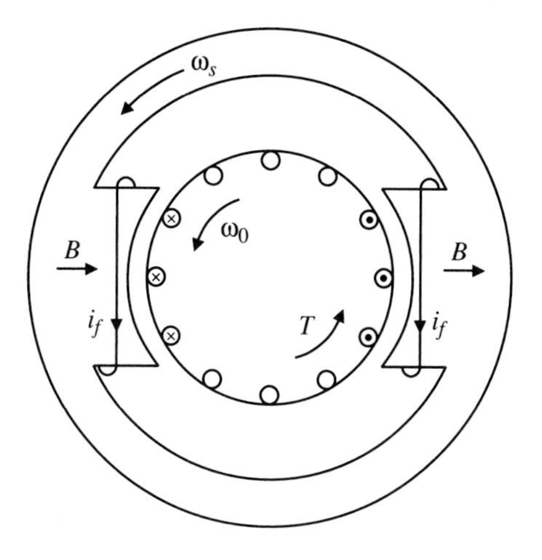

(a)

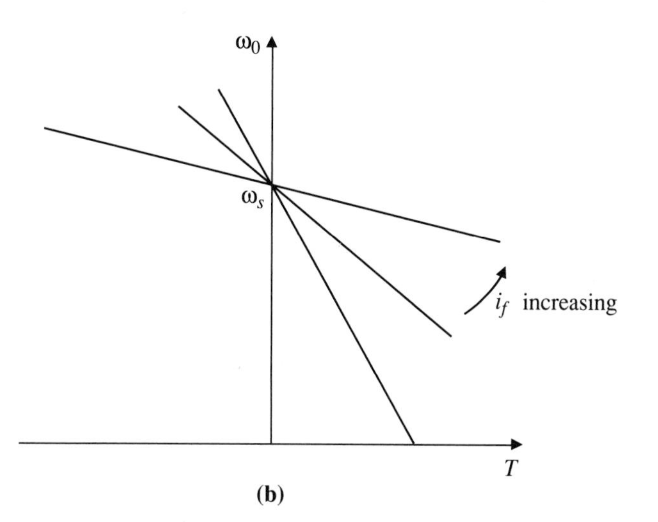

(b)

Figure 3.16 (a) Electrically controlled clutch and (b) speed-torque relation.

counterclockwise, as shown for time t_2 in Fig. 3.17(c). At time t_3, the field is rotated an additional 60°. For one complete cycle of the three-phase current, the magnetic field rotates through one complete revolution. Thus, a three-phase sinusoidal current supply of angular frequency ω_s flowing in a set of three symmetrically displaced windings produces a magnetic field that rotates at angular velocity ω_s.

This production of a rotating magnetic field using electric currents is the basis for the *induction machine*, invented by Tesla in 1883. This machine is capable of starting when connected to the available constant-frequency, three-phase,

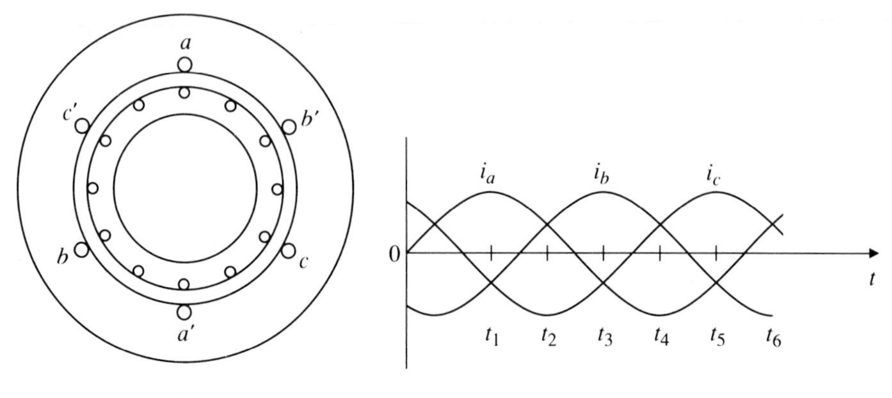

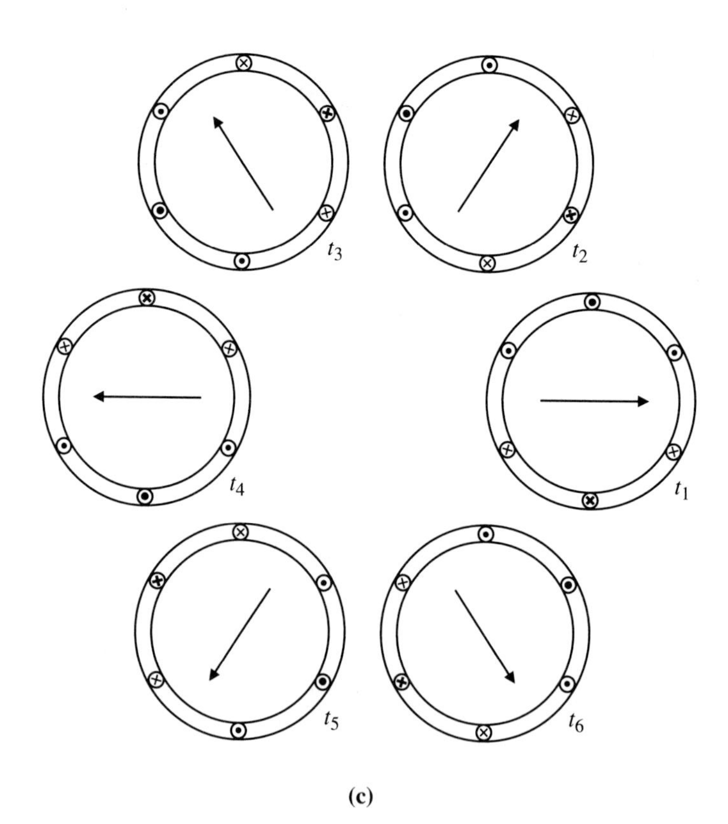

(c)

Figure 3.17 (a) Stator with three-phase windings, (b) three-phase stator winding currents, and (c) magnetic field produced by stator through one cycle.

alternating voltage supply. It normally runs as a motor at a speed slightly below the synchronous speed, i.e., the angular speed of the magnetic field. However, if it is connected to a three-phase voltage source and driven by a prime mover at a speed somewhat greater than synchronous speed, it can act as an induction generator. Induction machines are considered in more detail in Chapter 5.

Another class of electrical machines is known as *commutator machines*, or, more commonly, *direct-current machines*. An elementary form of commutator machine can be produced by replacing the slip rings in the single-loop machine of Fig. 3.8(a) by a two-segment switch or commutator mounted on the shaft, as shown in Fig. 3.18(a). The two ends of the coil are connected to two arcs of conductor insulated from each other and from the shaft. Stationary brushes are so placed that the polarity of the connections between the coil ends and the external terminals is reversed when the induced voltage in the coil is zero, i.e., twice per revolution when the coil is in a vertical plane. When the rotor is rotated at angular velocity ω_0, the terminal voltage has the waveform shown in Fig. 3.18(b). In contrast with the alternating voltage of the elementary slip ring machine shown in Fig. 3.9(a), this voltage is unidirectional. If a direct current is passed into the positive terminal, there will be an average electrical power input, and the machine will act as a motor. If a resistive load is connected to the terminals, an average direct current will flow out of the positive terminal, and the machine will act as a direct-current generator.

In a practical machine, the whole cylindrical rotor is used to accommodate coils located in slots. These coils are connected in series in a closed loop. Each connection point is attached to a commutator segment or bar. These bars come into play only when they are in contact with one of the two brushes. The result is an essentially constant terminal voltage, proportional to the rotor speed and the magnetic field. This machine, together with suitable control gear, can provide adjustable speed and torque over a wide range, as will be discussed in Chapter 4.

EXAMPLE 3.2 A machine has a stator winding distributed over two arcs each of 60°, as shown in Fig. 3.12. The slots are rectangular with a depth of 10 mm. At the stator inner surface, the width of each slot is equal to the width of the tooth. The conductor turns fill 55% of the slot area. The current density in the conductor material is 5 A/mm^2. The flux density averaged over the arc is 0.9 T. The inner radius of the stator is 70 mm, and its axial length is 120 mm. Determine the torque.

Solution The actual winding can reasonably be replaced by an equivalent solid conductor of depth

$$d_e = 0.01 \times \tfrac{1}{2} \times 0.55 = 0.00275 \qquad \text{m}$$

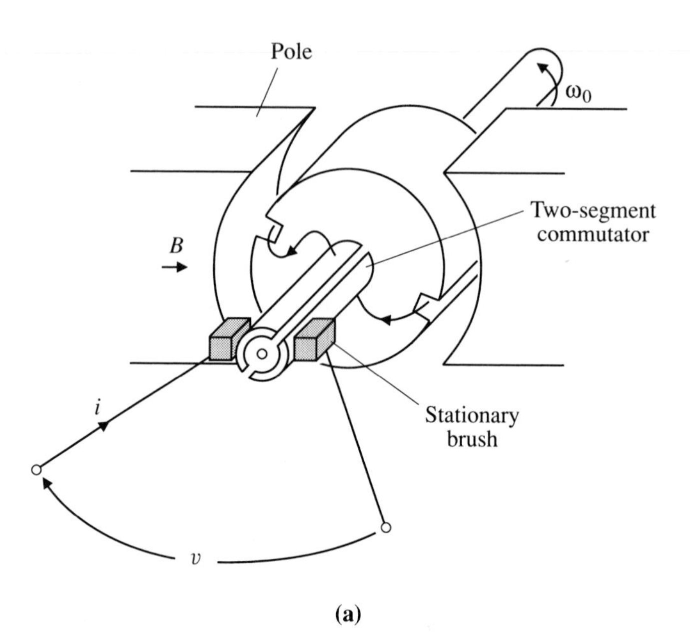

(a)

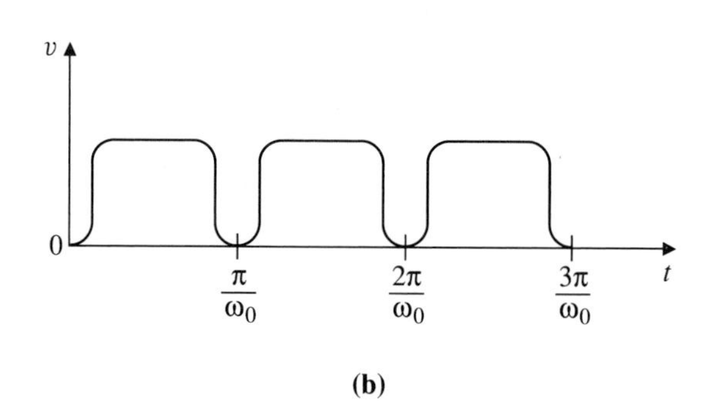

(b)

Figure 3.18 (a) Elementary commutator machine and (b) generated terminal voltage.

The linear current density is then

$$K = d_e J = 0.00275 \times 5 \times 10^6 = 13\,750 \qquad \text{A/m}$$

The force per unit of stator surface area is

$$\frac{F}{A} = BK = 0.9 \times 13\,750 = 12\,380 \qquad \text{N/m}^2$$

The effective area is

$$A_e = 2\gamma r\ell = 2 \times \frac{\pi}{3} \times 0.07 \times 0.12 = 0.0176 \qquad \text{m}^2$$

The torque is

$$T = \frac{F}{A} A_e r = 12\,380 \times 0.0176 \times 0.07 = 15.25 \qquad \text{N}\cdot\text{m}$$

3.6 Ferromagnetic Actuator

This section introduces a group of electromechanical devices that directly exploit the forces discussed in Section 3.2 that tend to close the air gap in a ferromagnetic system. An example is shown in Fig. 3.19. A coil on a fixed ferromagnetic core establishes flux across two air gaps to a ferromagnetic armature that is arranged so as to lift a mechanical load against either gravity or the force of a spring. Let us use this example to establish some relations for the force produced across an air gap by a magnetic field.

From the concept of conservation of energy, the following relation applies:

$$\begin{pmatrix}\text{Electrical}\\ \text{energy}\\ \text{from source}\end{pmatrix} = \begin{pmatrix}\text{Mechanical}\\ \text{energy}\\ \text{to load}\end{pmatrix} + \begin{pmatrix}\text{Energy}\\ \text{lost}\end{pmatrix} + \begin{pmatrix}\text{Increase of}\\ \text{energy stored in}\\ \text{magnetic field}\end{pmatrix} \qquad (3.26)$$

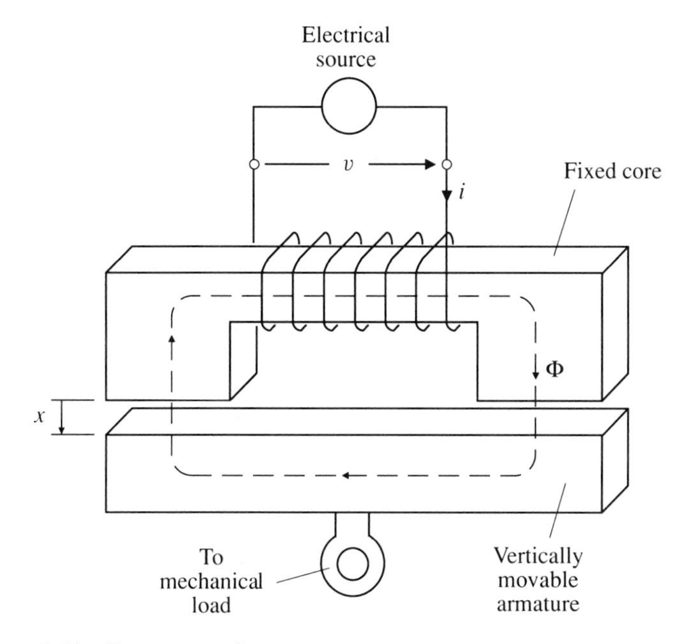

Figure 3.19 Ferromagnetic actuator.

Suppose we assume that the hysteresis and eddy current losses in the core are negligible and that any mechanical energy loss is included in the energy delivered to the load. Equation (3.26) can then be restated as

$$\begin{pmatrix}\text{Electrical energy}\\ \text{minus}\\ \text{resistance loss}\end{pmatrix} = \begin{pmatrix}\text{Mechanical energy}\\ \text{including}\\ \text{mechanical losses}\end{pmatrix} + \begin{pmatrix}\text{Increase in}\\ \text{stored energy}\\ \text{in field}\end{pmatrix} \tag{3.27}$$

or, in symbolic form, for a motion of the armature through a distance dx in the direction of positive x, occurring in a time dt,

$$dW_e = dW_0 + dW_f \qquad \text{J} \tag{3.28}$$

The electrical input energy minus resistance loss is

$$\begin{aligned} dW_e &= vi\,dt - Ri^2\,dt \\ &= (v - Ri)i\,dt \\ &= ei\,dt \qquad \text{J} \end{aligned} \tag{3.29}$$

The induced voltage e is related to the coil flux linkage λ by

$$e = \frac{d\lambda}{dt} \qquad \text{V} \tag{3.30}$$

Thus,

$$dW_e = i\,d\lambda \qquad \text{J} \tag{3.31}$$

Let us further assume that either the relative permeability of the core material can be regarded as constant or that the field intensity required in the core may be ignored. From Eq. (1.27), the stored energy in the magnetic field of this linear system may be expressed in terms of its inductance L as

$$W_f = \frac{\lambda^2}{2L} \qquad \text{J} \tag{3.32}$$

Suppose we consider first a very rapid displacement dx that occurs so quickly the coil flux linkage remains essentially constant. Note that any very rapid change in flux linkage would produce a very large induced voltage. With flux linkage constant, from Eq. (3.31), $d\lambda$ and dW_e will both be zero. Thus, from Eq. (3.28),

$$dW_m = -dW_f \qquad \text{J} \tag{3.33}$$

i.e., all of the mechanical energy comes out of the stored energy in the field.

The mechanical output energy is equal to the force F_x exerted in the x-direction multiplied by the distance moved dx. Thus,

$$
\begin{aligned}
F_x &= \frac{dW_m}{dx} \\
&= -\left(\frac{\partial W_f}{\partial x}\right)_{\lambda \text{ constant}} \qquad (3.34) \\
&= \frac{\lambda^2}{2L^2}\frac{dL}{dx} \quad \text{N}
\end{aligned}
$$

As λ^2 is always positive, the force on the armature in the x direction is positive only if inductance increases as x increases. In other words, the force acts in such a direction as to increase the inductance of the system. In the case of the actuator shown in Fig. 3.19, the coil flux linkage per unit of coil current increases as displacement x is reduced. Thus, the force tends to reduce x, i.e., to close the gap.

An alternative form of the force relation is obtained if the stored energy is expressed in terms of the flux and the reluctance. From Eq. (1.28),

$$
W_f = \frac{\mathcal{R}\Phi^2}{2} \quad \text{J} \qquad (3.35)
$$

It follows from Eq. (3.34) that the force is

$$
F_x = -\frac{\Phi^2}{2}\frac{d\mathcal{R}}{dx} \quad \text{N} \qquad (3.36)
$$

Thus, the force acts to decrease the reluctance of the magnetic system.

In the actuator of Fig. 3.19, let us assume that the flux is confined to an area A as it crosses each of the two air gaps of length x. The reluctance of the total flux path, assuming that negligible field intensity is required in the core material, is

$$
\mathcal{R} = \frac{2x}{\mu_0 A} \quad \text{A/Wb} \qquad (3.37)
$$

By inserting this into Eq. (3.36), the force can be expressed in terms of the air-gap flux density as

$$
\begin{aligned}
F_x &= -\frac{(BA)^2}{2}\frac{2}{\mu_0 A} \\
&= -\frac{B^2}{2\mu_0}(2A) \quad \text{N}
\end{aligned} \qquad (3.38)
$$

This is the same form of expression as stated earlier without derivation in Eq. (3.7) for the force between two ferromagnetic members carrying flux density B.

Consider now a very slow armature displacement dx. For this condition, the induced voltage e will be very small because the flux linkage is changing only slowly with time. Thus, the current i will be essentially constant at $i = v/R$. From Eq. (1.26), the stored energy can be expressed in terms of the current as

$$W_f = \tfrac{1}{2} Li^2 \qquad \text{J} \tag{3.39}$$

and

$$dW_f = \frac{i^2}{2}\, dL \qquad \text{J} \tag{3.40}$$

From Eq. (3.31), the electrical input energy increment is

$$dW_e = i\, d\lambda = i^2\, dL \qquad \text{J} \tag{3.41}$$

Thus, from Eq. (3.25), the force is

$$\begin{aligned} F &= \frac{dW_m}{dx} \\ &= \frac{\partial}{\partial x}\left(i^2\, dL - \frac{i^2}{2}\, dL\right)\Bigg|_{i=constant} \\ &= \frac{i^2}{2}\frac{dL}{dx} \qquad \text{N} \end{aligned} \tag{3.42}$$

Because $\lambda = Li$, this is identical with the expression of Eq. (3.34). The force is always in the direction that leads to an increase in inductance.

The expressions for force in Eqs. (3.34) and (3.42) are valid only in a magnetically linear system in which the inductance is dependent only on the dimensions of the structure and is independent of the magnitude of the current or flux. For situations in which the nonlinearity and hysteresis of the magnetic material is significant, a fuller treatment is given in Section 7.8.

EXAMPLE 3.3 The dimensions of an actuator are given in Fig. 3.20. The ferromagnetic core material may be assumed to have a nearly constant relative permeability of 1800 up to a flux density of 1.3 T. The coil has 2000 turns. Leakage flux and flux fringing at the air gaps may be ignored.

(a) Develop a magnetic equivalent circuit for this device and use it to find the coil current required to produce an air-gap flux density of 1.1 T with air gaps each of length of $g = 2$ mm.

(b) Find the force on the armature for the condition of (a).

(c) Find the force with the same current as in (a) but with an air-gap length of 4 mm.

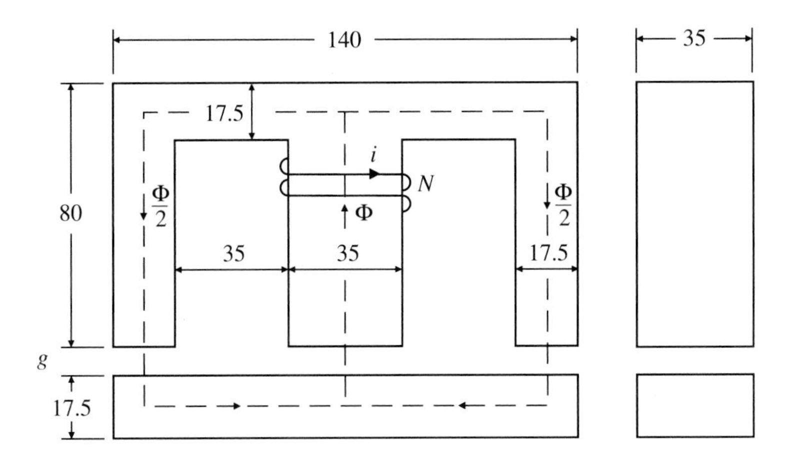

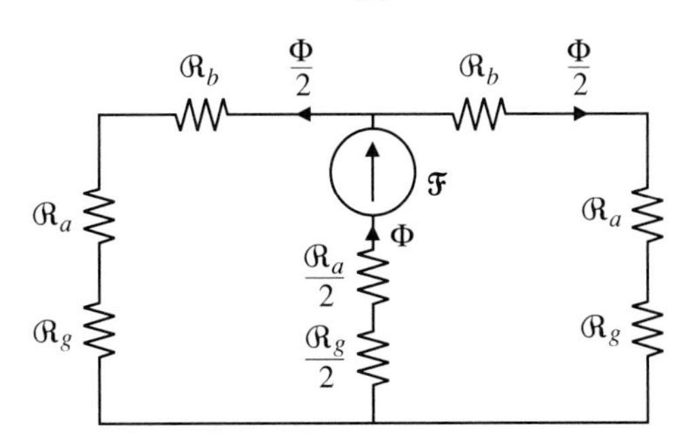

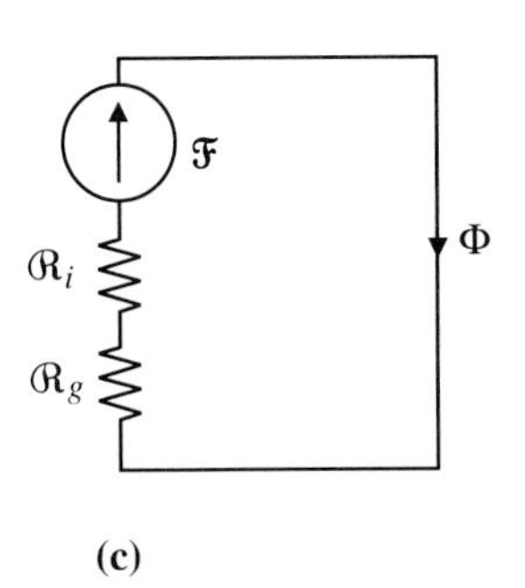

Figure 3.20 (a) Dimensions (in millimeters) of actuator, (b) equivalent magnetic circuit, and (c) simplified circuit.

(d) Suppose that the steel core is effectively saturated at a flux density of 2 T where $H_{steel} = 3$ kA/m. What is the maximum holding force when the gap is closed and approximately what coil current is required to maintain this force?

Solution

(a) With a flux density of 1.1 T, the flux is

$$\Phi = BA = 1.1 \times 0.035^2 = 1.35 \qquad \text{mWb}$$

A magnetic equivalent circuit is shown in Fig. 3.20(b). Because the areas of the side branches are each one-half that of the central branch, this circuit may be simplified by paralleling the two side branches leading to the equivalent circuit of Fig. 3.20(c), where

$$\mathcal{R}_i = \mathcal{R}_a + \mathcal{R}_b$$

The area of the path in both sections a and b is

$$A_i = 0.0175 \times 0.035 = 0.613 \times 10^{-3} \qquad \text{m}^2$$

The total steel path length in sections a and b is

$$\ell_i = 80 + (70 - \tfrac{17.5}{2}) = 141.3 \qquad \text{mm}$$

Thus, the reluctance of the steel path is

$$\mathcal{R}_i = \frac{\ell_i}{\mu_r \mu_0 A_i} = \frac{0.1413}{1800 \times 4\pi \times 10^{-7} \times 0.613 \times 10^{-3}}$$

$$= 1.02 \times 10^5 \qquad \text{A/Wb}$$

The reluctance of one of the side gaps is

$$\mathcal{R}_g = \frac{g}{\mu_0 A_i} = \frac{g}{4\pi \times 10^{-7} \times 0.613 \times 10^{-3}} = 1.3 \times 10^9 g \qquad \text{A/Wb}$$

The total reluctance of the flux path is

$$\mathcal{R} = \mathcal{R}_i + \mathcal{R}_g$$

For $g = 0.002$ m,

$$\mathcal{F} = (\mathcal{R}_i + \mathcal{R}_g)\Phi$$

$$= (1.02 \times 10^5 + 1.3 \times 10^9 \times 0.002)1.35 \times 10^{-3}$$

$$= 3524 \qquad \text{A}$$

$$i = \frac{\mathcal{F}}{N} = \frac{3524}{2000} = 1.76 \qquad \text{A}$$

(b) The force tending to close the air gaps of length g can be expressed from Eq. (3.36) as

$$F_g = \frac{\Phi^2}{2} \frac{d(\mathcal{R}_i + \mathcal{R}_g)}{dg}$$

$$= \frac{(1.35 \times 10^{-3})^2}{2} \frac{d}{dg} (1.02 \times 10^5 + 1.3 \times 10^9 g)$$

$$= 1185 \qquad \text{N}$$

An alternative approach would be to use Eq. (3.7). The total area of all three gaps is

$$A = 2 \times 0.035^2 = 2.45 \times 10^{-3} \quad \text{m}^2$$

$$F_g = \frac{B^2 A}{2\mu_0} = \frac{1.1^2 \times 2.45 \times 10^{-3}}{8\pi \times 10^{-7}} = 1180 \quad \text{N}$$

(c) With a known coil current of $i = 1.76$ A, a convenient form of force relation is that of Eq. (3.42). The coil inductance is

$$L = \frac{N^2}{\mathcal{R}} = \frac{2000^2}{1.02 \times 10^5 + 1.3 \times 10^9 g} \quad \text{H}$$

The force, then, for $g = 0.004$ m, is

$$F = \frac{i^2}{2}\frac{dL}{dg} = \frac{1.76^2}{2}\,\frac{2000^2 \times 1.3 \times 10^9}{(1.02 \times 10^5 + 1.3 \times 10^9 g)^2} = 286.5 \quad \text{N}$$

Alternatively, the flux Φ could be calculated for this gap length and coil current and used in Eq. (3.36) to obtain the force. Note that the flux density with this gap length will be much less than 1.3 T. Thus, the linear model is applicable.

(d) With a gap flux density of 1.8 T,

$$F = \frac{B^2 A}{2\mu_0} = \frac{1.8^2 \times 2.45 \times 10^3}{8\pi \times 10^{-7}} = 3158 \quad \text{N}$$

$$i = \frac{\mathcal{F}}{N} = \frac{H(2\ell_i)}{N} = \frac{3 \times 10^3 \times 2 \times 0.1413}{2000} = 0.42 \quad \text{N}$$

3.7

Reluctance Machines

The forces between flux-carrying sections of ferromagnetic material can be used to produce a rotating actuator. In the structure shown in Fig. 3.21(a), the rotor has two cylindrical poles, each extending over an arc of $\pi/2$ rad. The stator also has two poles, each with the same arc and radially displaced from the rotor by a gap g. Let us first develop a physical picture of how this device produces rotary action or torque.

Assuming ideal iron, i.e., $H_i = 0$, there will be a radially directed flux density in the overlap portions of the air gaps:

$$B = \mu_0 \frac{Ni}{2g} \quad \text{T} \qquad (3.43)$$

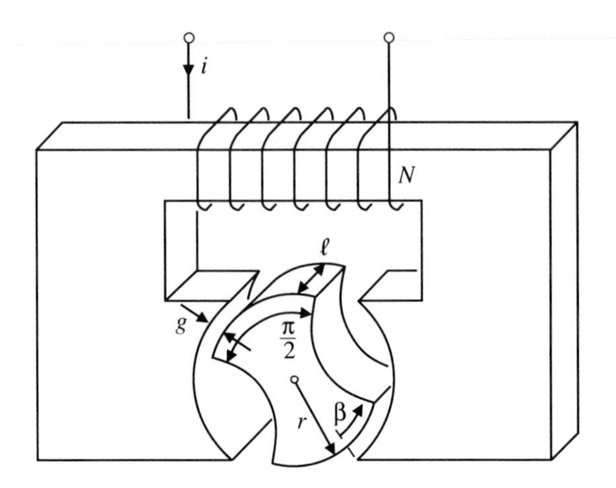

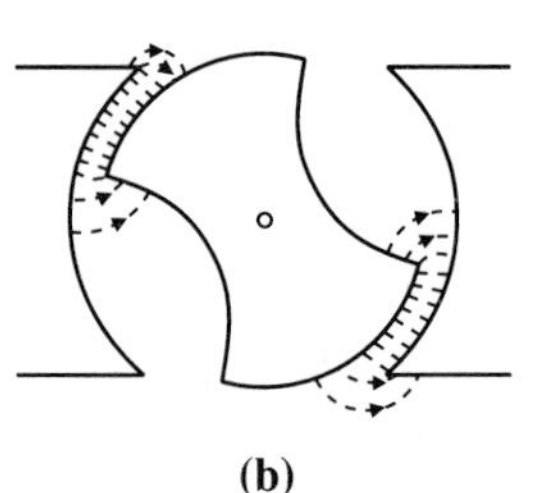

Figure 3.21 (a) Rotary actuator and (b) flux pattern.

This flux density, being radial as shown in the flux plot of Fig. 3.21(b), produces only radially directed forces on the overlap portions of the rotor. If there is to be a torque on the rotor, it must arise from the fringing flux shown in Fig. 3.21(b) at the ends of the overlapping regions. Qualitatively, we might expect the forces in these regions to be such as to shorten the length of the leakage flux paths, thus producing clockwise torque on the rotor. Further, we might note that the pattern of fringing flux will remain essentially constant as the overlap angle β changes, at least until the overlap angle approaches $\pi/2$ rad. Thus, we might expect that the torque would be nearly constant and independent of angle β for a given coil current.

Calculation of the radial force would be reasonably straightforward using the flux density of Eq. (3.43) in

$$\frac{F}{A} = \frac{B^2}{2\mu_0} \qquad \text{N/m}^2 \tag{3.44}$$

However, this approach is of little value in calculating the circumferential forces producing torque because of the complex distribution of the fringing flux density.

As an alternative, consider deriving the torque using an expression analogous to that of Eq. (3.42):

$$T = \frac{dW_m}{d\beta} = \frac{i^2}{2}\frac{dL}{d\beta} \qquad \text{N}\cdot\text{m} \tag{3.45}$$

Ignoring any leakage flux, the inductance of the coil can be expressed as

$$L = \frac{\lambda}{i} = \frac{N\Phi}{i} \qquad \text{H} \tag{3.46}$$

The flux in the overlap region is the flux density of Eq. (3.43) multiplied by the overlap area. The flux in the fringing region can be assumed to be proportional to the coil current, but the constant of proportionality that is designated as k_f is unknown. Thus, the flux is

$$\Phi = \mu_0 \frac{Ni}{2g} \ell r\beta + k_f Ni \qquad \text{Wb} \tag{3.47}$$

Inserting this in Eq. (3.46), gives

$$L = \mu_0 \frac{N^2 \ell r\beta}{2g} + N^2 k_f \qquad \text{H} \tag{3.48}$$

Then, using Eq. (3.45), the torque is given by

$$T = \frac{i^2}{2}\frac{dL}{d\beta} = \frac{\mu_0 N^2 \ell r}{4g} i^2 \qquad \text{N}\cdot\text{m} \tag{3.49}$$

This torque is seen to be constant for a given coil current and independent of overlap angle β. The expression is valid only over the range of β where it can be assumed that the fringing flux pattern remains constant.

It is interesting to note that if the fringing flux had been ignored in deriving the coil inductance of Eq. (3.48), the same torque expression would have resulted. We would have obtained a numerically correct value while ignoring the only part of the magnetic field that was effective in producing torque. This example emphasizes the need to have an understanding of the relevant physical mechanisms before making simplifying assumptions. Expressions for force and torque such as those given in Eqs. (3.42) and (3.49) are powerful computational tools, but their intelligent application requires a good appreciation of the physical phenomena that are acting.

The machine of Fig. 3.21(a) can produce continuous rotation as a motor by appropriate control of the coil current. Suppose the machine stator is fitted with a sensor, such as a Hall effect device, that detects the edge of the rotor as it passes. This signal can be used to control a current source so that it supplies the

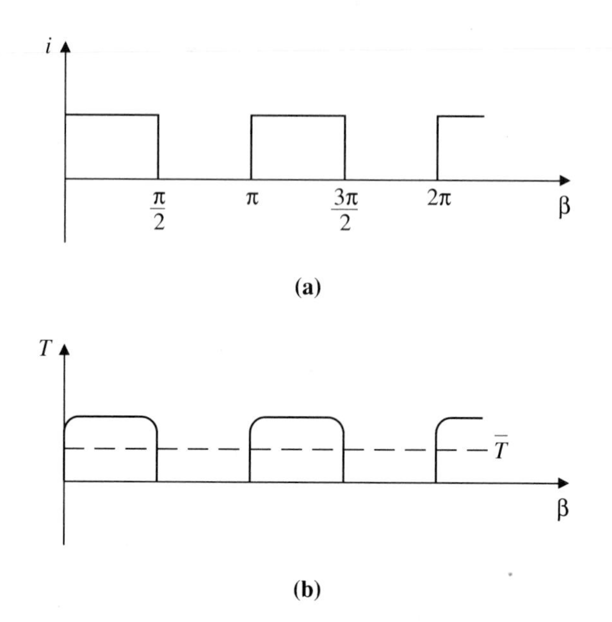

Figure 3.22 (a) Coil current for switched reluctance motor and (b) reluctance motor torque.

coil current waveform shown in Fig. 3.22(a). Because the torque depends on the square of the current, the polarity of the current is not significant. For the range $0 < \beta < \pi/2$, the torque will be counterclockwise with the value given by Eq. (3.49). The torque will be zero for $\pi/2 < \beta < \pi$ and will again be counterclockwise for $\pi < \beta < 3\pi/2$. The average counterclockwise torque will be approximately half of the value of Eq. (3.49), as suggested by Fig. 3.22(b). A more nearly constant torque can be obtained by inserting two other poles in the open arcs around the stator and providing these with current in the intervals when the first poles are unexcited. This elementary machine is the basis for a class of machines known as *switched reluctance machines*, which are discussed more fully in Section 11.7.

EXAMPLE 3.4 The switched reluctance motor shown in cross section in Fig. 3.23(a) has six stator poles, each covering an arc of $\pi/6$ rad. The rotor has four poles, each with an arc of $\pi/4$ rad. The air-gap radius is 50 mm and the axial length of the motor is 80 mm. The air-gap length is 0.5 mm. The stator coils, each having 200 turns, are supplied with a current i which is switched to the series-connected coils a-a', b-b' and c-c' in turn, as shown in Fig. 3.23(b).

(a) Determine the source current i required to establish a flux density of 1.2 T in the overlap air gap.

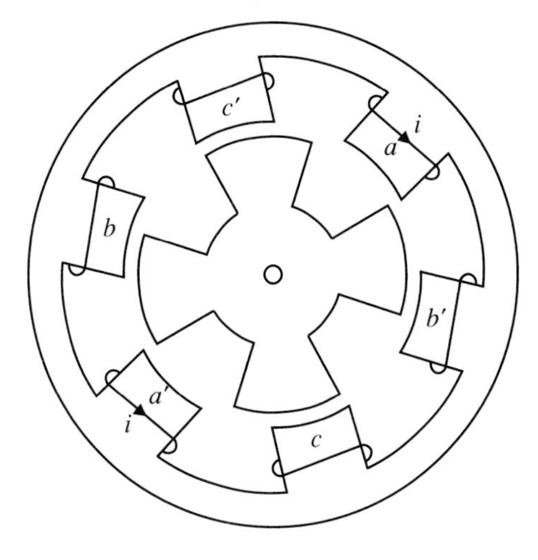

(a)

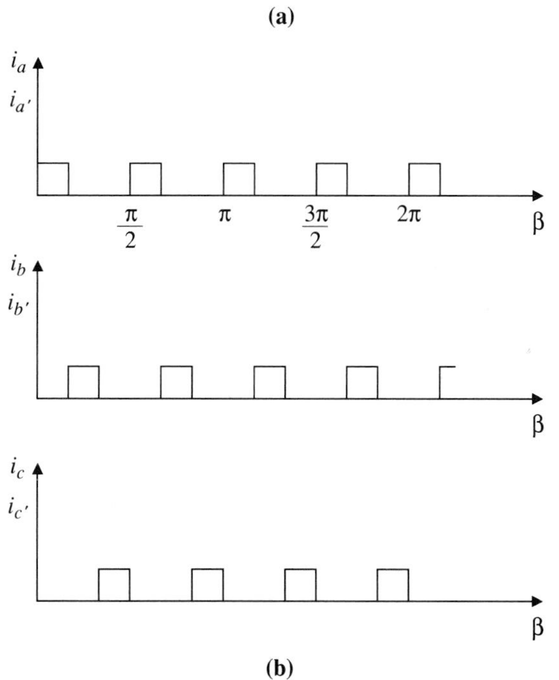

(b)

Figure 3.23 (a) Switched reluctance motor and (b) currents in coils.

(b) With the current found in (a), find the average torque on the rotor.

(c) If the motor is rotating at 120 rad/s, estimate the required voltage of the electrical supply. Conductor resistance may be ignored, and the switches may be assumed to be perfect.

Solution

(a) For a flux density of $B = 1.2$ T in the two air gaps, each of length g, the required total mmf of the two coils a and a' in series is

$$\mathfrak{F} = 2gH = 2g\frac{B}{\mu_0} = \frac{2 \times 0.0005 \times 1.2}{4\pi \times 10^{-7}} = 955 \qquad \text{A}$$

Half of this is supplied by each coil. Thus, the coil current is

$$i = \frac{955}{2 \times 200} = 2.4 \qquad \text{A}$$

(b) If coils a and a' are supplied with current i at the rotor position shown in Fig. 3.23(a), the torque for a counterclockwise angular motion of $\pi/6$ rad can be found by deriving the inductance of the coils in series. The flux linking these coils is, by analogy with Eq. (3.47), for overlap angle β:

$$\Phi = BA = \mu_0\left(\frac{2Ni}{2g}\right)\ell r\beta + k_f(2Ni)$$

$$L_{a\text{-}a'} = \frac{2N\Phi}{i} = 2\mu_0\frac{N^2\ell r\beta}{g} + k_f(2N)^2$$

Then, the torque is

$$\begin{aligned} T &= \frac{i^2}{2}\frac{dL}{d\beta} \\ &= \mu_0\frac{N^2\ell r}{g}i^2 \\ &= 4\pi \times 10^{-7} \times 200^2 \times \frac{0.08 \times 0.05}{0.0005} \times 2.4^2 \\ &= 2.32 \qquad \text{N}\cdot\text{m} \end{aligned}$$

With ideal switching, this also will be the torque when phases b-b' and c-c' are energized. Thus, it is also the average torque.

(c) The mechanical power output is

$$P_0 = T\omega_0 = 2.32 \times 120 = 278 \qquad \text{W}$$

If losses are ignored, there is also the electrical input power. Thus, the voltage of the supply must be

$$v = \frac{P_0}{i} = \frac{278}{2.4} = 115.8 \qquad \text{V}$$

PROBLEMS

3.1 Figure 3.24 shows the movement of a direct-current instrument. The permanent magnet and the central cylinder of iron are shaped so as to produce an essentially radial flux density of 0.4 T in the two air gaps. The rotating coil is 15 mm in height and 20 mm in width. It has 2000 turns of very fine wire. The electrical connections to the coil are brought out through helical springs which together exert a restoring torque of 10^{-5} N·m/rad. Determine the coil current required to produce the 70° rotation that corresponds to full-scale deflection.

(Section 3.1)

3.2 A 13.2-kV alternating-current transmission line consists of two conductors, each 20 mm in diameter and spaced 1.5 m apart. The span between supporting poles is 100 m. During a short-circuit condition, a current of 8000 A rms flows in the line. Determine the average magnitude and the direction of the force on each conductor over one span.

(Section 3.1)

3.3 The pair of parallel bus bars shown in Fig. 3.25 have a height of 150 mm and a spacing of 5 mm. They supply a direct current of 10 kA to a load. Estimate the magnitude and direction of the force on each bar per meter of its length. Because the height is large in comparison with the spacing, the magnetic field produced by each bar can be considered to be uniform over its height, up one side and down the other, and the field in the end regions can be ignored.

(Section 3.1)

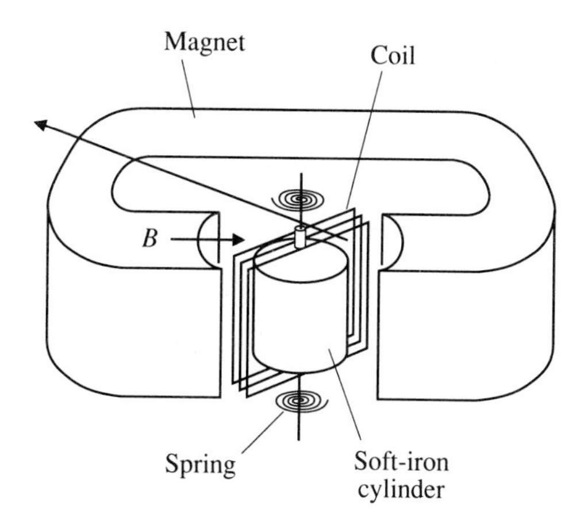

Figure 3.24 Instrument for Problem 3.1.

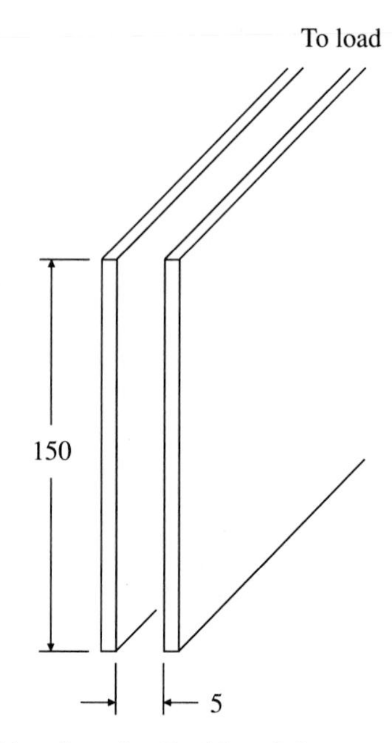

Figure 3.25 Parallel bus bars for Problem 3.3.

3.4 Figure 3.26 shows a rectangular loop carrying a current i in a region having a uniform flux density B.

(a) Show that the net translational force on the loop is zero.

(b) Show that the torque on the loop can be expressed as

$$T = p_m \times B \qquad \text{N·m}$$

where p_m is the magnetic moment of the loop.

(Section 3.2)

3.5 A toroidal core with a mean diameter of 250 mm and a cross-sectional area of 1000 mm^2 has an air gap of 1.25 mm cut in it. Fringing of flux around the air gap may be ignored. The core material is of the type shown in Fig. 1.17.

(a) Suppose the current in a coil wound around the core is increased just enough to produce a flux density of 1.4 T. Determine the force tending to close the air gap.

(b) Suppose the core material is completely aligned or saturated at a flux density of 1.4 T. Will the force tending to close the air gap be increased if the coil current is increased beyond the level of (a)?

(Section 3.2)

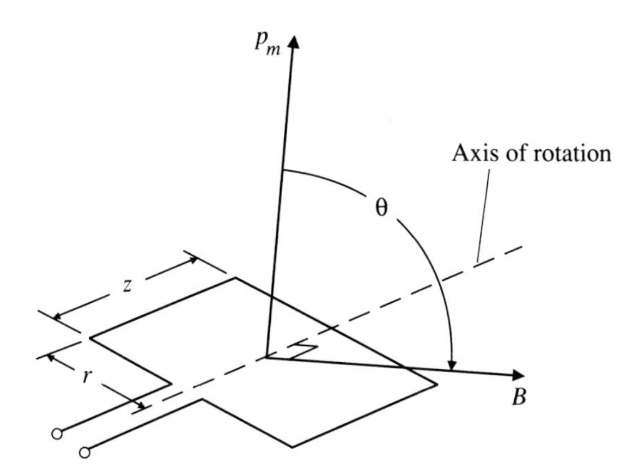

Figure 3.26 Rectangular loop for Problem 3.4.

3.6 Figure 3.27 shows a magnetic device with a steel movable wedge located between two poles. The cross-sectional area of the left pole is twice that of the right pole. Fringing of the flux around the pole edges may be ignored.

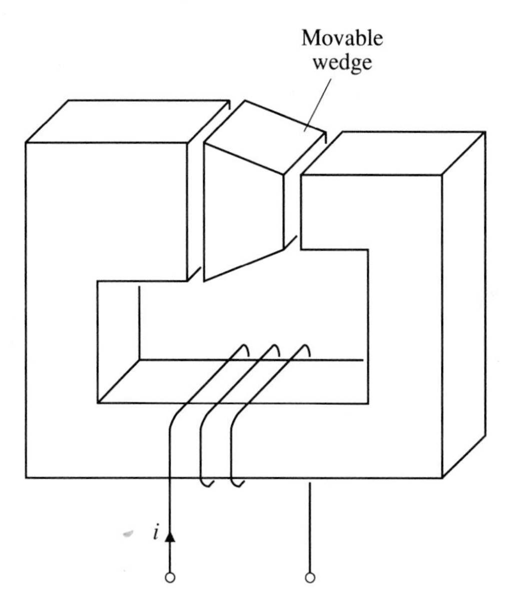

Figure 3.27 Device with movable wedge for Problem 3.6.

(a) What is the direction of the force on the wedge when a flux is established?

(b) Is the force dependent on the position of the wedge if the flux is kept constant?

(c) Is the force dependent on position if the current is kept constant?

(d) If the area of the right pole is 100 mm^2 and both air gaps are 1 mm in length, find the force for a coil mmf of 2000 A.

(Section 3.2)

3.7 The magnetic holding device shown in Fig. 3.28 consists of two ferrite magnets each having a length of 8 mm in the direction of magnetization and a cross-sectional area of 700 mm^2. The magnets are fixed to a high-permeability iron backing plate and are faced by a high-permeability iron moving member. Ignoring fringing of flux around the air-gap edges, determine the force on the moving member when it is 5 mm from the magnets.

(Section 3.2)

3.8 In the elementary rotating machine shown in Fig. 3.5, the magnets produce a radially directed flux density of 0.95 T. The radius r of the conductor is 50 mm and its length in the gap is 100 mm.

(a) For a current of 40 A, find the torque on the conductor loop when it is in the air gap.

(b) If the loop revolves at 20 r/s, determine the generated voltage in the loop when in the air gap.

(c) For the condition of (a) and (b), determine the rate at which energy is being converted.

(Section 3.3)

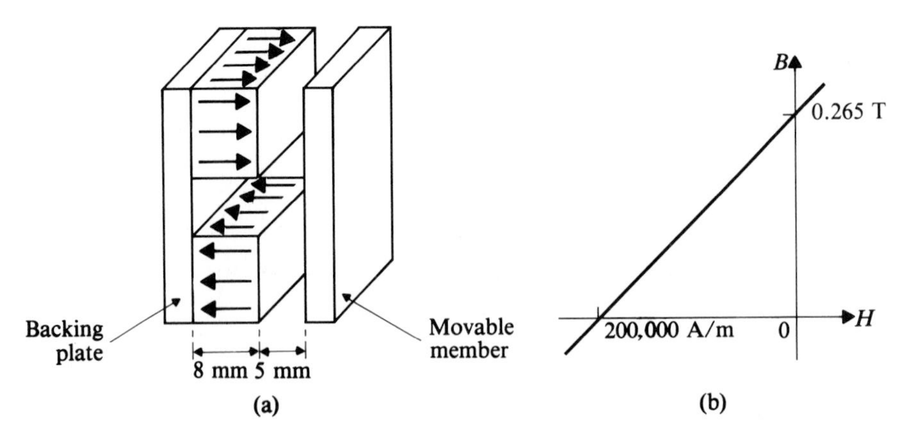

Figure 3.28 Magnetic holding device for Problem 3.7.

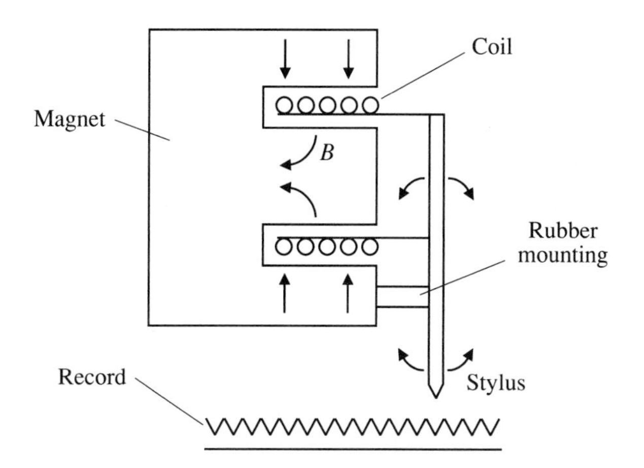

Figure 3.29 Phonograph pickup for Problem 3.9.

3.9 The phonograph pickup shown in Fig. 3.29 consists of a 50-turn cylindrical coil of 3 mm diameter attached to the stylus. The coil is situated in the air gap of a cylindrical permanent magnet. The radially directed flux density is 0.3 T. The stylus-coil assembly is attached to the magnet by a flexible coupling so that the horizontal motion of the coil is approximately equal in magnitude to that of the stylus. If the tip of the stylus oscillates sinusoidally with a peak-to-peak amplitude of 0.1 mm at a frequency of 600 Hz, determine the rms output voltage of the pickup coil.

(Section 3.3)

3.10 Figure 3.30 shows a pump used to move liquid metal along a rectangular channel. The channel is made of insulating material except in the region of the two electrodes that pass a direct current through the metal. A uniform flux density of 0.8 T is produced across the channel by use of a permanent magnet.

(a) Supposing the metal flows at a velocity of 12 m/s, determine the voltage between the electrodes when the current is zero.

(b) If the current is 2000 A, what will be the pressure difference between the entrance and exit of the pump? Assume the friction is negligible.

(c) If the resistivity of the metal is $10^{-6}\ \Omega \cdot \text{m}$ and the current is uniformly distributed, determine the voltage required between the electrodes for a current of 2000 A and a metal velocity of 12 m/s.

(d) Show that the electrical power input is equal to the mechanical power output plus the loss in the resistance of the metal.

(Section 3.3)

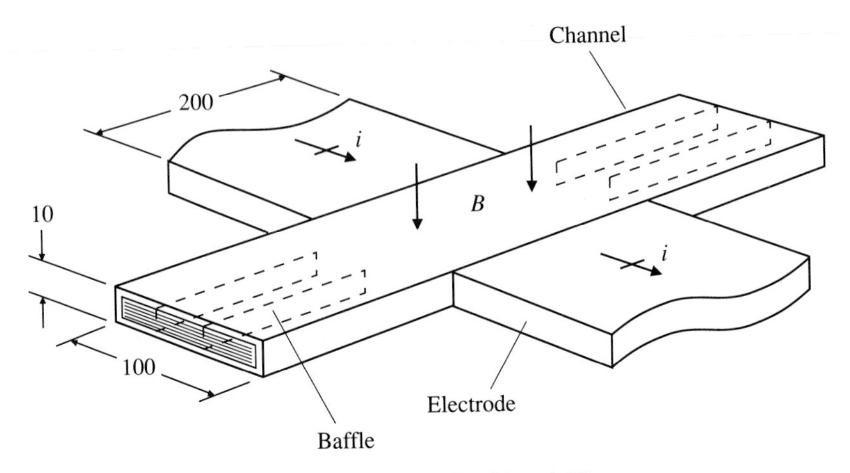

Figure 3.30 Electromagnetic pump for Problem 3.10.

3.11 Figure 3.31 shows a cross section of a low-voltage, high-current machine known as a homopolar generator. A thin metal cylinder of radius r, length x, and thickness c is rotated at angular velocity ω in a radial magnetic field of density B. Current is passed through the cylinder by means of liquid metal brushes around the inside and the outside rims. If the current density in the cylinder is not to exceed 8 A/mm^2 and the flux density is 1.2 T,

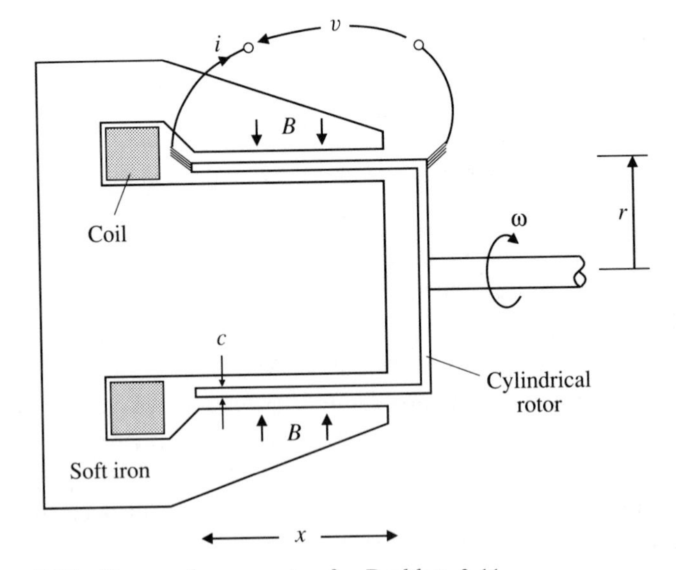

Figure 3.31 Homopolar generator for Problem 3.11.

find the maximum electrical power output of a generator with a cylinder having a radius of 125 mm, length 100 mm, and thickness 10 mm and with an angular velocity of 1500 rad/s. Losses may be ignored.

(Section 3.3)

3.12 A circular metal disk of radius r_0 is rotated with angular velocity ω rad/s, as shown in Fig. 3.32. A permanent magnet produces a flux density B which can be assumed to be uniform in the region of the disk and perpendicular to its plane. Electrical connections are made to the disk by stationary brushes distributed uniformly around its outer rim and on the conducting shaft of radius r_s.

(a) Derive an expression for the voltage produced between the brushes.

(b) Suppose a current i is circulated through the disk by way of the brushes. Derive an expression for the torque produced.

(c) Ignoring losses, show that the electrical input power equals the mechanical output power.

(Section 3.3)

3.13 To examine some of the practical features and limitations of the machine of Problem 3.12 (shown in Fig. 3.32), assume that the peripheral veloc-

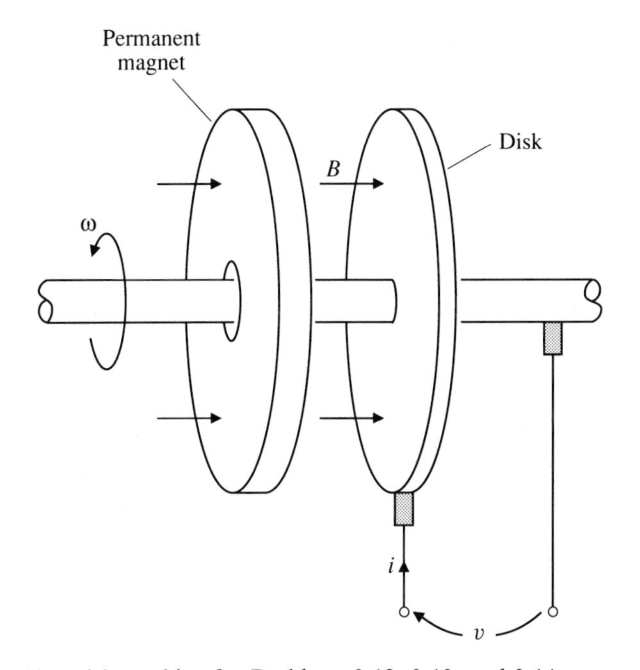

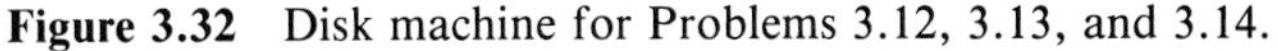
Figure 3.32 Disk machine for Problems 3.12, 3.13, and 3.14.

ity of the disk is limited to about 200 m/s to prevent mechanical overstressing. An average flux density of 0.6 T may be assumed.

(a) If the outer radius of the disk is 0.2 m and the shaft radius is 20 mm, determine the maximum voltage that can be generated.

(b) If the thickness of the disk is 10 mm and the current density is not to exceed 10 A/mm^2 at any point, find the maximum current that the machine can carry.

(c) Suppose the disk is made of copper having a resistivity of 2×10^{-8} $\Omega \cdot$m at the operating temperature. Find the maximum output power of the machine operating as a generator.

(Section 3.3)

3.14 In the disk machine shown in Fig. 3.32, suppose the magnet is attached to the disk and rotates with it. Will there be a generated voltage v at the terminals of the machine when the shaft is rotated? Does your answer depend on where the brush is placed on the shaft?

(Section 3.3)

3.15 The machine shown in Fig. 3.14 has three concentrated stator coils, each of eight turns. The coils are arranged symmetrically around the stator at 120° intervals. The poles of the rotor are so shaped that the air-gap flux density due to the rotor mmf is nearly sinusoidally distributed. The total rotor flux is 0.040 Wb. The rotor is driven at 3000 r/min.

(a) Determine the frequency of the voltages generated in the coils.

(b) If the stator terminals a', b', and c' are connected to form a neutral point o, find the rms voltage between any two of the terminals a, b, and c.

(Section 3.5)

3.16 Figure 3.33 shows a cross section of a simple electric brake. The rotor winding consists of two shorted loops, each having a resistance of 0.005 Ω. Each stator pole covers an arc of 90°. The axial length of the rotor is 50 mm. The magnetic material may be considered to be ideal, fringing may be neglected, and the magnetic effect of the rotor currents may be ignored.

(a) Derive expressions for the generated voltage and the current in each of the loops while the loop sides are under the poles.

(b) Derive an expression for the torque acting on the rotor.

(c) What field current is required to provide a braking power of 200 W when the speed is 400 rad/s?

(Section 3.5)

3.17 A toroidal coil consists of 250 circular turns each of 20 mm radius. The length of the coil is 0.3 m. It is self supporting and contains only air.

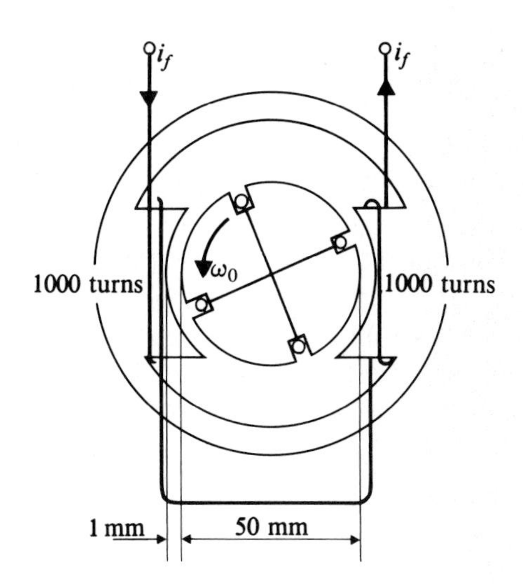

Figure 3.33 Electric brake for Problem 3.16.

(a) Determine the inductance of the coil, assuming the field to be uniform inside the coil and zero elsewhere.

(b) Find the force tending to change the length of the coil when its current is 18 A.

(c) Find the force tending to change the radius of the turns when the coil current is 18 A.

(Section 3.6)

3.18 Figure 3.34 shows a pair of parallel conducting plates. A cloud of ionized gas or plasma is established between the plates at a mean distance z from the ends to which a current source of 10 kA is connected.

(a) Determine the inductance of the conducting loop, assuming the magnetic field to be uniform between the plates and zero elsewhere.

(b) Determine the force acting on the plasma cloud.

(c) Assuming the cloud is bounded by the plates, determine the pressure on the cloud.

(Section 3.6)

3.19 A 13.2-kV transmission line consists of two conductors, each 20 mm in diameter and spaced 1.5 m apart. The span between poles is 100 m.

(a) Derive an expression for the inductance of the line for a length z.

(b) Determine the average force acting on each conductor over a span when a sinusoidal current of 8000 A (rms) flows in the line.

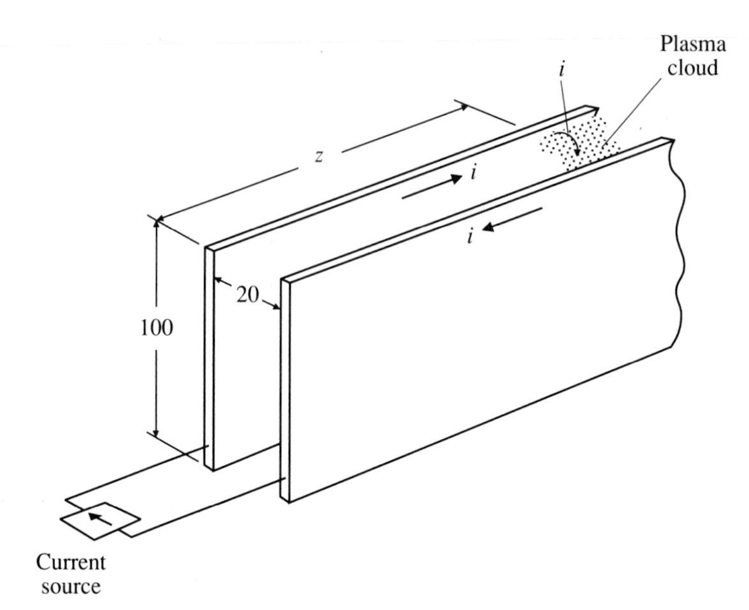

Figure 3.34 System for Problem 3.18.

(c) Suppose the transmission line is shorted by a round bar placed on top of the conductors and free to roll along them. For a current of 8000 A in the line and the bar, find the force on the bar.

(Section 3.6)

3.20 The magnetic system of Fig. 3.35 has a square cross section 30×30 mm. When the two sections of the core are fitted together, air gaps, each of length $x = 1.0$ mm, separate them. The coil has 250 turns and a resistance

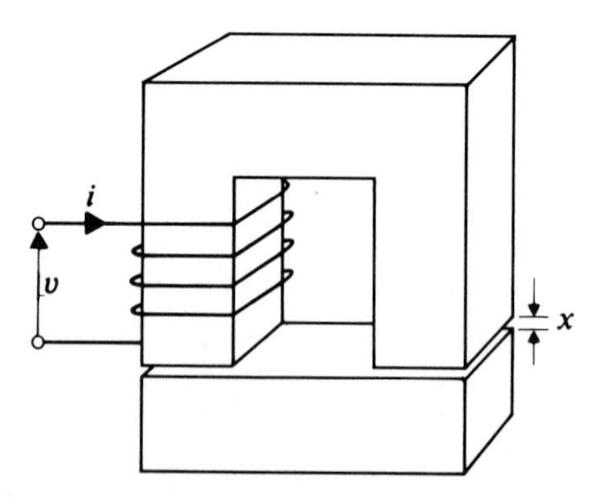

Figure 3.35 Magnetic system for Problem 3.20.

of 7.5 Ω. The magnetic field intensity required by the core material is negligible.

(a) If a constant voltage source of 40 V is connected to the coil terminals, what will be the total force holding the two sections together?

(b) If a sinusoidal voltage of 100 V at 60 Hz is connected to the coil terminals, what is the average force holding the two sections together?

(c) The effect of flux fringing can be allowed for by assuming that the air-gap flux density is uniform over a gap with dimensions $(30 + x)$ in each direction for a gap length x. Repeat (b) and (c) including this effect.

(Section 3.6)

3.21 The actuator shown in Fig. 3.36 is used to raise a mass m through a distance y. The coil has 500 turns and can carry a current of 20 A without overheating. The magnetic material can support a flux density of 1.5 T with negligible field intensity. Fringing may be ignored.

(a) Determine the maximum air gap y for which a flux density of 1.5 T can be established with a current of 20 A.

(b) What is the force exerted by the actuator for the condition of (a)?

(c) The mass density of the core material is 7800 kg/m^3. Determine the approximate value of the net mass that can be lifted against the force of gravity by the actuator at the air gap determined in (a).

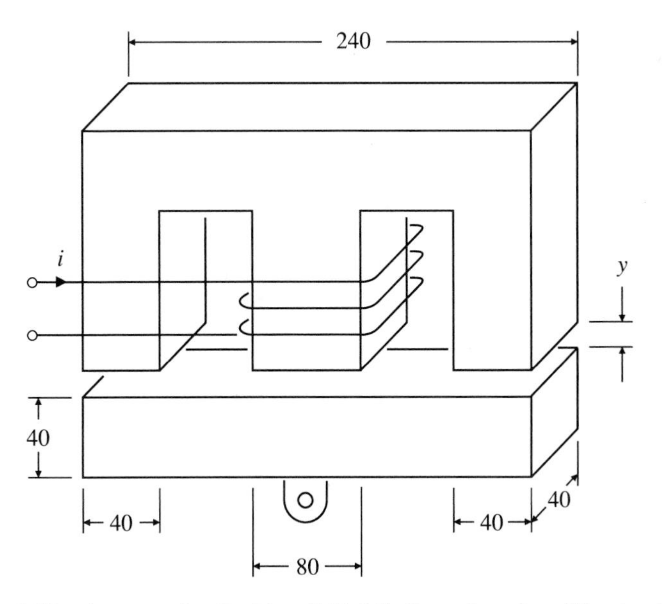

Figure 3.36 Actuator for Problem 3.21 (all dimensions in millimeters).

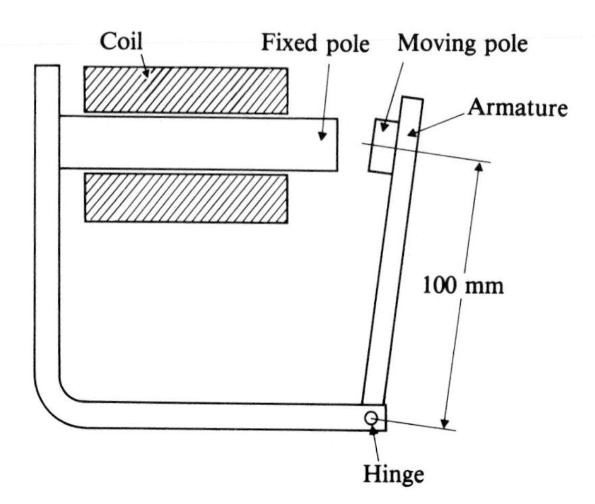

Figure 3.37 Relay for Problem 3.22.

(d) Find the initial acceleration of the unloaded actuator if it is released at the air gap of (a) with a coil current of 2 A.

(Section 3.6)

3.22 In the relay mechanism illustrated in Fig. 3.37, the two steel poles are 30 × 30 mm in cross section. When they meet, the center line of the armature is vertical. Assume the permeability of the steel is infinite, ignore the gap at the hinge, and ignore fringing. Determine the force on the armature when the coil carries a direct current of 0.5 A in its 1000 turns and the armature is at an angle of 2° from the vertical position.

(Section 3.6)

3.23 Figure 3.38 shows a cross section of a cylindrical magnetic actuator. The plunger of cross-sectional area 1500 mm^2 is free to slide vertically through a circular hole in the outer magnetic casing, the gap between the two being negligible. The coil has 3000 turns and a resistance of 8 Ω. A source of 12 V is applied to the coil terminals. The magnetic material can be regarded as perfect up to a flux density of 1.6 T. Fringing may be ignored.

(a) Determine the static force on the plunger as a function of the gap length y.

(b) Over what range of gap length y will the force on the plunger be essentially constant because the saturation flux density (assumed to be 1.6 T) has been reached?

(c) Suppose the plunger is constrained to move slowly from a gap of 10 mm to the fully closed position. What mechanical energy will be produced?

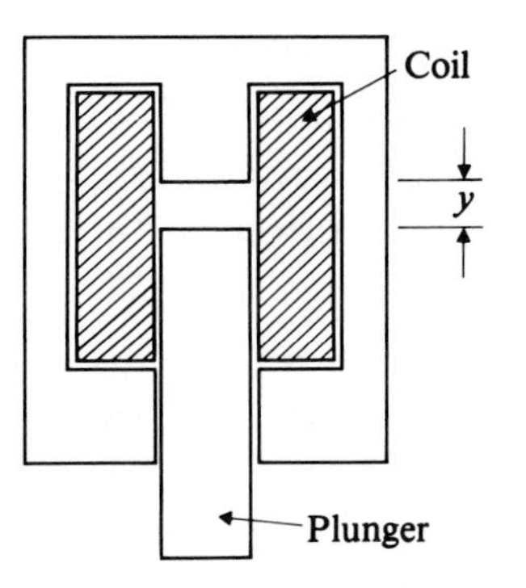

Figure 3.38 Actuator for Problem 3.23.

(d) Suppose the plunger moves so quickly from its original gap of 10 mm to complete closure that the coil flux linkage does not change appreciably during the motion. How much mechanical energy will be produced?

(Section 3.6)

3.24 Figure 3.39 illustrates the movement of a moving-iron ammeter in which a curved ferromagnetic rod is drawn into a curved solenoid against the torque of a restraining spring. The inductance of the coil can be expressed as

$$L = 5 + 20\Theta \qquad \mu\text{H}$$

where Θ is the deflection angle in radians. The coil resistance is 0.01 Ω. The spring constant is 7×10^{-4} N·m/rad.

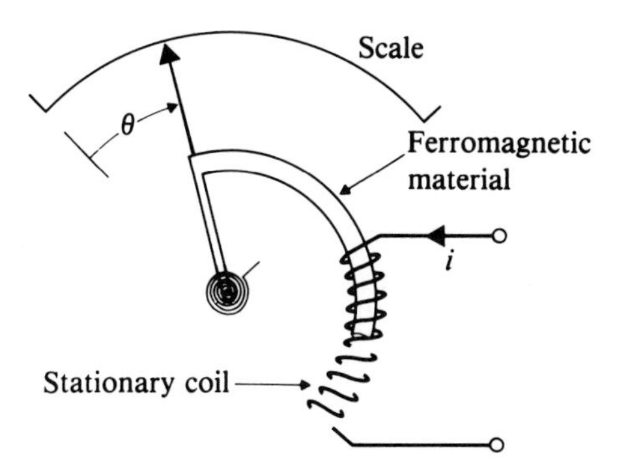

Figure 3.39 Moving-iron instrument for Problem 3.24.

(a) Show that the instrument measures the rms value of the coil current.

(b) What will be the full-scale deflection for a rated current of 10 A?

(c) What will be the voltage across the coil terminals for a sinusoidal coil current of 5 A (rms) at a frequency of 180 Hz?

(Section 3.6)

3.25 A rotating actuator of the form shown in Fig. 3.40 has the following dimensions: $g = 1$ mm, $r = 20$ mm, $z = 40$ mm. The coil has 400 turns. The magnetic material may be considered perfect up to a flux density of 1.2 T.

(a) Determine the maximum coil current if the flux density is to be limited to 1.2 T in the material bounding the air gaps.

(b) With the current of (a), determine the torque produced at an overlap angle Θ.

(c) What mechanical work will be done as the shaft moves from $\Theta = 0$ to $\Theta = \pi/2$?

(Section 3.7)

3.26 In some control applications, a rotating actuator with reversible torque is required. Figure 3.41 illustrates a device that has this feature. Each of the four stator poles subtends an angle of 45° at the axis of rotation, and each of the rotor poles subtends an angle of 90°. The axial lengths of the poles is z. Each stator pole has two N-turn coils, one carrying current i_1 and the other carrying current i_2. The coils are connected so that the mmf on each horizontal pole is $N(i_1 + i_2)$ while that on each vertical pole is $N(i_1 - i_2)$. The permeability of the steel may be assumed to be infinite.

(a) Derive an expression for the torque on the rotor as a function of the coil currents.

(b) Let $r = 25$ mm, $z = 50$ mm, $g = 1$ mm, and $N = 1000$ turns. If the air-gap flux density is not to exceed 1.5 T and the currents i_1, and i_2 are

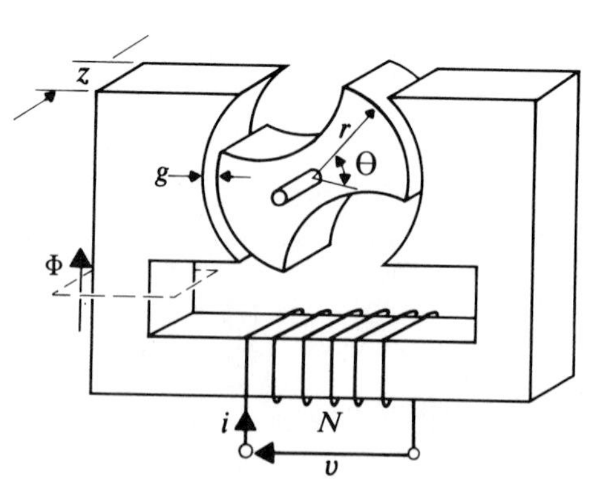

Figure 3.40 Rotating actuator for Problem 3.25.

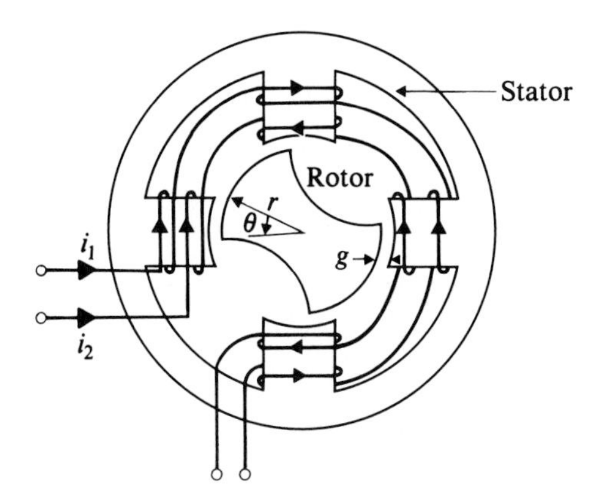

Figure 3.41 Reversible rotating actuator for Problem 3.26.

equal in magnitude, what is the largest permissible value of the coil currents?

(c) If i_1 is maintained at the value of (b), what is the torque per ampere of current i_2?

(Section 3.7)

3.27 A reluctance motor of the form shown in Fig. 3.40 has a magnetic flux path whose reluctance can be expressed approximately as

$$R = 5.06 \times 10^4 \ (2.5 + 1.5 \cos 2\Theta) \qquad \text{A/Wb}$$

over the range $0 < \Theta < \pi/2$. The coil has 15 turns of negligible resistance. A sinusoidal voltage of 110 V rms at 60 Hz is applied to the machine.

(a) What is the magnetic flux in the machine?

(b) At what values of angular velocity of rotation does the machine develop an average unidirectional torque?

(c) What is the maximum value of average torque that this motor can produce?

(d) What is the maximum mechanical power for (c)?

(Section 3.7)

3.28 A rotating machine of the form shown in Fig. 3.40 has a coil inductance that can be approximated by

$$L = 0.01 - 0.03 \cos 2\Theta - 0.02 \cos 4\Theta \qquad \text{H.}$$

A current of 5 A (rms) at 50 Hz is passed through the coil, and the rotor is driven at a speed of ω_0 rad/s.

(a) At what values of speed ω_0 can this machine develop useful torque?

(b) Determine the maximum torque at each of the speeds in (a).

(c) Determine the maximum power output for each of the speeds in (a).

(Section 3.7)

GENERAL REFERENCES

Fitzgerald, A.E., Kingsley, C., and Umans, S.D., *Electric Machinery*, 5th ed., McGraw-Hill, New York, 1990.

Nasar, S.A. and Unnewehr, L.E., *Electromechanics and Electric Machines*, 2nd ed., John Wiley and Sons, New York, 1983.

Say, M.G., *Alternating Current Machines*, 4th ed., Pitman, London, 1976.

Sen, P.C., *Principles of Electric Machines and Power Electronics*, John Wiley and Sons, New York, 1989.

CHAPTER

4

Commutator Machines

Commutator machines normally operate with direct voltage supplies and are usually referred to as *direct-current machines*. They are widely used in vehicles that have electric storage batteries, as motors for auxiliary equipment, and as starters. A major area of use is in variable-speed drives such as required in steel and paper manufacture where the ability to control speed and position is important. A use of particular interest to students is as controllable drivers or loads for the testing of electrical machines in a laboratory.

The basic principles of commutator machines have been introduced in Section 3.5. In this chapter, their behavior will be discussed in more detail, and means of predicting some performance characteristics will be developed.

4.1 Magnetic System

In a commutator machine, the magnetic field is produced by permanent magnets or field coils on the stator. The current-carrying coils that interact with this magnetic field are placed on the rotor so that they can be switched by a rotor-mounted commutator as it passes stationary brushes that lead to the machine terminals.

Figure 4.1 shows the magnetic system of a simple two-pole machine. A field coil encircles each pole. The flux passes along the pole core and is then distributed over an arc of the rotor periphery by a pole shoe. It then crosses the air gap to the teeth on a slotted rotor which contains the rotor or armature winding. After passing around the rotor core inside the teeth and crossing the other gap and the other pole, the flux returns around two magnetically parallel paths that are known as the *yoke.*

The field coils are normally connected in series and supplied with a controllable direct current from an external source. The relation between the field current i_f and the pole flux Φ can be developed using the magnetic equivalent circuit methods described in Section 1.12. Over a considerable range of flux density, the ferromagnetic material in the flux paths may be regarded as ideal. Each field coil of N_f turns produces an mmf source of

$$\mathfrak{F}_p = N_f i_f \qquad \text{A} \tag{4.1}$$

These two magnetomotive forces establish a flux around the closed path and in particular across the two air-gap reluctances. Thus, the pole flux can be approximated by

$$\Phi = \frac{2\mathfrak{F}_p}{2\mathcal{R}_g} = \frac{N_f i_f}{\mathcal{R}_g} \qquad \text{Wb} \tag{4.2}$$

The gap reluctance is inversely proportional to the gap area and directly proportional to the effective air gap length g_e taking into account the effect of rotor slotting:

$$\mathcal{R}_p = \frac{g_e}{\mu_0 A_p} \qquad \text{A/Wb} \tag{4.3}$$

As the flux is increased, the required magnetic field intensity in various ferromagnetic parts of the flux path will become significant, particularly in the ro-

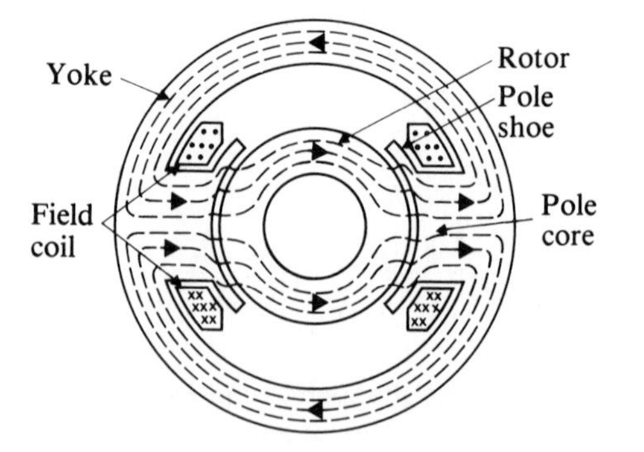

Figure 4.1 Magnetic system of a two-pole machine.

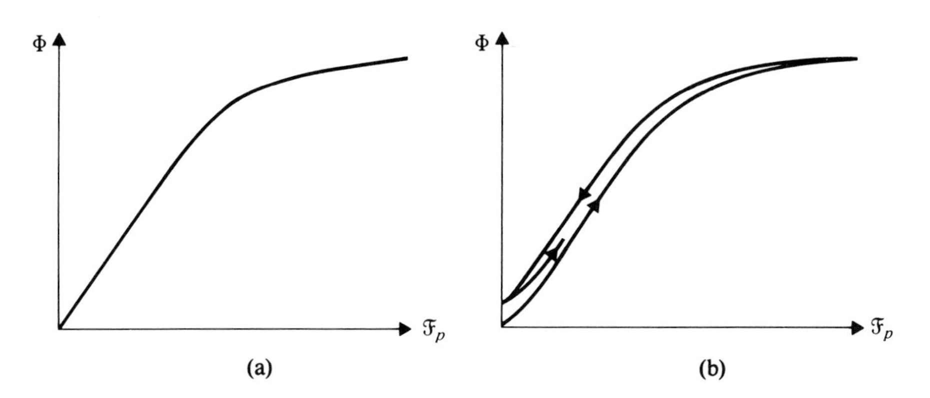

Figure 4.2 Relation between air-gap flux Φ_p and magnetomotive force $\mathfrak{F}_p$ per pole. (a) Neglecting hysteresis. (b) Including hysteresis.

tor teeth. The flux-mmf relation will then tend toward saturation, as shown in Fig. 4.2(a). It is normal to operate most machines somewhat into this nonlinear saturating region because more flux can be obtained for a relatively small increase in the power required to provide the field current.

In some applications of commutator machines, particularly when used as generators, magnetic hysteresis may be significant, even though its effect appears to be small. Figure 4.2(b) shows, in somewhat accentuated form, the multivalued relationship between flux and field mmf that may occur with hysteresis. From this, it will be appreciated that the flux for a given field current will be predictable only within an accuracy of a few percent.

In a two-pole machine such as that illustrated in Fig. 4.1, much of the available rotor surface is not used effectively to produce torque. Also, the yoke contains a considerable mass of magnetic material. By use of a larger even number of poles, as shown in Fig. 4.3, a more compact design requiring less magnetic material per unit of torque may be achieved. Because of the symmetry of the

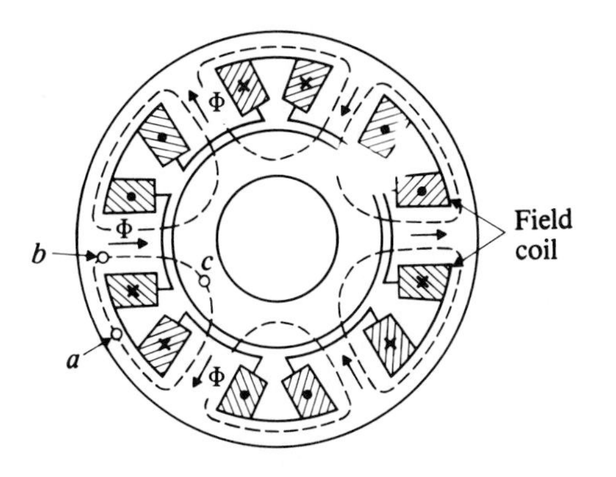

Figure 4.3 Magnetic system of a six-pole machine.

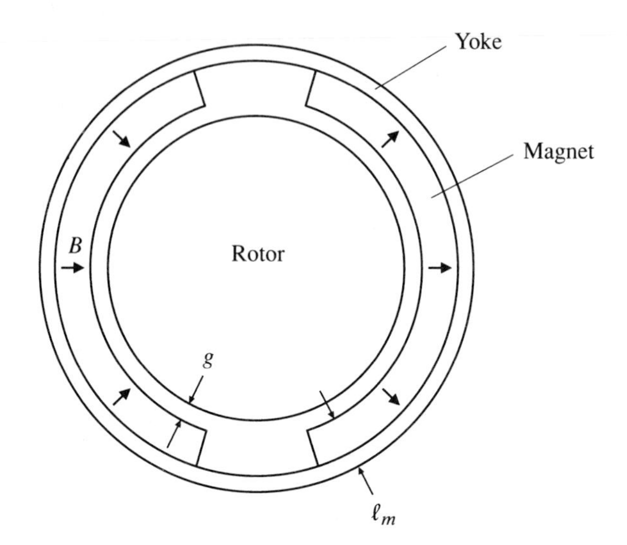

Figure 4.4 Magnetic system of a permanent magnet motor.

magnetic system, the Φ-$\mathfrak{F}_p$ relation may be determined considering only the reluctances in one flux path such as *a-b-c* in Fig. 4.3.

The magnetic flux in a commutator machine also may be established using permanent magnet poles, as shown in Fig. 4.4. The magnets are typically made of materials such as those for which the *B-H* characteristics are shown in Fig. 1.18. The ferrite material has a relatively low residual flux density B_r of about 0.3 to 0.4 T, but its cost per unit of magnet volume is also low. The neodymium (Nd-Fe-B) material has a higher residual density of about 1.1 to 1.2 T and is more expensive. In some machines, magnets made of samarium cobalt with a residual flux density of about 0.85 T are used.

The analysis of a magnetic system with a permanent magnet and an air gap has been discussed in Section 1.11. Ignoring any flux fringing and assuming the relative permeability of the magnet material to be unity, the air-gap flux density of the machine in Fig. 4.4 is given approximately by

$$B_g = \frac{\ell_m}{\ell_m + g_e} B_r \quad \text{T} \tag{4.4}$$

4.2

Rotor Windings and Commutator

In order to make efficient use of the available rotor or armature surface, conductors are distributed in slots over the complete circumference, as shown in

Fig. 4.5 for a two-pole machine. To obtain continuous torque relatively free of ripple, it is desirable to have, at any instant in rotation, the pattern of conductor current directions shown. This is achieved by use of coils which are usually multi-turn. A coil that has one of its sides in slot a in Fig. 4.5 has its other side in slot a' in an approximately similar location under the next pole. At the front or commutator end of the machine, a connection goes from the end conductor emerging from slot a' to the beginning conductor of the coil in slot b next to slot a. This sequential connection of coils continues until all the coils are connected in series in a closed loop, the last one connecting onto the start of the coil in slot a. In large machines that have somewhat rigid coils, it is usual to have one side of the coil in the bottom of a slot while the return side is in the top of a slot.

In order to maintain continuously the pattern of currents under the poles, as shown in Fig. 4.5, each-front end connection between sequential coils is attached to an insulated segment of a mechanical switch or commutator. This commutator is fitted to the shaft and consists of a number of segments equal to the number of rotor coils. The connections for a simple two-pole, six-slot machine are shown in Fig. 4.6. Stationary brushes are fitted to slide on the commutator surface and are located so as to feed current from the external electric system into two parallel paths through the closed loop of series-connected rotor coils. Each conductor carries half the terminal or armature current i_a. The width of the brush is such that it causes a short circuit between the two adjacent commutator bars to which a coil is connected; however, at that time, this coil is located in an area of the rotor between the poles and thus does not have an induced voltage.

In a two-pole machine, the rotor coils are always connected in two parallel paths between the two brushes. As the machine rotates, the induced voltages in the series of coils of each path are added to provide two parallel and equal voltages at the commutator terminals that are in contact with the two brushes.

For a p-pole machine, the coils normally span an arc of about $2\pi/p$ rad. The coils of the armature winding are all connected in a series loop with one commutator bar per coil. There are several ways of making the front connections between coils. These differ in the number of parallel paths into which the

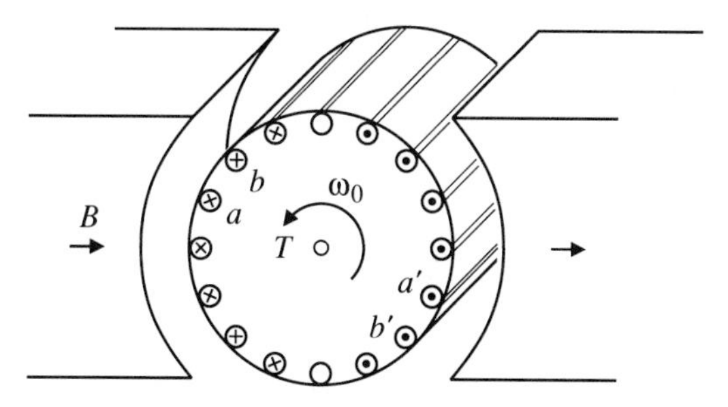

Figure 4.5 Desired pattern of armature conductor currents.

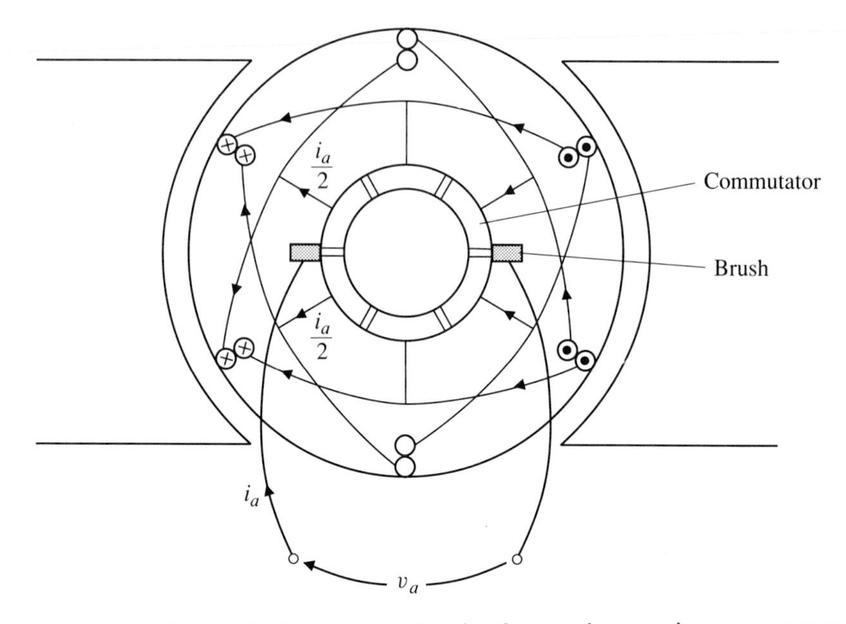

Figure 4.6 Simple six-slot armature showing front-end connections, commutator, and brushes.

armature coils are arranged. In small machines with more than two poles, the number of parallel paths, designated as a, is usually two. In larger, high-current machines, the number of parallel paths is made equal to p, or sometimes to a multiple of p. Normally, the number of brushes is equal to the number of poles, with every other brush being connected together to form two external armature terminals. A sketch of a part of the winding of a machine with $a = p$ is shown in Fig. 4.7. This winding is known as a *lap winding*.

4.3

Torque and Generated Voltage

The armature winding of the two-pole machine shown in Fig. 4.8 has a total of N_a turns in its series-connected coils. Each of the $2N_a$ conductors seen in the cross section carries half the armature terminal current i_a (with the exception of the turns in the coil being commutated from one path to the other). The linear current density K per unit of the rotor periphery is then given by

$$K = \frac{2N_a}{2\pi r}\frac{i_a}{2} \qquad \text{A/m} \tag{4.5}$$

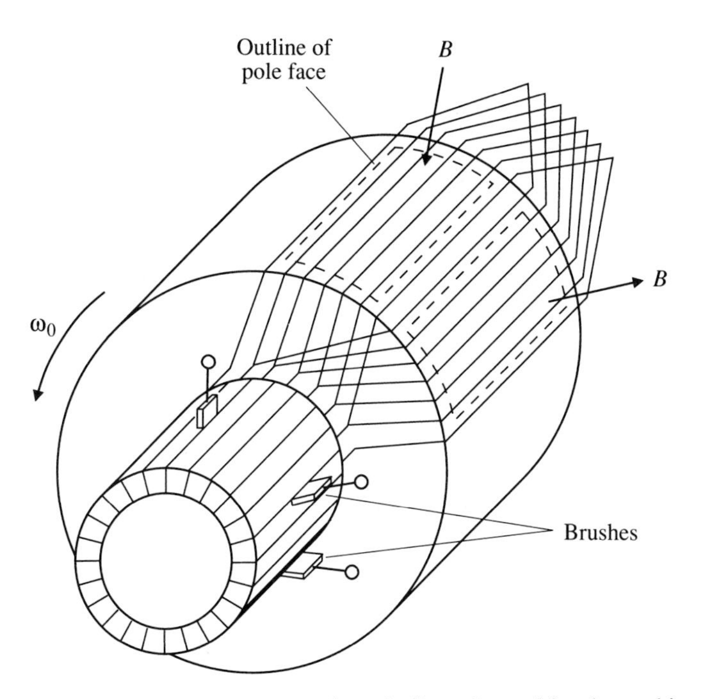

Figure 4.7 Part of the armature lap winding of a multi-pole machine.

Because the flux density B is radially directed, the force per unit area of armature surface under the poles is BK, from Eq. (3.22). Thus, the torque exerted on the rotor is then

$$\begin{aligned} T &= 2w\ell rBK \\ &= \frac{N_a}{\pi}(Bw\ell)i_a \\ &= k\Phi i_a \qquad \text{N}\cdot\text{m} \end{aligned} \tag{4.6}$$

where w is the pole width, ℓ is the axial length of both pole and rotor, and Φ is the flux per pole. In a two-pole machine, the constant k depends only on the number of armature winding turns. In a p-pole machine, with the winding in a parallel paths,

$$k = \frac{N_a p}{\pi a} \tag{4.7}$$

Ignoring losses, the electrical power input to the armature winding equals the mechanical power output. Denoting e_a as the induced or generated voltage in each parallel path of the winding,

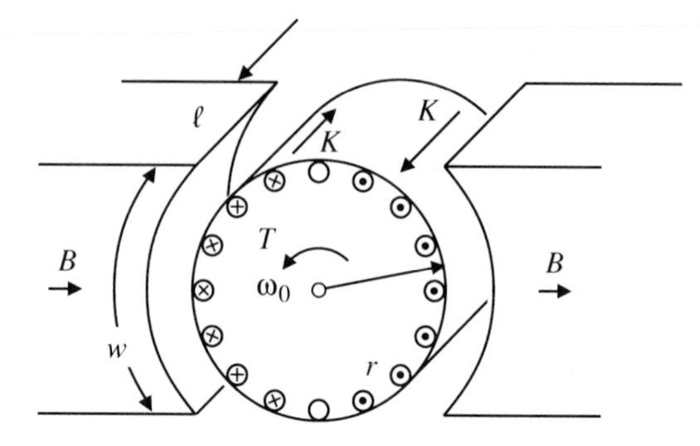

Figure 4.8 Armature and pole dimensions.

$$\begin{aligned} p_{elec} &= e_a i_a \\ &= T\omega_0 \\ &= k\Phi i_a \omega_0 \qquad \text{W} \end{aligned} \tag{4.8}$$

Thus,

$$e_a = k\Phi\omega_0 \qquad \text{V} \tag{4.9}$$

An alternative way of deriving this expression for the generated voltage is to consider that the flux linkage of a single turn on the rotor changes from $+\Phi$ to $-\Phi$ as the turn moves across the face of a pole. The time taken for the turn to rotate through π rad is π/ω_0 s. Thus, the average voltage induced in the $N_a/2$ series turns in a path is

$$\begin{aligned} e_a &= \frac{N_a}{2}\left(\frac{d\lambda}{dt}\right)_{av} \\ &= \frac{N_a}{2}\,\frac{2\Phi}{\pi/\omega_0} \\ &= \frac{N_a}{\pi}\,\Phi\omega_0 \qquad \text{V} \end{aligned} \tag{4.10}$$

The quantity $k\Phi$ can be found as a function of field current i_f from measurements. Suppose the machine is driven by a prime mover at speed ω_0 and the no-load terminal voltage v_t is measured for a range of values of field current i_f. Because the armature current is zero, the terminal voltage v_t is equal to the generated voltage e_a. Thus,

$$k\Phi = \frac{v_t}{\omega_0} \qquad \text{V}\cdot\text{s or Wb} \tag{4.11}$$

Ignoring any hysteresis effect, the result is a curve of the form of Fig. 4.9. Within the linear region, the relation can be expressed as

$$k\Phi = k_f i_f \qquad \text{Wb} \tag{4.12}$$

and

$$e_a = k_f i_f \omega_0 \qquad \text{V} \tag{4.13}$$

An alternate way of obtaining the $k\Phi$-i_f characteristic of Fig. 4.9 is to measure the shaft torque at standstill when a current i_a is passed through the armature terminals. If any friction torque is ignored, from Eq. (4.6),

$$k\Phi = \frac{T}{i_a} \qquad \text{N·m/A} \tag{4.14}$$

EXAMPLE 4.1 A small two-pole commutator machine is used as a tachometer, being driven from the shaft of which speed is to be measured. The flux is established by permanent magnets that produce an air-gap flux of 0.3 mWb for each pole. The armature winding has 400 turns. What speed in radians per second will be indicated by each one-volt division on the scale of a high-resistance voltmeter connected to the armature terminals?

Solution The armature terminal voltage may be assumed to be equal to the generated voltage in view of the negligible current in the voltmeter. From Eq. (4.9),

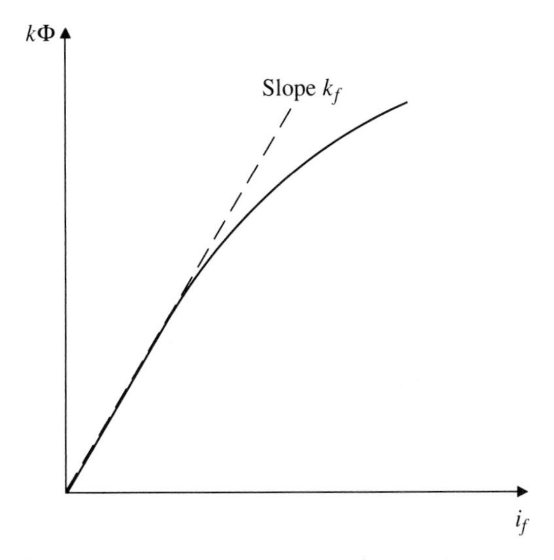

Figure 4.9 Flux quantity $k\Phi$ as a function of field current i_f.

$$\frac{\omega_0}{e_a} = \frac{1}{k\Phi} \qquad \text{Wb}^{-1}$$

From Eq. (4.7),

$$k = \frac{N_a p}{\pi a} = \frac{400}{\pi}$$

$$\frac{\omega_0}{e_a} = \frac{\pi}{400} \times \frac{10^3}{0.3} = 26.2 \qquad \text{rad/V·s}$$

EXAMPLE 4.2 In the speed-measuring system of Example 4.1, suppose the high-resistance voltmeter is replaced by a meter of 1000 Ω resistance. The resistance of the armature circuit of the machine is 250 Ω. If the tachometer is now driven at 3000 r/min, determine the torque and mechanical power required to drive it, ignoring friction losses.

Solution From Example 4.1,

$$k\Phi = \frac{1}{26.2} = 0.0382 \qquad \text{Wb}$$

$$\omega_0 = \frac{3000}{60} \times 2\pi = 314.2 \qquad \text{rad/s}$$

$$e_a = k\Phi\omega_0 = 0.0382 \times 314.2 = 12 \qquad \text{V}$$

$$i_a = \frac{e_a}{R_{total}} = \frac{12}{1000 + 250} = 0.0096 \qquad \text{A}$$

$$T = k\Phi i_a = 0.0382 \times 0.0096 = 0.367 \qquad \text{mN·m}$$

$$P = T\omega_0 = 0.115 \qquad \text{W}; \qquad P = e_a i_a = 12 \times 0.0096 = 0.115 \qquad \text{W}$$

4.4

Equivalent Circuit

For the purposes of analysis, the commutator machine can be visualized as an ideal electromechanical energy converter with added parameters that represent its various imperfections. The ideal machine is described by the two expressions

$$e_a = k\Phi\omega_0 \qquad \text{V} \tag{4.15}$$

and

$$T = k\Phi\omega_0 \qquad \text{N·m} \tag{4.16}$$

This ideal machine is analogous to an ideal transformer.

Next, let us consider the imperfections. When armature current flows, there is power loss in the armature coils. The resistance R_a of the armature circuit may be derived from the measured voltage between the commutator segments under entry and exit brushes when the rotor is stationary and carrying armature current.

There is also an electric power loss in the brushes. These are normally made of carbon, sometimes with a metal such as copper in solution. A voltage v_b occurs between the brush and the commutator surface. This voltage varies with the current density J_b at the contact surface, as shown for a typical brush in Fig. 4.10. The total brush drop of about 2 V for two brushes in series is significant in low-voltage machines but may often be ignored in higher-voltage machines. Alternatively, the resistance R_a may include a term to represent the brush resistance.

The armature circuit also has an inductance L_a representing the flux linkage of the armature winding per ampere of armature current. The pattern of this armature flux is suggested by the two-pole cross section of a permanent magnet machine shown in Fig. 4.11. This flux is small relative to the field pole flux because of the large effective air gap between the rotor and the yoke along the axis between the poles. The armature inductance is significant in situations where the armature current is changing rapidly. Including these parameters, the armature circuit may be represented as shown in the equivalent circuit of Fig. 4.12, where

$$v_t = R_a i_a + L_a \frac{di_a}{dt} + k\Phi\omega_0 \qquad \text{V} \tag{4.17}$$

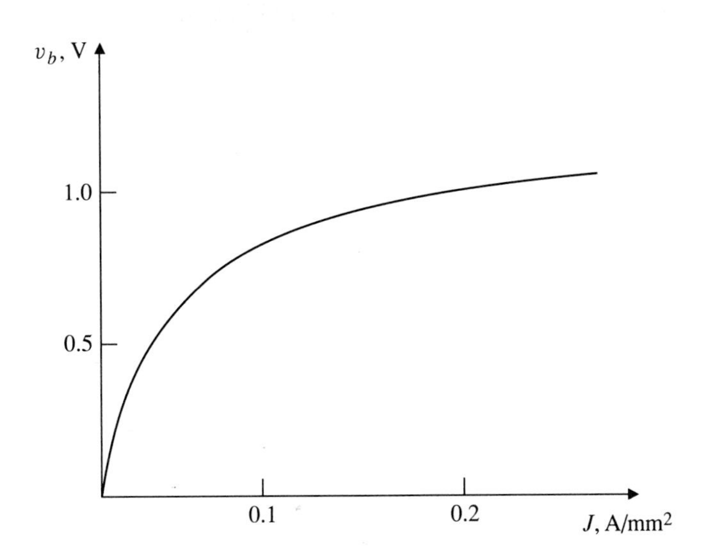

Figure 4.10 Voltage across a typical brush-commutator contact.

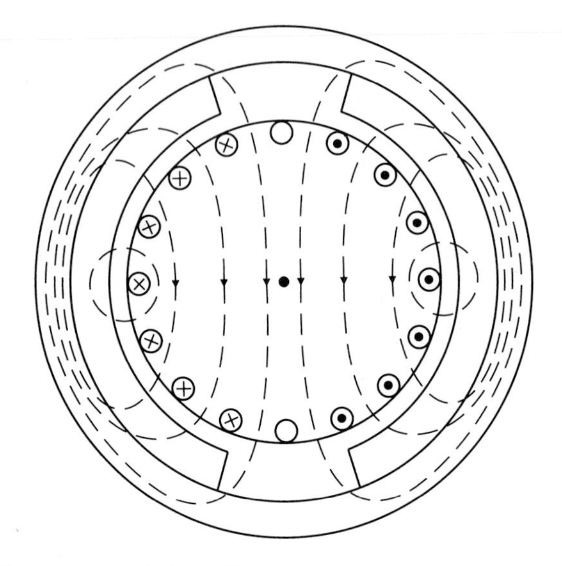

Figure 4.11 Armature leakage flux pattern.

The series coils of a wound-field machine will have a resistance R_f. The terminal voltage and current of the field circuit are then related by

$$v_f = R_f i_f + \frac{d\lambda_f}{dt} \qquad \text{V} \tag{4.18}$$

The field flux linkage λ_f is related to the field current i_f by a graph of the same form as shown in Fig. 4.9. In the linear range, this λ_f-i_f relation may be modeled as the field circuit inductance

$$L_f = \frac{\lambda_f}{i_f} \qquad \text{H} \tag{4.19}$$

Equation (4.18) then may be stated as

$$v_f = R_f i_f + L_f \frac{di_f}{dt} \qquad \text{V} \tag{4.20}$$

In the mechanical system, the torque exerted on the rotor by the interaction of the armature current with the field flux is given in Eq. (4.16). This torque provides the output shaft torque T_0 plus any loss torque due to friction and windage T_{loss} plus the torque required to accelerate the inertia J of the rotor. Thus,

$$\begin{aligned} T &= k\Phi i_a \\ &= J\frac{d\omega_0}{dt} + T_{loss} + T_0 \qquad \text{N} \end{aligned} \tag{4.21}$$

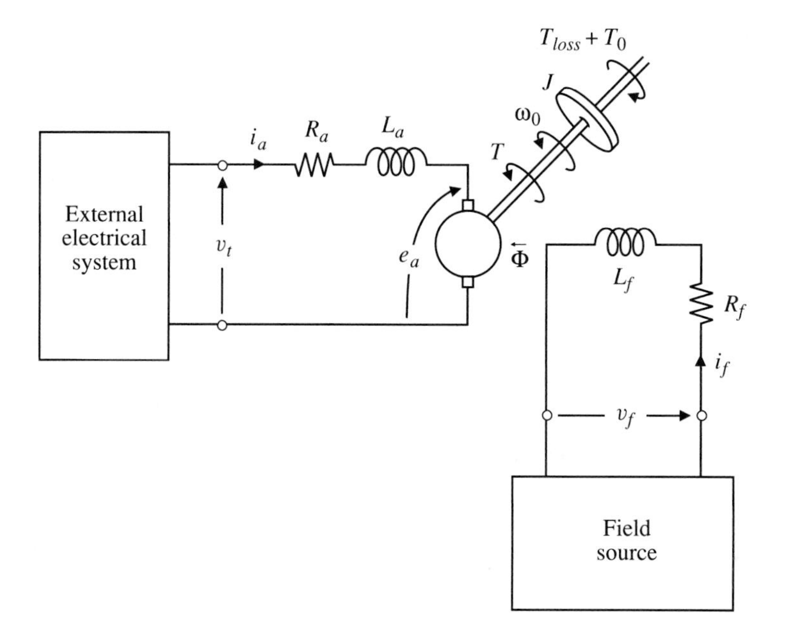

Figure 4.12 Equivalent circuit of a commutator machine.

The transient behavior of the machine is described by the three differential Eqs. (4.17), (4.20), and (4.21), together with a flux-field current relation such as shown in Fig. 4.9. If any of the variables i_a, i_f, or ω_0 is constant, the term in these equations involving its derivative with time becomes zero. If all three variables are constant, the resultant steady-state equations are

$$v_t = R_a i_a + k\Phi\omega_0 \qquad \text{V} \tag{4.22}$$

$$v_f = R_f i_f \qquad \text{V} \tag{4.23}$$

$$k\Phi i_a = T_{loss} + T_0 \qquad \text{N}\cdot\text{m} \tag{4.24}$$

To determine the performance of the machine, the above set of three equations, either transient or steady state as required, must be combined with three other equations representing the two external electrical systems connected to the armature and the field and with the external mechanical system attached to the shaft.

EXAMPLE 4.3 A fan motor for an automobile has a ferrite permanent magnet field. When tested at standstill, an applied terminal voltage of 4 V produced an armature current of 6.5 A and a torque of 0.23 N·m. Determine the armature circuit resistance and the flux constant for this machine. Friction and windage torque may be ignored.

Solution From the test data, the armature resistance is

$$R_a = \frac{v_t}{i_a} = \frac{4}{6.5} = 0.615 \qquad \Omega$$

From Eq. (4.16), the flux constant is

$$k\Phi = \frac{T}{i_a} = \frac{0.23}{6.5} = 0.0354 \qquad \text{N}\cdot\text{m/A}$$

4.5 Steady-State Performance

Because of the wide variety of applications of commutator machines, it is convenient to classify them as generators or motors, depending on the dominant direction of power flow in the application. However, each commutator machine is capable of both generating and motoring action.

Consider a machine in which the field current i_f has been set, providing a constant pole flux Φ, usually near its maximum design value. Suppose the armature terminals are connected to a supply of either fixed or controllable direct voltage such as a battery, another commutator machine, or an electronically controlled voltage source. Suppose the shaft is connected to a mechanical system that has a known relation between its steady-state torque and its speed:

$$T_0 = f(\omega_0) \qquad \text{N}\cdot\text{m} \tag{4.25}$$

A typical characteristic for a fan or pump load is shown in Fig. 4.13. In this load, the torque is approximately proportional to the square of the speed. Other mechanical loads may have a nearly constant torque or may have a torque nearly proportional to speed. For convenience, the loss torque in the machine may be considered as part of the mechanical load.

In some situations, it is necessary to predict how the speed will change when the load torque changes or as the supply voltage is changed. The relation between terminal voltage v_t and load torque T_0 may be found by rearranging Eqs. (4.16) and (4.22) to give

$$\begin{aligned} \omega_0 &= \frac{v_t - R_a i_a}{k\Phi} \\ &= \frac{v_t}{k\Phi} - \frac{R_a T_0}{(k\Phi)^2} \qquad \text{rad/s} \end{aligned} \tag{4.26}$$

This relation is shown for a set of values of terminal voltage v_t as the straight lines with downward slope in Fig. 4.13. The intersection of a line and the load characteristic represents the solution of the two steady-state Eqs. (4.25) and (4.26) for a particular supply voltage, giving the speed and torque of operation.

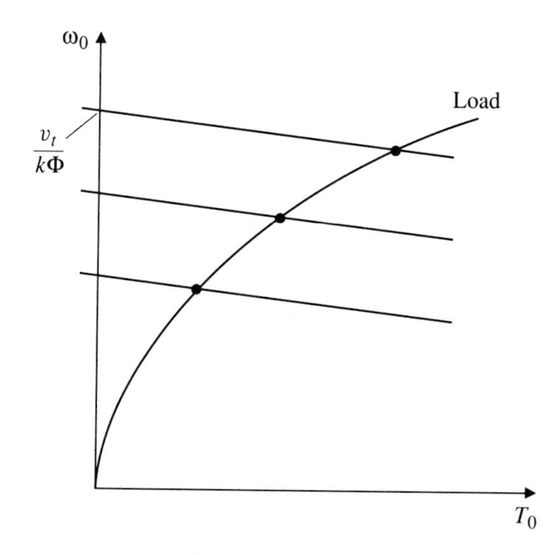

Figure 4.13 Graphical solution for steady-state speed and torque.

A commutator machine with a controlled voltage supply to its armature may operate in all four quadrants of the speed-torque plane. Consider the system shown in Fig. 4.14 where an elevator cage is driven up and down by a motor coupled through a gear box to a drum on which the suspension cable is wound. When the mass of the cage is greater than the mass of the counter-

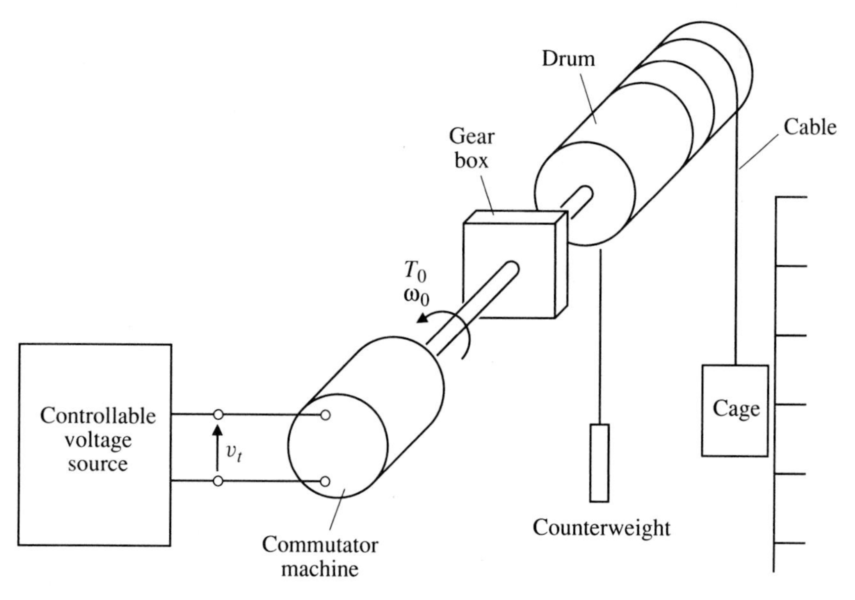

Figure 4.14 Elevator drive.

weight, the load torque T_0 on the motor will be positive. If the cage is to be raised with this value of torque, a positive voltage v_t is required to provide positive motor speed ω_0, i.e., operation is in the first quadrant of Fig. 4.15. If the cage is to be lowered with this torque, the voltage must be made negative, giving operation in the fourth quadrant. When the cage mass is less than that of the counterweight, the load torque is negative, and operation is then in quadrants two or three.

The maximum torque that can be provided by the machine is limited by the maximum armature current for which the machine is designed. The armature voltage must be controlled so that this armature current limit is not exceeded.

The speed-torque characteristics of Fig. 4.15 have been drawn as straight lines on the assumption that the flux Φ is constant. At high values of armature current, the magnetomotive force of the armature winding may distort the flux pattern across the poles to the extent that the net flux Φ in each pole will be somewhat reduced. This may lead to a small increase in the speed above that which would otherwise be predicted at high values of torque. This effect on the flux is known as *armature reaction* and is discussed further in Section 9.8.

A variable-speed drive using a commutator machine normally operates at or near its maximum design value of flux density in the rotor, poles, and yoke because this gives the maximum torque per ampere of armature current. However, the controllable voltage source that supplies the armature has a maximum voltage $\hat{v}_t$. Thus, there is a limit to the speed that can be achieved with the maximum value of the flux constant $k\Phi$. Higher speed operation can be achieved by reducing the field current and thus reducing the flux constant $k\Phi$. From Eq. (4.26), it can be seen that this increases the no-load speed to a value

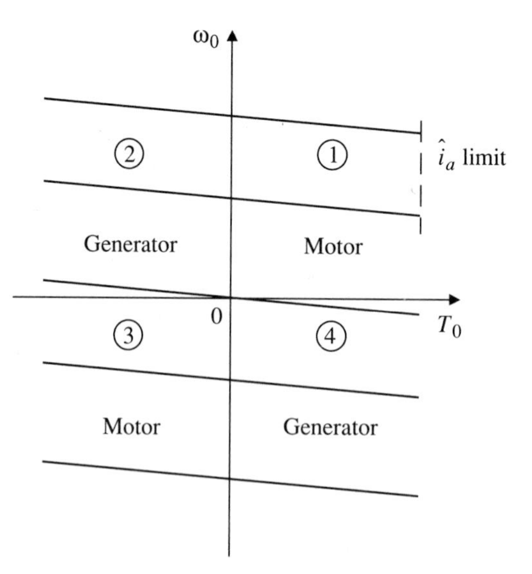

Figure 4.15 Four-quadrant operation of elevator-type drive.

$$\omega_0 = \frac{\hat{v}_t}{k\Phi} \qquad \text{rad/s} \tag{4.27}$$

At the same time, it reduces the torque per unit of armature current. If the armature current is limited to some design value $\hat{i}_a$, the maximum available torque will be reduced as $k\Phi$ is reduced:

$$\hat{T} = k\Phi\hat{i}_a \qquad \text{N}\cdot\text{m} \tag{4.28}$$

From Eq. (4.26), the maximum torque can be related to the speed by

$$\begin{aligned}\hat{T} &= k\Phi\hat{i}_a \\ &= \left(\frac{\hat{v}_t - R_a\hat{i}_a}{\omega_0}\right)\hat{i}_a \\ &= \frac{\hat{v}_t\hat{i}_a - R_a\hat{i}_a^2}{\omega_0} \qquad \text{N}\cdot\text{m}\end{aligned} \tag{4.29}$$

The electrical input power is limited by the maximum voltage of the source and the maximum allowable armature current. Thus, the mechanical output power is limited to this value minus the constant losses in the armature resistance. Therefore, the maximum mechanical output power is essentially constant in this region of flux weakening.

Figure 4.16 shows typical speed-torque characteristics within this field-weakening region of operation. It is noted that the downward slopes of these characteristics increase as the flux is decreased, as would follow from Eq. (4.26).

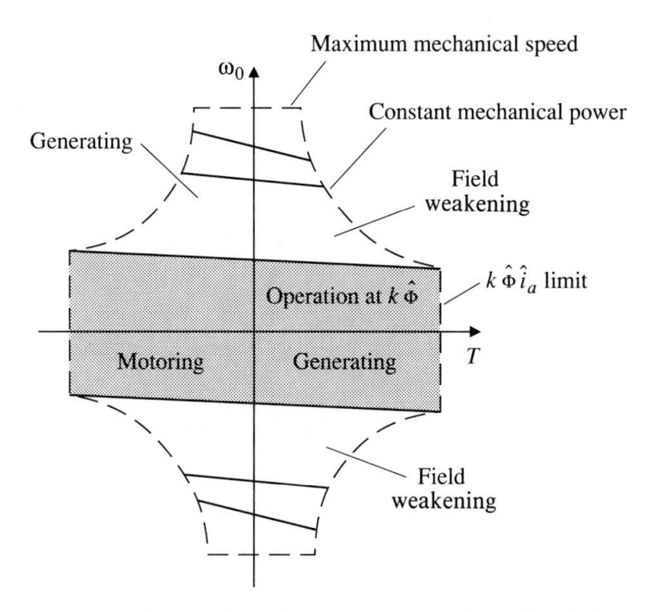

Figure 4.16 Operating region of commutator machine with field weakening.

Field weakening also applies in the generating region where, with positive speed ω_0, the current i_a is negative. The mechanical power input limit from Eq. (4.29) then is equal to the electrical power output plus the losses.

For a commutator machine, the maximum permissible speed is limited, partly by mechanical considerations and partly by increasing difficulty in commutating the armature coils without sparking. Field weakening is used extensively in drives for traction vehicles such as electric trains, buses, subways, and street cars. At low speed, maximum acceleration and deceleration is achieved by use of maximum flux and controlling the armature voltage so as to achieve maximum allowable armature current and therefore maximum torque. This mode of operation is continued up to a speed at which the maximum available supply voltage is reached. For higher speed, the field is weakened gradually while the terminal voltage is held constant.

EXAMPLE 4.4 The motor of Example 4.3 is to drive a fan for which the torque varies as the square of the speed according to the relation

$$T_0 = 8 \times 10^{-6}\omega_0^2 \qquad \text{N}\cdot\text{m}$$

with ω_0 in rad/s. The motor is connected to the 12-V battery supply of the vehicle. At what steady-state speed will it drive the fan?

Solution In the steady state, the speed and torque of the motor are related by Eq. (4.26). This may be combined with the speed-torque relation for the load to obtain an expression for speed. From Example 4.3, the machine parameters are $R_a = 0.615\ \Omega$ and $k\Phi = 0.0354$ N·m/A with $v_t = 12$ V. Thus,

$$\begin{aligned}\omega_0 &= \frac{v_t}{k\Phi} - \frac{R_a T_0}{(k\Phi)^2} \\ &= \frac{12}{0.0354} - \frac{0.615 \times 8 \times 10^{-6}}{0.0354^2}\omega_0^2 \\ &= 339 - 0.00393\omega_0^2\end{aligned}$$

Solution of this quadratic equation gives the positive value of speed as

$$\omega_0 = 193 \qquad \text{rad/s}$$

EXAMPLE 4.5 An electric trolley bus has a mass of 12 500 kg. A drive system is required to provide a maximum acceleration or deceleration at low speed of 1 m/s^2. This system is to consist of a commutator motor coupled to the drive wheels through a gear box. An electronic controller is to provide a controllable voltage to the motor up to the maximum value of 600 V that is available on the trolley wire. The electric power taken from the supply is to be limited to a maximum of 100 kW. In the following estimates, all losses may be ignored, and the vehicle may be assumed to operate on a level street.

(a) Up to what speed can maximum acceleration be maintained?
(b) If the maximum cruising speed of the bus is to be 72 km/h and the maximum speed of the motor is to be 4500 r/min, what should be the minimum value of flux constant in the motor?
(c) What should be the maximum value of flux constant with full-field current?
(d) What should be the armature voltage at a speed of 10 km/h?

Solution

(a) The thrust force to produce an acceleration of 1 m/s^2 is

$$F = ma = 12\,500 \qquad \text{N}$$

If the power is to be limited to 100 kW, the velocity up to which this force can be achieved is

$$v = \frac{\hat{P}}{F} = \frac{100\,000}{12\,500} = 8 \qquad \text{m/s}$$

(b) The maximum velocity of the vehicle is specified as

$$\hat{v} = \frac{72\,000}{3600} = 20 \qquad \text{m/s}$$

The maximum speed of the motor is specified as

$$\hat{\omega}_0 = 4500 \times \frac{2\pi}{60} = 471.2 \qquad \text{rad/s}$$

The flux constant that will provide a generated voltage of 600 V at this speed is

$$k\Phi = \frac{600}{471.2} = 1.27 \qquad \text{V}\cdot\text{s/rad}$$

(c) Full torque is to be maintained to 8 m/s. This corresponds to a motor speed of

$$\omega_0 = 471.2 \times \frac{8}{20} = 188.5 \qquad \text{rad/s}$$

The flux constant at this speed should be

$$(\widehat{k\Phi}) = \frac{600}{188.5} = 3.18 \qquad \text{V}\cdot\text{s/rad}$$

(d) At 10 km/h, the motor speed, using the data of (b), is

$$\omega_0 = \frac{10}{72} \times 471.2 = 65.4 \qquad \text{rad/s}$$

Then, the terminal voltage for operation at this speed should be

$$v_t = (\widehat{k\Phi})\omega_0 = 3.18 \times 65.4 = 208.3 \qquad \text{V}$$

4.6

Motors Operating on Constant Voltage Supply

In many situations, only a constant direct-voltage supply is available. For example, this may be a 12-, 24-, or 48-V battery in a vehicle or a 100- to 240-V supply in a laboratory. It may be permissible to start some small motors by connecting the armature terminals directly to the supply using a simple switch. Examples will be found in the starter motor and other auxiliary motors in automobiles. In these motors, when the switch is closed, the armature current quickly rises to approximately

$$i_a = \frac{v_t}{R_a} \qquad \text{A} \tag{4.30}$$

This current may be several times the continuous current rating of the armature winding. However, the torque produced quickly accelerates the motor to its full speed where the armature current is reduced below its permissible continuous or rated value.

Motors in this category are inherently inefficient. For example, if the maximum current during starting is not to exceed four times rated value, the armature circuit resistance must be such that the voltage drop across it will be one quarter of the supply voltage when rated current flows in the armature. Therefore, the efficiency at rated current is limited to 75%.

Larger motors with field windings may have both their armature and field windings connected to the constant voltage supply through series resistances, as shown in the circuit diagram of Fig. 4.17. The field winding of the machine is normally designed with the number of turns and the resistance R_f which limits the field current i_f to an acceptable maximum value when the series resistor R_{fe} in the field circuit is set to zero. This is the normal setting of the external field resistor for starting because it produces maximum flux and thus maximum torque per ampere of armature current.

To start the motor, sufficient external resistance R_{ae} is inserted in series in the armature circuit to limit the maximum armature current to a value that can be tolerated by the commutator without excessive sparking at the brushes. Typically, this maximum value may be two to three times the rated armature current.

The steady-state speed-torque characteristic of the drive can be controlled by variation of the series resistor R_{ae}, as shown in Fig. 4.18. If the constant supply voltage is v_s, the speed is given, by analogy with Eq. (4.26), by

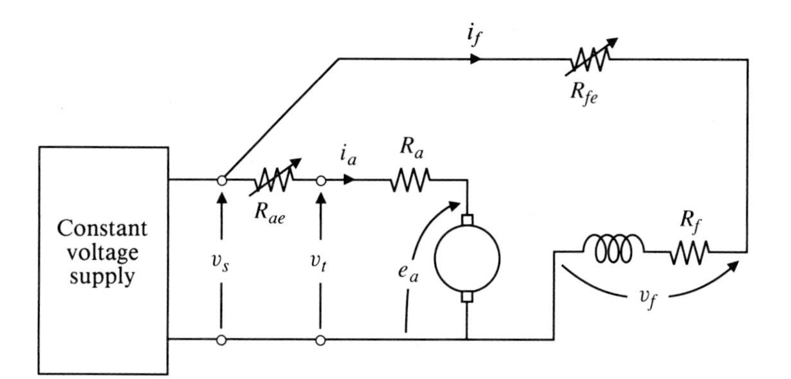

Figure 4.17 Equivalent circuit of a motor operating on a constant voltage supply.

$$\begin{aligned} \omega_0 &= \frac{v_s - (R_a + R_{ae})i_a}{k_a\Phi} \\ &= \frac{v_s}{k_a\Phi} - \frac{(R_a + R_{ae})T}{(k_a\Phi)^2} \qquad \text{rad/s} \end{aligned} \tag{4.31}$$

The steady-state speed of a load at any resistance setting is given by the intersection of its speed-torque relation (including friction and windage losses in the motor) with the appropriate speed-torque line of the motor for that resistance.

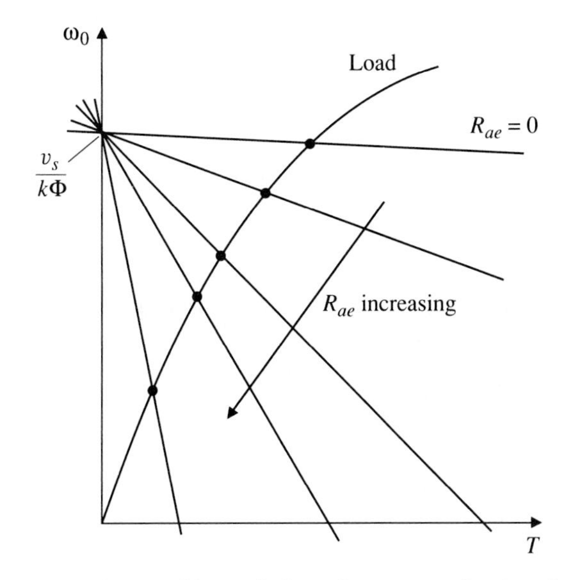

Figure 4.18 Speed control by variation of armature circuit resistance.

Transiently, the horizontal difference between these two curves at any speed represents the accelerating torque, bringing the motor toward the point of intersection.

This method of speed control is effective only if the motor is loaded. If load torque is removed, it can be seen from Eq. (4.31) that the speed tends to rise to the value $v_s/k\Phi$.

Speed control by variation of armature circuit resistance is seldom employed for continuous operation of a motor because much energy is wasted in the external resistor R_{ae}. However, this method of speed control may be applied in special circumstances such as an occasionally used hoist.

Armature circuit resistance control is the standard method of starting a commutator motor from a constant voltage source. A rapid start can be achieved by reducing the resistance in steps as the speed increases. A typical manual starter, as shown in Fig. 4.19, has a spring-loaded contact arm that is rotated clockwise. This provides initial connection of the field circuit and gradual reduction of the armature circuit resistance as the speed increases toward its steady-state value. A hold-on electromagnetic actuator holds the arm in the zero-resistance position as long as field current flows through its coil. Loss of supply or opening of the field circuit releases the arm. The series field resistor should be set at zero during starting to give maximum field current and maximum flux.

Automatic starters perform essentially the same functions as the manual starter of Fig. 4.19. Electromagnetic contactors short out sections of the starting resistor either in a predetermined time sequence or when the armature current has dropped to a predetermined value.

When the starting sequence is complete, the motor will have a speed-torque characteristic such as that shown for maximum field current i_f in Fig. 4.20. If

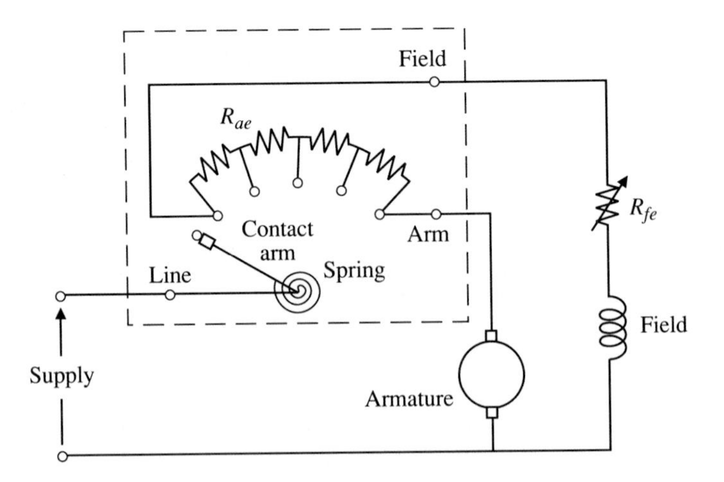

Figure 4.19 Manual starter.

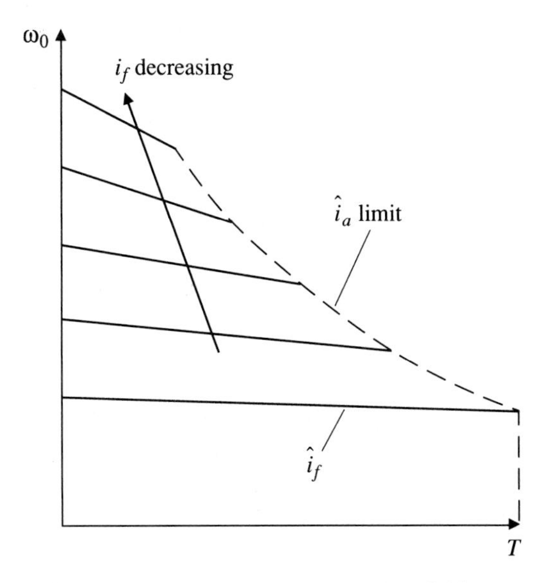

Figure 4.20 Speed-torque relations with decreasing field current.

desired, the speed can now be increased by increasing the series field resistance R_{fe}, thus reducing the field current and the flux Φ. It should be noted that if the armature current is to be limited to its rated value, the available torque at any value of speed will be reduced as the quantity $k\Phi$ is reduced. Figure 4.20 shows a set of characteristics, each terminating at a torque corresponding to rated current. The maximum speed is limited either by mechanical considerations or by an increased difficulty in commutating as the flux is decreased.

Motors that are designed for operation from a constant voltage supply and connected as in Fig. 4.17 are commonly known as *shunt motors* because the field circuit is in shunt, or parallel with, the armature circuit. The field current is usually less than 5% of the rated armature current. Such motors provide a limited range of speed control through variation of the field resistor from a minimum to a maximum value. A particular application of interest is in the testing of other electric machines in a laboratory.

EXAMPLE 4.6 A commutator motor has a rated armature voltage of 125 V and rated armature current of 36 A. The armature resistance is 0.37 Ω. With full field and rated armature voltage and current, its speed is known to be 1150 r/min.

(a) Operating from a 120-V supply, what series resistance should be inserted in the armature circuit to limit the starting current to 2.5 times rated value?

(b) Suppose the motor is driving a load that requires a torque of 0.3 N·m per rad/s of speed. With full field and a 120-V supply, what series resistance is required to drive this load at 600 r/min?

Solution

(a) The armature current is to be limited to

$$\hat{i}_a = 2.5 \times 36 = 90 \quad \text{A}$$

The required resistance is

$$R_{ae} = \frac{120}{90} - 0.37 = 0.96 \quad \Omega$$

(b) The flux constant at full field current can be obtained from the data given. Rearranging Eq. (4.26),

$$k\Phi = \frac{v_t - R_a i_a}{\omega_0} = \frac{125 - 0.37 \times 36}{1150 \times \frac{2\pi}{60}} = 0.927$$

For the stated load condition,

$$v_s = 120 \quad \text{V}$$

$$\omega_0 = 600 \times 2\pi/60 = 62.8 \quad \text{rad/s}$$

$$T_0 = 0.3\omega_0 \quad \text{N·m}$$

Inserting these values into Eq. (4.31) gives

$$\omega_0 = \frac{v_s}{k\Phi} - \frac{(R_a + R_{ae})T_0}{(k\Phi)^2}$$

$$62.8 = \frac{120}{0.927} - \frac{(0.37 + R_{ae})0.3 \times 62.8}{0.927^2} = 129.5 - (0.37 + R_{ae})21.9$$

$$R_{ae} = \frac{129.5 - 62.8}{21.9} - 0.37 = 2.68 \quad \Omega$$

EXAMPLE 4.7 An automatic starter is required for the machine described in Example 4.6. During starting, the armature current is allowed to vary over the range $40 < i_a < 90$ A. The supply voltage is 125 V. How many sections are required in the starter resistance, what are their values, and at what speeds are they to be switched out?

Solution Initially, the total resistance R_{T1} in the armature circuit should be set at

$$R_{T1} = \frac{125}{90} = 1.39 \qquad \Omega$$

As the machine speeds up, the generated voltage e_a increases. When the armature current i_a has been reduced to 40 A, the generated voltage will be

$$e_{a1} = 125 - 1.39 \times 40 = 69.4 \qquad \text{V}$$

At this point, the armature circuit resistance can be reduced to

$$R_{T2} = \frac{125 - 69.4}{90} = 0.62 \qquad \Omega$$

When the armature current again has been reduced to 40 A, the generated voltage will be

$$e_{a2} = 125 - 0.62 \times 40 = 100.3 \qquad \text{V}$$

At this point, the armature circuit resistance can be reduced to

$$R_{T3} = \frac{125 - 100.3}{90} = 0.274 \qquad \Omega$$

Because this is less than $R_a = 0.37$, only two steps are required. The external armature circuit resistance should consist of two resistances in series, the second being

$$R_2 = R_{T1} - R_a = 0.62 - 0.37 = 0.25 \qquad \Omega$$

and the first being

$$R_1 = R_{T1} - R_2 - R_a = 1.39 - 0.62 = 0.77 \qquad \Omega$$

The speed at which R_1 is switched out is, using $k\Phi$ from Example 4.6,

$$\omega_0 = \frac{e_{a1}}{k\Phi} = \frac{69.4}{0.927} = 74.9 \qquad \text{rad/s}$$

The second resistor R_2 is switched out at

$$\omega_0 = \frac{e_{a2}}{k\Phi} = \frac{100.3}{0.927} = 108.2 \qquad \text{rad/s}$$

EXAMPLE 4.8 The motor of Example 4.6 has been started on a 125-V supply at full field. The field current is now reduced until the no-load speed is 1800 r/min. Ignore rotational losses. At this setting of field current, what torque can be produced by the motor at rated armature current, and what will be the corresponding speed?

Solution The desired speed value on no load is

$$\omega_0 = 1800 \times \frac{2\pi}{60} = 188.5 \qquad \text{rad/s}$$

To achieve this speed, the flux constant must be

$$k\Phi = \frac{v_t}{\omega_0} = \frac{125}{188.5} = 0.66$$

With rated armature current of 36 A, the available torque is

$$T = k\Phi i_a = 0.66 \times 36 = 23.9 \qquad \text{N·m}$$

The speed will be

$$\omega_0 = \frac{v_t - R_a i_a}{k\Phi} = \frac{125 - 0.37 \times 36}{0.66} = 169.2 \text{ rad/s} = 1615 \text{ r/min}$$

4.7

Operation as a Generator

When a commutator machine is operated as a generator, its speed is determined by the characteristics of its prime mover. This speed may be essentially constant if the prime mover is a well-governed engine or a synchronous motor. With some prime movers, the speed may decrease significantly as shaft torque is increased.

When a wide range of output voltage is required from a generator, the field current is supplied from a separate, controllable source, as shown in Fig. 4.21. This source may be another generator, a controlled rectifier, or simply a battery with a variable series resistor. The combined system is known as a *separately excited generator*. For convenience, the positive direction of the armature current has been reversed in comparison with the equivalent circuit of Fig. 4.12.

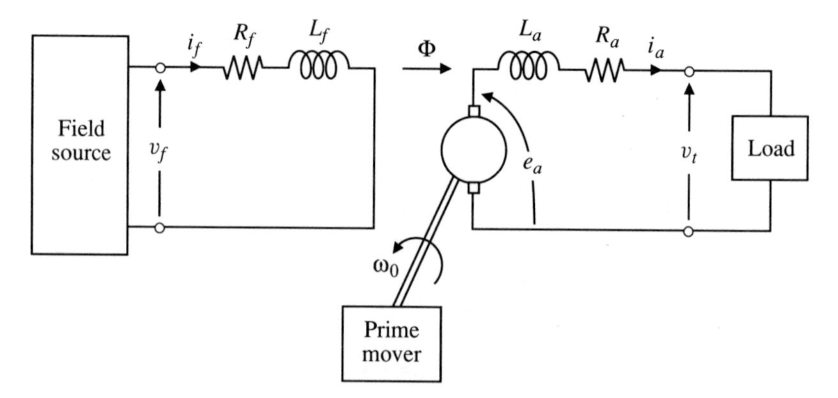

Figure 4.21 Connections for separately excited generator.

In steady-state operation, the field current is set to give the desired value of the flux factor $k\Phi$. The relation between the armature voltage and current is then

$$v_t = k\Phi\omega_0 - R_a i_a \qquad \text{V} \tag{4.32}$$

The intersection of this relation with the voltage-current characteristic of the load defines the point of steady-state operation.

When certain conditions are fulfilled, the generator's own armature circuit may be employed as a source of field excitation. Consider the generator system shown in Fig. 4.22, commonly known as a *shunt generator*. The first condition for self excitation is that there must be some residual magnetism from hysteresis in the ferromagnetic system of the machine. The direction of the residual flux must be such as to produce a small generated voltage with the desired polarity. If no residual flux exists or if it is in the wrong direction, an external source can be used momentarily to pass a current through the field winding.

The field circuit in Fig. 4.22 is connected across the armature circuit so that the small initial generated voltage e_a due to hysteresis will cause a field current i_f in such a direction as to increase the flux, thus increasing e_a. With no load connected to the armature, $i_a = i_f$ and the field current is governed by the relation

$$e_a = (R_a + R_{fe} + R_f)i_f + \frac{d\lambda_f}{dt} \qquad \text{V} \tag{4.33}$$

The generated voltage e_a and the field current i_f are also related by the no-load saturation curve that applies at the speed ω_0 at which the generator is being driven. This curve, shown in Fig. 4.23, includes the effects of both residual flux and magnetic saturation. This figure also shows a straight line representing the voltage-current relation for the total resistance $R_a + R_{fe} + R_f$ in the circuit. The condition for the generated voltage to increase is that the saturation curve be above this resistance line. When this is true, from Eq. (4.33), the field flux linkage λ_f will be increasing with time, and the flux Φ and the generated voltage e_a

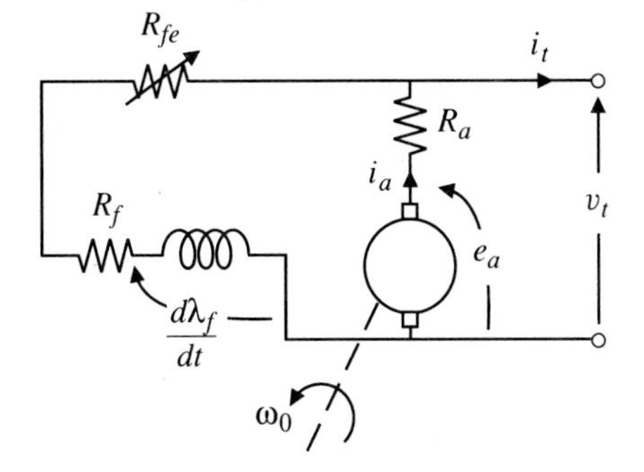

Figure 4.22 Connections for a shunt generator.

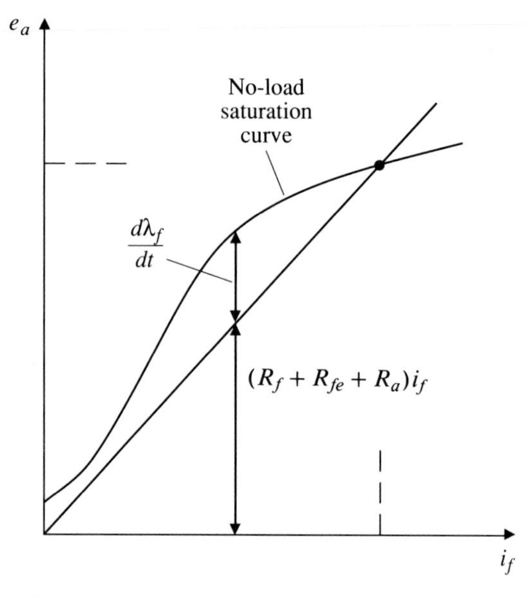

Figure 4.23 Build-up of generated voltage and field current in a shunt generator.

will increase. The stable operating point, at no load, will occur when these two relations intersect due to the approach to saturation in the magnetic system.

If the external field resistance R_{fe} is set at too high a value or if the speed of the prime mover is too low, the intersection of the two relations in Fig. 4.23 may occur near the origin, and the machine will be virtually useless as a generator.

When an electrical load is taken from the shunt generator, the steady-state operating condition is defined by the relation

$$\begin{aligned} v_t &= e_a - R_a I_a \\ &= (R_{fe} + R_f)i_f \qquad \text{V} \end{aligned} \tag{4.34}$$

and by the saturation curve. For a given value of armature current i_a, the corresponding terminal voltage v_t is shown graphically in Fig. 4.24(a). The vertical distance between the saturation curve and the field circuit resistance line is equal to the voltage across the armature resistance. Figure 4.24(b) shows the corresponding relation between the terminal voltage v_t and load current $i_t = i_a - i_f$. Note that the drop in terminal voltage as the load current is increased is greater than for a separately excited generator. If necessary, the generator may be fitted with control gear which automatically adjusts the resistance R_{fe} to maintain v_t approximately constant.

EXAMPLE 4.9 A 4.5-kW, 125-V, 1150-r/min generator has an armature-circuit resistance of 0.37 Ω. When the machine is driven at rated speed, its

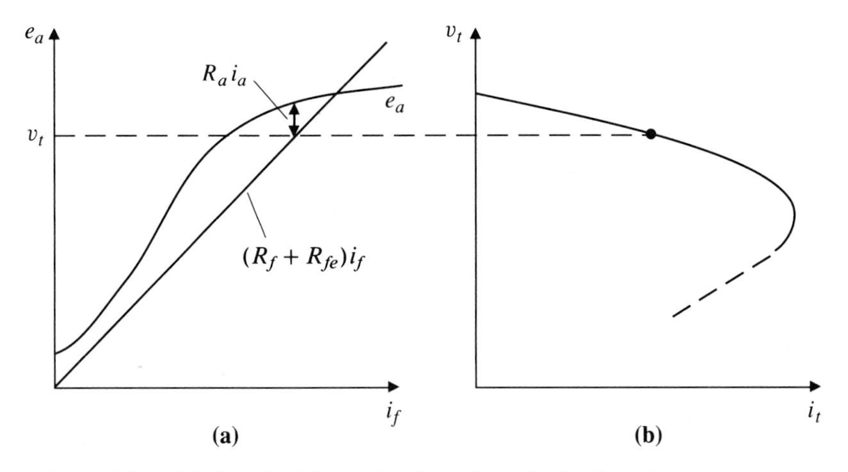

Figure 4.24 (a) Graphical determination of terminal voltage for a given armature current and (b) terminal voltage-current relation.

no-load saturation curve is as shown in Fig. 4.25. Suppose the machine is operated as a separately excited generator and that the field current is adjusted to 2 A. If the machine is driven at 1000 r/min and loaded to deliver its rated current, what will be the terminal voltage?

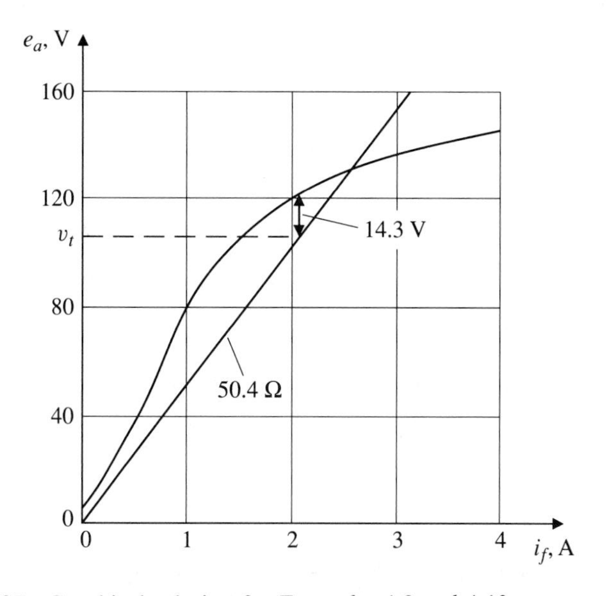

Figure 4.25 Graphical solution for Examples 4.9 and 4.10.

Solution At rated speed, the generated voltage e_a at $i_f = 2$ A is 120 V. At a speed of 1000 r/min and with the same value of $k\Phi$, the generated voltage will be

$$e_a = \frac{1000}{1150} \times 120 = 104.4 \quad \text{V}$$

Rated armature current is

$$i_a = \frac{4500}{125} = 36 \quad \text{A}$$

and the terminal voltage is

$$v_t = e_a - R_a i_a = 104.4 - 0.37 \times 36 = 91 \quad \text{V}$$

EXAMPLE 4.10 The machine of Example 4.9 is driven as a shunt generator at rated speed, and the field circuit resistor is adjusted to give a no-load terminal voltage of 130 V. What will be the approximate terminal voltage when the load current is 36 A?

Solution From Figure 4.25, the generated voltage is 130 V at a field current of 2.5 A. Thus, the field circuit resistor might be set so that the resistance line in Fig. 4.25 would have a slope

$$R = \frac{130}{2.5} = 52 \quad \Omega$$

Actually, this would result in an open-circuit terminal voltage of

$$v_t = 130 - 0.37 \times 2.5 = 129.1 \quad \text{V}$$

so a slightly lower resistance might more properly be chosen to give an intersection at about $e_a = 131$ V, $i_f = 2.6$ A, and $R = 50.4\ \Omega$. With a load current of 36 A, the armature current will be approximately

$$i_a = i_t + i_f \approx 36 + 2.5 \approx 38.5 \quad \text{A}$$

The voltage across the armature circuit resistance will then be

$$R_a i_a \approx 0.37 \times 38.5 = 14.3 \quad \text{V}$$

Referring now to Fig. 4.25, the vertical distance between the saturation curve and the field circuit resistance line is about 14.3 V at $e_a = 121$ V and $v_t = 107$ V. At this point, we could go back and correct the value of field current used in calculating i_a and repeat the process. However, refinement is not justified in this type of problem because we do not know the saturation curve to a high degree of accuracy. Being influenced by hysteresis, it will depend on its past history, as demonstrated in Fig. 4.2(b).

PROBLEMS

4.1 A commutator machine has four poles of 0.025 m^2 cross-sectional area and an air gap of effective length 6 mm. Each field coil has 1500 turns, and the coils are connected in series. Assuming the ferromagnetic materials to have infinite permeability, determine the air-gap flux per pole per unit of field current.

(Section 4.1)

4.2 A demagnetizing characteristic for a ferrite permanent magnet material is shown in Fig. 4.26. This material is to be used to provide an air-gap flux density of 0.28 T in the permanent-magnet motor shown in Fig. 4.4. If the effective length of the air gap is 1 mm, what thickness of magnet is required?

(Section 4.1)

4.3 In the machine illustrated in Fig. 4.1, suppose each pole shoe has an arc length of 0.15 m and an axial length of 0.12 m. The air-gap length is 5 mm.

(a) Ignoring the slotting of the rotor and assuming ideal iron, determine the flux per pole if each of the 2500-turn, series-connected field coils carries a field current of 1.2 A.

(b) Suppose the width of the rotor teeth is equal to the width of the slots. If the magnetic field intensity in the iron can be considered zero up to a flux density of 1.4 T, at what value of flux per pole and field current would the relationship between flux and field current be expected to depart from linearity?

(Section 4.1)

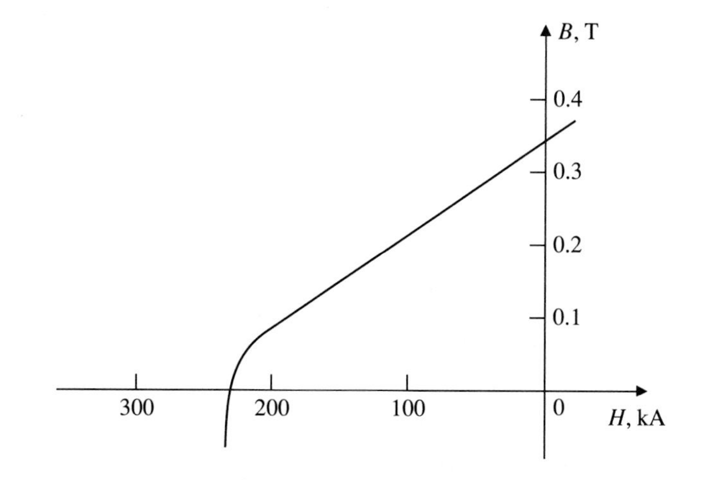

Figure 4.26 Ferrite demagnetization characteristic for Problem 4.2.

4.4 The armature of a motor has a radius of 0.12 m and an effective length of 0.25 m. The armature winding consists of 37 coils each with six turns accommodated in 37 slots. The field structure has four poles which cover 70% of the armature periphery. The average flux density in the air gap under each pole shoe is 0.8 T. The armature winding is connected in two parallel paths.

(a) Determine the constant k for this machine.

(b) Determine the generated voltage when the armature is rotated at a speed of 20 r/s.

(c) Determine the torque produced when the armature current is 40 A.

(Section 4.3)

4.5 A 1500-kW, ten-pole motor has a rated open-circuit armature voltage of 250 V when driven at its rated speed of 300 r/min. The armature winding has 220 turns connected with one parallel path for each pole.

(a) Find the flux per pole under these conditions.

(b) Find the torque when operating at rated conditions.

(Section 4.3)

4.6 A 20-kW, four-pole commutator generator has 279 armature turns in two parallel paths. The flux per pole is 0.03 Wb. At what speed must this generator be driven to produce a generated voltage of 440 V?

(Section 4.3)

4.7 A direct-current machine has no nameplate to indicate its rating. Examination of the machine yields the following information: It has four poles each with a pole face area of 250 × 200 mm, 150 bars on the commutator, armature coils with two turns per coil, and four parallel paths. The cross-sectional area of each armature conductor appears to be adequate to carry about 50 A. It is estimated that a reasonable value of flux density at each pole face would be about 0.9 T. Assuming the machine to be operated at 1000 r/min, estimate:

(a) the rated voltage,

(b) the rated torque, and

(c) the rated power.

(Section 4.3)

4.8 A commutator machine has the following physical properties: six poles, 2000 turns per field coil, 10-Ω resistance per field coil, all field coils connected in series, armature radius 50 mm, axial length 50 mm, pole face area 2000 mm^2, effective air-gap length 2 mm, 300 armature turns in two parallel paths, and each turn with a resistance of 8 mΩ. The polar moment of inertia of the rotor is 0.015 $kg \cdot m^2$. Saturation in the magnetic material is negligible, and friction and windage may be ignored. It is estimated that the pole flux is about 15% greater than the air-gap flux because of the leakage between poles.

(a) Determine the resistance and inductance of the field circuit, the armature circuit resistance, and the constant k_f.

(b) Write three equations relating the instantaneous values of the six terminal variables of this machine: the terminal voltage v_t, field voltage v_f, armature current i_a, field current i_f, shaft torque T_0, and speed ω_0. Include the numerical values of the machine parameters.

(Section 4.4)

4.9 When a commutator machine is driven at a speed of 1170 r/min, its open-circuit armature voltage (v_t) is measured for a set of values of field current (i_f) giving the following results:

i_f (A)	0.1	0.2	0.3	0.4	0.5	0.6	0.7	0.8
v_t (V)	55	110	161	218	265	287	303	312

What torque will this machine develop with a field current of 0.7 A and an armature current of 25 A?

(Section 4.4)

4.10 A permanent magnet commutator motor produces a torque of 8 N·m with an armature current of 10 A. The armature circuit resistance is 0.16 Ω. If this motor is to drive a mechanical load requiring a torque of 20 N·m at a speed of 1800 r/min, what must be the voltage applied to the armature terminals?

(Section 4.5)

4.11 The machine of Problem 4.10 is to drive a load that requires a torque that is proportional to its speed. One point on the mechanical torque-speed relation is 16 N·m at 400 r/min. If the armature terminal voltage is 50 V, at what steady-state speed will the load be driven?

(Section 4.5)

4.12 An industrial drive consists of a commutator motor with its armature supplied from a variable-voltage source. As this voltage is varied from 0 to 600 V, the drive speed is to vary from 0 to 1600 r/min, the field flux being maintained constant. All losses may be ignored.

(a) Determine the armature current if the load torque is held constant at 420 N·m.

(b) A load is to be driven in the speed range from 1600 to 4000 r/min by weakening the field flux while the armature voltage is held constant at 600 V. Determine the torque available at maximum speed if the armature current of (a) is not to be exceeded.

(Section 4.5)

4.13 A permanent magnet commutator motor has an air gap flux that may be assumed to be constant under all operating conditions. With no shaft torque, the motor operates at a speed of 6000 r/min when its armature is connected to a 120-V supply. The armature-circuit resistance is 1.5 Ω. Rotational losses may be ignored. Determine the motor speed when connected to a 60-V direct-voltage supply and providing a torque of 0.5 N·m.

(Section 4.5)

4.14 A trolley bus is driven by a commutator motor connected through a gear box of ratio 8:1 to the 0.9-m diameter wheels. The motor armature is supplied by a source for which the voltage can be varied from 0 to 600 V. All losses may be ignored.

(a) Determine the required value of $k\Phi$ to give a vehicle speed of 12.5 m/s at an armature voltage of 600 V.

(b) With the value of $k\Phi$ from (a), determine the armature current to provide a thrust of 8 kN on the trolley bus.

(c) The speed range above 12.5 m/s is provided by field weakening with constant armature voltage of 600 V. Determine the required value of $k\Phi$ at a speed of 25 m/s.

(Section 4.5)

4.15 A four-pole commutator machine is rated for 230 V and 5 kW at a speed of 1200 r/min.

(a) If the speed is increased to 1500 r/min by reduction of the field current, determine new values for the rated voltage and power of this machine.

(b) If the armature of the machine is rewound with its coils in two parallel paths rather than four, what would be the appropriate values of rated voltage and power at 1200 r/min?

(Section 4.5)

4.16 An automatic starter is required for a 50-hp, 230-V, 1750-r/min shunt motor. The armature circuit resistance is 0.12 Ω. When delivering rated output, the armature current is 214 A. The starter is to consist of resistors connected in series with the armature with contactors to short out the resistances in sequence. The armature current is not to exceed 500 A, and a resistor can be switched out when the armature current drops to 250 A. Determine the required number of resistors and the resistance of each.

(Section 4.6)

4.17 A shunt motor has the following nameplate data: 230 V, 40 A, 1000 r/min, $R_a = 0.5\ \Omega$.

(a) Find the value of $k\Phi$ for operation of this motor under rated conditions.

(b) What resistance would be required in series with the armature circuit of this machine to provide a torque of 62.5 N·m at a speed of 500 r/min?

(Section 4.6)

4.18 A shunt motor takes an armature current of 75 A from its 550-V supply when driving a load at 1350 r/min. The armature resistance is 0.46 Ω.

(a) Ignoring mechanical losses, calculate the torque applied to the load.

(b) Suppose the flux is reduced by 5%. Find the motor speed if the torque remains at the value found in (a).

(Section 4.6)

4.19 When a load torque of 50 N·m is applied to a shunt motor, its speed drops by 5% of its no-load value. The flux is then reduced to one-half of its original value, and the same 50 N·m torque load is applied. What will now be the speed drop as a percentage of no-load speed?

(Section 4.6)

4.20 The open-circuit voltage of a commutator generator driven at 200 rad/s with rated field current is 85 V. The resistance of the armature circuit is 4.3 Ω. If a resistive load of 50 Ω is connected to the armature terminals, what power will be delivered to this load when the motor is driven at 260 rad/s with rated field current?

(Section 4.7)

4.21 A 4.5-kW, 125-V, 1150-r/min generator has the open-circuit voltage-field current curve of Fig. 4.25 when driven at rated speed. A field current regulator is required to vary the field current so that rated voltage can be maintained from no load up to rated armature current. The prime mover drives the generator at rated speed at no load, but its speed falls by 5% with full load. The armature circuit resistance is 0.37 Ω. Determine the range of field current that the regulator must provide.

(Section 4.7)

4.22 The machine of Problem 4.21 is connected as a shunt generator and driven at 900 r/min. The resistor in series with the field is adjusted to give a no-load terminal voltage of 110 V. Determine the terminal voltage when the armature current is 40.5 A.

(Section 4.7)

GENERAL REFERENCES

Fitzgerald, A.E., Kingsley, C., and Umans, S.D., *Electric Machinery*, McGraw-Hill, New York, 1990.

Match, L., *Electromagnetic and Electromechanical Machines*, 2nd ed., Harper and Row, New York, 1977.

Nasar, S.A., *Electric Machines and Transformers*, Macmillan, New York, 1984.

Say, M.G. and Taylor, E.O., *Direct Current Machines*, Halsted Press, New York, 1980.

CHAPTER

5

Induction Machines

For many years, induction motors have provided the most common form of electromechanical drive for those industrial, commercial, and domestic applications that can operate at essentially constant speed. More recently, with the introduction of electronic power converters, induction machines are being used increasingly in variable-speed drives. The objective of this chapter is first to develop an understanding of the basic properties of induction machines and then to show how the performance of these machines can be predicted through the use of models incorporating these properties.

In its most commonly encountered form, an induction machine has a cylindrical stator with distributed windings, usually three, displaced symmetrically in slots around its inner periphery. These are connected to a polyphase sinusoidal voltage supply to produce a revolving magnetic field. This field sweeps past the shorted conductors that are placed in slots on the cylindrical rotor, inducing currents in them which in turn interact with the magnetic flux to produce torque in the direction of field rotation.

The elementary concepts of induction machine action were introduced in Section 3.5. In this chapter, the structure of the machine required to produce relatively ripple-free torque is discussed, and an equivalent circuit model is developed for prediction of steady-state performance. Special features such as transient performance and application in variable-speed drives are treated in Chapter 10.

5.1

Sinusoidally Distributed Windings

Figure 3.17 showed conceptually how three windings displaced around a stator and supplied by a set of three-phase sinusoidal currents could produce a rotating magnetic field. In a practical induction machine, it is desirable to have a field that has constant magnitude and shape and rotates at a constant angular velocity. The clue to accomplishing this is the unique property of a sine wave, which, when added to another displaced sine wave of the same period, produces a wave that is also of sine shape. Suppose we could arrange each phase winding so that its current would produce an air-gap magnetic field that would be sinusoidally distributed in space around the stator periphery. Then, the sum of the fields produced by currents in all three phase windings would also be sinusoidally distributed in angular space. The magnitude and position of this resultant sinusoidally distributed field then could be regulated by control of the three phase currents.

Ideally, it would be preferable to have the turns of each phase winding distributed sinusoidally in angular space around the stator. Figure 5.1 suggests such

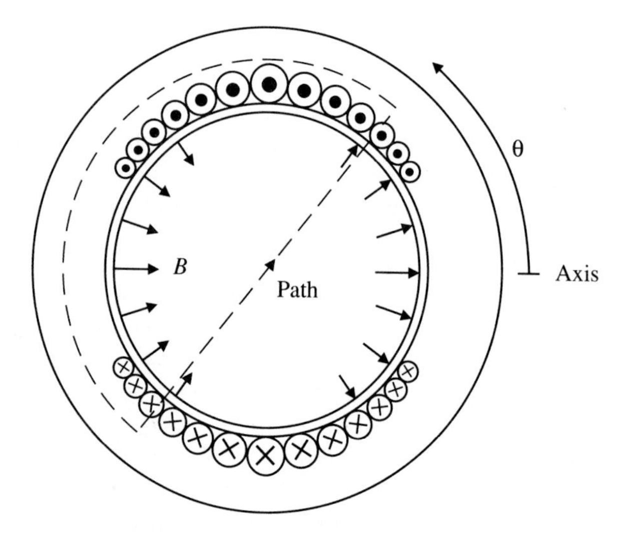

Figure 5.1 Sinusoidally distributed winding.

an arrangement where the size of each circle denotes the density of conductors in that region. For the phase a winding with N_s turns, the conductor density is ideally given by

$$n_a = \frac{N_s}{2} \sin \theta \tag{5.1}$$

A negative density can be interpreted as the return path of the turns.

Suppose we pass a current i_a through this phase a winding and examine the mmf that it produces. For the path shown around the machine at angle θ from the axis of winding a, the mmf is

$$\begin{aligned} \mathfrak{F}_{\theta(path)} &= \int_{\theta}^{\theta+\pi} n_a i_a \, d\theta \\ &= N_s i_a \cos \theta \qquad \text{A} \end{aligned} \tag{5.2}$$

Because of symmetry, it is convenient to associate half of this path mmf with each side of the machine. Thus, the mmf per air gap is

$$\mathfrak{F}_{a\theta} = \frac{N_s i_a}{2} \cos \theta \qquad \text{A} \tag{5.3}$$

If the magnetic field intensity in the iron sections of the path is negligible, the air-gap flux density due to the phase current i_a is

$$\begin{aligned} B_{ga\theta} &= \frac{\mu_0 \mathfrak{F}_{a\theta}}{g_e} \\ &= \frac{\mu_0 N_s i_a \cos \theta}{2 g_e} \qquad \text{T} \end{aligned} \tag{5.4}$$

where g_e is the effective air gap length.

The winding for phase b is similar to that of a but is displaced $2\pi/3$ rad forward in angle, i.e., the conductor density for winding b is

$$n_b = \frac{N_s}{2} \sin\left(\theta - \frac{2\pi}{3}\right) \tag{5.5}$$

The forward displacement for the phase-c winding is a further $2\pi/3$ rad.

Practically, a winding that produces close to a sinusoidal distribution of mmf can be achieved with a finite number of slots in a number of ways. Figure 5.2(a) shows a simple 12-slot stator in which the four identical coils of phase a span an arc of only 5 slots rather than 6, one side being in the bottom of a slot while the other side is in the top. This is known as a *short-pitched winding*. The corresponding mmf produced by this winding is shown in Fig. 5.2(b) together with its fundamental component. Even in this very simple winding, the approximation to a sine wave is reasonably good. With the larger number of slots that occur in most machines, the sinusoidal form can be more closely approximated.

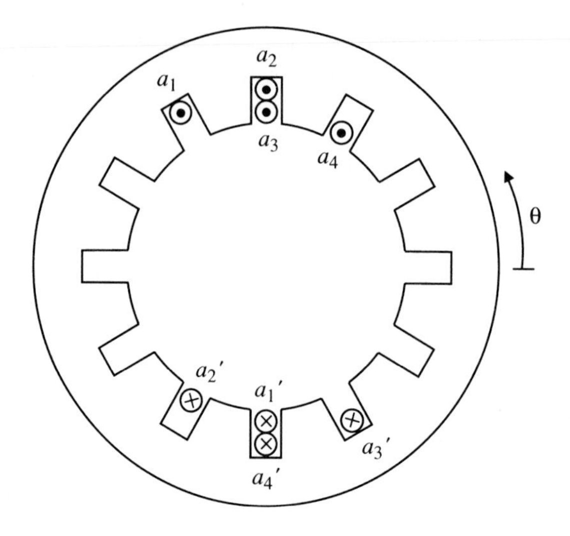

(a)

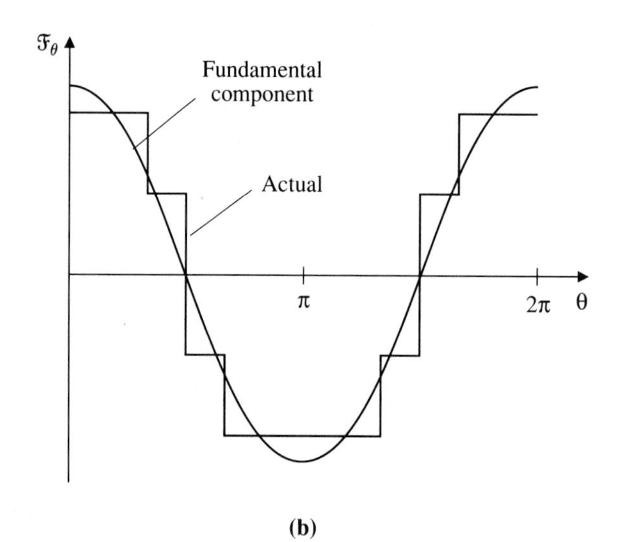

(b)

Figure 5.2 (a) Short-pitched winding (phase *a*) and (b) magnetomotive force due to phase *a*.

Another approach to a sinusoidal distribution of turns is discussed in Example 5.1, and a fuller description of windings is given in Section 10.7. In the analysis to follow, the symbol N_{se} will be designated as the equivalent number of sinusoidally distributed turns of a practical distributed winding.

EXAMPLE 5.1 Consider the 12-slot stator of Fig. 5.3(a) where the winding consists of two sets of three concentric coils. The number of turns in each of these series-connected coils is to be such as to approximate as closely as possible a sinusoidal distribution. If the total number of series turns is 1000, determine the appropriate number of turns in each of the coils. Also,

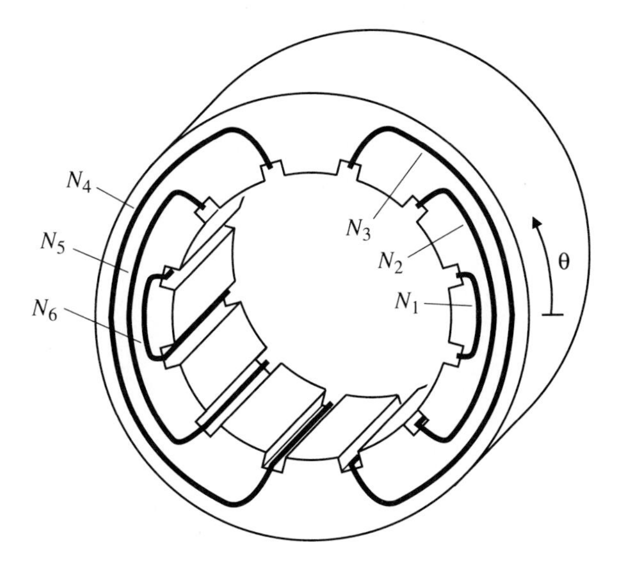

(a)

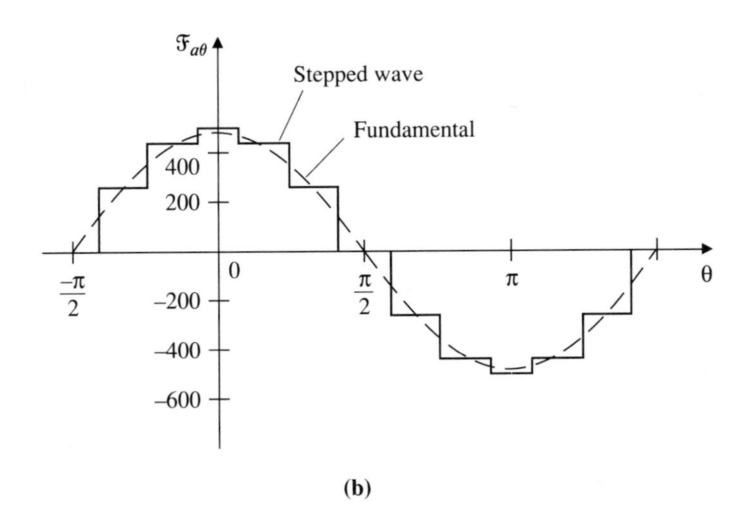

(b)

Figure 5.3 (a) Concentric winding in 12-slot stator and (b) magnetomotive force for Example 5.1.

derive the magnitude of the fundamental component of the mmf at each air gap and the magnitude of the two most significant harmonics. What is the equivalent number of sinusoidal turns for this winding?

Solution In general, for an s-slot stator, the appropriate number of conductors in slot q can be found by gathering together the ideal sinusoidal distribution over an arc of $2\pi/s$ rad around slot q. From Eq. (5.1),

$$N_q = \frac{N_s}{2}\int_{2\pi(q-1)/s}^{2\pi q/s} \sin\theta\, d\theta \qquad q = 1, 2 \ldots 6.$$

For $s = 12$, $N_s = 1000$, the required numbers of turns are

$$N_1 = \frac{1000}{2}\int_0^{\pi/6} \sin\theta\, d\theta = 67 = N_6$$

$$N_2 = \frac{1000}{2}\int_{\pi/6}^{\pi/3} \sin\theta\, d\theta = 183 = N_5$$

$$N_3 = \frac{1000}{2}\int_{\pi/3}^{\pi/2} \sin\theta\, d\theta = 250 = N_4$$

With a current of $i_a = 1$ A, the mmf at each air gap due to coils N_3 and N_4 consists of a flat-topped wave of magnitude 250 A and width $5\pi/6$ rad. Coils N_2 and N_5 produce a wave of magnitude 183 A over $\pi/2$ rad, and coils N_1 and N_6 produce a wave of 67 A over $\pi/6$ rad, as shown in Fig. 5.3(b).

Because of the quarter-wave symmetry, only odd cosine terms can occur in the Fourier series of the mmf $\mathfrak{F}_{a\theta}$:

$$\mathfrak{F}_{a\theta} = \frac{4}{\pi}\int_0^{\pi/2} \mathfrak{F}\cos(h\theta)\, d\theta \qquad h = 1, 3, 5, \ldots.$$

The fundamental ($h = 1$) component has the peak magnitude

$$\mathfrak{F}_{1a} = \frac{4}{\pi}[250\sin(5\pi/12) + 183\sin(\pi/4) + 67\sin(\pi/12)] = 494 \qquad \text{A}$$

The third harmonic ($h = 3$) magnitude is

$$\mathfrak{F}_{3a} = \frac{4}{3\pi}[250\sin(15\pi/12) + 183\sin(3\pi/4) + 67\sin(\pi/4)]$$

$$= \frac{4}{3\pi}[-176.8 + 129.4 + 47.4] = 0$$

Similarly the fifth, seventh, and ninth harmonics can be shown to be zero. The 11th and 13th harmonics are

$$\mathfrak{F}_{11a} = 45 \quad \text{A}; \qquad \mathfrak{F}_{13a} = 38 \quad \text{A},$$

i.e., less than 10% of the value of the fundamental component. The equivalent number of sinusoidal turns on the winding is

$$N_{se} = 1000 \times \frac{494}{500} = 988$$

5.2 Rotating Magnetic Field

This section demonstrates the rotating magnetic field analytically. Figure 5.4 shows a cross section of a stator with three sinusoidally distributed phase windings, each of N_{se} equivalent turns. For simplicity, only the center conductor is shown for each winding. Suppose we pass a balanced three-phase set of magnetizing currents, each varying sinusoidally with time at angular frequency ω_s, through these windings. The subscript M will be used to denote this as a magnetizing component of current:

$$\begin{aligned} i_a &= \hat{i}_M \cos(\omega_s t + \alpha_M) \\ i_b &= \hat{i}_M \cos(\omega_s t + \alpha_M - 2\pi/3) \\ i_c &= \hat{i}_M \cos(\omega_s t + \alpha_M - 4\pi/3) \quad \text{A} \end{aligned} \tag{5.6}$$

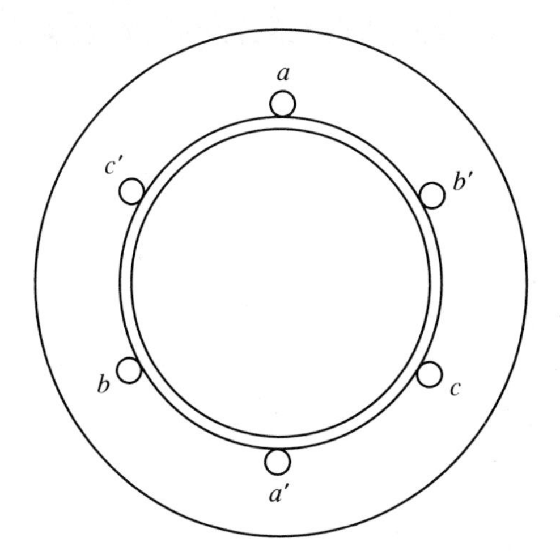

Figure 5.4 Three-phase stator winding showing only the center conductors of each phase.

Each of these winding currents will produce an mmf. The total mmf at angle θ can be expressed, from Eq. (5.3), as

$$\mathfrak{F}_{M\theta} = \mathfrak{F}_{a\theta} + \mathfrak{F}_{b\theta} + \mathfrak{F}_{c\theta}$$

$$= \frac{N_{se}}{2}\,\hat{i}_M\left[\cos(\omega_s t + \alpha_M)\cos\theta + \cos\left(\omega_s t + \alpha_M - \frac{2\pi}{3}\right)\cos\left(\theta - \frac{2\pi}{3}\right)\right.$$

$$\left. + \cos\left(\omega_s t + \alpha_M - \frac{4\pi}{3}\right)\cos\left(\theta - \frac{4\pi}{3}\right)\right] \quad \text{A} \tag{5.7}$$

Expansion of the terms in Eq. (5.7) yields

$$\mathfrak{F}_{M\theta} = \frac{N_{se}\hat{i}_M}{4}\left[3\cos(\omega_s t + \alpha_M - \theta) - \cos(\omega_s t + \alpha_M + \theta)\right.$$

$$\left. - \cos\left(\omega_s t + \alpha_M + \theta - \frac{4\pi}{3}\right) - \cos\left(\omega_s t + \alpha_M + \theta - \frac{2\pi}{3}\right)\right] \quad \text{A} \tag{5.8}$$

The last three terms in this equation have a sum of zero. Thus,

$$\mathfrak{F}_{M\theta} = \frac{3N_{se}\hat{i}_M}{4}\cos(\omega_s t + \alpha_M - \theta) \quad \text{A} \tag{5.9}$$

This expression for $\mathfrak{F}_{M\theta}$ describes an mmf that is sinusoidally distributed in space and rotating with time. Its peak magnitude depends on the peak winding current $\hat{i}_M$. This positive peak value of magnetic field will occur at the angular position,

$$\theta = \omega_s t + \alpha_M \tag{5.10}$$

With the windings *a*, *b*, and *c* arranged as in Fig. 5.4, the direction of rotation of the field is counterclockwise. Clockwise rotation of the field can be arranged by having the winding currents rise to their maximum values in the sequence *a-c-b* rather than *a-b-c*. It will be convenient in some instances to regard this sequence as a negative value of the angular velocity ω_s of the three-phase supply.

Another useful concept is to consider the revolving field to be produced by a revolving current sheet on the inner surface of the stator. The angular current density for any phase, in amperes per radian, is the phase current multiplied by the conductor density as given in Eqs. (5.1) or (5.5). Because the arc length in meters is the radius r times the angle in radians, the linear current density due to the three magnetizing currents can be expressed as

$$
\begin{aligned}
K_{M\theta} &= \frac{1}{r}\,(n_a i_a + n_b i_b + n_c i_c) \\
&= \frac{N_{se}\hat{i}_M}{2r}\left[\sin\theta\cos(\omega_s t + \alpha_M)\right. \\
&\qquad + \sin\left(\theta - \frac{2\pi}{3}\right)\cos\left(\omega_s t + \alpha_M - \frac{2\pi}{3}\right) \\
&\qquad \left. + \sin\left(\theta - \frac{4\pi}{3}\right)\cos\left(\omega_s t + \alpha_M - \frac{4\pi}{3}\right)\right] \\
&= \frac{3N_{se}}{4r}\,\hat{i}_m\cos\left(\omega_s t + \alpha_M + \frac{\pi}{2} - \theta\right) \qquad \text{A/m}
\end{aligned}
\tag{5.11}
$$

Figure 5.5 is a representation of this sinusoidally distributed current sheet at time $t = 0$. An advance in the phase angle α_M of the magnetizing current produces an equal advance in the angular position of the current sheet.

When dealing with steady-state sinusoidal currents and voltages, it is convenient to use a phasor representation. This is particularly true of three-phase balanced systems. Instead of writing out the expressions for the instantaneous values of all three-phase quantities, as for the currents in Eq. (5.6), the essential information for steady-state conditions is retained in one complex number

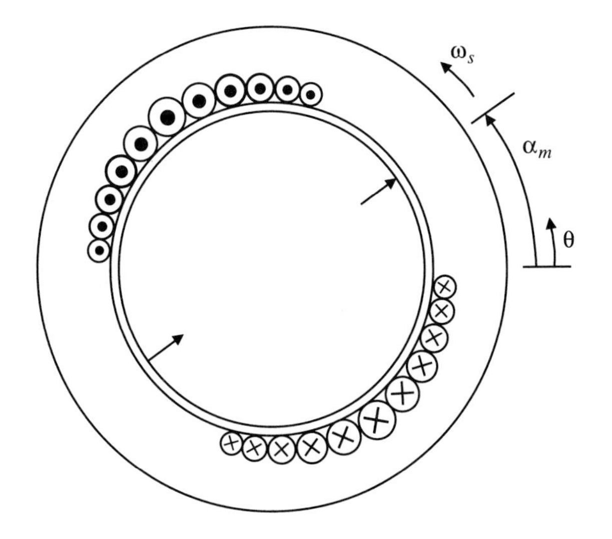

Figure 5.5 Revolving current sheet.

or phasor. Phase a is taken as the representative phase. The magnitude of the phasor is the rms value of each of the sines waves, i.e., the peak value divided by $\sqrt{2}$. Because all sine waves in a steady-state situation have the same frequency, it is not necessary to repeat the ωt term for each variable. The angle of the phasor is the angle of the cosine term for phase a at $t = 0$, i.e., α_M in Eq. (5.6). It is understood that the angles for the other two phases are delayed by 120 and 240 degrees, respectively. The phase a magnetizing current of Eq. (5.6) thus can be denoted as the phasor

$$\vec{I}_M = \frac{\hat{i}_M}{\sqrt{2}} \epsilon^{j\alpha_M} = I_M \angle \alpha_M \qquad \text{A} \tag{5.12}$$

This phasor is shown in Fig. 5.6. We have chosen to locate the axis of phase a at $\theta = 0$. The instantaneous value of the current in phase a is then given as $\sqrt{2}$ times the projection of this phasor on the a axis. The instantaneous values of the currents in the other two phases may be found by the projection of $\sqrt{2}\vec{I}_M$ on the b and c axes.

The magnetizing current produces distributions of mmf and linear current density that are sinusoidally distributed around the stator periphery as given in Eqs. (5.9) and (5.11). Each of these also may be represented by a complex number. Because each of these variables represents a quantity that is sinusoidally distributed in angular space, its complex number notation is known as a *space vector*. For steady-state studies, we will define the space vector of a variable as having a magnitude equal to the peak value of the variable divided by $\sqrt{2}$, i.e., its rms value, and an angle equal to the angular position of the maximum value of the variable at time $t = 0$. Thus, the space vector of the mmf of Eq. (5.9) can be written as

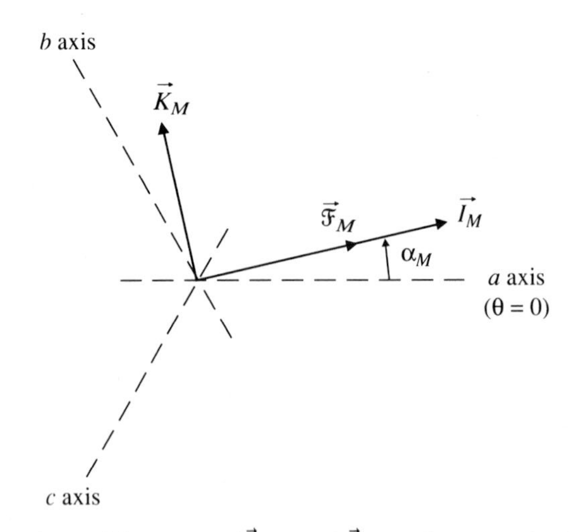

Figure 5.6 Magnetizing current $\vec{I}_M$, mmf $\vec{\mathfrak{F}}_M$, and current density $\vec{K}_M$.

$$\vec{\mathfrak{F}}_M = \frac{\hat{\mathfrak{F}}_M}{\sqrt{2}} \epsilon^{j\alpha_m} \qquad \text{A} \tag{5.13}$$

As shown in Fig. 5.6, this space vector is directed along $\theta = \alpha_M$ in line with the $\vec{I}_M$ phasor.

The linear current density of Eq. (5.11) can similarly be expressed as a space vector:

$$\vec{K}_M = \hat{K}_M \epsilon^{j(\alpha_M + \pi/2)} \qquad \text{A/m} \tag{5.14}$$

This space vector conveys all the necessary steady-state information of the distribution shown in Fig. 5.5, i.e., the maximum current density $\hat{K}_M$ at $(\alpha_M + \pi/2)$, and the understood cosinusoidal distribution. It is also understood that the rotation is at angular velocity ω_s just as the angular frequency ω_s is understood for phasor notation.

Using this notation, space vectors of field quantities can be readily linked to the corresponding phasors of time variables. From Eqs. (5.9) and (5.13), the mmf space vector may be related to the magnetizing current phasor by

$$\vec{\mathfrak{F}}_M = \frac{3N_{se}}{4} \vec{I}_M \qquad \text{A} \tag{5.15}$$

Similarly, the linear current density can be related to the magnetizing current by

$$\vec{K}_M = j\,\frac{3N_{se}}{4r} \vec{I}_M \qquad \text{A/m} \tag{5.16}$$

as shown in Fig. 5.6.

5.3

Flux Linkage and Magnetizing Inductance

The next stage in our analysis of the induction machine is to determine the flux linkages that the revolving field produces in the stator windings. The time rate of change of these flux linkages then will give us the induced voltages in the windings.

Suppose a balanced set of magnetizing currents has set up a sinusoidally distributed mmf $\mathfrak{F}_{M\theta}$ in the air gap, as given in Eq. (5.9), or in space vector form $\vec{\mathfrak{F}}_M$ in Eq. (5.13). If the iron in the stator and rotor is considered to be ideal, the flux density in an air gap of effective length g_e is

$$\vec{B}_g = \mu_0 \frac{\vec{\mathfrak{F}}_M}{g_e} = \frac{3N_{se}\mu_0}{4g_e} \vec{I}_M \qquad \text{T} \tag{5.17}$$

Consider the representative phase a. To find the maximum flux linkage in this phase winding, let us consider the instant of maximum current in phase a, i.e., $i_a = \hat{i}_M$ at $\omega_s t = -\alpha_M$. At this instant, the magnetic field is oriented along the axis of phase a and, from Eqs. (5.9) and (5.17), is given by

$$B_{g\theta} = \frac{3N_{se}\mu_0}{4g_e}\,\hat{i}_M \cos\theta = \hat{B}_g \cos\theta \qquad \text{T} \tag{5.18}$$

Note that both the flux density and the winding turns (Eq. [5.1]) are distributed in angular space. Consider first a single turn of winding a located at θ and $\theta + \pi$. For a stator of radius r and axial length ℓ, the flux linkage of this turn is

$$\lambda_{turn} = \int_{\theta}^{\theta+\pi} B_{g\theta}\,\ell r\, d\theta = 2\hat{B}_g \ell r \sin\theta \qquad \text{Wb} \tag{5.19}$$

To find the total peak flux linkage of winding a, this linkage per turn can be integrated over the turns distribution of Eq. (5.1):

$$\begin{aligned} \hat{\lambda} &= \int_{-\pi/2}^{\pi/2} n_a \lambda_{turn}\, d\theta \\ &= \int_{-\pi/2}^{\pi/2} \frac{N_{se}}{2} \sin\theta\,(2\hat{B}_g \ell r \sin\theta)\, d\theta \\ &= \frac{\pi}{2} N_{se} \hat{B}_g \ell r \qquad \text{Wb} \end{aligned} \tag{5.20}$$

In general, the flux linkage $\vec{\Lambda}_s$ now can be expressed as a phasor using its rms value. From Eqs. (5.15), (5.17), and (5.20),

$$\begin{aligned} \vec{\Lambda}_s &= \frac{\pi}{2} N_{se} \ell r \vec{B}_g \\ &= \frac{3\pi}{8} N_{se}^2 \frac{\mu_0 \ell r}{g_e} \vec{I}_M \qquad \text{Wb} \end{aligned} \tag{5.21}$$

The flux linkage is seen to be proportional to the magnetizing current and to be in phase with it. Therefore, the proportionality can be expressed as the magnetizing inductance L_M where

$$L_M = \frac{3\pi}{8} N_{se}^2 \frac{\mu_0 \ell r}{g_e} \qquad \text{H} \tag{5.22}$$

To make efficient use of the magnetic material in the machine, it is usual to operate at reasonably high values of maximum flux density where the magnetic field intensity integrated over the flux paths in the teeth, yoke, and rotor core are significant. While the mmf distribution around the machine is always sinusoidal because of the turns distribution, the flux density distribution can become somewhat flat topped, mainly due to saturation in the teeth.

One of the interesting features of using sinusoidally distributed windings is that sinusoidally varying magnetizing currents in the phase windings will still produce a sinusoidally varying flux linkage in the windings, even though magnetic saturation has occurred. The reason for this can be appreciated if the flat-topped flux density space wave is represented as a Fourier series. Only the fundamental component of this space wave can couple with each of the sinusoidally distributed windings. None of the odd harmonics in the flux density space wave, third, fifth, etc., will produce any net flux linkage with the sinusoidally distributed stator winding.

In the saturating region, the flux linkage-magnetizing current relation is of the form shown in Fig. 5.7. The slope of the linear portion of this curve is as given in Eq. (5.22).

The induced voltage in the phase-*a* winding is given by the phasor

$$\vec{E}_s = j\omega_s \vec{\Lambda}_s = j\omega_s L_M \vec{I}_M \qquad \text{V} \tag{5.23}$$

This induced voltage will differ from the terminal voltage of the stator winding $\vec{V}_s$ by the voltage drop across the winding resistance R_s. The stator now can be represented by the single-phase equivalent circuit model of Fig. 5.8.

Typical values of the magnetizing current for an induction machine operated at rated voltage are in the range 0.15 to 0.6 pu based on the rated current of the machine. Conversely, the values of the magnetizing inductance and the magnetizing reactance range between about 1.5 and 7 pu, i.e., the inverse of the pu magnetizing current.

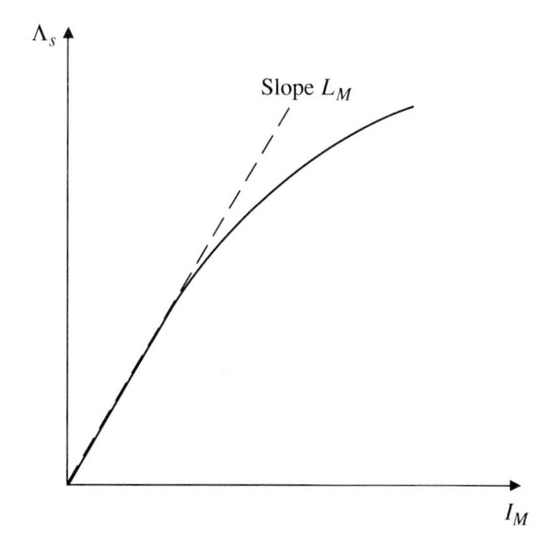

Figure 5.7 Stator flux linkage Λ_s as a function of magnetizing current I_M.

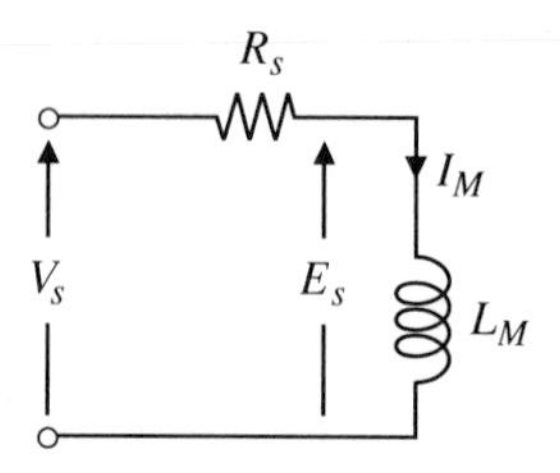

Figure 5.8 Equivalent circuit of induction machine without rotor winding.

EXAMPLE 5.2 A three-phase, two-pole induction motor has a rotor length of 80 mm, an air-gap radius of 60 mm, and an effective air-gap length of 0.5 mm. The effective number of sinusoidally distributed turns per phase is 78.6. The motor is connected to a three-phase, 60-Hz supply with a line-to-line voltage of 230 V. Note that it is conventional to state supply voltage as line to line (L-L) for three-phase systems rather than line to neutral (L-N).

(a) Determine the maximum value of the air-gap flux density.
(b) Determine the magnetizing inductance.
(c) Determine the rms value of the magnetizing current.

Solution

(a) Let us ignore the resistance of the stator winding. Normally, its effect on the voltage-flux linkage relation is negligible. The rms phase voltage applied to the motor is

$$V_s = \frac{230}{\sqrt{3}} = 132.8 \qquad \text{V}$$

The angular frequency of the supply is

$$\omega_s = 2\pi \times 60 = 377 \qquad \text{rad/s}$$

From Eq. (5.23), the flux linkage of each phase winding has an rms value of

$$\Lambda_s = \frac{V_s}{\omega_s} = \frac{132.8}{377} = 0.352 \qquad \text{Wb}$$

From Eq. (5.20), the peak value of air-gap flux density is

$$\hat{B}_g = \frac{2\sqrt{2}\Lambda_s}{\pi N_{se}\ell r} = \frac{2\sqrt{2} \times 0.352}{\pi \times 78.6 \times 0.08 \times 0.06} = 0.84 \qquad \text{T}$$

This is a typical value. The steel laminations in the stator teeth can have a maximum flux density in the range 1.6 to 1.8 T. If the tooth width is approximately equal to the slot width, the average flux density at the air gap will be in the range 0.8 to 0.9 T.

(b) From Eq. (5.22), the magnetizing inductance is

$$L_M = \frac{3\pi}{8} N_{se}^2 \frac{\mu_0 \ell r}{g_e}$$

$$= \frac{3\pi}{8} \times 78.6^2 \frac{4\pi \times 10^{-7} \times 0.08 \times 0.06}{0.0005} = 87.8 \qquad \text{mH}$$

(c) The magnetizing current is

$$I_M = \frac{V_s}{\omega_s L_M} = \frac{132.8}{377 \times 0.0878} = 4.01 \qquad \text{A}$$

5.4 Torque Production with Squirrel Cage Rotor

The most common form of rotor for an induction motor is known as the *squirrel cage rotor.* The rotor core is made of a stack of sheet-steel laminations. Aluminum, copper, or bronze conductors are placed in slots around the outer periphery of the rotor core, and these conductors are shorted together by circular end rings at each end of the rotor, as shown in Fig. 5.9. The conductors need not be insulated from the rotor core. In many rotors, the conductors and end rings are of cast aluminum, and projections are cast into the end rings to provide for ventilation fans for the machine.

In most rotors, the slots are not parallel to the axis but are skewed in angle, as shown in Fig. 5.9. This is done to avoid a cogging torque which might otherwise arise due to a varying ferromagnetic attractive force between rotor teeth and stator teeth as the rotor rotates.

Suppose we have created a rotating magnetic field with radially directed flux density across the air gap of

$$B_{g\theta} = \hat{B}_g \cos(\omega_s t + \alpha_M - \theta) \qquad \text{T} \tag{5.24}$$

This can be represented by the space vector

$$\vec{B}_g = \frac{\hat{B}_g}{\sqrt{2}} \epsilon^{j\alpha_M} \qquad \text{T} \tag{5.25}$$

This field is rotating counterclockwise at angular velocity ω_s. Suppose the rotor is rotating at angular velocity ω_0 measured in the same direction. The linear velocity ν of the field relative to the rotor conductors will then be

$$\nu = r(\omega_s - \omega_0) \qquad \text{m/s} \tag{5.26}$$

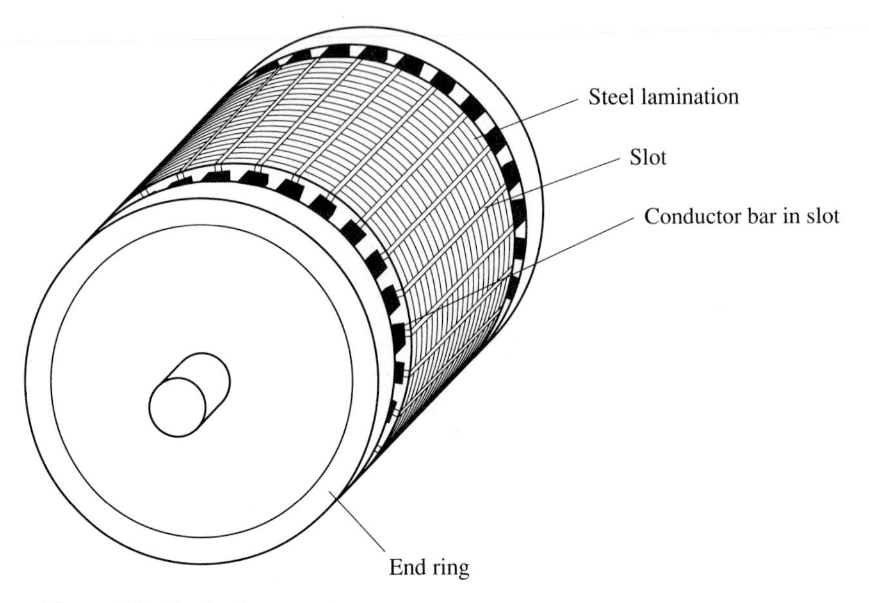

Figure 5.9 Squirrel cage rotor.

Because this velocity is perpendicular to the flux density, the electric field intensity in the rotor conductors can be represented by the space vector

$$\vec{\mathcal{E}} = \vec{v} \cdot \vec{B}_g \qquad \text{V/m} \tag{5.27}$$

If we assume that the end rings produce an ideal short circuit, the current density that will be established in the rotor conductors will then be

$$\vec{J} = \frac{\vec{\mathcal{E}}}{\rho} \qquad \text{A/m}^2 \tag{5.28}$$

where ρ is the resistivity of the rotor conductor material. Note that all the space vectors $\vec{J}$, $\vec{\mathcal{E}}$, and $\vec{B}_g$ are colinear or in space phase. They all represent space distributions of the sinusoidal type as given for $B_{g\theta}$ in Eq. (5.24).

Whatever the shape of the rotor bars, they will have a total cross-sectional area. The angular current distribution will be essentially the same as would occur in a conductor sheet of uniform depth d_e around the curved rotor surface and having the same cross-sectional area. Therefore, the rotor can be regarded as a sinusoidally distributed current sheet with a linear current density as described by the space vector

$$\vec{K}_r = d_e \vec{J} = \frac{d_e r}{\rho} (\omega_s - \omega_0) \vec{B}_g \qquad \text{A/m} \tag{5.29}$$

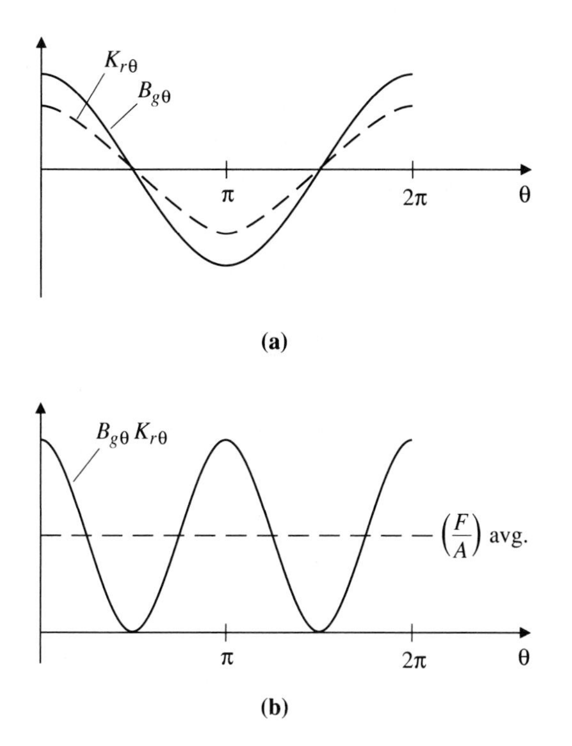

Figure 5.10 (a) Air-gap flux density and rotor linear current density distribution. (b) Tangential force distribution.

This current density is in space phase with the revolving flux density, as shown in Fig. 5.10(a). Its magnitude is seen, from Eq. (5.29), to be dependent on the difference between the angular velocity of the field and that of the rotor.

From Eq. (3.22), the force exerted per unit area at any point on the rotor periphery due to the interaction of flux density and current density is given by

$$\frac{F}{A} = BK \qquad \text{N/m}^2 \tag{5.30}$$

The flux density and the current density are directed perpendicular to each other and thus produce a tangential force. The average value of the product $B_{g\theta} K_{r\theta}$ around the machine can be seen from Fig. 5.10(b) to be equal to the product of the rms values of the space vectors B_g and K_r. Thus,

$$\left(\frac{F}{A}\right)_{avg} = B_g K_r \qquad \text{N/m}^2 \tag{5.31}$$

The torque exerted on the rotor may now be found from the average force density of Eq. (5.31) acting over the area $2\pi r\ell$ at a radius arm r:

$$T = \frac{2\pi r^3 \ell d_e}{\rho} B_g^2(\omega_s - \omega_0) \qquad \text{N}\cdot\text{m} \tag{5.32}$$

A more convenient expression would be one relating the torque to the supply voltage. When the stator terminals are connected to a three-phase supply of phase voltage V_s and angular frequency ω_s, alternating magnetizing currents flow to establish changing flux linkages in the windings. Ignoring the voltage across the winding resistance, the rms flux linkage must be

$$\Lambda_s = \frac{E_s}{\omega_s} \approx \frac{V_s}{\omega_s} \qquad \text{Wb} \tag{5.33}$$

From Eq. (5.21), the flux linkage can be related to the air-gap flux density. Thus, substituting into Eq. (5.32), the torque can be related to the stator flux linkage by

$$\begin{aligned} T &= \frac{2\pi r^3 \ell d_e}{\rho}\left(\frac{2\Lambda_s}{\pi N_{se}\ell r}\right)^2(\omega_s - \omega_0) \\ &= \frac{8}{\pi}\frac{r d_e}{N_{se}^2 \rho \ell}\Lambda_s^2(\omega_s - \omega_0) \qquad \text{N}\cdot\text{m} \end{aligned} \tag{5.34}$$

When operated at constant terminal voltage and frequency, the torque-speed relation has the simple form shown in Fig. 5.11. When the rotor rotates synchronously with the revolving field, no torque is produced. As a mechanical load is applied, the rotor slows down, inducing currents in the rotor conductors. The frequency of the current in any rotor conductor is

$$\omega_r = \omega_s - \omega_0 \qquad \text{rad/s} \tag{5.35}$$

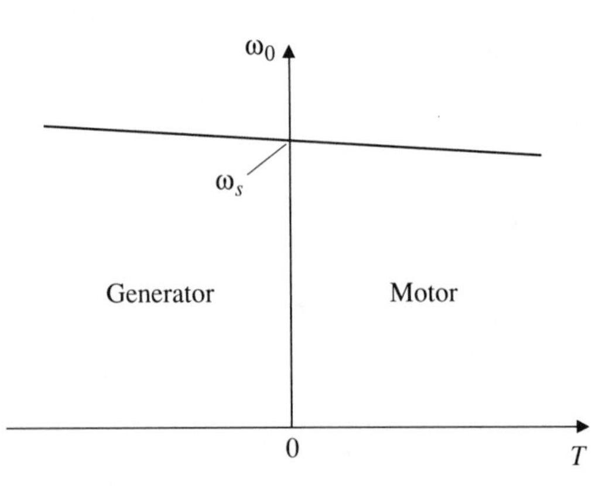

Figure 5.11 Speed-torque relation of induction machine near synchronous speed.

As seen by an observer on the rotor, these rotor currents produce an mmf wave rotating counterclockwise at angular velocity ω_r. When added to the rotor velocity ω_0, the net effect as seen by a stationary observer on the stator is a rotor mmf rotating synchronously with the stator field.

If the induction machine is connected to a three-phase supply and is driven at a speed greater than the synchronous speed of the field, i.e., where $\omega_0 > \omega_s$, the torque will be negative, and the machine will act as a generator, i.e., in the second quadrant of Fig. 5.11.

The rotor frequency ω_r is usually small, being only large enough to produce current in the shorted rotor conductors. For small machines, it may be as high as 3 Hz; for large machines, it may be lower than 0.3 Hz. Thus, in normal operation on a fixed frequency supply, an induction machine is a nearly constant-speed drive.

The ratio ω_r/ω_s is frequently referred to as the *slip s*. If the supply frequency ω_s is used as the base quantity, the slip is the rotor frequency in per unit. Because we are interested in variable- as well as constant-frequency operation, we will retain the identity of the rotor frequency in the analysis to follow.

EXAMPLE 5.3 The induction motor of Example 5.2 has a cast aluminum squirrel cage rotor. The cross-sectional area of the aluminum bars is 1250 mm^2. The resistivity of the aluminum at operating temperature is 3.85×10^{-8} Ω·m. Determine the rotor frequency, the speed of the machine and the mechanical power when it is producing a torque of 28.5 N·m.

Solution From Example 5.2, $r = 0.06$ m, $\ell = 0.08$ m, $B_g = 0.84$ T, and $\omega_s = 377$ rad/s. The rms flux density is

$$B_g = \frac{0.84}{\sqrt{2}} = 0.594 \qquad \text{T}$$

The equivalent depth of a conductor sheet having the same area as the rotor bars is

$$d_e = \frac{A}{2\pi r} = \frac{0.000125}{2\pi \times 0.06} = 3.32 \qquad \text{mm}$$

From Eqs. (5.32) and (5.35), the rotor frequency for the required torque is

$$\omega_r = \frac{\rho T}{2\pi r^3 \ell d_e B_g^2} = \frac{3.85 \times 10^{-8} \times 28.5}{2\pi \times 0.06^3 \times 0.08 \times 0.00332 \times 0.594^2} = 8.63 \qquad \text{rad/s}$$

i.e., the rotor frequency is $8.63/2\pi = 1.37$ Hz. The speed of the rotor is

$$\omega_0 = \omega_s - \omega_r = 377 - 8.63 = 368.4 \text{ rad/s} = 3518 \text{ r/min}$$

The mechanical power developed by the motor is

$$P_0 = T\omega_0 = 28.5 \times 368.4 = 10.5 \qquad \text{kW}$$

5.5

Equivalent Circuit Near Synchronous Speed

Suppose a supply of phase voltage V_s at frequency ω_s is applied to the stator. Because this voltage is fixed, it is usually convenient to use it as a reference for phasors. Thus,

$$\vec{V}_s = V_s \epsilon^{j0} \qquad \text{V} \tag{5.36}$$

Ignoring the stator resistance for the present, this also will be the induced voltage $\vec{E}_s$. From Eq. (5.23), the magnetizing current phasor will be

$$\vec{I}_M = \frac{\vec{V}_s}{j\omega_s L_M} \qquad \text{A} \tag{5.37}$$

i.e., lagging $\vec{V}_s$ by $\pi/2$ rad, as shown in the phasor diagram of Fig. 5.12(a). This magnetizing current produces the sinusoidally distributed current sheet $\vec{K}_M$ of Eq. (5.16), as shown in distributed form in Fig. 5.12(b), and in the space vector diagram of Fig. 5.12(c).

The flux density $\vec{B}_g$ is a space vector colinear with $\vec{I}_M$, from Eq. (5.17), as shown in Fig. 5.12(c). Motion of this flux density field past the rotor conductors produces the rotor current sheet $\vec{K}_r$ (Eq. [5.29]), which is colinear with $\vec{B}_g$, as shown in distributed form in Fig. 5.12(d) and in the space vector diagram of Fig. 5.12(e).

If the flux density $\vec{B}_g$ is to be preserved, there must now be an additional component of stator current, denoted as $\vec{I}_R$, which will produce a current sheet $\vec{K}_R$ equal in magnitude and opposite in direction to the rotor current sheet $\vec{K}_r$, as shown in Figs. 5.12(d) and (e). As might be expected, the phasor $\vec{I}_R$ is in phase with the stator-induced voltage $\vec{E}_s$, as seen in Fig. 5.12(a), and is known as the power component of stator current, in contrast to the magnetizing component $\vec{I}_M$.

The current $\vec{I}_R$ now can be related to the induced voltage $\vec{E}_s$. From Eq. (5.21),

$$\vec{E}_s = j\omega_s \vec{\Lambda}_s$$

$$= j\omega_s \frac{\pi}{2} N_{se} \ell r \vec{B}_g \qquad \text{V} \tag{5.38}$$

From Eq. (5.29), the linear current density $\vec{K}_r$ in the rotor is related to $\vec{B}_g$ by

$$\vec{B}_g = \frac{\rho \vec{K}_r}{d_e r(\omega_s - \omega_0)} \qquad \text{T} \tag{5.39}$$

The power component of stator current density $\vec{K}_R$ is equal to $-\vec{K}_r$. By analogy with Eq. (5.16), this current density is related to its stator current phasor by

$$\vec{K}_R = j\frac{3N_{se}}{4r}\vec{I}_R \qquad \text{A/m} \tag{5.40}$$

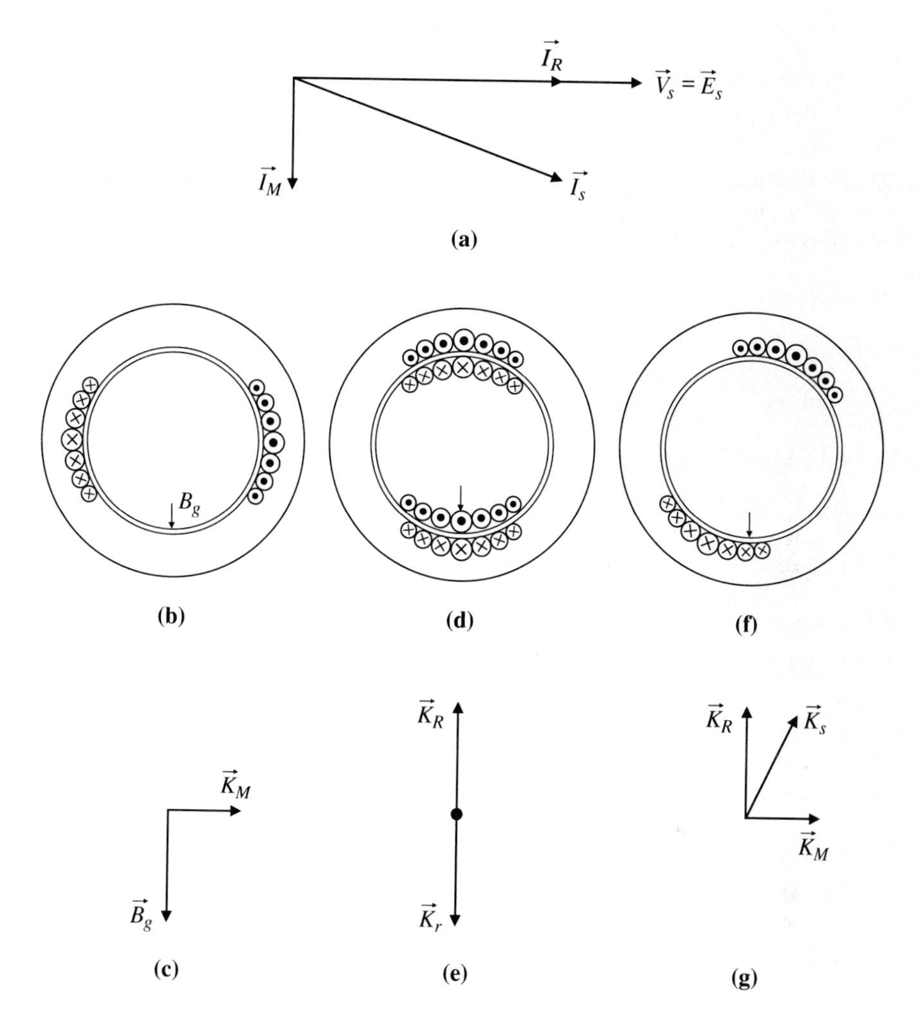

Figure 5.12 Phasors, current distributions, and space vectors of induction machine.

Combining Eqs. (5.38) and (5.40) gives

$$\begin{aligned}\vec{E}_s &= j\omega_s \frac{\pi}{2} N_{se} \frac{\ell r \rho}{d_e r(\omega_s - \omega_0)} \left(-j \frac{3N_{se}}{4r}\right) \vec{I}_R \\ &= R_R \frac{\omega_s}{(\omega_s - \omega_0)} \vec{I}_R \qquad \text{V}\end{aligned} \tag{5.41}$$

where

$$R_R = \frac{3\pi}{8} N_{se}^2 \frac{\rho \ell}{d_e r} \qquad \Omega \tag{5.42}$$

Thus, the rotor system as seen by the stator can be modeled as a resistance per phase of $R_R\omega_s/\omega_r$ where ω_r is the rotor frequency.

Of the power that crosses the air gap, a part is used in the losses in the rotor resistance. From the principle of conservation of energy, the rest is mechanical power output. The resistance R_R is the effect of the resistance of the rotor as seen by the stator. The remaining part of the total resistance $R_R\omega_s/\omega_r$ represents the effect of the mechanical output power per phase, i.e.,

$$\begin{aligned} R_R\frac{\omega_s}{\omega_r} - R_R &= R_R\frac{(\omega_s - \omega_r)}{\omega_r} \\ &= R_R\frac{\omega_0}{\omega_r} \quad \Omega \end{aligned} \tag{5.43}$$

The induction motor may now be modeled by the equivalent circuit of Fig. 5.13.

The torque produced by the machine can be determined from the equivalent circuit by calculating the power in the resistance $R_R\omega_0/\omega_r$, multiplying by 3 for three phases, and dividing by the rotor speed ω_0. Examination of the equivalent circuit of Fig. 5.13 shows that the power in the resistance $R_R\omega_0/\omega_r$ is the fraction ω_0/ω_s of the power crossing the air gap. Thus, the torque also may be expressed in terms of the air-gap power as

$$\begin{aligned} T &= \frac{3P_g}{\omega_s} \\ &= 3\frac{E_s^2}{\omega_s}\left(\frac{\omega_r}{R_R\omega_s}\right) \\ &= 3\frac{\Lambda_s^2\omega_r}{R_R} \quad \text{N}\cdot\text{m} \end{aligned} \tag{5.44}$$

Inserting the expression for R_R in Eq. (5.42), this can be shown to be identical with the torque expression of Eq. (5.34). Typical values of the equivalent rotor resistance vary from greater than about 0.05 pu (based on machine rating) for small machines to as low as 0.005 pu for very large machines.

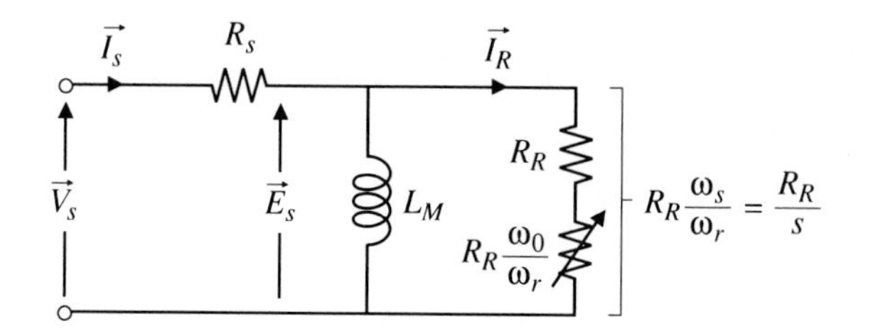

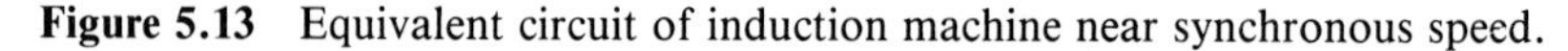
Figure 5.13 Equivalent circuit of induction machine near synchronous speed.

EXAMPLE 5.4 Using the induction machine data of Examples 5.2 and 5.3, determine the equivalent rotor resistance. Also, determine the stator current when the stator voltage is 230 V (L-L) at 60 Hz and the load torque is 28.5 N·m.

Solution From the earlier examples, $N_{se} = 78.6$, $\rho = 3.85 \times 10^{-8}\ \Omega\cdot\text{m}$, $\ell = 0.08$ m, $r = 0.06$ m, and $d_e = 3.32$ mm. Thus, from Eq. (5.42),

$$R_R = \frac{3\pi}{8}\frac{N_{se}^2 \rho \ell}{d_e r} = \frac{3\pi}{8} \times \frac{78.6^2 \times 3.85 \times 10^{-8} \times 0.08}{0.00332 \times 0.06} = 0.1125 \qquad \Omega.$$

The rms flux linkage, ignoring any stator resistance, is

$$\Lambda_s = \frac{V_s}{\omega_s} = \frac{132.8}{377} = 0.352 \qquad \text{Wb}$$

From Eq. (5.44), the rotor frequency can be found for the required torque:

$$\omega_r = \frac{R_R T}{3\Lambda_s^2} = \frac{0.1125 \times 28.5}{3 \times 0.352^2} = 8.63 \qquad \text{rad/s}$$

The component I_R of stator current is given by

$$I_R = \frac{V_s \omega_r}{R_R \omega_s} = \frac{132.8 \times 8.63}{0.1125 \times 377} = 27 \qquad \text{A}$$

With a magnetizing current $I_M = 4.01$ A from Example 5.2,

$$\vec{I}_s = \vec{I}_R + \vec{I}_M = 27 - j4.01 = 27.3\angle{-8.45} \qquad \text{A}$$

5.6

Machines with Multiple Pole Pairs

Relatively few induction machines are built with two poles. With a 60-Hz supply, the synchronous speed of 60 r/s or 3600 r/min is higher than required for many drives. A further reason for more than two poles is that the two-pole machine has a rather massive iron yoke because half of the flux must go around each side of the stator. The most common induction motors have four or six poles. Eight and higher numbers of poles are used when lower synchronous speeds are required.

The structure of a p-pole stator winding is essentially similar to that of the two-pole sinusoidal distribution of conductors shown in Fig. 5.1 except that the sine distribution is repeated $p/2$ times around the periphery, as shown in Fig. 5.14 for a six-pole machine. One cycle of the supply rotates the magnetic

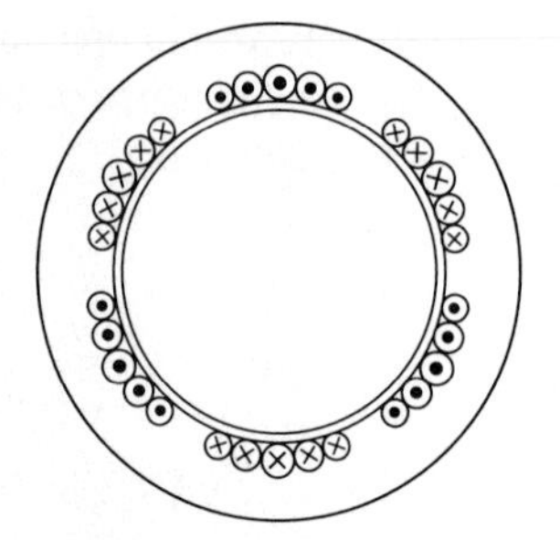

Figure 5.14 Six-pole induction machine showing current distribution in stator winding.

field through a mechanical angle of $4\pi/p$ rad. The synchronous mechanical speed then becomes

$$\omega_{mech} = \frac{2}{p}\,\omega_s \qquad \text{rad/s} \tag{5.45}$$

For simplicity, the analysis of the previous section has been presented for a two-pole machine. In order to interpret this previous analysis for p-pole machines, it is convenient to introduce the concept of electrical angle θ where

$$\theta = \frac{p}{2}\,\theta_{mech} \qquad \text{rad} \tag{5.46}$$

and to consider the previous analysis as representative of a typical pole pair.

In a p-pole machine, the maximum value of the air-gap flux density will be essentially as in the two-pole machine because it is limited by tooth saturation. Similarly, the load components of stator and rotor linear current density will be the same because these are limited by conductor heating. Thus, from Eq. (5.31) and Fig. 5.10(b), it will be seen that, for the same rotor radius and length, the available torque of the machine will be determined by the flux density limit and the current density limit. It will be essentially independent of the number of poles. However, with the fixed available line frequency, the speed will be reduced as the number of poles is increased, and the mechanical power available from a given frame size will also be proportionately reduced.

The turns of all poles of a phase winding are normally connected in series. For a total, sinusoidally distributed number of turns N_{se}, the conductor density (analogous to Eq. [5.1]) is

$$n_a = \frac{N_{se}}{p}\sin\theta \tag{5.47}$$

The flux linkage vector is given, by analogy to Eq. (5.21), as

$$\vec{\Lambda}_s = \frac{\pi}{p} N_{se} \ell r \vec{B} \qquad \text{Wb} \tag{5.48}$$

The current phasors are still related to their corresponding linear current densities by

$$\vec{K} = j\,\frac{3N_{se}}{4r}\,\vec{I} \qquad \text{A/m} \tag{5.49}$$

In the equivalent circuit of Fig. 5.13, the magnetizing inductance and the reflected rotor resistance are given by

$$L_M = \frac{3\pi}{2p^2} N_{se}^2 \frac{\mu_0 \ell r}{g_e} \qquad \text{H} \tag{5.50}$$

and

$$R_R = \frac{3\pi}{4p} N_{se}^2 \frac{\rho \ell}{d_e r} \qquad \Omega \tag{5.51}$$

The torque expression of Eq. (5.44) is adapted by noting that the torque is the power crossing the air gap divided by the synchronous mechanical speed. Thus,

$$\begin{aligned} T &= \frac{3P_g}{\left(\dfrac{2}{p}\right)\omega_s} \\ &= \frac{3p}{2}\,\frac{\Lambda_s^2 \omega_r}{R_R} \qquad \text{N}\cdot\text{m} \end{aligned} \tag{5.52}$$

EXAMPLE 5.5 An induction motor drive is required for a pump. The speed is to be in the range 300 to 325 r/min, and the mechanical power is to be about 100 kW. The electrical supply is at 60 Hz.

(a) Choose an appropriate number of poles.

(b) Suppose the motor has an rms air-gap flux density of 0.65 T and that the linear current density of the power component of stator current is 30 kA/m. If the poles of the machine are to be approximately square, estimate the rotor radius and length of the machine.

Solution

(a) The values of synchronous speed for a 60-Hz induction machine are

$$\frac{2}{p} \times 60 \times 60 \qquad \text{r/min}$$

The operating speed will be somewhat below the synchronous value (possibly about 1 to 2%). The choice $p = 24$ provides a synchronous speed at the bottom of the required range, and, therefore, the mechanical operating speed will be too low. The choice $p = 22$ gives a synchronous speed of 327.3 r/min, which should give an operating speed of 320 to 325 r/min.

(b) Assuming a mechanical speed of

$$\omega_{mech} = 325 \text{ r/min} = 34 \text{ rad/s}$$

the required torque is

$$T = \frac{P_{mech}}{\omega_{mech}} = \frac{10^5}{34} = 2938 \qquad \text{N}\cdot\text{m}$$

From Eq. (5.30), the tangential force per unit area of air gap is

$$\frac{F}{A} = BK = 0.65 \times 30\,000 = 19\,500 \qquad \text{N}\cdot\text{m}$$

The torque can be expressed as

$$T = (BK)2\pi r^2 \ell \qquad \text{N}\cdot\text{m}$$

To have square poles,

$$\ell = \frac{2\pi r}{p} \qquad \text{m}$$

Thus,

$$r = \frac{Tp}{4\pi^2 BK} = \left(\frac{2938 \times 22}{4\pi^2 \times 19\,500}\right)^{1/3} = 0.44 \qquad \text{m}$$

$$\ell = 2\pi \times 0.44/22 = 0.13 \qquad \text{m}$$

5.7

Leakage Inductance and Maximum Torque

Our analysis of the induction machine up to this point would suggest that the torque produced by the machine will continue to increase as the speed is reduced from its synchronous value, as shown in Fig. 5.11. While this is a useful approximation for normal operation, it is an oversimplification. The additional factor that must be considered now is the effect of leakage flux.

When an induction machine is connected to an alternating supply of constant voltage and frequency, the flux linking the stator winding must be such that its rate of change induces a voltage approximately equal to the supply voltage, the only difference being the voltage drop across the stator resistance. In the preceding analysis, it has been assumed that all of the flux crossed the air gap and interacted with the rotor conductors. This is a fair approximation with low values of rotor current. As the rotor current and the reflected power component $\vec{I}_R$ of stator current increase, the current distribution in the windings takes on the form as developed in Fig. 5.12 and shown in detail in Fig. 5.15(a). Some of the stator flux crosses the air gap and gives rise to the rotor current in response to the velocity difference between the field and the rotor conductors. However, part of the flux avoids the opposing mmf of the rotor conductors and passes along a circumferential path around rather than across the air gap. This is leakage flux, similar to the leakage that occurs in a transformer, as discussed in Section 2.2.4.

The path of this leakage flux is demonstrated in Fig. 5.15(b). This shows a one-pole sector of a multi-pole machine. Only the leakage flux pattern is shown. While some of the leakage flux is directed around the air gap, the most important leakage paths occur from tooth tip to tooth tip around both the stator and the rotor. Some of the leakage flux extends up into the stator and rotor slots, linking part but not all of the windings. An additional component of leakage flux occurs around the end windings outside the stator and rotor laminations.

The space vector of stator flux linkage $\vec{\Lambda}_s$ represents a sinusoidally distributed flux in the stator rotating at a velocity dependent on the stator frequency. In the region of the air gap, this space vector can be considered as dividing into two components, a flux linkage $\vec{\Lambda}_L$ representing the leakage flux and a flux linkage $\vec{\Lambda}_R$ representing the flux that crosses into the rotor core. The reluctance of the leakage flux path is dominated by the air paths between tooth tips, along the air gap, and between the end windings, and thus can be regarded as linear. Thus, the leakage flux linkage $\vec{\Lambda}_L$ will be directly proportional to the mmf causing it, i.e., to the load component of current $\vec{I}_R$. The constant of proportionality is the leakage inductance L_L between stator and rotor windings:

$$\vec{\Lambda}_L = L_L \vec{I}_R \qquad \text{Wb} \tag{5.53}$$

This leakage inductance now can be included in an equivalent circuit, as shown in Fig. 5.16. Of the total induced voltage $\vec{E}_s$ in the stator, a phasor component $j\omega_s L_L \vec{I}_R$ is due to leakage. The remaining phasor component $j\omega_s \vec{\Lambda}_R$ represents the effect of the rotor resistance and the mechanical load as seen by the stator.

Using the concepts developed in Section 5.5, the torque produced by the machine may be expressed in terms of the active power crossing the air gap. From Eq. (5.52),

$$T = \frac{3p}{2\omega_s} P_g \qquad \text{N} \cdot \text{m} \tag{5.54}$$

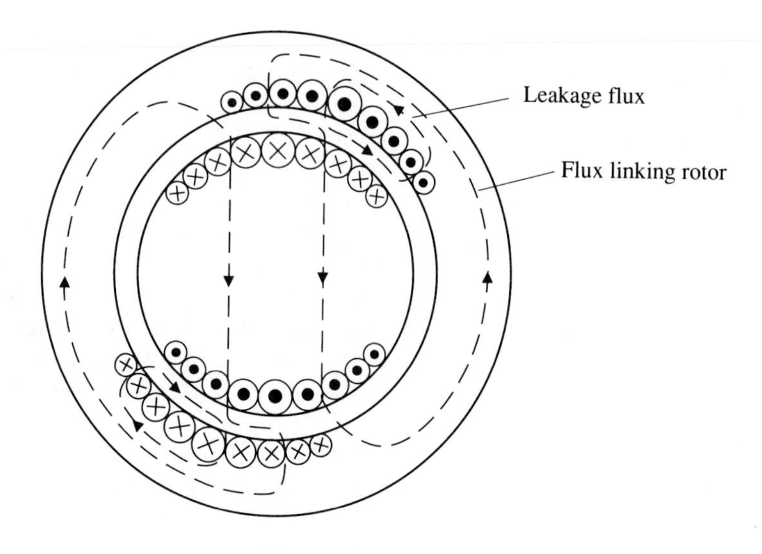

(a)

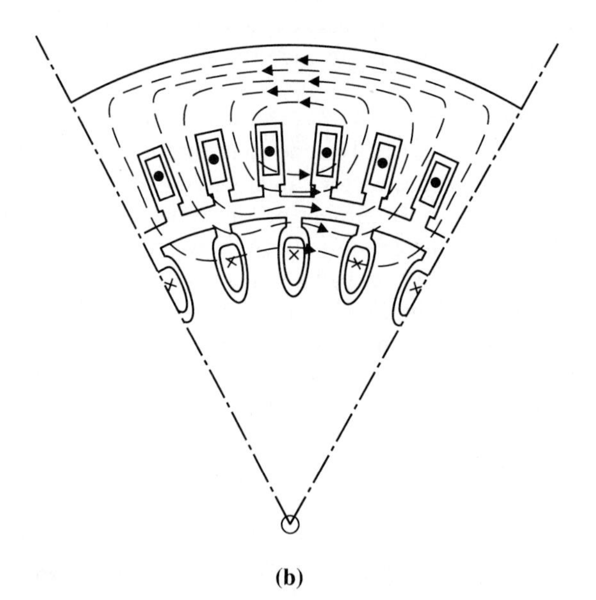

(b)

Figure 5.15 (a) Leakage and mutual flux and (b) leakage flux pattern.

It is convenient to express the torque in terms of the induced voltage E_s. When the stator resistance can be regarded as negligible, the induced voltage can be assumed to be the known supply voltage V_s. An appropriate correction can be made when the stator resistance is significant:

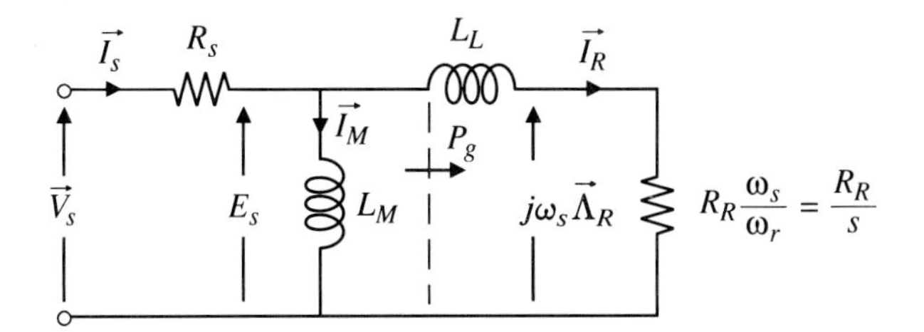

Figure 5.16 Equivalent circuit including leakage inductance.

$$\begin{aligned} T &= \frac{3p}{2\omega_s} P_g \\ &= \frac{3p}{2\omega_s}\left(\frac{R_R\omega_s}{\omega_r}\right)(I_R)^2 \\ &= \frac{3pR_R}{2\omega_r} \frac{E_s^2}{\left(\dfrac{R_R\omega_s}{\omega_r}\right)^2 + (\omega_s L_L)^2} \quad \text{N}\cdot\text{m} \end{aligned} \tag{5.55}$$

The torque can also be expressed as a function of the stator flux linkage. Because $E_s = \omega_s \Lambda_s$,

$$T = \frac{3pR_R}{2\omega_r} \frac{\Lambda_s^2}{\left(\dfrac{R_R}{\omega_r}\right)^2 + L_L^2} \quad \text{N}\cdot\text{m} \tag{5.56}$$

Note that this torque is independent of the stator frequency but dependent only on the rotor frequency. For operation near synchronous speed, where the rotor frequency ω_r is much less than the ratio R_R/L_L, the torque may be approximated, from Eq. (5.56), by

$$T \approx \frac{3p}{2} \frac{\Lambda_s^2 \omega_r}{R_R} \quad \text{N}\cdot\text{m} \tag{5.57}$$

which is identical with the previously developed expression of Eq. (5.52) for the case where leakage inductance was ignored.

As the machine is loaded, decreasing the speed and increasing the rotor frequency, the speed-torque relation ceases to be linear as was shown in Fig. 5.11, but tends toward a maximum torque $\hat{T}$, as shown in Fig. 5.17. The condition for maximum torque can be determined by differentiating Eq. (5.56) with respect to ω_r and equating to zero. If this is done, the condition for $dT/d\omega_r = 0$ is found to be

$$\omega_r = R_R/L_L \quad \text{rad/s} \tag{5.58}$$

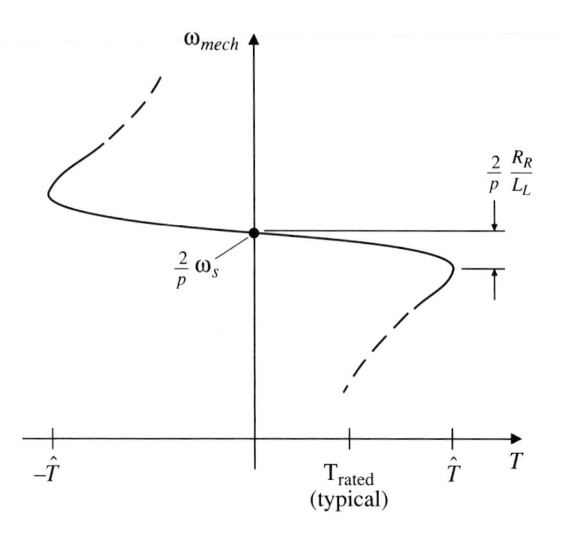

Figure 5.17 Relation between mechanical speed and torque including effect of leakage.

This result might have been expected from examination of the equivalent circuit of Fig. 5.16 because it is known that maximum power is transferred to the variable resistance $R_R\omega_s/\omega_r$ when it is equal to the constant series impedance $\omega_s L_L$.

Insertion of the condition of Eq. (5.58) into Eq. (5.56) gives the maximum torque as

$$\hat{T} = \frac{3p}{4}\frac{E_s^2}{\omega_s^2 L_L} = \frac{3p}{4}\frac{\Lambda_s^2}{L_L} \quad \text{N}\cdot\text{m} \tag{5.59}$$

When operated as a generator with a negative value of rotor frequency ω_r, i.e., negative phase sequence in the rotor, the maximum torque is given by the same expression (Eq. [5.59]), and it occurs at $\omega_r = -R_R/L_L$.

Figure 5.18 shows a phasor diagram of an induction machine operating with constant stator voltage and near-rated load. As the speed changes, changing the rotor frequency ω_r, the rotor current $\vec{I}_R$ follows a circular locus of diameter Λ_s/L_L. Maximum torque occurs when the in-phase component of the stator current is a maximum. It may be noted from Fig. 5.18 that maximum stator power factor typically occurs in the region of rated load.

Most induction motors are operated at essentially constant stator voltage and frequency, i.e., at essentially constant stator flux linkage. Thus, the max-

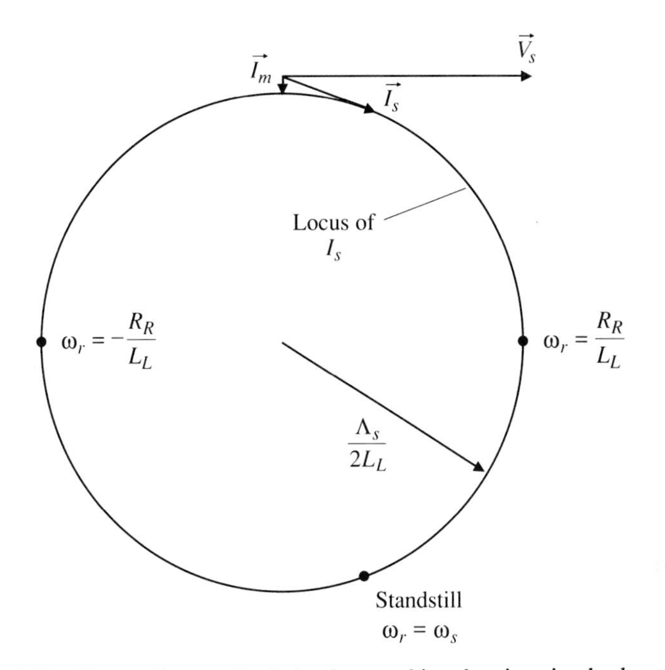

Figure 5.18 Phasor diagram for induction machine showing circular locus of stator current.

imum torque obtainable from the machine is seen, from Eq. (5.59), to be inversely proportional to the leakage inductance L_L. Standard induction motors are designed to have a value of leakage inductance such that the maximum or breakdown torque is in the range 1.75 to 3 times the rated continuous torque. For most machines, this ratio is approximately 2. Typical values for the leakage inductance vary from about 0.15 to 0.3 pu.

A variable-speed drive can be produced using an induction machine supplied from a three-phase, variable-frequency source. As frequency is varied, the speed torque curve of Fig. 5.17 is shifted up or down. The shape of the speed-torque characteristic, including the value of peak torque, will remain the same if the stator flux linkage Λ_s is maintained constant. To achieve this, the supply voltage V_s must be made proportional to the supply frequency. Drives based on this approach are discussed further in Chapter 10.

EXAMPLE 5.6 Suppose the induction motor of Examples 5.2, 5.3, and 5.4 has a leakage inductance of 3.25 mH. It is connected to a 230-V (L-L), 60-Hz supply. Determine the value of the breakdown torque and the speed at which it will occur.

Solution From Example 5.4, the stator flux linkage at the given stator voltage and frequency is

$$\Lambda_s = 0.352 \qquad \text{Wb}$$

The equivalent rotor resistance is

$$R_R = 0.1125 \qquad \Omega$$

The maximum torque for this two-pole machine is then given, from Eq. (5.59), as

$$\hat{T} = \frac{3p}{4}\frac{\Lambda_s^2}{L_L} = \frac{3}{2} \times \frac{0.352^2}{0.00325} = 57.2 \qquad \text{N}\cdot\text{m}$$

This will occur at a rotor frequency ω_r given, from Eq. (5.57), as

$$\omega_r = \frac{R_R}{L_L} = \frac{0.1125}{0.00325} = 34.6 \qquad \text{rad/s}$$

The mechanical speed for maximum torque, as a motor, is

$$\omega_{mech} = \omega_s - \omega_r = 377 - 34.6 = 342.6 \text{ rad/s} = 3270 \text{ r/min}$$

EXAMPLE 5.7 A 230-V (L-L), 60-Hz, three-phase induction motor has a nameplate power rating of 10 hp and rated speed of 1745 r/min. The parameters of the equivalent circuit are $R_s = 0.20\ \Omega$, $R_R = 0.191\ \Omega$, $L_L = 2.23$ mH, and $L_M = 34.2$ mH. When operating at rated speed, determine the torque developed by the machine and its stator current.

Solution For simplicity, let us first ignore the effect of the stator resistance and correct for it later if necessary. The stator voltage per phase is

$$V_s \approx E_s = \frac{230}{\sqrt{3}} = 132.8 \qquad \text{V}$$

The stator flux linkage is

$$\Lambda_s = \frac{E_s}{\omega_s} = \frac{132.8}{2\pi \times 60} = 0.352 \qquad \text{Wb}$$

Using the induced voltage E_s as a reference phasor, the magnetizing current is

$$\vec{I}_M = \frac{\vec{E}_s}{j\omega_s L_M} = \frac{132.8\angle 0}{j(377 \times 0.0342)} = -j10.3 \qquad \text{A}$$

From the rated speed and the supply frequency, it is evident that this is a four-pole machine. At rated speed, the rotor frequency is

$$\omega_r = \left(\frac{1800 - 1745}{1800}\right) 2\pi \times 60 = 11.52 \qquad \text{rad/s}$$

From Eq. (5.56), the developed torque is

$$T = \frac{3pR_R}{2\omega_r} \frac{\Lambda_s^2}{\left(\frac{R_R}{\omega_r}\right)^2 + L_L^2} = \frac{3 \times 4 \times 0.191}{2 \times 11.52} \frac{0.352^2}{\left(\frac{0.191}{11.52}\right)^2 + 0.00223^2}$$

$$= 44.04 \qquad \text{N} \cdot \text{m}$$

The load component of stator current is

$$\vec{I}_R = \frac{\vec{E}_s}{\frac{R_R \omega_s}{\omega_r} + j\omega_s L_L} = \frac{132.8 \angle 0}{\frac{0.191 \times 377}{11.52} + j377 \times 0.00223} = 20.86 - j2.81 \qquad \text{A}$$

The stator current is then

$$\vec{I}_s = \vec{I}_R + \vec{I}_M = 20.86 - j13.1 = 24.64 \angle -32.1.$$

Let us now correct for the voltage drop across the stator resistance. If $E_s = 132.8$ V,

$$\vec{V}_s = \vec{E}_s + R_s \vec{I}_s = 132.8 + 0.2(24.64 \angle -32.1) = 137 \angle -1.1 \qquad \text{V}$$

Because V_s is actually 132.8 V, the torque can be corrected in proportion to the square of the voltage:

$$T = 44.04 \left(\frac{132.8}{137}\right)^2 = 41.38 \qquad \text{N} \cdot \text{m}$$

The stator current can be corrected in proportion to the voltage:

$$I_s = 24.64 \times \frac{132.8}{137} = 23.88 \qquad \text{A}$$

The nameplate rating indicates a shaft torque of

$$T = \frac{10 \times 746}{1745 \times 2\pi/60} = 40.83 \qquad \text{N} \cdot \text{m}$$

The difference between this value and the value computed previously is due to windage and friction torque.

It might be of interest to determine the inaccuracy that would have occurred if both the stator resistance and the leakage inductance were ignored. From Eq. (5.57),

$$T = \frac{3p}{2} \frac{\Lambda_s^2 \omega_r}{R_R} = \frac{3 \times 4}{2} \frac{0.352^2 \times 11.52}{0.191} = 44.84 \qquad \text{N} \cdot \text{m}$$

This is 1.8% larger than the value with leakage inductance included.

5.8

Rotor Design for Starting Torque

The dotted section of the speed-torque relation shown in Fig. 5.17 represents a region where the rotor frequency ω_r is greater than R_R/L_L. In this region, the torque expression of Eq. (5.56) can be approximated by

$$T \approx \frac{3pR_R}{2\omega_r}\left(\frac{\Lambda_s}{L_L}\right)^2 \quad \text{N}\cdot\text{m} \tag{5.60}$$

i.e., the torque is inversely proportional to the rotor frequency. The torque available to start the motor then will be the value from Eq. (5.60) with $\omega_r = \omega_s$. As seen from Fig. 5.17, this may be less than the rated torque of the motor. With some types of mechanical load, the motor may fail to start. The condition would be worse if the supply voltage was reduced during starting because of the high current drawn from the supply.

The starting torque can be increased by increasing the rotor resistance R_R, as can be seen from Eq. (5.60). However, this will increase the power loss of $R_R I_R^2$ W/phase in the rotor, thus reducing machine efficiency.

The design approach used in most induction motors to avoid this compromise between starting torque and efficiency is to use a somewhat more complex shape for the rotor conductors, such as that shown in Fig. 5.19(a). This double-cage rotor consists of an outer cage with a conductor of relatively small cross section and a more deeply buried cage with a larger conductor. The upper cage, with an equivalent resistance R_{R1} as seen by the stator, can be modeled as in the equivalent circuit of Fig. 5.16 with its leakage inductance L_{L1} being dependent mainly on the tooth-to-tooth air gaps above the cage conductors. However, there is a further path for leakage flux between the two cages. This component of leakage flux links the outer cage but not the inner one. This path can be modeled by an additional leakage inductance L_{L2}. The inner cage can be modeled by its equivalent resistance R_{R2}. The overall equivalent circuit model then has the form shown in Fig. 5.19(b).

At very low values of motor speed, the rotor frequency ω_r is approximately equal to the supply frequency ω_s. Under this condition, the impedance $\omega_s L_{L2}$ of the inner leakage inductance is much higher than the resistance R_{R1} of the outer cage. As a result, most of the current $\vec{I}_R$ flows through the outer cage, giving the effect of a high rotor resistance as seen by the stator. In contrast, at the normal operating condition near synchronous speed, the rotor frequency ω_r is small. With $\omega_r \ll R_{R2}/L_{L2}$, the resistances of the two rotor cages can be considered as effectively in parallel, thus giving a low value of effective resistance as seen by the stator.

By varying the shapes of the cages and the air gaps between them, various shapes of speed-torque relation can be achieved, as shown in Fig. 5.19(c). The most widely used motor is a Class B design, providing starting torque usually

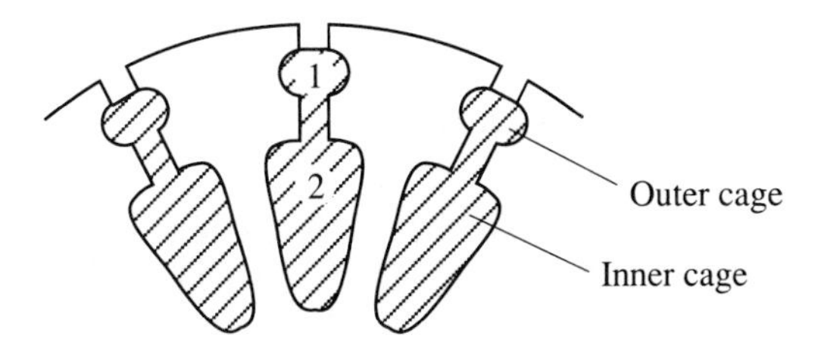

(a)

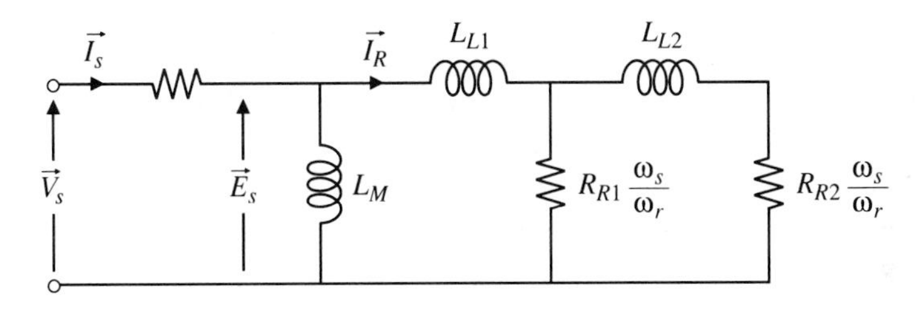

(b)

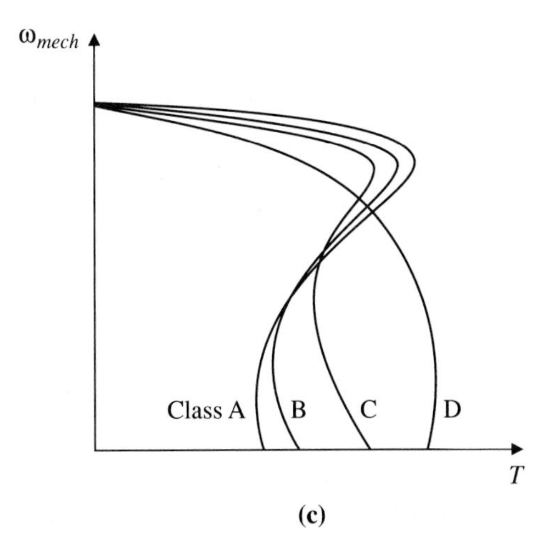

(c)

Figure 5.19 (a) Double-cage rotor, (b) equivalent circuit, and (c) typical speed-torque curves for standard classes of induction motors.

in the range 1.0 to 1.75 times rated continuous torque. A Class A design may be chosen for a large motor, providing high efficiency with a somewhat lower starting torque. Where higher starting torque is required, a Class C or D design may be chosen.

5.9

Single-Phase Induction Motors

Production of a rotating field in an induction machine requires a set of alternating currents, displaced in phase, flowing in a set of stator windings that are displaced around the stator periphery. This was demonstrated for a three-phase supply in Section 5.2. However, most commercial and domestic supplies are only of a single phase, typically with a voltage of 120 or 240 V. There are several ways in which a revolving field can be produced from such a single-phase supply.

The arrangement of a capacitor induction motor is shown in Fig. 5.20(a). The stator has two windings, each approximately sinusoidally distributed and displaced 90° from each other for a two-pole machine. Winding *a-a'*, known as the *main winding*, is connected directly to the single-phase supply. For starting, winding *b-b'*, known as the *auxiliary winding*, is connected through a series capacitor to the same supply. The effect of the capacitor is to make the current entering winding *b-b'* lead the current in winding *a-a'* by approximately 90°. Thus, a rotating field and starting torque are produced.

As the rotor speed approaches its rated value, it is no longer necessary to use the auxiliary winding to preserve the rotating field. The currents induced in the rotor conductors as they pass winding *a-a'* are retained with negligible decay as they rotate past winding *b-b'*. Thus, the rotor can maintain the rotating field with only the main winding connected. The auxiliary winding is usually disconnected by a centrifugal switch which is attached to the shaft and which opens when the speed is about 80% of its rated value.

The efficiency of the motor can be somewhat increased and the line current decreased by use of two capacitors, only one of which is switched out. This arrangement, also shown in Fig. 5.20(a), is known as a *capacitor-start, capacitor-run induction motor.*

Power ratings for capacitor induction motors are usually restricted to a maximum of about 2 kW for a 120-V supply and 10 kW for a 230-V supply because of the need to limit the voltage drop on the supply lines that occurs on starting. The most usual synchronous speeds for a 60-Hz supply are 1800 and 1200 r/min for four and six-pole machines, respectively.

An alternative means of providing a rotating field for starting is to use two stator windings, as in Fig. 5.20(b), where the auxiliary winding *b-b'* is made of more turns of smaller conductor than winding *a-a'*. At starting, both windings are connected to the supply. The effect of the higher resistance of winding *b-b'* is that its current leads that of *a-a'*, but only by 20 to 30° at standstill. While the magnetic field in the machine for this condition is mainly pulsating, it contains enough rotating component to provide a starting torque usually greater than the rated value. To prevent overheating, the auxiliary winding is switched out when the speed reaches 75 to 80% of rated value. This motor is commonly known as a *split-phase motor.*

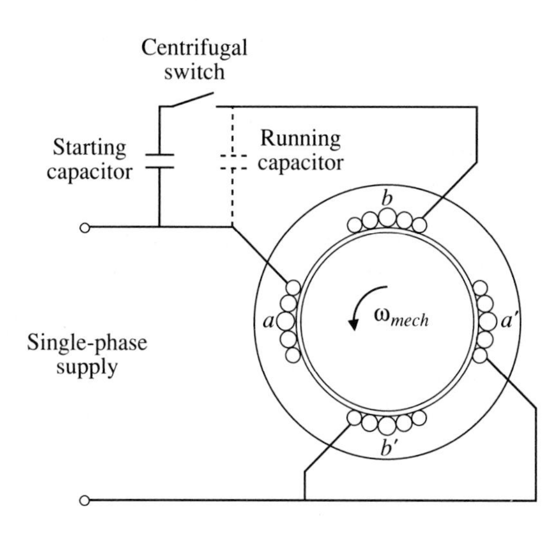

(a)

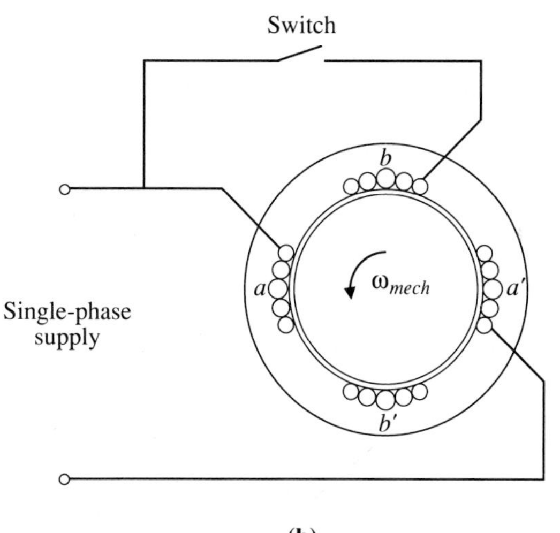

(b)

Figure 5.20 (a) Capacitor induction motor and (b) split-phase induction motor.

PROBLEMS

5.1 A three-phase, two-pole induction motor has the following physical properties: rotor radius = 35 mm, rotor length = 50 mm, effective air-gap length = 0.35 mm, effective number of sinusoidally distributed turns per phase = 420. The motor is to be operated from a 50-Hz three-phase supply. Stator resistance may be ignored.

(a) Determine the appropriate value of rms stator voltage per phase if the peak air-gap flux density is to be 0.9 T.

(b) Determine the magnetizing inductance for this machine.

(c) Determine the magnetizing current with the stator voltage found in (a).

(Section 5.4)

5.2 Suppose the induction machine in Problem 5.1 can support a linear rms current density of 2 kA/m of its rotor periphery without overheating. The motor is to be supplied with 230 V line to neutral, three phase at 50 Hz. Determine an appropriate value for its rated torque.

(Section 5.4)

5.3 The rotor of the machine in Problem 5.1 has 23 cast aluminum bars, each having a cross-sectional area of 35 mm^2. The resistivity of the aluminum may be taken as 0.03 $\mu\Omega \cdot m$ at the operating temperature.

(a) Determine the depth of the rotor conductor sheet that is equivalent to the rotor bars.

(b) Ignoring the resistance of the end rings, determine the rotor resistance per phase as seen from the stator winding.

(c) When the motor is operating from a 50-Hz, three-phase supply with 230 V per phase, find the rotor frequency to provide a torque of 12 $N \cdot m$.

(d) Find the mechanical output power for the condition of (c).

(Section 5.5)

5.4 A three-phase, four-pole, 440-V line to line, 60-Hz induction motor has a magnetizing inductance of 90 mH/phase and an effective rotor resistance as seen from the stator of 0.4 Ω/phase. The stator resistance may be ignored.

(a) Draw an approximate equivalent circuit for this machine.

(b) When connected to a 440-V, 60-Hz supply, at what speed will this drive a load requiring a torque of 45 $N \cdot m$?

(c) What will be the supply current per phase for the condition of (b)?

(Section 5.5)

5.5 A six-pole induction motor has a rated voltage of 230 V line to line at 60 Hz. Its leakage inductance is 14 mH, and its rotor resistance reflected to the stator turns is 1.2 Ω. The stator resistance may be ignored.

(a) When operating on a supply of rated voltage and frequency, what is the maximum torque that this machine can produce?

(b) If the machine were operated with a 50-Hz supply while maintaining the same stator flux linkage, what would be its maximum torque?

(c) At what speed would the maximum torque occur for the condition of (b)?

(Section 5.7)

5.6 A 460-V, four-pole, 60-Hz, 50-kW, 1770 r/min induction motor has a no-load current of 24 A. All losses except rotor losses can be ignored. Also, the leakage inductance can be ignored. Estimate the stator current and power factor when this motor is delivering rated torque.

(Section 5.7)

5.7 A 230-V, six-pole, 60-Hz induction motor has the following parameters: $R_s = 3.1\ \Omega$, $R_R = 6.6\ \Omega$, $X_M = 190\ \Omega$, and $X_L = 14.5\ \Omega$. For a rotor speed of 1125 r/min, determine the line current, the power factor, the torque, and the shaft power. Rotational losses may be ignored.

(Section 5.7)

5.8 Using the motor parameters given in Problem 5.7 but ignoring the stator resistance, draw a phasor circle diagram of the form shown in Fig. 5.18 for operation at rated stator voltage and frequency. Mark the condition for standstill, maximum torque, and maximum stator power factor. Determine the maximum stator power factor and estimate the stator current at which it occurs.

(Section 5.7)

5.9 A three-phase induction motor has a no-load current of 8.0 A when connected to its rated supply of 440 V line to line at 60 Hz. When driving a particular load, its input current is 20 A, its input power is 13.15 kW, and its speed is 1750 r/min. The resistance measured at standstill between a pair of stator terminals is 1.32 Ω.

(a) Determine the approximate value of the equivalent circuit parameters R_s, L_M, L_L, and R_R.

(b) Estimate the maximum torque that the machine can produce when operated on rated voltage and frequency.

(c) At approximately what speed will the maximum torque of (b) occur?

(Section 5.7)

5.10 A three-phase, two-pole, 110-V, 60-Hz induction motor has a leakage reactance of 1.6 Ω and an effective rotor resistance of 0.42 Ω. Stator resistance is negligible. This motor is used on a small hoist. When the load is being lowered, the motor is driven at more than synchronous speed. Determine the maximum braking torque that the motor can develop under this condition and the speed at which that maximum torque occurs.

(Section 5.7)

5.11 A four-pole squirrel-cage induction motor has a rated voltage of 440 V (L-L) at a frequency of 60 Hz. Its rated torque of 60 N·m is produced at a speed of 1760 r/min. This motor is to be used in a variable-speed drive and is to be supplied by a variable-voltage, variable-frequency three-phase source. The stator resistance and the leakage inductance may be ignored.

(a) What should be the stator voltage and frequency to supply rated torque output at a speed of 600 r/min?

(b) Suppose the mechanical system connected to the motor shaft acts as a power source driving the motor with a torque equal to its rated value. What should be the stator voltage and frequency to maintain a speed of 600 r/min?

(c) What power is fed back to the supply for the condition of (b)?

(Section 5.7)

5.12 A variable-speed drive consists of a squirrel-cage induction motor supplied with near-sinusoidal, three-phase, controllable voltage and frequency. The motor rating is 460 V, 140 A, 60 Hz, 100 kW at 1770 r/min. The mechanical load requires a torque of 3 N·m per rad/s of speed. Stator resistance, leakage inductance, and losses may be ignored. To drive this load at 1200 r/min, what should be the voltage and frequency of the supply?

(Section 5.7)

5.13 A small two-pole, 110-V (L-L), three-phase, 60-Hz induction motor has the following equivalent circuit parameters: $R_s = 0.22\ \Omega$, $R_R = 0.42\ \Omega$, $X_L = 1.6\ \Omega$, and $X_M = 34\ \Omega$. To simplify computation, the magnetizing branch may be moved out to the stator terminals of the equivalent circuit.

(a) Determine the mechanical power developed at a speed of 3500 r/min.

(b) Determine the input power factor for the condition of (a).

(c) Assuming that the rotor resistance and the leakage inductance are independent of rotor frequency, determine the starting torque.

(Section 5.8)

5.14 For an induction motor operated at constant stator flux linkage, Eq. (5.58) gives an expression for the maximum or breakdown torque $\hat{T}$ which occurs at a rotor frequency ω_r.

(a) Assuming that the rotor resistance is independent of rotor frequency, show that the torque T produced at any rotor frequency ω_r is given by the expression

$$T = \frac{2\hat{T}}{\dfrac{\hat{\omega}_r}{\omega_r} + \dfrac{\omega_r}{\hat{\omega}_r}} \qquad \text{N}\cdot\text{m}$$

At what value of the ratio $\omega_r/\hat{\omega}_r$ does the assumption of neglecting the leakage inductance L_L result in an error of 5% in predicting the torque?

(Section 5.8)

5.15 A three-phase, four-pole, 550-V, 60-Hz induction motor has a maximum or breakdown torque of 100 N·m which occurs at a speed of 1350 r/min. Assume that the stator resistance is negligible and that the rotor resistance does not vary with rotor frequency. The machine is to be operated on a 400-V, 50-Hz supply.

(a) Determine the new value of breakdown torque and the speed at which it occurs.

(b) Determine the new value of starting torque and compare it with that for 550-V, 60-Hz operation.

(Section 5.8)

5.16 A 440-V, three-phase, 60-Hz, four-pole induction motor has a double-rotor cage. The parameters for an equivalent circuit of the form shown in Fig. 5.19 are $L_{L1} = 5$ mH, $L_{L2} = 12$ mH, $L_M = 140$ mH, $R_{R1} = 4.5\ \Omega$, and $R_{R2} = 1.1\ \Omega$. The stator resistance may be ignored. Calculate the torque developed by the machine at speeds of 0, 80, and 96% of synchronous speed.

(Section 5.8)

GENERAL REFERENCES

Alger, P.L., *Induction Machines*, Gordon and Breach, New York, 1970.

Fitzgerald, A.E., Kingsley, C., and Umans, S.D., *Electric Machinery*, 5th ed., McGraw-Hill, New York, 1990.

Miller, T.J.E., *Brushless Permanent-Magnet and Reluctance Motor Drives*, Clarendon Press, Oxford, 1989.

Nasar, S.A., *Handbook of Electric Machines*, McGraw-Hill, New York, 1987.

Say, M.G., *Alternating Current Machines*, 5th ed., Pitman, London, 1983.

Vienott, C.B., *Fractional and Subfractional-Horsepower Electric Motors*, 3rd ed., McGraw-Hill, New York, 1970.

CHAPTER

6

Synchronous Machines

The concept of a rotating field produced by a set of three-phase sinusoidal currents flowing in a set of distributed stator windings was introduced in Section 3.5 and was further developed for the induction machine in Section 5.2. Synchronous machines use this same concept of a stator-produced rotating field. In contrast with the induction machine, the rotor of a synchronous machine consists of electromagnets or permanent magnets that rotate synchronously with the stator field. Thus, for a p-pole synchronous machine, the steady-state speed must be related to the stator frequency by

$$\omega_{mech} = \frac{2}{p}\,\omega_s \qquad \text{rad/s} \tag{6.1}$$

The largest and most significant synchronous machines are those used as generators to provide power for transmission and distribution over electric power networks. The power is derived from hydraulic turbines at waterfalls, from steam turbines whose energy comes from the burning of gas, oil, coal, or nuclear fuel, or occasionally from reciprocating diesel or gasoline engines. Steam turbines operate at relatively high speed. Therefore, the generators to which they are coupled are usually of two- or four-pole construction to provide the usual power system frequencies of 50 or 60 Hz. Hydraulic turbines, particularly those operating with a low pressure and a large volume of water flow, operate at low speed. The number of poles may be as high as 120. Power ratings for steam turbine-driven generators may extend up to about 2000 MW.

Synchronous generators are also used for stand-alone supplies of electric power such as in vehicles, aircraft, and remote installations and for emergency power supplies.

A number of synchronous machines are employed as motors operating at constant speed from the constant-frequency electric supply. Usually, these tend to be large machines, up to about 10 MW rating. They are used where the control of stator power factor provides an advantage over induction motors.

Synchronous motors have a wide and expanding field of application as variable speed drives. For such applications, they are supplied from electronic converters that provide a controllable frequency to produce the desired speed.

6.1

Cylindrical Wound-Field Machines

In principle, the stator of a synchronous machine is identical with that of an induction machine. If, for the moment, we visualize a machine with a cylindrical iron rotor without windings, all of the discussion of Sections 5.1 to 5.3 relating to the revolving field and the magnetizing inductance of the induction machine applies. Electrically, the machine can be modeled by the equivalent circuit of Fig. 5.8 consisting of its stator resistance R_s and its magnetizing inductance L_M (Eq. [5.22]). To complete the equivalent circuit as a synchronous machine, it is now necessary to model its rotor properties.

A typical rotor for a two-pole synchronous machine is shown in Fig. 6.1. Because the rotor rotates synchronously with the revolving field, its flux is normally constant. With constant flux in the rotor, there will be no significant hysteresis or eddy current losses. Thus, for large synchronous machines, the rotor body is frequently made of a solid steel forging.

Slots are cut in the cylindrical rotor, and series-connected coils are placed in these slots so as to approximate a sinusoidal distribution of conductors in much the same way as shown for a stator in Fig. 5.3(a). If the equivalent number of sinusoidally distributed turns is N_{re}, the conductor density can be expressed, by analogy with Eq. (5.1), as

$$n_r = \frac{N_{re}}{2} \sin \theta_r \tag{6.2}$$

If we now pass a field current i_f through this winding, using insulated slip rings on the rotor shaft and stationary brush connections to an external source, the mmf produced, on a per-pole basis, will be

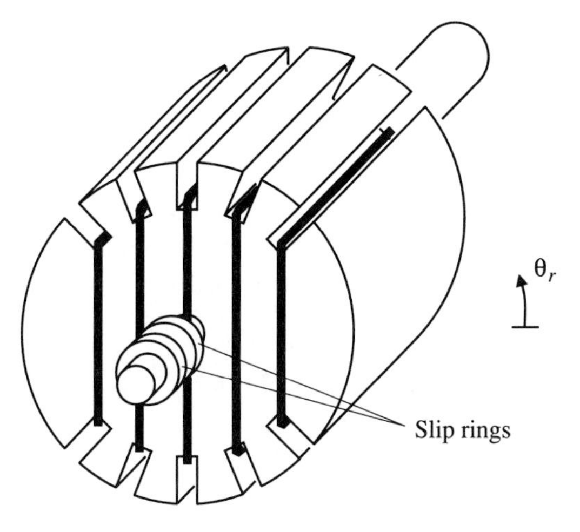

Figure 6.1 Cylindrical two-pole rotor for synchronous machine.

$$\mathfrak{F}_{r\theta_r} = \frac{N_{re}}{2} i_f \cos\theta_r \qquad \text{A} \tag{6.3}$$

by analogy with Eq. (5.3), where θ_r is the angular position on the rotor relative to the rotor axis, as shown in Fig. 6.1.

Let us now rotate the rotor so that the angle β of its axis relative to the axis of phase a on the stator is

$$\beta = \omega_0 t + \beta_0 \qquad \text{rad} \tag{6.4}$$

An angular position θ_r on the rotor is then related to a position θ measured from the axis of stator phase a by

$$\theta_r = \beta - \theta \qquad \text{rad} \tag{6.5}$$

As seen by the stator, the mmf per pole produced by the rotor is then given by

$$\begin{aligned} \mathfrak{F}_{r\theta} &= \frac{N_{re}}{2} i_f \cos(\beta - \theta) \\ &= \frac{N_{re}}{2} i_f \cos(\omega_0 t + \beta_0 - \theta) \qquad \text{A} \end{aligned} \tag{6.6}$$

In Section 5.2 (Eq. [5.9]), it was shown that a set of three-phase sinusoidal stator currents of peak value $\hat{i}$, angular frequency ω_s, and phase angle α in the three stator windings each of N_{se} equivalent turns produced an mmf of

$$\mathfrak{F}_{\theta} = \frac{3N_{se}}{4} \hat{i} \cos(\omega_s t + \alpha - \theta) \qquad \text{A} \tag{6.7}$$

Thus, as seen by the stator, the rotor looks like a three-phase sinusoidal current source. For a two-pole machine, its frequency is the rotational speed ω_0 which, in the steady state, will be equal to the stator angular frequency ω_s. Comparison of Eqs. (6.6) and (6.7) shows that the peak magnitude of this current source, designated by the subscript F, can be related to the field current by

$$\hat{i}_F = \frac{2}{3}\frac{N_{re}}{N_{se}} i_f \qquad \text{A} \tag{6.8}$$

From the same equations, the phase angle of this current source is β_0. Therefore, the effect of the field on the stator is that of an alternating current source which can be expressed as the rms phasor quantity

$$\begin{aligned} \vec{I}_F &= \frac{\hat{i}_F}{\sqrt{2}} \epsilon^{j\beta_0} \\ &= \frac{\sqrt{2}}{3}\frac{N_{re}}{N_{se}} i_f \angle \beta_0 \\ &= n i_f \angle \beta_0 \qquad \text{A} \end{aligned} \tag{6.9}$$

where n is a current ratio relating the direct-field current to its rms equivalent in stator terms.

The magnetomotive forces due to the stator phase currents denoted by the phasor $\vec{I}_s$ and this equivalent rotor current $\vec{I}_F$ combine to produce magnetic flux in the machine. Because this flux is rotating at the same speed as the rotor, the rate of change of flux linkages in the rotor winding in the steady state will be zero. Thus, in the steady state, the field current will be related to the direct-voltage field supply v_f and the field circuit resistance R_f by

$$i_f = v_f/R_f \qquad \text{A} \tag{6.10}$$

It follows that the magnitude of the alternating source current I_F as seen by the stator is also constant.

6.2

Steady-State Equivalent Circuit

A distinctive feature of a synchronous machine is that it can be magnetized either from the stator or from the rotor. The mmf space vector $\vec{\mathfrak{F}}_s$ produced by the alternating stator current $\vec{I}_s$ combines with a space vector $\vec{\mathfrak{F}}_r$ produced by direct-field current in a rotating rotor to establish a net magnetizing mmf vector $\vec{\mathfrak{F}}_M$, which, in turn, produces an air-gap flux density space vector $\vec{B}_g$ and an rms stator flux linkage $\vec{\Lambda}_s$. The relations among these quantities have already been derived for the induction machine in Eqs. (5.13), (5.18), and (5.21). Thus, as in the induction machine, the synchronous machine has a magnetizing inductance L_M relating the stator flux linkage $\vec{\Lambda}_s$ to the net magnetizing current $\vec{I}_M$, where

$$\vec{I}_M = \vec{I}_s + \vec{I}_F \qquad \text{A} \tag{6.11}$$

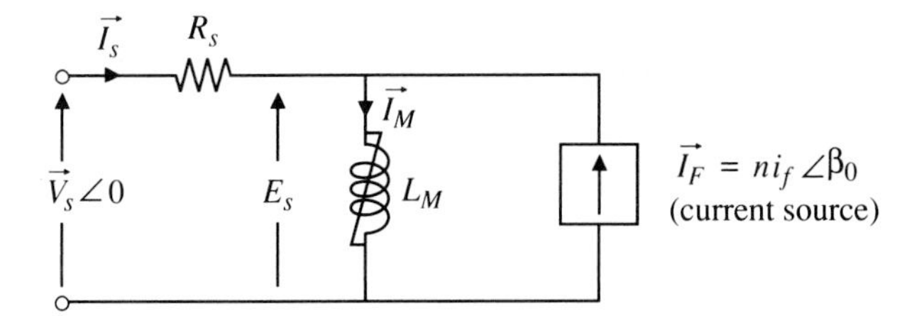

Figure 6.2 Equivalent circuit of synchronous machine.

The time rate of change of the stator flux linkage $\vec{\Lambda}_s$ produces the induced stator voltage phasor $\vec{E}_s$ which differs from the stator terminal voltage $\vec{V}_s$ by the drop across the stator resistance. Therefore, the cylindrical-rotor synchronous machine may be represented in the steady state by the simple equivalent circuit of Fig. 6.2. While there may be significant leakage flux which links the field winding but not the stator winding, the corresponding leakage parameter, the leakage inductance, is in series with a current source $\vec{I}_F$ and thus does not affect the steady-state performance.

If the iron in both stator and rotor can be regarded as a perfect magnetic conductor, the magnetizing inductance can be related to the dimensions of the machine in the region of the air gap by Eq. (5.22). However, most synchronous machines operate some sections of the iron path, particularly the stator and rotor teeth, at levels of flux density such that the mmf required for the iron is significant. Thus, the magnetizing inductance in Fig. 6.2 has been labeled as nonlinear and has the typical form shown in the Λ_s-I_M graph of Fig. 5.7.

Much of the literature on synchronous machines makes use of an alternate form of equivalent circuit which is that obtained by applying Thevenin's theorem to the circuit of Fig. 6.2. The result is a voltage source $\vec{E}_F = j\omega_s L_M \vec{I}_F$ in series with the inductive reactance $\omega_s L_M$; this latter parameter is generally referred to as the *synchronous reactance*. This voltage source model is not favored, partly because it loses the identity of the magnetizing current $\vec{I}_M$ on which the value of the inductance L_M depends, partly because the nonlinearity in inductance L_M is introduced twice in the circuit, and partly because the fundamental property of the synchronous machine as a near-ideal alternating-current source is lost.

6.3

Steady-State Operation

Consider a synchronous machine connected to a three-phase supply of constant voltage V_s and constant frequency ω_s, and rotating at synchronous speed $\omega_0 = \omega_s$ electrical rad/s where, for a p-pole machine, $\omega_0 = \omega_{mech}\, p/2$. Because the voltage V_s is known, it is convenient to use it as the reference phasor. For simplicity, the usually small effect of the stator resistance R_s will be ignored in the following analysis.

From the equivalent circuit of Fig. 6.2, the stator current phasor is

$$\vec{I}_s = \frac{\vec{V}_s}{j\omega_s L_M} - \vec{I}_F \qquad \text{A} \tag{6.12}$$

The complex power per phase entering the stator terminals is then

$$\vec{S}_s = P_s + jQ_s = \vec{I}_s \vec{V}_s^* \qquad \text{VA} \tag{6.13}$$

where $\vec{V}_s^*$ is the conjugate of the phasor $\vec{V}_s$. Inserting the expression for $\vec{I}_s$ from Eq. (6.12) gives

$$\vec{S}_s = -j\frac{V_s^2}{\omega_s L_M} - I_F V_s \angle \beta_0 \qquad \text{VA} \tag{6.14}$$

Because the losses in the stator are being ignored at this stage, the real component of $\vec{S}_s$ in Eq. (6.14) is also the power per phase crossing the air gap:

$$P_s = P_g = -I_F V_s \cos\beta_0 \qquad \text{N}\cdot\text{m} \tag{6.15}$$

By analogy with Eqs. (5.44) and (5.52), the torque is equal to the total air-gap power divided by the synchronous mechanical speed. Thus, the torque is

$$\begin{aligned} T &= \frac{3P_g}{\omega_{mech}} \\ &= -\frac{3p}{2\omega_s} I_F V_s \cos\beta_0 \qquad \text{N}\cdot\text{m} \end{aligned} \tag{6.16}$$

The relationship between torque T and rotor angle β_0 is shown in Fig. 6.3. Consider the condition where the torque is zero at $\beta_0 = -\pi/2$ rad. With this rotor angle, the equivalent rotor current $\vec{I}_F$ is in phase with the magnetizing current $\vec{I}_M$. The rotor mmf vector is rotating in space phase with the flux linkage vector $\vec{\Lambda}_s$. If a mechanical load with a torque T_L now is gradually applied, the machine will momentarily slow down, thus reducing the angle β_0. With β_0 more negative than $-\pi/2$, it is seen from Fig. 6.3 that the motor will produce a positive torque. In the transient state, an oscillation of the angle β_0 will occur for a time. Eventually, this will be damped out, and the steady-state condition will occur when the angle β_0 is such that the motor torque equals the load torque.

If the synchronous machine is coupled to a prime mover that applies a shaft torque in the same direction as the rotation, the effect will be to accelerate the machine momentarily, thus increasing the angle β_0 to a value greater than $-\pi/2$. As a result, the machine produces a negative torque, i.e., in the direction opposite to rotation and such as to balance the applied torque. In this condition, the machine operates as a synchronous generator.

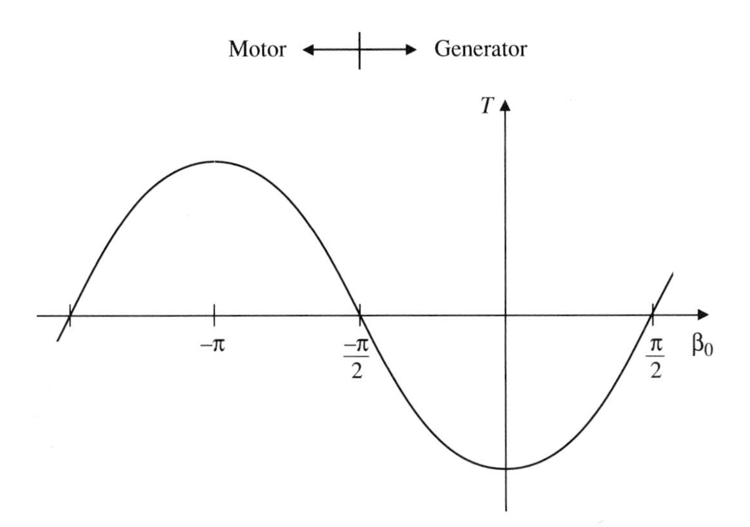

Figure 6.3 Torque-angle characteristic of synchronous machine.

From Eq. (6.16), it is seen that the peak value of available torque for a synchronous machine, operating on a constant voltage system either as a motor or as a generator, is directly proportional to the field current. Usually, the field current is set such that the rated torque of the machine is considerably lower than this maximum value. If, for any reason, the load torque as a motor exceeded the maximum available torque, the motor would stall. If a prime-mover torque greater than the maximum available torque as a generator was applied, the generator would run away.

To ensure a sufficiently large value of maximum torque, synchronous machines are normally designed with somewhat larger air gaps than induction machines. This is done to decrease the value of the magnetizing inductance and increase the magnetizing current, thus requiring a large value of the source current I_F. Typical values of the magnetizing inductance or reactance at rated frequency are in the range 0.6 to 2 pu based on the rated stator quantities. The magnetizing current in per unit at rated voltage and frequency then will be the inverse of the per unit magnetizing inductance.

A major factor influencing the setting of the field current magnitude is the stator power factor. From Eq. (6.14), the reactive power taken from the supply is

$$Q = -\frac{V_s^2}{\omega_s L_M} - I_F V_s \sin \beta_0 \qquad \text{VA} \tag{6.17}$$

Consider a machine operating with zero torque, i.e., with $\beta_0 = -\pi/2$. Then

$$Q = -\frac{V_s^2}{\omega_s L_M} + I_F V_s \qquad \text{VA} \tag{6.18}$$

This shows that the reactive power taken by the machine can be controlled by regulating the magnitude of the field current. For example, the reactive power at zero torque will be zero if the field current is set at

$$I_F = \frac{V_s}{\omega_s L_M} \quad \text{A} \tag{6.19}$$

For this condition, the machine is said to be floating on the electric supply system or bus with, theoretically, zero stator current. Increase of the field current causes the machine to act as a capacitive load on the system while reduction of the field current causes the machine to act as an inductive load. The theoretical limit of this latter condition of inductive load occurs with zero field current where the machine loads the system by the value of its magnetizing inductance. The relation between the stator current and the field current is of the form shown in Fig. 6.4. In some installations, a synchronous machine is used for the sole purpose of improving the power factor of a load system. In this case it is referred to as a *synchronous phase modifier* or *synchronous capacitor.*

When operated as a motor, the field current at any load is adjusted to maintain the desired power factor. To take the minimum current from the supply for any given torque load, the stator power factor should be made unity. In many applications, a synchronous motor is installed in an industrial plant to drive a large load and also to take a leading power factor so that the overall power factor of the plant will be within acceptable limits. For this condition, the field current is increased beyond the unity-power-factor condition and the motor is said to be overexcited.

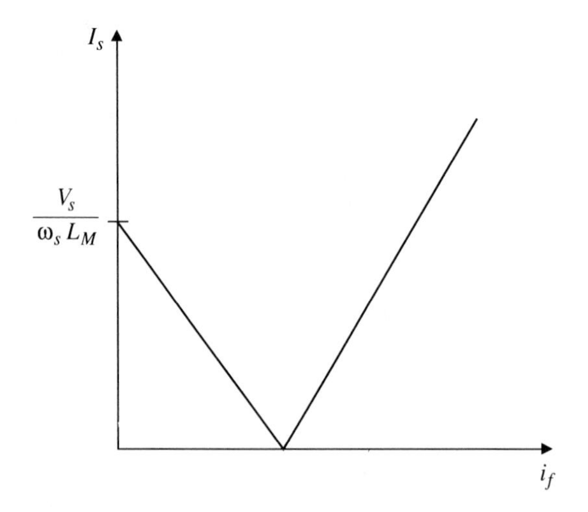

Figure 6.4 Relation between stator and field currents for operation as a synchronous phase modifier.

A phasor diagram for a synchronous motor is shown in Fig. 6.5(a) for a somewhat leading power factor, i.e., $\vec{I}_s$ leading $\vec{V}_s$ by the angle α_s. If the load torque is constant and machine losses are ignored, the component of the stator current in phase with the stator voltage will be constant. If the field current is increased, the stator current phasor $\vec{I}_s$ will follow the vertical locus as shown,

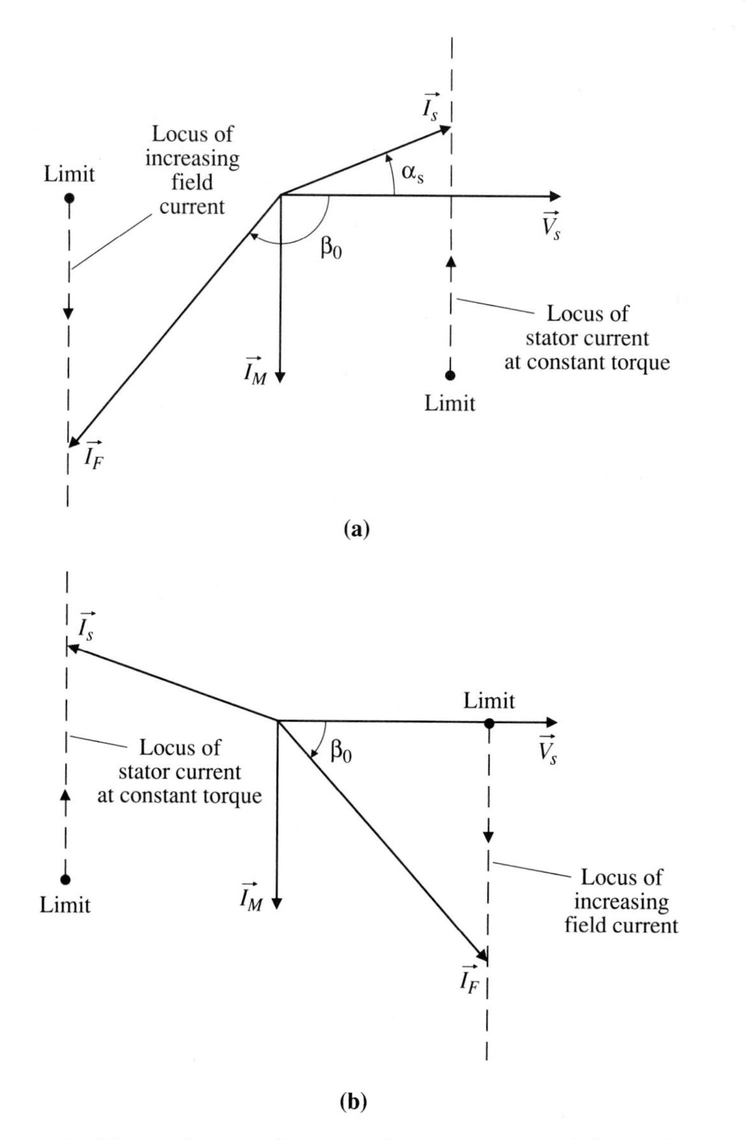

Figure 6.5 Phasor diagrams for (a) synchronous motor and (b) synchronous generator.

increasing the capacitive component of the stator current. At the same time, the load angle β_0 will be decreased somewhat. Conversely, a decrease in field current will eventually cause the stator current to become inductive. The theoretical limit for decrease of field current is reached when $\beta_0 = -\pi$ because, from Eq. (6.16), at this condition the maximum motor torque is just equal to the load torque. It would not be normal to operate in this region near $\beta_0 = -\pi$ because it is close to the stability limit. When the torque load on a synchronous motor varies over a wide range, it is usually necessary to provide means of detecting the stator power factor and regulating the field current to keep it within appropriate limits.

A synchronous generator connected to a power system will have its torque and, therefore, the in-phase component of its stator current determined by the setting of its prime mover. Figure 6.5(b) shows a phasor diagram for a generator operating at a constant value of output active power. In the condition shown, it is also acting as a capacitive load on the power system. Again, as with the synchronous motor, an increase in field current causes the generator to increase its capacitive load on the system, thus supplying the inductive reactive power required by the system's electrical loads and its transmission lines. Decrease of the field current eventually causes the machine to act as an inductive load on the system, the stability limit occurring at the rotor angle $\beta_0 = 0$. In situations where the system voltage is not substantially constant at the point of generator connection, increase of the generator's field current will tend to increase the system voltage.

So far in this section, synchronous generators have been considered as connected to a large electrical system. However, a number of synchronous generators are required to supply independent loads. An example would be an emergency power supply at a hospital. In such an installation, the frequency is maintained by the speed regulator on the prime mover. The field current of the generator is controlled so as to maintain a reasonably constant terminal voltage as the load changes.

EXAMPLE 6.1 A three-phase, 2300-V (L-L), 60-Hz, 100-kVA synchronous machine has a magnetizing inductance of 0.171 H. The ratio n of equivalent rms source current to field current is 1.62. All losses in the machine may be ignored. The machine is to be operated as a synchronous motor on a rated voltage supply delivering a mechanical power of 75 kW. Its power factor is to be 0.8 leading or capacitive. Determine the required field current and the load angle.

Solution The rated stator phase voltage is

$$\vec{V}_s = \frac{2300}{\sqrt{3}} = 1328\angle 0 \qquad \text{V}$$

With no losses, the input power per phase is 25 kW. Thus, for a power factor of 0.8, the stator current is

$$\vec{I}_s = \frac{25\,000}{1328 \times 0.8} \angle \cos^{-1} 0.8 = 23.53 \angle 36.87 \qquad \text{A}$$

The magnetizing reactance at 60 Hz is

$$X_M = \omega_s L_M = 377 \times 0.171 = 64.47 \qquad \Omega$$

The magnetizing current at rated voltage is

$$\vec{I}_M = \frac{\vec{V}_s}{jX_M} = \frac{1328 \angle 0}{j64.47} = -j20.60 \qquad \text{A}$$

The source or field current phasor is

$$\vec{I}_F = \vec{I}_M - \vec{I}_s = -j20.60 - 23.53 \angle 36.87 = 39.49 \angle -118.5 \qquad \text{A}$$

The field current then is obtained as

$$i_f = \frac{I_F}{n} = \frac{39.49}{1.62} = 24.38 \qquad \text{A}$$

The load angle is

$$\beta_0 = -118.5°$$

EXAMPLE 6.2 Suppose the machine in Example 6.1 is to be used as a synchronous phase modifier or capacitor connected to a rated voltage supply. What range of field current is required to provide a full range of capacitive load on the system within the rating of the machine?

Solution From Example 6.1,

$$\vec{V}_s = 1328 \angle 0 \qquad \text{V}; \qquad \vec{I}_M = -j20.60 \qquad \text{A}$$

At the electrical rating of the stator of 100 kVA, the rated stator current is

$$\vec{I}_s = \frac{j100\,000}{3 \times 1328} = j25.1 \qquad \text{A}$$

With this stator current and with no load torque, the equivalent field source current is

$$\vec{I}_F = \vec{I}_M - \vec{I}_s = -j45.7 \qquad \text{A}$$

Then, for full capacitive load, the required maximum field current is

$$i_f = \frac{I_F}{n} = \frac{45.7}{1.62} = 28.21 \qquad \text{A}$$

The minimum capacitive load is zero, and this occurs at $I_s = 0$. Thus, for this case,

$$\vec{I}_F = \vec{I}_M = j25.1 \qquad \text{A}$$

and

$$i_f = \frac{25.1}{1.62} = 15.49 \qquad \text{A}$$

EXAMPLE 6.3 The synchronous machine of Example 6.1 is to be used as a generator supplying an active power of 80 kW and an inductive reactive power of 60 kVA at 60 Hz to an electric bus with a line-to-line voltage of 2450 V. Determine the required field current and the load angle.

Solution The stator phase voltage is given as

$$\vec{V}_s = \frac{2450}{\sqrt{3}} = 1414.5\angle 0 \qquad \text{V}$$

The complex load power per phase is

$$S_L = \tfrac{1}{3}(80\,000 - j60\,000) \qquad \text{VA}$$

Because the assigned direction of the stator current $\vec{I}_s$ in the equivalent circuit of Fig. 6.2 is into the machine, the stator current phasor is given, from Eq. (6.13), as

$$\vec{I}_s = \frac{\vec{S}_s}{\vec{V}_s^*} = -\frac{\vec{S}_L}{\vec{V}_s^*} = -18.85 + j14.14 \qquad \text{A}$$

The magnetizing current is

$$\vec{I}_M = \frac{\vec{V}_s}{jX_M} = \frac{1414.5\angle 0}{j64.47} = -j21.94 \qquad \text{A}$$

The source current is then

$$\vec{I}_F = \vec{I}_M - \vec{I}_s = 40.71\angle -62.42^\circ \qquad \text{A}$$

Thus, the field current is

$$i_f = \frac{I_F}{n} = \frac{40.71}{1.62} = 25.13 \qquad \text{A}$$

and the load angle is

$$\beta_0 = -62.42^\circ$$

6.4

Starting and Excitation of Synchronous Machines

A source of direct current is required for synchronous machines with field windings. In small synchronous machines, this current may be supplied from an external source through a pair of brushes on two slip rings, as shown in Fig. 6.1. The power required by the field winding is dissipated as heat. In a small machine, this power may be of the order of 5% of the machine rating. For a large machine, the field power is usually less than 1% of rating. However, in a 1000-MVA machine, this will still be several megawatts and will be difficult to provide through slip rings. Thus, for large machines, the field current is normally provided by another generator mounted on the same shaft. In some systems, this may be a direct-current generator. In most modern installations, another synchronous generator is used as an exciter. The field windings of the exciter are placed on the stator and the phase windings on the rotor. A rectifier mounted on the rotating shaft of the exciter converts the alternating current to direct current. The field current of the main machine then can be regulated by controlling the exciter field current.

A synchronous generator is started with its stator terminals disconnected from the system that it is to supply. The prime mover controls are used to accelerate it up to the speed that is synchronous with the electric system. At the same time, the field current is adjusted so that the open-circuit generated voltage at the stator terminals is approximately equal to that of the system. Finally, the prime mover controls are used to bring the phase angles of the stator and the system voltages to approximate equality, whereupon the switch connecting the generator to the system is closed. This process is called *synchronization.* In small generators, it may be manual with the aid of instruments; in large generators, it is automated.

A synchronous motor with only a field winding carrying a direct current would not be self starting if connected to a constant frequency supply. At any speed other than synchronous speed, its rotor would experience an oscillating torque of zero average value as the rotating magnetic field produced by the stator passed the stationary or slower-moving rotor field. To provide for starting torque, synchronous motors are normally provided with a squirrel cage winding somewhat similar to that employed in induction motors. Figure 6.6 shows such a winding, sometimes referred to as a *damper winding* or *amortisseur,* fitted to a salient-pole synchronous motor with shaped poles and concentrated field windings. The motor is started by connecting the stator to its supply, either at full or at reduced voltage, and accelerated toward synchronous speed. During this acceleration, the field winding is usually short circuited to protect it from excessive induced voltage. When the speed reaches about 95% of synchronous speed, the field current is applied, and the rotor pulls into synchronism with the revolving field.

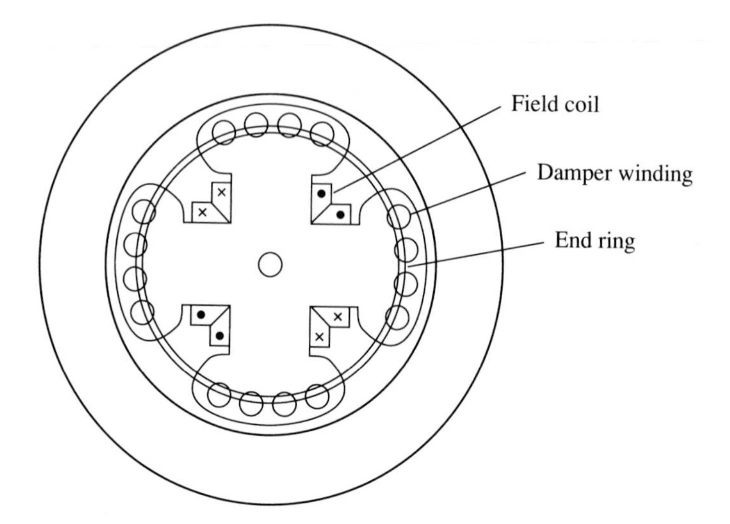

Figure 6.6 Four-pole salient-pole motor with starting (damper) winding.

6.5

Measurement of Parameters

The parameters of the equivalent circuit of Fig. 6.2 may be determined with an accuracy adequate for most purposes by performing two simple tests. For these tests, the machine is normally driven by a prime mover at rated speed, although any other available speed is satisfactory provided it is not too small. With the stator terminals open circuited and the speed constant, the stator voltage is measured for a range of values of field current. This range should extend up into the saturating region, as shown in Fig. 6.7(a). In the second test, also operated at any reasonable speed, the stator terminals are short circuited through ammeters when the field current is zero. The stator current is then measured for a range of field current, as shown in Fig. 6.7(a).

Assuming that the stator resistance R_s in Fig. 6.2 can be ignored, the stator current on short circuit will be equal to the source current I_F. Then, from Eq. (6.9), the current ratio is

$$n = \frac{I_s}{i_f} \tag{6.20}$$

i.e., it is the slope of the short-circuit test graph of Fig. 6.7(a). Normally, this is a straight line because the flux in the machine during short-circuit conditions is much too low to produce saturation.

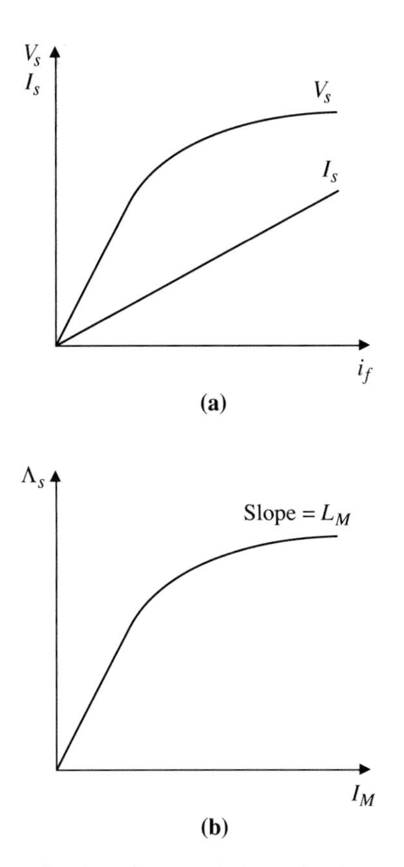

Figure 6.7 (a) Open circuit voltage and short circuit current. (b) Magnetization curve.

In the unsaturated region, the magnetizing inductance can be found from the open-circuit test results by

$$L_M = \frac{V_s}{\omega_s n i_f} \qquad \text{H} \tag{6.21}$$

where ω_s is the frequency at which the test was made and V_s is the phase voltage corresponding to a field current i_f.

In the saturating region, Eq. (6.21) is still valid but the flux linkage-magnetizing current relation then is nonlinear and is usually best represented by rescaling the open-circuit curve of Fig. 6.7(a) to produce the magnetization curve relating flux linkage Λ_s to magnetizing current I_M of Fig. 6.7(b). This form is particularly useful if the machine is operated at other than rated frequency. As with other machine types, the stator resistance can be measured using direct current between two stator terminals.

EXAMPLE 6.4 The following test results were obtained for a three-phase, 60-Hz, six-pole, 15-kVA, 220-V synchronous generator, driven at rated speed: open-circuit test: $i_f = 6.7$ A, $V_s = 220$ V; short-circuit test: $i_f = 4.6$ A, $I_s = 39.1$ A.

(a) Determine the values of the equivalent circuit parameters for operation at rated voltage.

(b) Determine the field current required to supply an independent rated load at a leading power factor of 0.8 and at rated frequency.

Solution

(a) The factor n relating field current to its sinusoidal equivalent current source is, from Eq. (6.20) and the short-circuit data,

$$n = \frac{I_s}{i_f} = \frac{39.1}{4.6} = 8.50$$

The magnetizing reactance is obtained, from Eq. (6.21), as

$$X_M = \frac{V_s}{ni_f} = \frac{220}{\sqrt{3} \times 8.5 \times 6.7} = 2.23 \quad \Omega$$

$$L_M = \frac{X_M}{\omega_s} = \frac{2.23}{377} = 5.92 \quad \text{mH}$$

(b) The rated stator voltage is

$$\vec{V}_s = 127\angle 0 \quad \text{V}$$

Full-load stator current at 0.8 leading power factor is

$$\vec{I}_s = -\frac{15\,000}{\sqrt{3} \times 220} \angle \cos^{-1} 0.8 = 39.4\angle 143.1^\circ \quad \text{A},$$

The magnetizing current is

$$\vec{I}_M = \frac{\vec{V}_s}{jX_M} = \frac{127\angle 0}{j2.23} = -j57.0 \quad \text{A}$$

The equivalent source current is

$$\vec{I}_F = \vec{I}_M - \vec{I}_s = 86.6\angle -68.7^\circ \quad \text{A}$$

Thus, the required field current is

$$i_f = \frac{86.6}{8.5} = 10.2 \qquad \text{A}$$

6.6

Permanent Magnet Synchronous Machines

As in the case with commutator machines, the flux in a synchronous machine can be provided by use of permanent magnets. These magnets can be placed on the rotor, thus eliminating the need for a source of direct current for excitation. The result is a simple, rugged machine. A cross section of one type of permanent magnet machine is shown in Fig. 6.8(a). The stator is the same as in a synchronous machine with a field winding. The rotor has a cylindrical steel core with radially directed permanent magnets on its surface. The permanent magnets might be of the neodymium material for which the *B-H* relation is shown in Fig. 1.18. In small, low-cost machines, they might be made of ferrite material.

The incremental relative permeability of a good permanent magnet material is approximately unity. As a result, a magnet of length ℓ_m can be regarded as a source of mmf equal to $\ell_m B_r/\mu_0$ A where B_r is the residual flux density of the permanent magnet material. Thus, a two-pole permanent magnet rotor rotating at angular velocity ω_0 can be visualized, as seen by the stator, as a source of alternating current at frequency $\omega_s = \omega_0$. The equivalent circuit model will be of the same form as shown for the wound rotor synchronous machine in Fig. 6.2.

To assess the parameters of the equivalent circuit, suppose each magnet in Fig. 6.8(a) covers an angle of 2γ rad. The mmf $\mathfrak{F}_{\theta r}$ produced by the magnet will have the rectangular form shown in Fig. 6.8(b). The stator windings are approximately sinusoidally distributed in angular space around the machine. As a result, only the fundamental space component of the flux density produced by the magnets can link with the stator windings and only the space fundamental component $\mathfrak{F}_{1\theta r}$ of the magnet mmf will be significant in producing interaction between the rotor and the stator. The peak value of this fundamental mmf per magnet is given by

$$\hat{\mathfrak{F}}_{1\theta_r} = \frac{4}{\pi}\frac{B_r \ell_m}{\mu_0}\sin\gamma \qquad \text{A} \tag{6.22}$$

This may be compared with the peak mmf produced by a sinusoidal set of currents of maximum value $\hat{i}$ in the stator windings, which is, from Eq. (6.7),

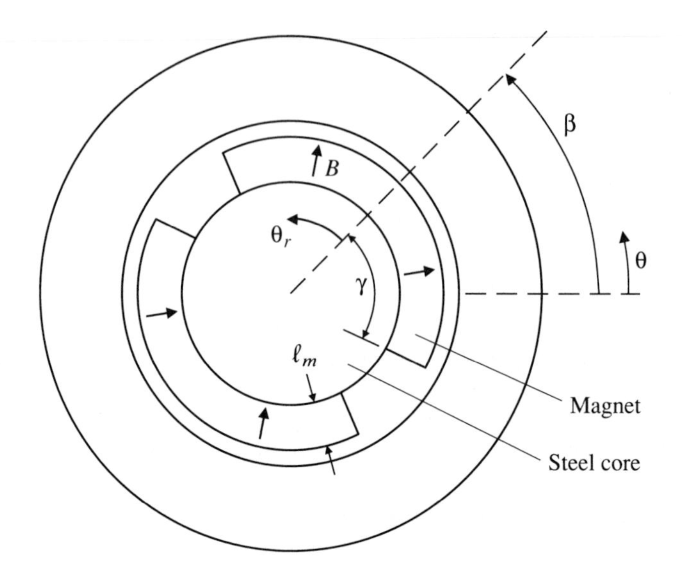

(a)

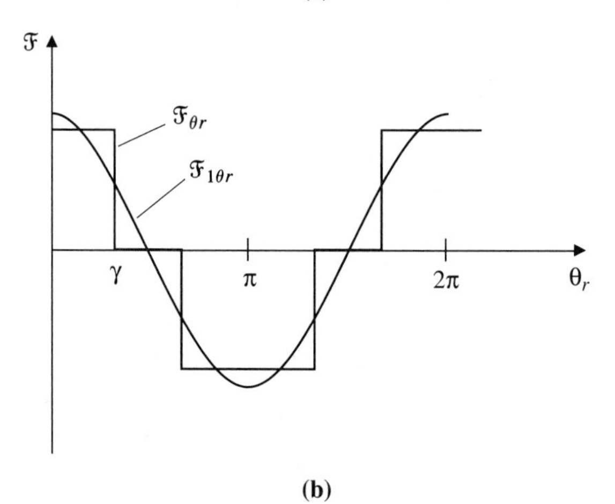

(b)

Figure 6.8 (a) Cross section of a two-pole permanent magnet machine. (b) Magnet mmf $\mathfrak{F}_{\theta r}$ and its space fundamental component $\mathfrak{F}_{1\theta r}$.

$$\hat{\mathfrak{F}}_\theta = \frac{3N_{se}}{4}\,\hat{i} \qquad \text{A} \tag{6.23}$$

By analogy with Eq. (6.9), the rotating magnet mmf can be represented as a current source $\vec{I}_F$. From Eqs. (6.22) and (6.23), the rms current source phasor is

$$\vec{I}_F = \frac{8\sqrt{2}}{3\pi}\,\frac{B_r \ell_m \sin\gamma}{\mu_0 N_{se}}\,\angle\beta_0 \qquad \text{A} \tag{6.24}$$

While the structure of the equivalent circuit for the permanent magnet machine is similar to that shown for the wound field machine in Fig. 6.2, the value of the magnetizing inductance is usually somewhat smaller. The reason for this is that the effective air-gap length to insert in the magnetizing inductance expression of Eq. (5.22) is the distance between the stator and the steel core of the rotor. The magnet, having a relative permeability of approximately unity, can be regarded as free space in determining the air-gap flux produced by a stator magnetizing current. Therefore, typical values of the magnetizing inductance may be in the range 0.25 to 0.5 pu based on rating. Similarly, typical values of the equivalent source current I_F will be the inverse of the magnetizing inductance values in per unit, i.e., 2 to 4 pu so that an open-circuit voltage of one per unit or rated value can be produced at rated speed.

The operating characteristics of permanent magnet machines, both generators and motors, are constrained by the lack of ability to adjust the magnitude of the equivalent source current. As a free-standing generator, the terminal voltage will vary with the load. To keep this voltage within reasonable limits, it may be necessary sometimes to control the load power factor by the use of adjustable shunt capacitors. As a motor, the permanent magnet machine is most widely used as a variable-speed drive supplied by a source of controllable voltage and frequency.

EXAMPLE 6.5 A permanent magnet synchronous machine is rated at 400 V, three phase, 50 Hz, four pole, 50 kVA. Its magnetizing inductance is 2.5 mH, and its equivalent source current is 310 A (rms). The stator resistance may be ignored.

(a) Suppose this machine is operated as a generator at rated frequency. Determine the maximum and minimum values of the stator phase voltage as the load current is varied from zero to rated value at unity power factor.

(b) Suppose the machine is operated as a variable-speed drive motor supplied by an electronic inverter that provides controllable three-phase voltage at controllable frequency. What should be the frequency and line-to-line voltage of the inverter to drive a load requiring a torque of 300 N·m at a speed of 600 r/min if it is also desired that the load on the inverter be at a unity power factor?

Solution

(a) The rated phase voltage is

$$V_s = \frac{400}{\sqrt{3}} = 231 \qquad \text{V}$$

and the rated stator current is

$$I_s = \frac{50\,000}{3 \times 231} = 72.2 \qquad \text{A}$$

With no load on the generator and with rated frequency, the phase voltage is

$$V_s = \omega_S L_M I_F = 2\pi \times 50 \times 0.0025 \times 310 = 243.5 \qquad \text{V}$$

With rated stator current at unity power factor, the stator current and the magnetizing current will be in quadrature, i.e., 90° from each other. The phasor sum must equal I_F. Thus,

$$I_M = (310^2 - 72.7^2)^{1/2} = 301.5 \qquad \text{A}$$

Therefore, the stator voltage is

$$V_s = 314 \times 0.0025 \times 301.5 = 236.8 \qquad \text{V}$$

Because of the small value of the magnetizing inductance and because of the high power factor of the load, the variation in the terminal voltage with load is small. It would be larger with a lagging-power-factor load.

(b) With four poles, the frequency for a speed of 600 r/min is

$$\omega_S = \frac{4}{2} \times 600 \times \frac{2\pi}{60} = 125.6 \text{ rad/s} = 20 \text{ Hz}$$

The torque can be expressed as the ratio of the power entering the machine divided by the synchronous mechanical speed because losses are to be ignored. Thus,

$$T = \frac{3p}{2\omega_s} V_s I_s = \frac{3p}{2} L_M I_M I_S = 300 \qquad \text{N}\cdot\text{m}$$

With unity power factor, the stator current and the magnetizing current will be in quadrature. Their phasor sum must equal I_F. Thus,

$$I_M^2 + I_s^2 = I_F^2 = 310^2$$

These two relations in I_s and I_M can be solved to give the condition

$$I_M = 302.9 \qquad \text{A}; \qquad I_s = 66.0 \qquad \text{A}$$

Thus, the required phase voltage will be

$$V_s = \omega_S L_M I_M = 125.6 \times 0.0025 \times 302.9 = 95.1 \qquad \text{V}$$

The line voltage will be

$$V_{S(L\text{-}L)} = \sqrt{3} V_s = 164.7 \qquad \text{V}$$

As a check, the input power is

$$P_e = \sqrt{3}V_{S(L\text{-}L)}I_S = 18.84 \qquad \text{kW}$$

The mechanical power is

$$P_0 = \frac{2\omega_s T}{p} = 18.84 \qquad \text{kW}$$

6.7

Electronically Switched Permanent Magnet Motors

The availability of efficient semiconductor switches has provided means for eliminating the mechanical switching on commutator machines while retaining many of their useful properties. This section introduces a typical example.

Figure 6.9(a) shows a cross section of a two-pole motor. The rotor has two surface-mounted permanent magnets, each covering approximately 180° of the rotor periphery and thereby producing a nearly rectangular space wave of flux density in the air gap. The stator has three phase windings but differs from the windings for three-phase alternating machines in that the conductors of each phase are distributed approximately uniformly in slots over two arcs of 60° for each phase.

The motor can be supplied from a system such as that shown in Fig. 6.9(b). The motor phases are connected to a source of controllable direct current i through six electronic switches. At any instant, one upper and one lower switch are closed, connecting two phases in series to the current supply. As seen in Fig. 6.9(a), each rotor magnet interacts with two 60° arcs of stator conductors carrying current with the polarity shown. When the rotor magnet edges reach the boundary between stator phases, a detector, such as a Hall device mounted on the stator, detects the reversal of the air-gap field and causes an appropriate opening and closing of the switches. At the position shown in Fig. 6.8(a) and with the rotor rotating in a counterclockwise direction, switch 1 will open and switch 3 will close, thus energizing phases b and c and continuing the torque.

Suppose the air-gap flux density is B_g. If there are N_s turns per phase, the linear current density over an energized phase winding for a rotor of radius r is

$$K = \frac{3N_s i}{\pi r} \qquad \text{A/m} \tag{6.25}$$

The tangential force per unit of area is $B_g K$ from Eq. (3.22). Because two-thirds of the total surface area of the stator is effective in producing force at any instant, the torque for an effective axial length ℓ is

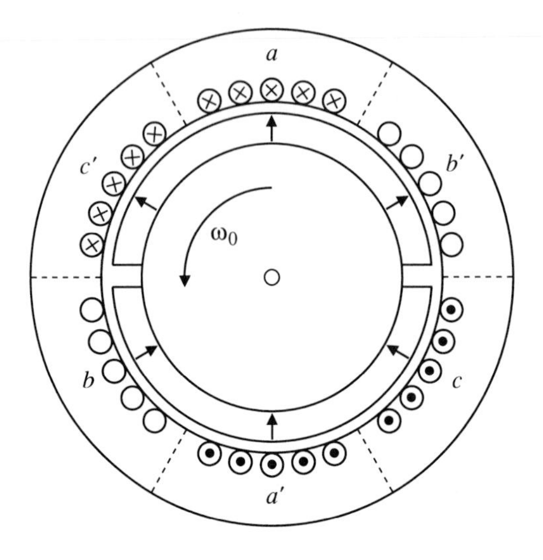

(a)

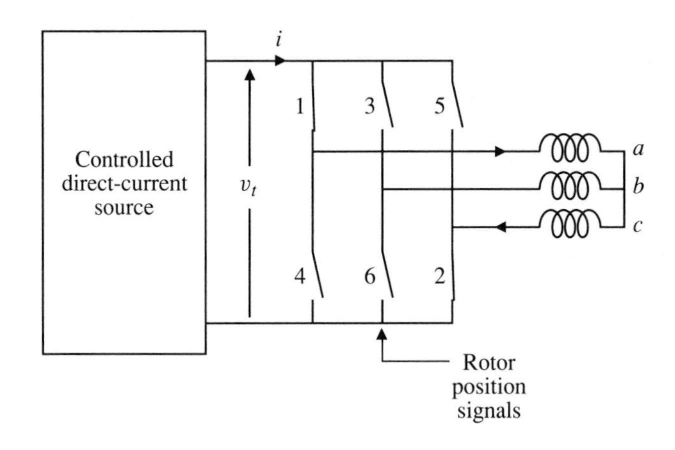

(b)

Figure 6.9 (a) Cross section of an electronically switched permanent magnet motor. (b) Electric supply and switching system.

$$
\begin{aligned}
T &= \tfrac{2}{3}\,(2\pi r)\,\ell r B_g K \\
&= 4 r \ell B_g N_s i \qquad\qquad (6.26) \\
&= ki \qquad \text{N}\cdot\text{m}
\end{aligned}
$$

Thus, the motor produces a torque directly proportional to its supply current. Reversal of the torque direction is usually achieved by appropriate change in the switching signals from the field detectors.

To determine the generated voltage produced by two phase windings in series, consider the machine to be rotating at angular velocity ω_0. In each of the $4N_s$ series conductors, the electric field intensity will be νB_g where $\nu = r\omega_0$. Thus, the generated voltage is

$$e = 4N_s r\ell B_g \omega_0 = k\omega_0 \qquad \text{V} \tag{6.27}$$

If the current source in Fig. 6.9(b) now is replaced by a controllable voltage source v_t, the motor will operate at a speed approximately proportional to the source voltage. By analogy with a commutator motor,

$$\omega_0 = \frac{v_t - Ri}{k} = \frac{v_t}{k} - \frac{RT}{k^2} \qquad \text{rad/s} \tag{6.28}$$

where R is the resistance of two phases in series. Switched permanent-magnet motors of the general type shown in Fig. 6.9(a) are used in a wide variety of drive applications such as in robots, machine tools, and disk drives.

PROBLEMS

6.1 A source of three-phase, 400-Hz voltage is required for an industrial operation. The only available supply is three-phase, 60 Hz. The frequency conversion is to be accomplished by use of a synchronous motor driving a synchronous generator. A variation of ±3% about the 400-Hz frequency is permissible. Determine a suitable number of poles for each of the synchronous machines and the speed at which the system will run.

(Section 6.1)

6.2 A three-phase, 440-V (L-L), 10-kVA, 60-Hz, 1200 r/min synchronous machine has a synchronous reactance X_M of 50.8 Ω, negligible stator resistance, and negligible rotational losses. A field current of 6 A is required to produce rated stator voltage on open circuit. Saturation is negligible.

(a) Draw an equivalent circuit for the machine, inserting values for the parameters L_M and n.

(b) Determine the magnetizing reactance in per unit of the rated impedance of the machine.

(Section 6.2)

6.3 The synchronous machine of Problem 6.2 is to be used as a motor connected to a 440-V, 60-Hz supply.

(a) If the mechanical power output of the motor is 10 kW, what field current is required to give a stator power factor of unity?

(b) If the mechanical load is now removed from the motor and the field current is unchanged, what is the stator current and power factor?

(c) What is the maximum torque that the motor can deliver with the field current found in (a)?

(d) What is the supply current for the condition of (c)?

(Section 6.3)

6.4 The synchronous machine of Problem 6.2 is used as a generator driven by an internal combustion engine. Its stator is connected to a 60-Hz, 440-V supply.

(a) Determine the field current required when an electrical power of 10 kW is being delivered to the supply at unity power factor.

(b) Using the field current of (a), determine the maximum torque that the engine can exert before the generator pulls out of synchronism.

(Section 6.3)

6.5 A three-phase, 60-Hz, two-pole, 60-MVA, 13.2-kV, synchronous generator is connected to a distribution system that has a constant voltage of 13.2 kV (L-L). The magnetizing inductance of the generator is 1.2 pu based on the machine rating. Losses may be ignored. A field current of 1000 A is required to give unity power factor output when the power input at the prime mover shaft is 50 MW.

(a) Derive the parameters for an equivalent circuit for this machine.

(b) With a shaft power of 50 MW, to what value can the field current be reduced without loss of synchronism?

(c) Determine the line current for the condition of (b). Is it within the rated value?

(Section 6.3)

6.6 A 1000-hp, 2300-V, 60-Hz, 900 r/min, three-phase synchronous motor has a synchronous reactance of 1.9 Ω/phase. On a no-load test, driven at rated speed, a field current of 15 A produced an open-circuit stator voltage of 2300 V (L-L). Stator resistance may be neglected. When used as a motor with rated supply voltage and frequency, the load and the field current are adjusted to give a stator current of 200 A. Determine the field current and the load angle β_0 when the field current is adjusted to give a power factor of (a) unity (b) 0.8 leading or capacitive.

(Section 6.3)

6.7 With no load on the motor of Problem 6.6 and with rated voltage and frequency on the stator, the field current is adjusted to the value that gives the minimum stator current. Holding the field current at this value, the torque on the motor is increased until the stator current is 250 A. All losses may be ignored. Determine the angle through which the rotor is retarded when this load is applied.

(Section 6.3)

6.8 A 1-MW, 6600-V, ten-pole, 60-Hz, three-phase synchronous motor has a synchronous reactance of 25 Ω/phase. All losses may be ignored. The current ratio n is 6.8. The field current and the mechanical load are set so that the motor operates at a power factor of 0.8 leading and draws a stator current of 95 A.

(a) Determine the field current and the load angle.

(b) Determine the shaft torque and power.

(Section 6.3)

6.9 A factory has 900 hp of induction motors with an average power factor of 0.75 and average motor efficiency of 0.85. It also has 75 kW of lighting and heating load. To accommodate a new process requiring a mechanical power of 250 hp and at the same time improve the factory power factor to at least 0.9, it is decided to use a synchronous motor that may be assumed to have an efficiency of 0.9. Determine the appropriate kVA rating of this motor.

(Section 6.3)

6.10 A 300-kW, 6600-V, eight-pole, 60-Hz, three-phase synchronous motor has a synchronous reactance of 88 Ω/phase. Neglecting losses in the motor, determine the pull-out torque when the motor is initially set to operate at rated power and with the field current adjusted to give (a) unity power factor and (b) a power factor of 0.8 leading.

(Section 6.3)

6.11 An emergency power supply consists of a synchronous generator driven by a diesel motor. The generator has a rating of 220 V (L-L), 20 kVA, 60 Hz, three phase, 1200 r/min. Its synchronous reactance is 2.2 Ω/phase, and its current ratio n is 6.1. Losses may be ignored, and the diesel motor may be assumed to maintain constant speed.

(a) Suppose the generator field is adjusted to produce rated stator voltage while running at rated speed with no load in a standby condition. Suppose a three-phase resistive load of 15 kW which was previously supplied at 220 V (L-L) from the power utility is switched on to the generator following a failure of the utility supply. Assuming no change in field current, find the stator voltage that will be available.

(b) If the stator voltage is to be restored to rated value, by what factor should the field current be increased?

(Section 6.3)

6.12 When a 50-kVA, three-phase, 440-V (L-L), 60-Hz synchronous machine is driven at its rated speed, it is found that its open-circuit stator voltage is 440 V (L-L) with a field current of 7 A. When the stator terminals are short circuited, rated stator current is produced with a field current of 5.5 A.

(a) Determine the synchronous reactance in ohms per phase and the current ratio n.

(b) Determine the per unit value of the synchronous reactance based on the machine rating.

(c) Find the synchronous inductance L_M in ohms and in per unit.

(Section 6.5)

6.13 The following test results were obtained from a three-phase, 60-Hz, 15-kW, 220-V (L-L) synchronous motor: open-circuit test: field current = 6.7 A, stator voltage = 220 V; short-circuit test: field current = 6.7 A, stator current = 104 A. The motor is designed to operate at a power factor of 0.8 leading at rated load.

(a) Sketch a phasor diagram representing the full-load condition at rated power factor and determine the required field current.

(b) Ignoring losses, determine the reactive power that the motor could absorb when supplying half of its rated mechanical power at rated stator current. Determine the field current for this condition.

(Section 6.5)

6.14 A permanent magnet synchronous generator has a rating of 230 V (L-L), three-phase, 10 kVA, 400 Hz, two pole. Its magnetizing inductance is 0.6 mH. When driven at rated speed, its no-load stator voltage is 240 V. Determine the stator voltage when the stator current is at rated magnitude and 0.9 power factor lagging.

(Section 6.6)

6.15 A three-phase, four-pole permanent magnet motor has a magnetizing inductance of 3 mH, a stator resistance of 1.2 Ω and an equivalent source current of 21 A (rms). The motor is to be incorporated into a variable-speed drive with the stator supplied from a variable-voltage, variable-frequency source. Determine the stator voltage and frequency to drive a load requiring a torque of 1.5 N·m at a speed of 6000 r/min. The stator voltage should be adjusted to require minimum stator current.

(Section 6.6)

6.16 A switched permanent magnet motor has the structure shown in Fig. 6.9(a). Its rotor radius is 20 mm and its rotor length is 40 mm. The ferrite permanent magnets produce an air-gap flux density of 0.28 Wb over the complete periphery. The linear current density in the stator windings must be limited to 4000 A/m (rms) to prevent overheating.

(a) Find the maximum steady-state torque that this motor can produce.

(b) Suppose the maximum motor speed is to be 500 rad/s and the maximum direct voltage available from the electronic source is 48 V. What must be the maximum value of the source current? Stator resistances may be ignored.

(c) How many stator turns per phase are required?

(Section 6.7)

6.17 Suppose the motor in Problem 6.16 is driving a load that requires a torque of 0.2 mN·m per rad/s of speed. The stator winding has a resistance of 1.2 Ω/phase. At what steady-state speed will the load be driven when the source voltage is 32 V?

(Section 6.7)

GENERAL REFERENCES

Fitzgerald, A.E., Kingsley, C. and Umans, S.D., *Electric Machinery*, 5th ed., McGraw-Hill, New York, 1990.
Levi, E., *Polyphase Motors*, John Wiley and Sons, New York, 1984.
Nasar, S.A., *Handbook of Electric Machines*, McGraw-Hill, New York, 1987.
Say, M.G., *Alternating Current Machines*, 5th ed., Pitman, London, 1983.

CHAPTER 7

Magnetic Systems and Transformers—Revisited

Chapters 1, 2, and 3 of this book present an introduction to those concepts considered necessary for a basic understanding of ferromagnetic materials, simple transformer devices, and machines. The topics included are those that are viewed as appropriate for study by all students of electrical engineering. In this chapter, attention is focused on additional topics that extend the engineering application of the basic concepts:

modeling of more complex magnetic devices
an introduction to the design of such devices
some practical operating properties of transformers
forces in saturable magnetic devices.

7.1 Analysis of Complex Magnetic Systems

In Section 2.2, an electric equivalent circuit model was produced for the analysis of the operating properties of a relatively simple transformer with a single ferromagnetic core. Other devices, including some types of transformer, have more complex core and winding configurations. To develop analytical models for these, we will pick up the magnetic circuit concepts introduced in Section 1.12 and show how these can be converted into electric equivalent circuits from which the operating performance may be predicted using well-known techniques of electric circuit analysis.

First, let us examine the criteria that might be used in choosing an equivalent circuit model. It should contain enough information in its parameters for the prediction of the performance properties of concern to the user. Where magnetic nonlinearity influences the performance significantly, it should be included in the model. When transient phenomena or frequency responses are of interest, the model should be adequate for these analyses. On the other hand, the model should not be more complex than necessary for the problem at hand. Where the effects of certain properties are negligible from an engineering standpoint, it should be simple to ignore those parameters that might otherwise represent these properties.

7.1.1 ▪ Equivalent Circuits

The methodology for producing an adequate equivalent circuit can be best demonstrated using a typical and simple example. Consider the magnetic system shown in Fig. 7.1(a). This system consists of a three-legged magnetic core, two of the legs having windings and the third central leg having an air gap. Basically, this is a three-dimensional magnetic-field situation for which a precise analysis could be extremely complex. By the use of simplifying assumptions appropriate to the desired end result, the magnetic field can be reduced to a magnetic circuit of lumped reluctances. For example, suppose we assume that, except for the air gap, all magnetic flux is confined to the magnetic core material, i.e., the leakage flux in air paths around the windings is considered to be negligible. Later, we should reexamine this assumption to foresee the situations where it may not be appropriate.

The magnetic system of Fig. 7.1(a) now may be divided into four sections, each of which has a uniform flux over its length. Three of these sections represent magnetic paths in the core material and the fourth represents the air-gap path. Each section may be represented by a reluctance element that relates the path flux to the mmf required to establish that flux along the length of the section.

Figure 7.1(b) shows the magnetic equivalent circuit that results from the foregoing assumptions. Reluctances $\mathcal{R}_1$, $\mathcal{R}_2$, and $\mathcal{R}_3$ represent the paths in the magnetic material carrying fluxes Φ_1, Φ_2, and Φ_3, respectively. The air gap is represented by the linear reluctance $\mathcal{R}_4$, and its magnetic flux is Φ_3. The circuital law of Eq. (1.1) applies to any closed path in a magnetic-field system. In a magnetic equivalent circuit, this law is represented by the following relation: around any closed path, the total mmf of the current-carrying windings is equal to the sum of the products of reluctance and flux:

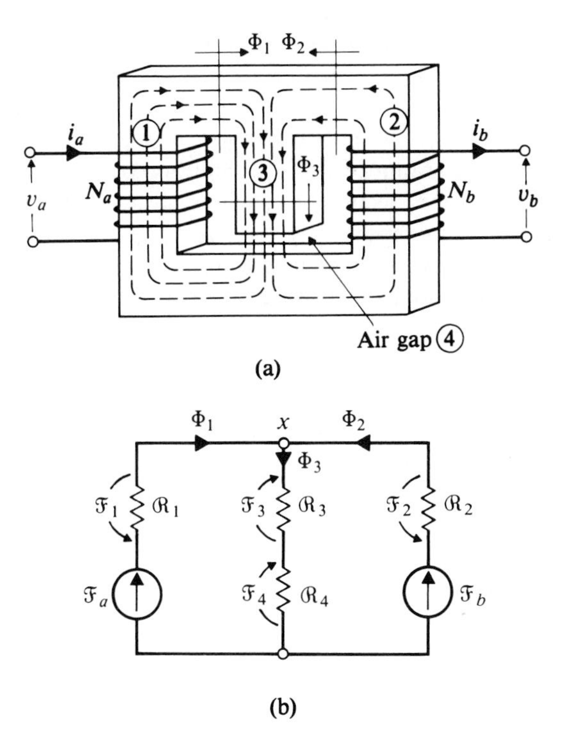

Figure 7.1 (a) A magnetic system. (b) A magnetic circuit for the system.

$$\sum \mathfrak{F} = \sum_{\text{around closed path}} \mathcal{R}\Phi \tag{7.1}$$

The continuity of flux in the magnetic field is represented by equating the sum of the fluxes entering any junction of magnetic paths in the equivalent circuit to zero:

$$\sum_{\text{into junction}} \Phi = 0 \tag{7.2}$$

If a solution in terms of the magnetic variables $\mathfrak{F}$ and Φ is desired, normal methods of circuit analysis may be employed to determine the fluxes in the circuit of Fig. 7.1(b) for a given set of magnetomotive forces. If the relative permeability μ_r of the magnetic material can be considered constant for the situation under study, the reluctances of the core sections can be expressed as

$$\mathcal{R} = \frac{\ell}{\mu_r \mu_0 A} \qquad \text{A/Wb} \tag{7.3}$$

where ℓ is the mean length of the flux path in the section and A is its cross-sectional area.

For this situation, the techniques of linear circuit analysis are applicable. If the nonlinearity of the B-H relation for the magnetic material is considered to be significant, each reluctance element may be represented by a graph or function relating its flux to its mmf. Examples of some useful approximate relations are shown in the graphs of Fig. 7.2. Solution of circuits containing such nonlinear elements requires techniques such as piecewise linear analysis, trial and error, or graphical methods.

Section 1.2 showed that a simple magnetic core represented by the magnetic circuit relation $\mathcal{F} = \mathcal{R}\Phi$ could also be represented by an inductance L in an elec-

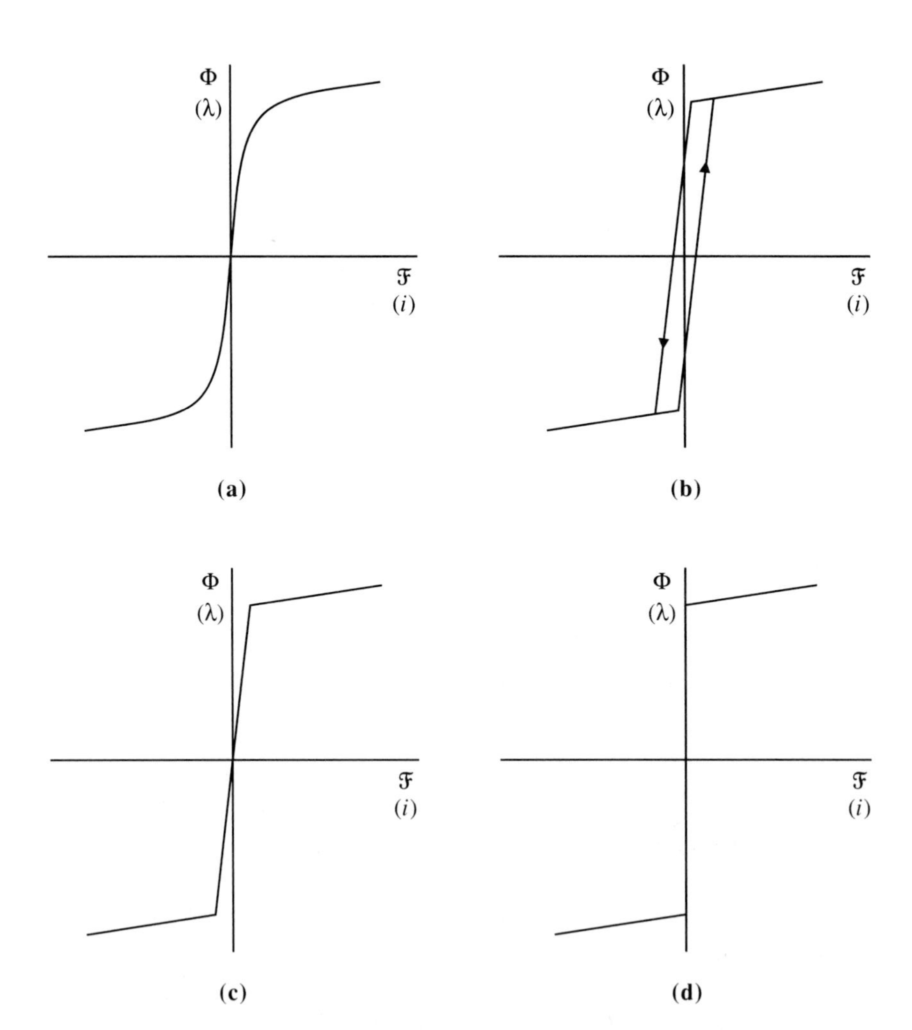

Figure 7.2 Models of flux-magnetomotive force relations (corresponding models for flux linkage-current relations are designated by the variables in parentheses).

tric equivalent circuit where, for a coil of N turns around the core carrying a current i,

$$L = \frac{\lambda}{i} = \frac{N^2}{\mathcal{R}} \qquad \text{H} \tag{7.4}$$

$$\lambda = N\Phi \qquad \text{Wb} \tag{7.5}$$

and

$$i = \mathfrak{F}/N \qquad \text{A} \tag{7.6}$$

Next, let us consider how to convert the magnetic circuit of Fig. 7.1(b) into an equivalent electric circuit. Suppose for the present that each of the fluxes Φ_1, Φ_2, and Φ_3 links an N-turn coil. The corresponding voltages induced in these coils will be

$$e_1 = N\frac{d\Phi_1}{dt} \qquad e_2 = N\frac{d\Phi_2}{dt} \qquad e_3 = N\frac{d\Phi_3}{dt} \qquad \text{V} \tag{7.7}$$

At node x in Fig. 7.1(b), the flux variables are related, from Eq. (7.2), by

$$\sum_{\text{into node } x} \Phi = \Phi_1 + \Phi_2 - \Phi_3 = 0 \qquad \text{Wb} \tag{7.8}$$

The corresponding induced-voltage variables in Eq. (7.7) therefore must be related by the expression

$$e_1 = e_2 - e_3 = 0 \qquad \text{V} \tag{7.9}$$

For the left-hand mesh of the magnetic circuit of Fig. 7.1(b), Eq. (7.1) gives the relation among the mmf variables as

$$\mathfrak{F}_a = \mathfrak{F}_1 + \mathfrak{F}_3 + \mathfrak{F}_4 \qquad \text{A} \tag{7.10}$$

Consider that each of these mmf components is produced by a corresponding component of current in an N-turn coil. These current components are then related by

$$i'_a = i_1 + i_3 + i_4 \qquad \text{A} \tag{7.11}$$

For the right-hand mesh, the corresponding relations are

$$\mathfrak{F}_b = \mathfrak{F}_2 + \mathfrak{F}_3 + \mathfrak{F}_4 \qquad \text{A} \tag{7.12}$$

and

$$i'_b = i_2 + i_3 + i_4 \qquad \text{A} \tag{7.13}$$

The prime symbols are added to i'_a and i'_b to distinguish them from the actual coil currents i_a and i_b.

Each reluctance in the magnetic circuit relates a flux variable Φ and an mmf variable $\mathfrak{F}$. The corresponding induced voltage variable e and the current variable i are related by an inductance parameter. For example, the relation $\mathfrak{F}_1 = \mathfrak{R}_1 \Phi_1$ in the magnetic circuit has a corresponding relation in the electric circuit of

$$e_1 = L_1 \frac{di_1}{dt} \quad \text{V} \tag{7.14}$$

Equations (7.9), (7.11), (7.13), and (7.14) describe the electric circuit shown in Fig. 7.3(a). For each of the two independent meshes of the magnetic circuit, there is an independent node in the electric circuit. The currents entering these two nodes, which are designated as a and b, are related by Eqs. (7.11) and (7.13). For each independent node in the magnetic circuit, there is a corresponding mesh in the electric circuit. The branch voltages around the lower central mesh of Fig. 7.3(a) are related by Eq. (7.9). For each reluctance branch in the magnetic circuit, there is a corresponding inductance branch in the electric circuit. For each mmf source in the magnetic circuit, there is a corresponding coil current in the electric circuit.

The form of the electric circuit of Fig. 7.3(a) may be derived directly from the magnetic circuit of Fig. 7.1(b) using the topological principle of duality. This technique is demonstrated in Fig. 7.3(b). A node is marked in each mesh of the magnetic circuit, i.e., a and b, and a reference node o is marked outside the circuit. These nodes are then joined by branches, one of which passes through each element of the magnetic circuit. It is observed that the structure of the resulting network is identical to the structure of the electric circuit of Fig. 7.3(a). For each reluctance in a mesh of the magnetic circuit, there is an inductance connecting the corresponding nodes of the electric circuit. Where a reluctance is common to two meshes in the magnetic circuit, the corresponding inductance connects the corresponding two nodes of the electric circuit. For each mmf source there is a corresponding driving current; for each flux there is a corresponding induced voltage between nodes.

When a reluctance in the magnetic circuit represents a nonlinear relation between flux and mmf, the corresponding inductance element in the electric circuit represents a similar nonlinear relation between the flux linkage in a coil of N turns encircling that branch of the magnetic system and the current in an N-turn coil that produces the mmf for that branch. The nonlinearities in the magnetic circuit therefore are preserved in the electric equivalent circuit. Models of Φ-$\mathfrak{F}$ relations such as those shown in Fig. 7.2 thus may be rescaled to provide appropriate λ-i models for the electric circuit. The choice of model depends on the problem under study.

The electric circuit of Fig. 7.3(a) has been developed assuming that all windings had N turns. Because the number of turns is generally different for the various windings, it is necessary to add ideal transformers at the terminals of the electric circuit to obtain the actual voltages and currents in the windings. The

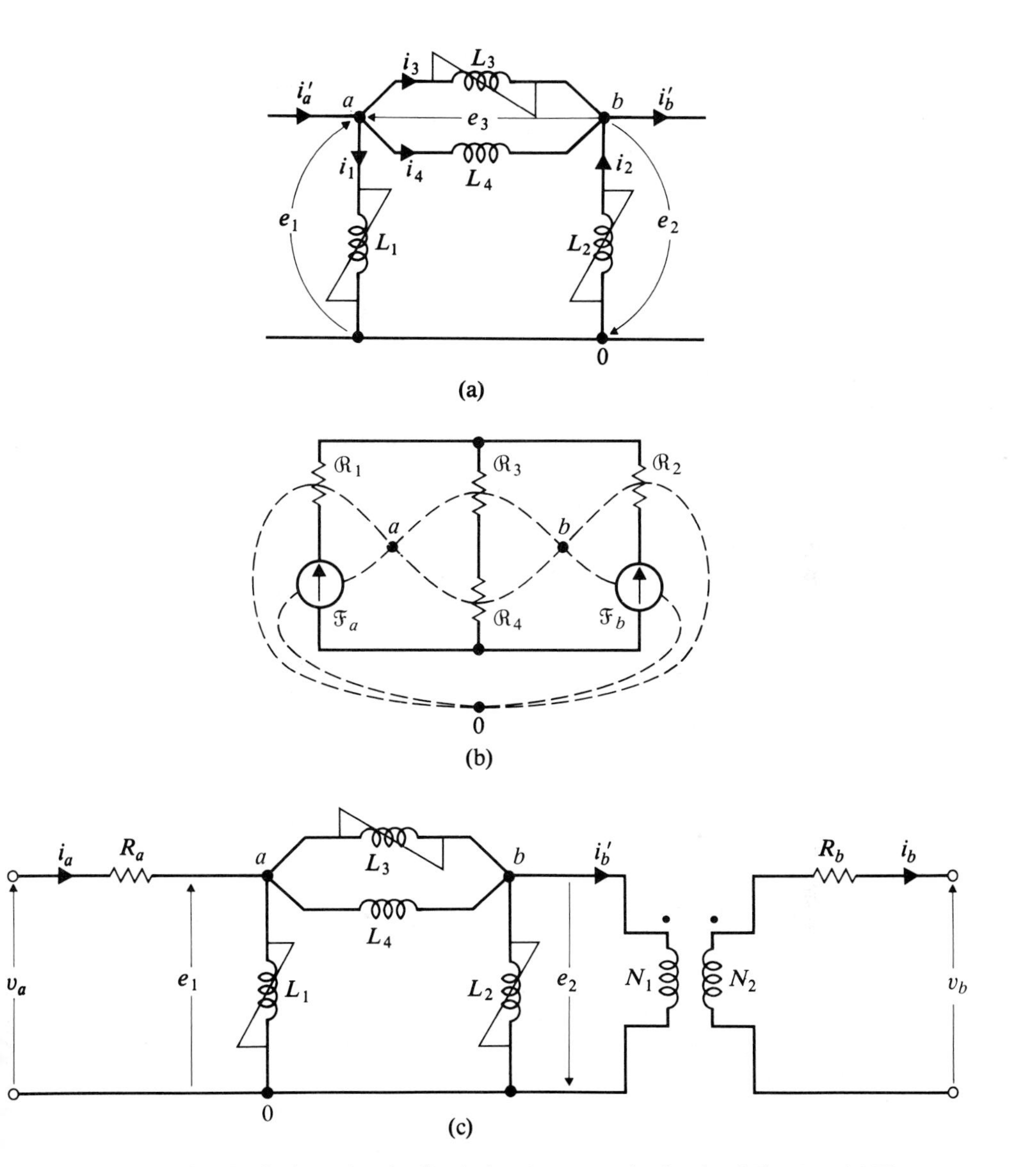

Figure 7.3 Equivalent electric circuit for the magnetic circuit of Fig. 7.1. (a) Elementary form of circuit. (b) Topological technique of derivation. (c) Circuit including ideal transformer and winding resistances.

reference number of turns is normally made equal to the number of turns in one of the windings. In Fig. 7.3(c), N is made equal to N_a and no ideal transformer is required for this winding. Note that even if $N_a = N_b$, an ideal transformer would be required for winding b to preserve the property that there is no conductive connection between the two windings. The winding resistances R_a and R_b now may be added to the equivalent circuit, providing the difference between the winding induced voltages and the terminal voltages v_a and v_b.

The equivalent circuit of Fig. 7.3(c) has been developed on the assumptions of no leakage flux, i.e., with all the flux confined to the three ferromagnetic core sections. Each of the inductances representing these core sections has been denoted as nonlinear and can be modeled by a variety of relations such as those shown in Fig. 7.2. If each core section is operated in a flux density range where the core mmf is negligible, all three of these inductances will be very large, and their effects can be ignored. In that case, the main property of the equivalent circuit of the device is as a series inductance L_4, the value of which can be adjusted, dependent on the air-gap dimensions.

Alternatively, a more detailed analysis of the device might justify a more complex circuit model. For example, the leakage fluxes in air paths around the two windings might be added to the magnetic-circuit model, resulting in added inductance elements to the electric equivalent circuit. Moreover, if these added leakage fluxes were large, it might be necessary to segregate the core into more than three sections. In situations where the core losses are significant, an appropriate loss resistance can be added in parallel with each of the inductances representing core sections, following the modeling approach developed in Section 1.10 and shown in Fig. 1.26. The main point is that there is no unique equivalent circuit for any device; a good choice of model is the one that includes the parameters needed for the problem at hand.

EXAMPLE 7.1 A cross section of a magnetic device is shown in Fig. 7.4(a). Coils A, B, and C are wound on three legs of the core, and coil D is wound around the outside. Develop the structure of an equivalent electric circuit for the device including leakage elements.

Solution A set of flux paths is shown by the dotted lines in Fig. 7.4(a). These provide for simple paths for leakage flux around each of the coils. It is assumed that the yoke sections of the core can be lumped into the vertical legs with negligible error.

The connection of reluctances of the magnetic equivalent circuit is shown in Fig. 7.4(b). Those representing core sections have been marked as nonlinear. Note that it is not obvious how the coil mmfs should be included in this model. While the mmfs of coils A, B, and C could be connected in series with $\mathcal{R}_3$, $\mathcal{R}_5$, and $\mathcal{R}_7$, respectively, the question remains as to where to insert the mmf of coil D. A practical approach is to proceed to produce the dual of the magnetic circuit without its mmf sources and later introduce the appropriate current sources into the electric circuit.

In Fig. 7.4(b), a node has been marked in each mesh with a reference node i outside. The dotted lines indicate the structure of the corresponding

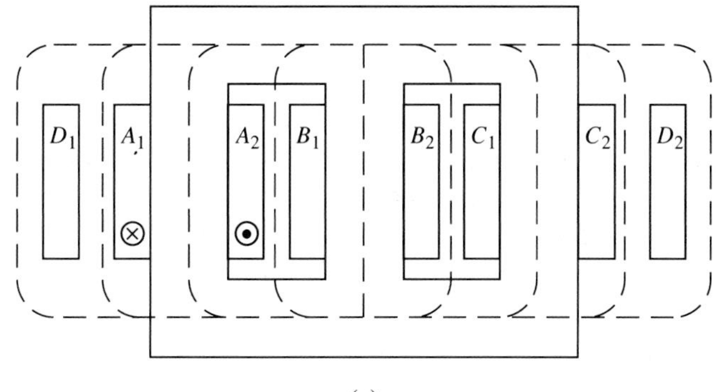

(a)

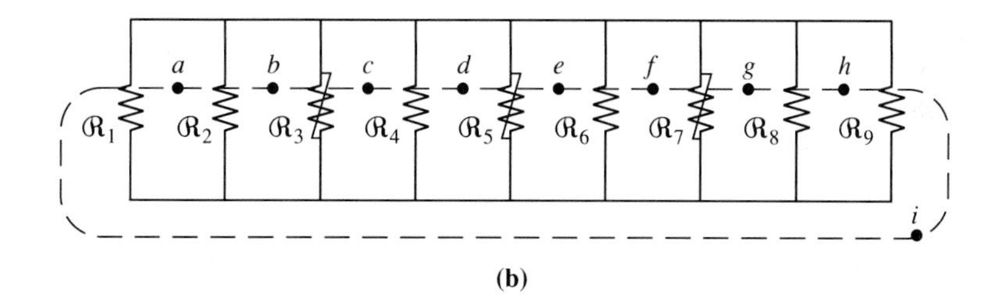

(b)

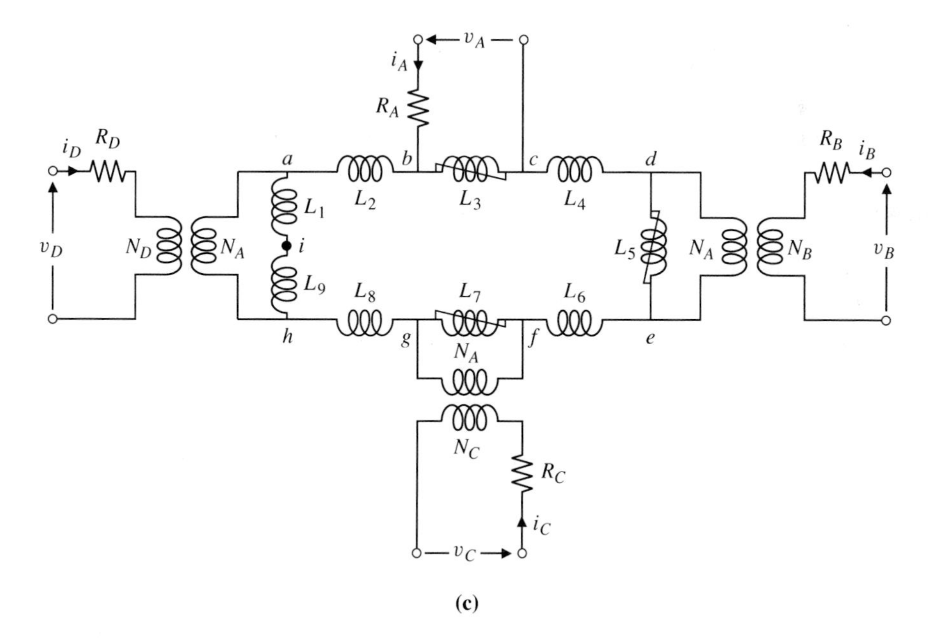

(c)

Figure 7.4 Development of an equivalent circuit for Example 7.1.

electric circuit, which is shown in Fig. 7.4(c). Note that the parallel arrangement of reluctances transforms into a series arrangement of inductances.

In the magnetic circuit, it is evident that coil side A has current entering at node b and leaving at node c. Therefore, its effect can be represented in the electric model by the current loop i_A shown connecting nodes b and c. For coil D, the corresponding branch in the electric circuit enters at node a and leaves at node h. The electric circuit model is completed with the addition of three ideal transformers and the four coil resistances.

7.1.2 ▪ Equivalent Circuit for a Two-Winding Shell-Type Transformer

Let us now apply the technique introduced in Section 7.1.1 to the development of an equivalent circuit model for one of the common types of single-phase transformers shown in cross section in Fig. 7.5(a). The first step is to choose the significant sections of the magnetic system to be represented by lumped reluctances. The pertinent assumptions are introduced at this stage. A particular choice is shown in the dotted flux paths of Fig. 7.5(a) and in the corresponding magnetic circuit of Fig. 7.5(b). The left and right outer-core sections including parts of the upper and lower yokes are represented by $\mathcal{R}_{1\ell}$ and $\mathcal{R}_{1r}$. The leakage paths between the A and B coils are represented by $\mathcal{R}_{L\ell}$ and $\mathcal{R}_{Lr}$. These should include the leakage paths between the front and back sections of the coils. The central leg is represented by reluctance $\mathcal{R}_2$, and this may incorporate the adjacent yoke sections for simplicity.

The basic form of the electric equivalent circuit now may be derived by designating the nodes a, b, c, d, and e in Fig. 7.5(b) and interconnecting them through the reluctance elements to give the structure of Fig. 7.5(c). Because of symmetry, this structure can be somewhat simplified as shown in Fig. 7.5(d). The series inductances $L_{1\ell}$ and L_{1r} have been combined to connect directly between nodes a and d. Also, the leakage inductance L_{Lr} has been deleted from the branch between nodes c and d and added to $L_{L\ell}$ to give a single leakage parameter L_L.

While the electric equivalent circuit produced by this approach is applicable for transient analysis, the circuit of Fig. 7.5(d) has been drawn in a form appropriate for steady-state analysis of the transformer with sinusoidal voltages at an operating frequency . The exciting current required by the central leg is that flowing in admittance Y_2 while the exciting current required for the outer legs is that flowing in admittance Y_1. In most operating situations, the voltage drop across the leakage impedance $j\omega L_L$ will be sufficiently small that Y_1 and Y_2 can be combined in parallel at whichever side of the leakage impedance is most convenient for analysis with negligible error. However, there are extreme operating conditions for which the model of Fig. 7.5(d) or even a more complex one including more core and leakage sections is required.

The parameters in the electric circuit of Fig. 7.5(d) have arbitrarily been referred to the turns N_A of the A winding. They can be transferred readily to the B-winding side using the square of the turns ratio.

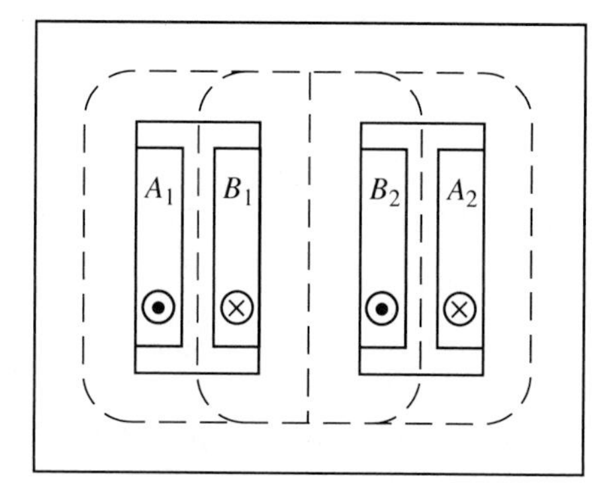

(a)

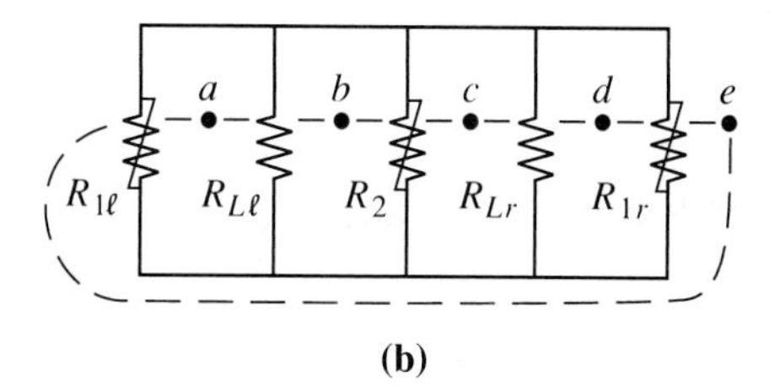

(b)

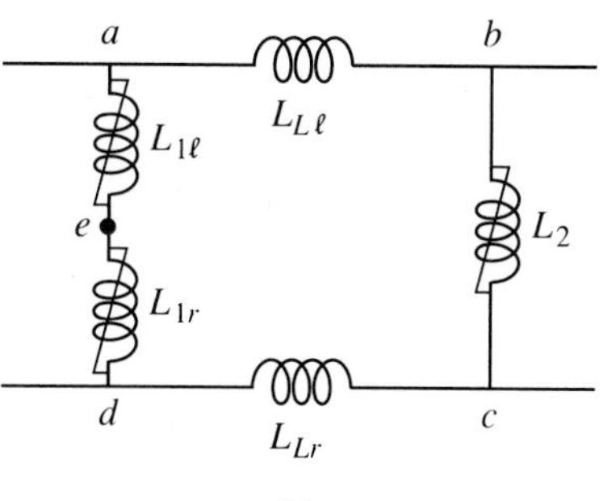

(c)

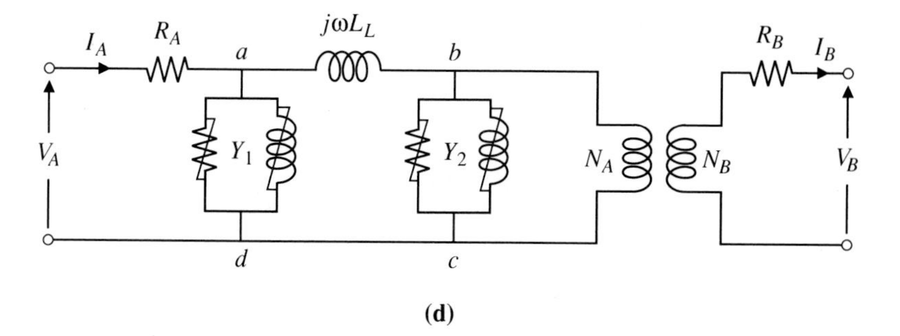

(d)

Figure 7.5 Development of an equivalent circuit for a two-winding shell-type transformer.

7.1.3 • Synthesis of Magnetic Devices

A technique for producing an electric equivalent circuit by developing the dual of the magnetic circuit has been introduced in Section 7.1.1. The converse process is equally feasible and is useful in devising the structure of a multi-coil magnetic device that will perform the same general functions as a given electric circuit. This process is best demonstrated by use of a typical example.

Suppose we wish to supply a nonlinear electrical load from a standard single-phase, 60-Hz supply. To limit the harmonic currents in the supply, an inductance-capacitance filter is to be inserted between the supply and the load. An additional requirement is that the load be coupled to the supply by a transformer, partly because its desired voltage is different from that of the supply and partly to provide conductive isolation. An electric circuit of the desired system is shown in Fig. 7.6(a).

Using the concept that the magnetic circuit is dual in form to the given electric circuit, the configuration of the required magnetic device can be derived. First, a node can be inserted into each independent mesh of the electric circuit, plus an external node. These nodes can now be joined by dotted lines as shown in Fig. 7.6(a) to give the general configuration of the magnetic circuit. From the earlier development of electric equivalent circuits, it will be appreciated that all elements in the electric circuit except inductances represent external elements that are connected to coils that in turn are wound around magnetic cores that have been assumed to be ideal, i.e., to have zero reluctance. Thus, the equivalent magnetic circuit has the form shown in Fig. 7.6(b). The one finite reluctance $\mathcal{R}$ is to be inversely proportional to the desired inductance L as in Eq. (7.4).

A physical structure for which Fig. 7.6(b) is a reasonable magnetic circuit model now can be devised. Such a structure is shown in cross section in Fig. 7.6(c). The finite reluctance $\mathcal{R}$ is produced by an air gap. The rest of the magnetic structure is of high-permeability material for which the reluctances may be considered negligible. Windings are provided to connect the source, the capacitor, and the load. The turns ratios can be chosen to provide the appropriate load voltage and also to provide a suitable voltage level for the most economical voltage rating of the capacitor. Normal methods of design may be employed to arrive at the core dimensions and actual numbers of turns. When a design has been produced, there may be reason to produce its electric equivalent circuit so that the effects of leakages around the coils may be assessed.

7.1.4 • Determination of Leakage Inductance from Dimensions

Up to this point, we have associated the leakage inductance mainly with the leakage flux path between the windings. In addition, there is leakage flux within each winding. In this section we will derive an expression for the leakage inductance in a simple case of two concentric windings arranged as shown in Fig. 7.5(a).

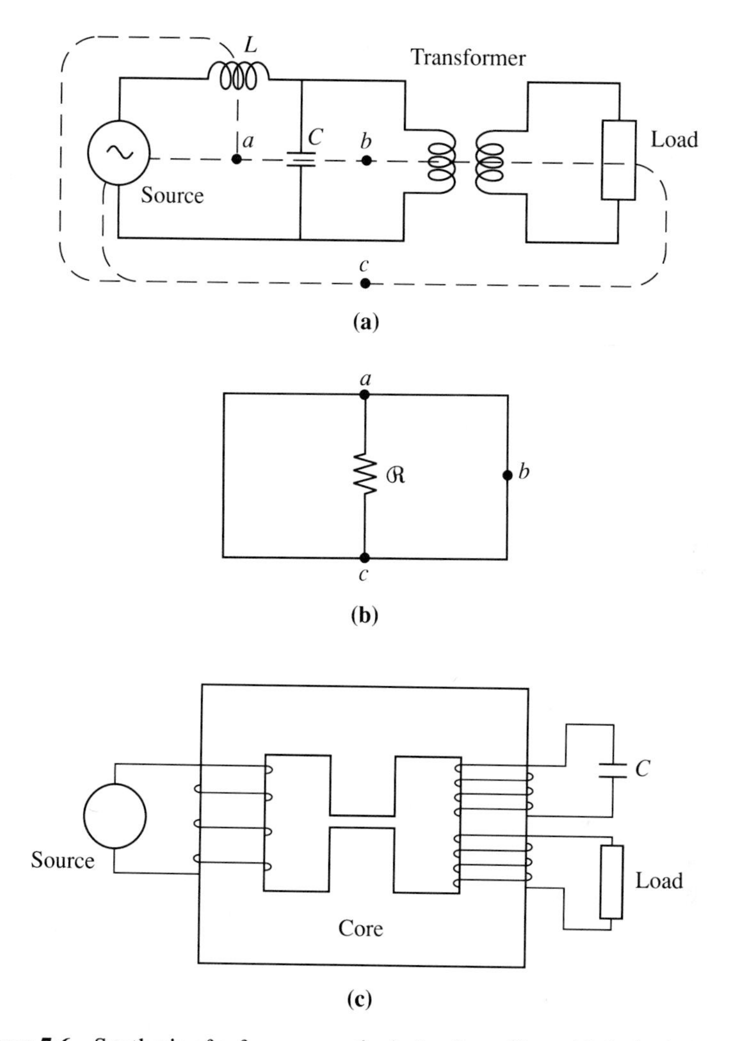

Figure 7.6 Synthesis of a ferromagnetic device for a filter: (a) desired electric circuit, (b) magnetic circuit, (c) device with system connections.

A section through the sides of a pair of such windings is shown in Fig. 7.7. For simplicity, suppose each winding has N turns and that the windings carry equal and opposite currents i. With the current direction shown, the magnetic field within and between the coils may be assumed to be approximately vertical in direction between the upper and lower core yokes, as shown in Fig. 7.7. If the energy stored in this magnetic field can be determined, the value of the leakage inductance can be found.

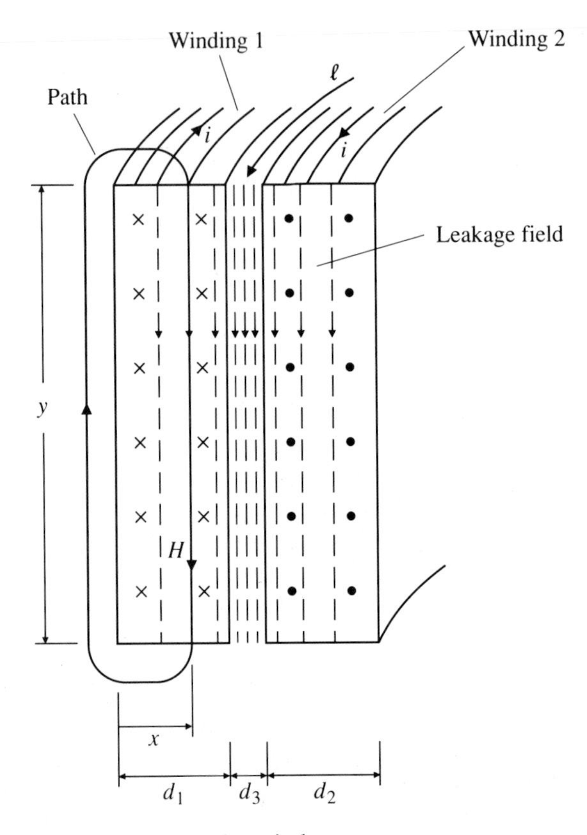

Figure 7.7 Calculation of leakage inductance.

Suppose we apply the circuital law of Eq. (1.1) to the path shown, recognizing that all but the right-hand side of this path is in a magnetic core where the magnetic field intensity is assumed to be negligible. Within winding 1,

$$Hy = Ni\,\frac{x}{d_1}$$

or

$$H = \frac{Nix}{yd_1} \qquad \text{A/m} \qquad \text{for } 0 < x < d_1 \tag{7.15}$$

Between the windings

$$H = \frac{Ni}{y} \qquad \text{A/m} \qquad \text{for } d_1 < x < (d_1 + d_3) \tag{7.16}$$

Within winding 2,

$$H = \frac{Ni}{yd_2}(d_1 + d_2 + d_3 - x) \qquad \text{A/m}$$
$$\text{for } (d_1 + d_2) < x < (d_1 + d_2 + d_3) \tag{7.17}$$

Let ℓ be the mean circumferential length of the winding turns. The energy stored in the volume V of the magnetic field then is

$$\begin{aligned} W &= \int \frac{1}{2}\mu_0 H^2\, dV \\ &= \int_0^{d_1+d_2+d_3} \frac{1}{2}\mu_0 H^2 y\ell\, dx \\ &= \frac{1}{2}\mu_0 \frac{N^2 i^2 \ell}{y}\left(\frac{d_1}{3} + d_2 + \frac{d_3}{3}\right) \qquad \text{J} \end{aligned} \tag{7.18}$$

Because $W = Li^2/2$, the leakage inductance for the pair of N-turn windings is

$$L = \mu_0 \frac{N^2\ell}{y}\left(\frac{d_1}{3} + d_2 + \frac{d_3}{3}\right) \qquad \text{H} \tag{7.19}$$

The expression of Eq. (7.19) indicates how the transformer designer can vary the leakage inductance by varying the arrangement and dimensions of the windings. To achieve a low value of leakage inductance, the height y should be large and the widths of the windings and the spacing between them should be small. A further reduction of leakage inductance may be achieved by interleaving layers of the two windings. In general, high-voltage transformers have relatively high values of leakage inductance because the insulation requirements prevent close proximity of the high and low voltage windings.

7.2

Design of Inductors and Transformers

The early phases of our study of magnetic devices have focused first on understanding the physical phenomena and then on developing analytical models that can be used to predict some important properties of performance. However, much of the essence of engineering is in the processes of design where we are required to devise a product that will meet the needs specified by a particular user. Designing for the commercial marketplace can be a highly detailed and specialized art involving criteria of economy, reliability, performance, and environmental impact. This section considers only a few of the major physical features of the design process as it applies to inductors and transformers.

7.2.1 ▪ Heat Transfer

Essentially all types of electrical apparatus considered in this book contain insulated conductors. The amount of current that can be carried in these conductors is limited by the ability to conduct away the heat produced by losses in the conductors without exceeding the maximum allowable operating temperature of the insulation on the conductors. For organic insulating materials such as cotton, paper, polyester film, and many varnishes (materials usually referred to as Class A insulation), this maximum temperature is about 105°C. For inorganic materials such as mica asbestos and glass fiber bonded with silicone varnishes (Class H), the maximum temperature is about 180°C. In between are Class B and Class F insulation types with limit temperatures of 130 and 155°C, respectively. If the safe operating temperature is exceeded, the insulation materials may oxidize, the chemical chains of polymers may break, and the insulation system may become brittle. An increase of only about 10°C in continuous operating temperature may reduce the useful life of insulation to half.

The heat produced in the conductors and the heat arising from core losses must be transferred to the cooling medium. This medium is usually air but oil and water may be used in some large devices. The ambient cooling air may have a temperature as high as 40°C for normal applications and may be higher depending on the local environment.

Most of the temperature drop between the heat sources in conductor and iron and the coolant is concentrated at the interface between the surfaces of the device and the cooling air. The amount of heat that can be removed from a surface of area A with a temperature difference $\Delta T = T_{surface} - T_{air}$ is denoted as the *heat transfer coefficient, h*. This coefficient depends on the velocity of the air over the surface and on the nature of the surface. Typical values for h are in the range 20 to 80 W/m²·°C for air cooling; for oil or water cooling, h may be of the order of 250 W/m²·°C.

To appreciate how the allowable current density in conductors is limited by cooling, consider the cross section of a winding of a transformer as shown in Fig. 7.8. Of the total cross-sectional area, only a fraction known as the *space factor*, k_s will be conductor area, the remainder being insulation space. Typical values of k_s are in the range 0.4 to 0.7. With a coil thickness d, height y, and circumference ℓ, the conductor volume is

$$V_c = dy\ell k_s \qquad \text{m}^3 \tag{7.20}$$

The power loss in the winding with a conductor current density J_c then is given by

$$P_L = \rho J_c^2\, dy\ell k_s \qquad \text{W} \tag{7.21}$$

where ρ is the conductor resistivity appropriate for the conductor temperature. For a typical coil, $y \gg d$, and the cooling surface A may be approximated by $A = 2y\ell$. Thus, the heat power produced per unit of cooling surface area is

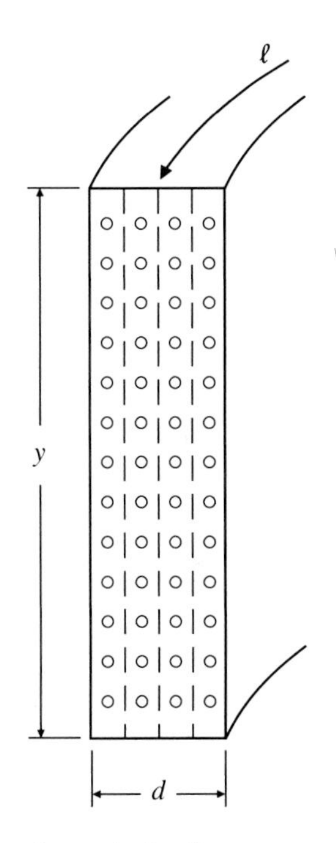

Figure 7.8 Cross section of a typical coil.

$$\frac{P_L}{A} = \frac{\rho}{2} J_c^2 \, dk_s \qquad \text{W/m}^2 \tag{7.22}$$

This must be equal to the cooling capability of air per unit of surface, which has been denoted as $h\Delta T$. Thus, the allowable current density in the conductor material can be expressed as

$$J_c = \left(\frac{2h\Delta T}{\rho \, dk_s} \right)^{1/2} \qquad \text{A/m}^2 \tag{7.23}$$

An example will give some appreciation of values typically encountered. Suppose Class B insulation is used with an operating temperature of 135°C. Let the air temperature be 40°C and allow 15°C for a temperature drop between the conductor and the outer insulation surface. Then, $\Delta T = 80$°C. At 135°C, the resistivity of copper is about 2.5×10^{-8} $\Omega \cdot$m. A typical heat transfer coefficient of 25 W/m$^2 \cdot$°C may be used. For a coil of 10 mm thickness with a space factor of 0.5,

$$J_c = \left(\frac{2 \times 25 \times 80}{2.5 \times 10^{-8} \times 0.01 \times 0.5} \right)^{1/2} = 5.7 \qquad \text{A/mm}^2 \tag{7.24}$$

Note that for a thicker coil of $d = 50$ mm, the allowable current density would be reduced to about 2.5 A/mm^2.

7.2.2 ▪ Inductor Design

The purpose of an inductor is to introduce inductance into an electric system. An example of the need for such a device is in a filter to control the flow of harmonic currents. Typical specifications for an inductor would include its rms operating current I, its inductance L, and its base frequency f or ω. For most applications, there will be a requirement for linearity, i.e., the inductance should be constant for all operating values of current. The following discussion will introduce some of the major considerations in arriving at an appropriate design.

In Section 1.1, Fig. 1.2, it was shown that a near-linear relation between flux linkage and coil current could be achieved using a magnetic core with an air gap, provided that the core was operated with limited flux density where its magnetic-field intensity could be ignored.

Our first consideration is the choice of a configuration or shape for the device. This will depend on the specific application and on what shapes of iron laminations are available. A frequently used configuration is shown in Fig. 7.9. The single coil is placed on the center leg. The cross section of the laminations in this center leg is normally made nearly square so that the maximum core area

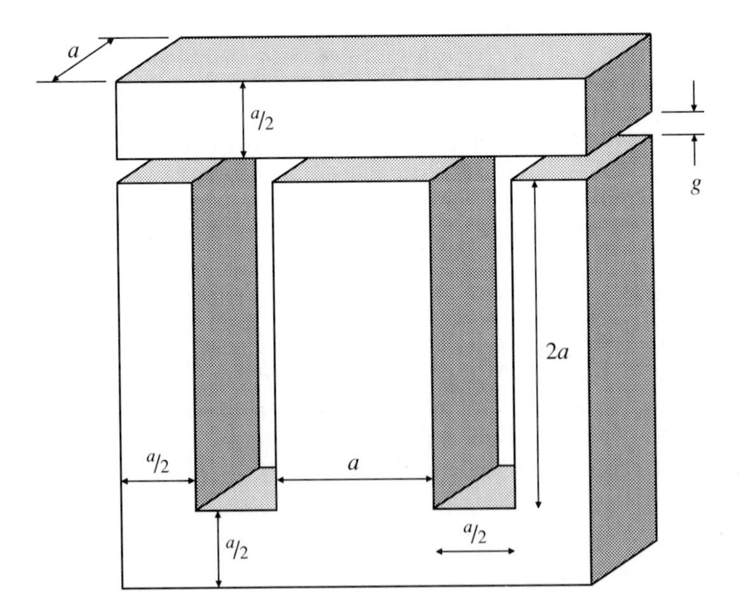

Figure 7.9 Inductor core.

can be enclosed by the minimum length of conductor. The core is proportioned to produce reasonably uniform flux density in each of its sections. The choice of the ratio of window height to width as shown in Fig. 7.9 is typical of available laminations.

Another consideration is the choice of the magnetic material. To preserve linearity in the device, a maximum flux density should be chosen, and the lamination thickness should be selected to limit the core losses. For example, a core operating at a frequency of 60 Hz might use the M-36 sheet-steel material for which the *B-H* characteristic is shown in Fig. 1.12. The maximum flux density B might be limited to about 1.4 T. Then, from Fig. 1.23, a lamination thickness of 0.356 mm might be chosen, resulting in a core loss of about 4 W/kg. The effects of these choices should be analyzed later in the design process.

Inductance is the ratio of flux linkage to current. The flux linkage in turn is the product of the flux and the number of turns. Therefore, the inductor could have a large core area with a large flux together with a small number of turns, or vice versa. The central design problem is to achieve a near-optimum choice of both the number of turns and the effective area of the core.

The cross-sectional area of the core cannot be completely made up of steel because of surface roughness and insulating coating on the laminations. The ratio of steel area to total core area is known as the *stacking factor*, k_i, a factor that is usually greater than 0.9. The peak flux linkage of the device then is given by the relation

$$\hat{\lambda} = \sqrt{2}LI = N\hat{B}k_i a^2 \qquad \text{Wb} \tag{7.25}$$

For the desired values of L and I, this provides one relation in the quantities N and a. A second relation can be developed from the constraint that the coil must not overheat. From the discussion in Section 7.2.1, appropriate values of conductor rms current density J_c and coil space factor k_s may be chosen. Then, for a window area of $2a \times a/2 = a^2$,

$$NI = k_s a^2 J_c \qquad \text{A} \tag{7.26}$$

The two relations, Eqs. (7.25) and (7.26), may now be solved for the turns N and the dimension a. At this point, it would be prudent to check back on the cooling capability for this coil thickness and adjust the current density if necessary.

The final step to be considered here is the choice of the air-gap length g. The maximum flux density B in the air gap as well as in the core (ignoring leakage flux) is found by applying the circuital law around a closed path through the coil:

$$\hat{B} = \mu_0 \frac{\sqrt{2}IN}{2g} \qquad \text{T} \tag{7.27}$$

The coil inductance then is given by

$$L = \frac{\hat{\lambda}}{\sqrt{2}I} = \frac{N\hat{B}k_i a^2}{\sqrt{2}I}$$
$$= \frac{N^2 \mu_0 a^2}{2g} \qquad \text{H} \tag{7.28}$$

from which the appropriate gap length g may be found.

In designing magnetic devices with air gaps, it is desirable to make some allowance for fringing of flux around the gap edges. A common approximation is to add the gap length g to each of the gap dimensions that make up its cross-sectional area.

The above concepts provide only a starting point for the design. Much further detail must be addressed to specify the conductor area, shape, and coating, the insulation, the structural and mounting arrangements, etc.

EXAMPLE 7.2 Provide a preliminary design for an inductor with an inductance of 2 mH capable of a continuous rms current of 80 A at a frequency of 60 Hz.

Solution Following the discussion in Section 7.2.2, consider use of M-36 steel with a thickness of 0.356 mm with a maximum flux density of 1.4 T. Use the lamination shape shown in Fig. 7.9 with a square central core with stacking factor of 0.95. From Eq. (7.25),

$$Na^2 = \frac{\sqrt{2}LI}{\hat{B}k_i} = \frac{\sqrt{2} \times 80 \times 0.002}{1.4 \times 0.95} = 0.17 \qquad \text{m}^2$$

Choosing an rms current density J_c of 2.5 A/mm^2 and a coil space factor k_s of 0.5, Eq. (7.26) gives

$$\frac{N}{a^2} = \frac{k_s J_c}{I} = \frac{0.5 \times 2.5 \times 10^6}{80} = 15\,600$$

Combining these two expressions gives

$$N = (0.17 \times 15\,600)^{1/2} = 52 \qquad \text{turns}$$

and

$$a = \left(\frac{0.17}{52}\right)^{1/2} = 57.4 \qquad \text{mm}$$

Reference back to the discussion of Section 7.2.1 suggests that, for a coil thickness of $a/2 = 29$ mm, a somewhat larger current density could have been chosen.

From Eq. (7.28), the appropriate gap length (ignoring fringing flux) is

$$g = \frac{N^2 \mu_0 a^2}{2L} = \frac{52^2 \times 4\pi \times 10^{-7} \times 0.0574^2}{2 \times 0.002} = 2.8 \qquad \text{mm}$$

If desired, this could now be adjusted to accommodate fringing.

Using the chosen value of current density, the required cross-sectional area of the conductor is 80/2.5 = 32 mm². A possible conductor arrangement could use rectangular conductors 8 × 4 mm arranged in 4 layers with 13 turns per layer.

7.2.3 ▪ Transformer Design

Many of the concepts developed for inductor design in Section 7.2.2 are applicable in modified form for transformer design. For a single-phase transformer, the basic requirement is for conversion of energy at voltage V_A and current I_A to voltage V_B and current I_B at an operating frequency ω. For a shell-type configuration such as that shown in Fig. 7.5(a), the basic relations required for determining dimensions and winding turns are

$$\frac{N_A}{N_B} = \frac{V_A}{V_B} = \frac{I_B}{I_A} \tag{7.29}$$

and

$$V_A = \omega N_A \frac{\hat{B}}{\sqrt{2}} k_i A_i \qquad \text{V} \tag{7.30}$$

where A_i is the cross-sectional area of the central core. The area A_w of the coil window is governed by the relation

$$N_A I_A + N_B I_B = 2 N_A I_A = k_s J_c A_w \qquad \text{A} \tag{7.31}$$

Thus, a core can be chosen that has the property

$$A_i A_w = \frac{2\sqrt{2}\, V_A I_A}{\omega \hat{B} k_i k_s J_c} \qquad \text{m}^4 \tag{7.32}$$

The criteria to use in a particular design depend very much on the application. For example, if energy conservation is a critical consideration, a high-quality core material would be chosen. The peak flux density and the lamination thickness would be chosen to provide low core loss. The conductor current density would be made low to provide low winding loss. Naturally, the initial cost of such a transformer will be relatively high. If the transformer is usually operated at rated load, it will be desirable to design it so that the core and winding losses are equal at that load, as shown in Eq. (2.51). If it is a distribution-type transformer in a residential area where its load is low for much of the time, particular attention must be paid to minimizing the core losses because these occur all the time. If the overriding criterion is low mass, as might be the case for airborne equipment, particular attention would be given to effective cooling means that would allow high values of core flux density and conductor current density. If low leakage inductance is required, consideration would be given to conductor arrangements that interleave the two windings.

EXAMPLE 7.3 A design is required for a transformer to be used in an electronic instrument. The voltage rating is 115 V : 24 V at 60 Hz and the transformer is to supply 50 W to a resistive load. A material having a core loss of 4.2 W/kg at a peak flux density of 1.5 T may be used. Estimate core dimensions and coil turns and find the approximate efficiency of the design.

Solution A manufacturer of core material lists a number of laminations having the general shape shown in Fig. 7.10. These E and I laminations are to be interleaved to provide the minimum effective corner gaps in the assembled core. The shape is an economical one because the I section comes out of the window of the E lamination.

Because this is a relatively small transformer, let us choose to use a current density of 4 A/mm^2 in the conductor. The winding space factor is estimated as 0.4 and the core stacking factor is about 0.9 for interleaved laminations according to the manufacturer. From Eq. (7.32),

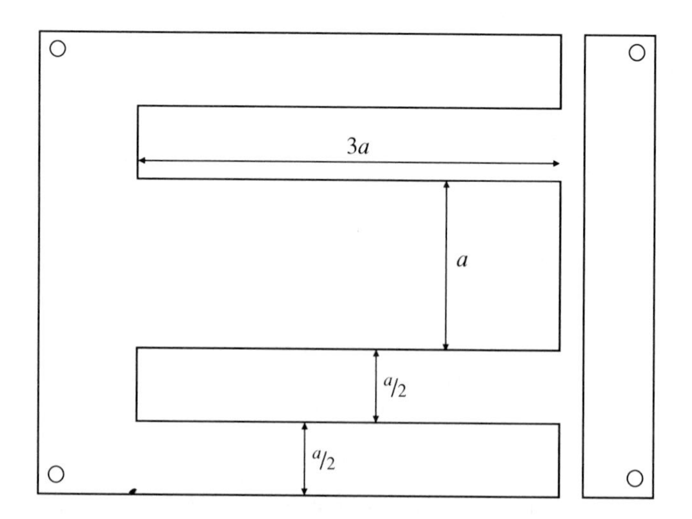

Figure 7.10 Laminations for Example 7.3.

$$A_i A_w = \frac{2\sqrt{2}\, V_A I_A}{\omega \hat{B} k_i k_s J_c}$$

$$= \frac{2\sqrt{2} \times 50}{2\pi \times 60 \times 1.5 \times 0.9 \times 0.4 \times 4 \times 10^6} = 0.174 \times 10^{-6} \qquad \text{m}^4$$

For the lamination shown in Fig. 7.10, $A_w = 1.5a^2$. With a square central core, $A_i = a^2$. Thus,

$$a = \left(\frac{0.174 \times 10^{-6}}{1.5}\right)^{1/4} = 18.4 \qquad \text{mm}$$

The nearest lamination listed by the manufacturer has $a = 19$ mm and $A_w = 1.5a^2 = 542$ mm^2. Thus, the required core area becomes

$$A_i = \frac{0.174 \times 10^{-6}}{542 \times 10^{-6}} = 322 \qquad \text{mm}^2$$

and the approximate stacking depth of the core is $322/19 = 17$ mm. The required number of primary turns is

$$N_p = \frac{\sqrt{2}\, V_p}{\hat{B} A_i k_i \omega}$$

$$= \frac{\sqrt{2} \times 115}{1.5 \times 322 \times 10^{-6} \times 0.9 \times 377} = 992$$

The number of secondary turns is

$$N_s = N_p \frac{V_s}{V_p} = 992 \times \frac{24}{115} = 207$$

The surface area of the pair of E and I laminations in Fig. 7.10 is $9a^2$. If the density of the core iron is 7600 kg/m^3 and the stacking factor is 0.9, the mass of the core is

$$m = 7600 \times 9 \times 0.019^2 \times 0.017 \times 0.9 = 0.38 \qquad \text{kg}$$

The estimated core loss is then

$$P_c = 0.38 \times 4.2 = 1.6 \qquad \text{W}$$

The mean length of a turn of the primary and secondary windings is approximately

$$\ell = 2(19 + 17) + 2\pi \times 9.5 = 131.7 \qquad \text{mm}$$

Using a resistivity of the copper windings of $2 \times 10^{-8}\ \Omega \cdot \text{m}$, the loss in the windings is approximately

$$
\begin{aligned}
P_w &= \rho J_c^2 V_{copper} \\
&= \rho J_c^2 (1.5a^2\ell) \\
&= 2 \times 10^{-8} \times (4 \times 10^6)^2 \times 1.5 \times 0.019^2 \times 0.1317 \times 0.4 \\
&= 9.13 \quad \text{W}
\end{aligned}
$$

The transformer efficiency is therefore

$$\eta = \frac{50}{50 + 1.6 + 9.13} = 0.823$$

In view of the voltage drop of about 16% which will be caused by the winding resistance, the process should now be repeated with an appropriate adjustment to the turns ratio to maintain the required output voltage.

It should be stressed that this is only a preliminary design. The efficiency may be unacceptably low for the application. The cooling may be inadequate. The available wire size may require a change in the core dimensions. The preliminary design now should be analyzed and the design process repeated as required.

7.3

Design of Permanent-Magnet Devices

Permanent-magnet materials are frequently used to produce a strong magnetic field in an air gap. Examples are found in audio speakers, moving coil instruments, and motors. Section 1.11 showed how the magnetic material could be modeled and how the air-gap flux density could be calculated for a given configuration. Let us here consider the converse problem of designing a device to produce a flux density B_g in an air gap of volume V_g using the minimum volume of permanent-magnet material V_m. This specification arises because of the relatively high cost per kilogram of the best permanent-magnet materials.

The basic principles can best be demonstrated by use of a typical configuration as shown in Fig. 7.11(a). The magnet material has been placed close to the air gap to minimize leakage. A soft iron section is used for the pole shoe to spread the flux over the required air-gap area, and other soft iron sections are used to close the magnetic path. The permanent-magnet material has a demagnetization characteristic of the form shown in Fig. 7.11(b) with a residual flux density B_r and a linear slope of permeability $\mu_r \mu_0$.

Let the cross-sectional areas of the air gap and the magnet material be A_g and A_m. From the circuital law applied around the magnetic path, and ignoring field intensity in the soft iron sections,

$$H_m \ell_m + H_g g = 0 \quad \text{A} \tag{7.33}$$

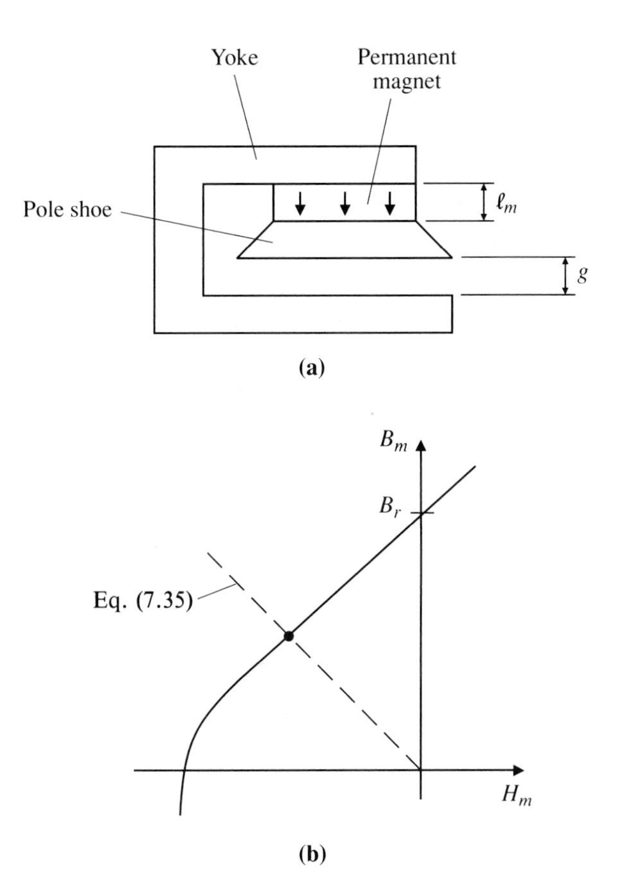

Figure 7.11 (a) Simple permanent-magnet device. (b) *B-H* characteristic of magnet material.

For flux continuity,

$$B_g A_g = B_m A_m \qquad \text{Wb} \tag{7.34}$$

Thus,

$$B_m = \frac{B_g A_g}{A_m} = -\mu_0 \frac{H_m \ell_m}{g} \frac{A_g}{A_m} \qquad \text{T} \tag{7.35}$$

This relation is shown as the dotted line in Fig. 7.11(b). The demagnetization curve provides a second relation between B_m and H_m. Thus, the intersection provides the operating point.

To design for minimum volume of magnet material, Eqs. (7.33) and (7.34) may be combined to give

$$\begin{aligned} V_m &= A_m \ell_m \\ &= \left(\frac{B_g A_g}{B_m}\right)\left(\frac{-H_g g}{H_m}\right) \\ &= \frac{B_g^2 V_g}{\mu_0 |B_m H_m|} \qquad \text{m}^3 \end{aligned} \tag{7.36}$$

From this, it is seen that the appropriate choice of operating point is the one that gives the maximum magnitude of the product $B_m H_m$ on the demagnetizing curve. Note that the energy stored in the magnetic field of the air gap is, from Eq. (1.24),

$$W_g = \frac{B_g^2}{2\mu_0} V_g \qquad \text{J} \tag{7.37}$$

This is seen, from Eq. (7.36), to be one-half of $B_m H_m V_m$. Thus, it is conventional for manufacturers to list the "maximum energy product," i.e., the maximum value of the product $(B_m H_m)$, for each permanent-magnet material.

Most of the modern good permanent-magnet materials can be characterized by the linear relation

$$B_m = B_r + \mu_r \mu_0 H_m \qquad \text{T} \tag{7.38}$$

over their useful range of operation. For these,

$$B_m H_m = B_r H_m + \mu_r \mu_0 H_m^2$$

The maximum value of the energy product is then found from

$$\frac{dB_m H_m}{dH_m} = B_r + 2\mu_r \mu_0 H_m = 0$$

from which

$$B_m = B_r + \mu_r \mu_0 \left(\frac{-B_r}{2\mu_r \mu_0}\right) = \frac{B_r}{2} \qquad \text{T} \tag{7.39}$$

Operation at this maximum energy product point is appropriate for permanent-magnet devices that contain no significant demagnetizing effects such as an mmf from an encircling current-carrying coil. However, where such applied mmf fields occur, the operating point is frequently determined by the requirement that operation must still be on the linear part of the curve for the maximum value of applied mmf to prevent permanent demagnetization of the magnet material.

A practical feature of a permanent-magnet device such as that shown in Fig. 7.11 is that there will be a substantial force tending to close the air gap. The magnitude of this force can be estimated using Eq. (3.7). The mechanical design must be such as to accommodate this force.

EXAMPLE 7.4 A facility is required in which to study the effects of intense magnetic fields on mice. A reasonably uniform flux density of 0.4 T is required over a circular chamber of 0.5 m diameter and 60 mm height. Produce a preliminary design using neodymium-iron-boron permanent-magnet material.

Solution From Fig. 1.8, Nd-Fe-B material has a residual flux density of about 1.2 T. From Eq. (7.39), the material should be operated at a flux density $B_m = B_r/2 = 0.6$ T if the minimum amount of permanent-magnet material is to be used. Suppose we consider the structure shown in Fig. 7.12. Because the air-gap flux density is to be lower than that of the magnets, outward flaring pole shoes are appropriate. The magnet material has been divided into two equal sections, both placed close to the air gap to minimize flux leakage.

Ignoring fringing flux around the gap, the air-gap area is

$$A_g = \frac{\pi}{4} \times 0.5^2 = 0.196 \qquad \text{m}^2$$

The air-gap flux is

$$\Phi = A_g B_g = 0.196 \times 0.4 = 0.079 \qquad \text{Wb}$$

Therefore, the required cross-sectional area of the magnets is

$$A_m = \frac{\Phi}{B_m} = \frac{0.079}{0.6} = 0.131 \qquad \text{m}^2$$

i.e., the diameter d_m of the circular magnet pieces should be 0.408 m.

The demagnetizing characteristic of Nd-Fe-B in Fig. 1.8 has a relative permeability μ_r of approximately 1.05. Thus,

$$H_m = \frac{B_m - B_r}{\mu_r \mu_0} \qquad \text{A/m}$$

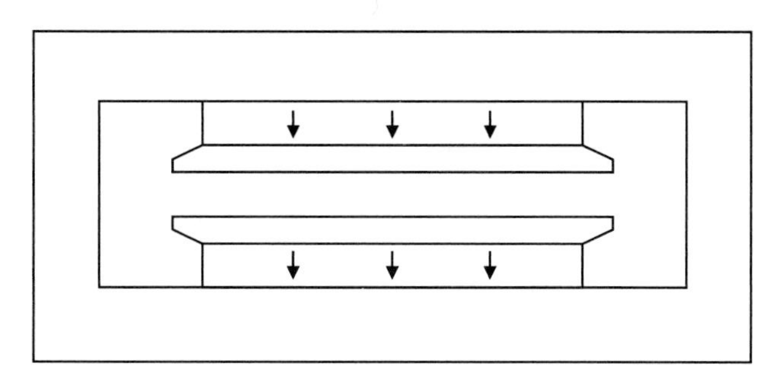

Figure 7.12 Cross section of magnet for Example 7.4 (approximately to scale).

From Eq. (7.33), ignoring field intensity in the soft iron sections, the total thickness of magnet material required is

$$\ell_m = -\frac{H_g g}{H_m} = \frac{B_g g \mu_R}{(B_r - B_m)} = \frac{0.4 \times 0.06 \times 1.05}{0.6} = 42 \quad \text{mm}$$

Thus, there will be two cylindrical magnets, each 21 mm thick.

The remainder of the magnetic system may be constructed using soft iron of reasonably high permeability. Because the flux will be constant, lamination is not necessary. Each of the two rectangular yoke sections must carry a flux $\Phi/2$. It is convenient to make the depth of the yoke equal to the magnet diameter. Suppose the flux density B_i in the iron is to be about 1.2 T. The required thickness of the yoke sections is then

$$\frac{\Phi}{2B_i d_m} = \frac{0.079}{2 \times 1.2 \times 0.408} = 81 \quad \text{mm}$$

The thickness of the pole shoes should be chosen such that the flux density does not exceed B_i as the flux spreads out from the magnet to the outer portion of the shoe. Thus, the thickness of the shoe should be

$$\frac{B_g(A_g - A_m)}{\pi d_m B_i} = \frac{0.4(0.196 - 0.131)}{\pi \times 0.408 \times 1.2} = 17 \quad \text{mm}$$

This preliminary design has been considerably simplified to demonstrate the main principles. A more detailed analysis of the flux density near the edges of the pole shoe would probably be required to meet a specification on the required uniformity of flux density over the experimental volume.

7.4

Variable Frequency Operation of Transformers

In electronic and communication systems, transformers are often operated over a wide range of frequencies. An example is the transformer used in an audio amplifier to couple the output stage to the speaker. Such a transformer should operate well over the audible range of about 20 to 20 000 Hz.

The various parameters of an equivalent circuit for a two-winding transformer have been introduced in Sections 2.2 and 7.1. For a variable-frequency transformer, the performance in its low-frequency range is dominated by its magnetizing inductance. In the mid-frequency range, winding resistance may be significant, particularly in small transformers. In the upper-frequency range, the leakage inductance limits performance. In addition, the capacitance between and across windings may also have a significant effect at high frequencies.

A set of equivalent circuit models for analysis is shown in Fig. 7.13. A variable-frequency source (for example, an electronic amplifier) is modeled by a voltage E_s in series with a resistance R_s. The load resistance is R_L. For the mid-frequency range, the transformer usually can be considered as ideal with turns ratio $a = N_1/N_2$, except for its winding resistances R_1 and R_2 as shown in Fig. 7.13(a). The ratio of the load voltage to the source voltage is

$$\frac{\vec{V}_L}{\vec{E}_s} = \frac{1}{a}\left(\frac{R_L}{R_L + R_2 + (R_s + R_1)/a^2}\right) = \frac{1}{a'} \tag{7.40}$$

To obtain maximum power delivery to the load for a given source voltage, the transformer parameters should be chosen so that the load resistance R_L is equal to the equivalent impedance as seen from the load terminals, i.e.,

$$R_L = R_2 + (R_s + R_1)/a^2 \quad \Omega \tag{7.41}$$

For this condition,

$$\frac{\vec{V}_L}{\vec{E}_s} = \frac{1}{2a} \tag{7.42}$$

As the frequency decreases, the magnetizing current becomes significant and the circuit of Fig. 7.13(b) applies. At low frequencies, core loss normally can be neglected. If the magnetizing inductance L_m can be regarded as con-

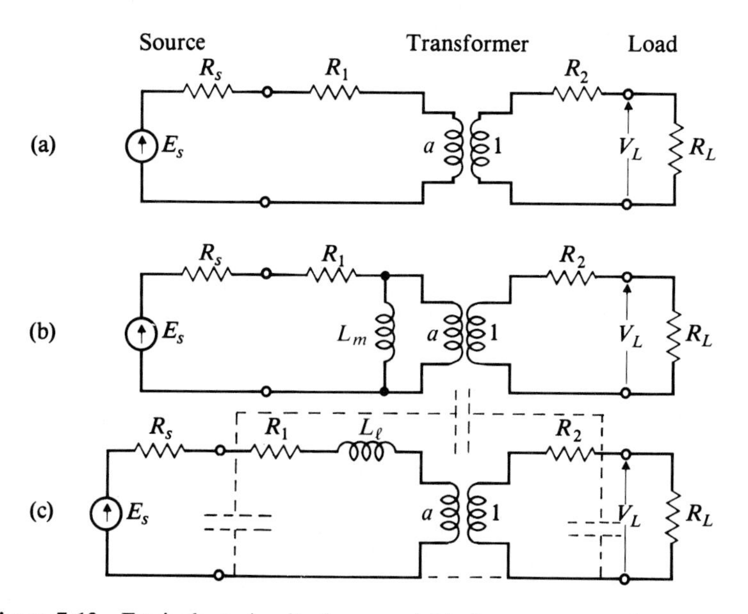

Figure 7.13 Equivalent circuits for a variable-frequency transformer. (a) Mid-frequency range, (b) low-frequency range, and (c) high-frequency range.

stant, the ratio of load voltage to source voltage can be expressed approximately as

$$\frac{\vec{V}_L}{\vec{E}_s} = \frac{1}{a'}\left(\frac{1}{1 - jR_\ell/\omega L_m}\right) \tag{7.43}$$

where R_ℓ is the circuit resistance as seen from the inductance L_m, i.e., $R_s + R_1$ in parallel with $a^2(R_2 + R_L)$. At the frequency $\omega_\ell = R_\ell/L_m$, the magnitude ratio $\bar{V}_L/\bar{E}_s$ is reduced to $1/\sqrt{2}$ of its value in the mid-frequency range and the load power is reduced to one half. With a constant value of source voltage, a frequency range will be reached where the core of the transformer is driven into saturation. The magnetizing inductance then cannot be assumed to be constant. In addition, the magnetizing current will contain significant harmonic components leading to distortion in the output waveform.

For the high-frequency range, the leakage inductance becomes significant, and the model of Fig. 7.13(c) is applicable. The load-to-source voltage ratio can be expressed as

$$\frac{\vec{V}_L}{\vec{E}_s} = \frac{1}{a'}\left(\frac{1}{1 + j\omega L_L/R_h}\right) \tag{7.44}$$

where R_h is the circuit resistance as seen from inductance L_L, i.e., $R_h = R_s + R_1 + a^2(R_2 + R_L)$. At the frequency $\omega_h = R_h/L_L$ rad/s, the magnitude ratio $\bar{V}_L/\bar{E}_s$ is again $1/\sqrt{2}$ of its mid-frequency value, and the load power is reduced to one-half.

The voltage magnitude ratio $\bar{V}_L/\bar{E}_s$ relative to its mid-frequency value is shown as a function of frequency on a logarithmic scale in Fig. 7.14. The bandwidth of the transformer is normally defined as the frequency range ω_ℓ to ω_h between the half-power points. In the high-frequency range, the effects of distributed capacitances may become significant. Both the turn-to-turn capaci-

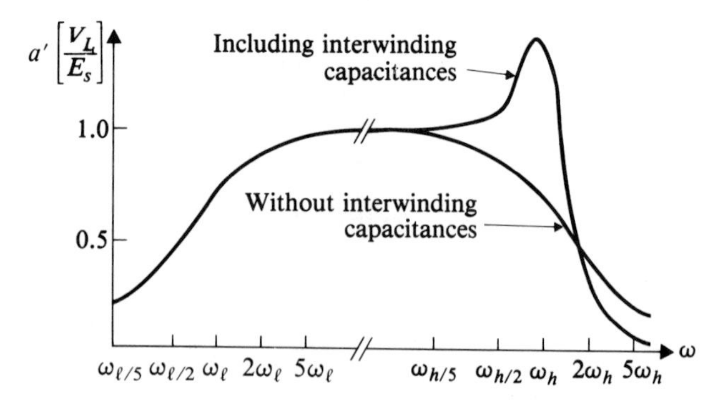

Figure 7.14 Normalized load-to-source voltage ratio of a transformer as a function of frequency ω.

tances in each of the windings and the capacitance between the two windings should be considered, as suggested by the dotted lines in Fig. 7.13. The values of these capacitances depend on the winding arrangements. For example, if the primary and secondary windings are interleaved to reduce the leakage inductance, the interwinding capacitance will be increased. However, if a conducting shield is placed between the primary and secondary windings and connected to a common terminal of the two windings, the interwinding capacitance can be effectively eliminated at the expense of increasing the capacitances across each of the windings. The interaction of these capacitances with the leakage inductance may cause a resonant condition such as that shown in Fig. 7.14. The effects of the winding capacitance are most pronounced in a step-up transformer where $a \ll 1$.

EXAMPLE 7.5 An audio transformer couples a variable-voltage, variable-frequency source to a speaker that has a resistive impedance of 8 Ω over its usable frequency range. The transformer can be modeled by the equivalent circuits of Fig. 7.13 with the following parameters:

$$R_1 = 4 \quad \Omega \qquad R_2 = 0.4 \quad \Omega$$

$$L_L = 0.95 \quad \text{mH} \qquad L_m = 35 \quad \text{mH}$$

$$a = 3.5$$

(a) Determine the source voltage required to supply a power of 60 W to the speaker in a mid-frequency range.

(b) If this source voltage is maintained constant, determine the frequencies at which the speaker power is reduced to 30 W.

Solution

(a) In the mid-frequency range, the load power is given by

$$P_L = V_L^2/R_L$$

Thus,

$$V_L = (60 \times 8)^{1/2} = 21.9 \quad \text{V}$$

From Eq. (7.40), the source voltage is

$$E_s = aV_L\left(\frac{R_L + R_2 + R_1/a^2}{R_L}\right)$$

$$= 3.5 \times 21.9\left(\frac{8 \times 0.4 \times 4/3.5^2}{8}\right) = 83.6 \quad \text{V}$$

(b) The low half-power frequency, from Eq. (7.43), is

$$\omega_\ell = R_\ell/L_m$$

where

$$\frac{1}{R_L} = \frac{1}{R_1} + \frac{1}{a^2(R_2 + R_L)} = \frac{1}{4} + \frac{1}{3.5^2 \times 8.4} = 0.26$$

giving $R_\ell = 3.85\ \Omega$. Thus, $\omega_\ell = 3.85/0.035 = 110$ rad/s, i.e., 17.5 Hz. The high half-power frequency, from Eq. (7.44), is $\omega_h = R_h/L_L$ where

$$R_h = R_1 + a^2(R_2 + R_L) = 4 + 3.5^2(8.4) = 106.9 \quad \Omega$$

Thus,

$$\omega_h = \frac{106.9}{0.95 \times 10^{-3}} = 112\,500 \quad \text{rad/s} \quad \text{or } 17.91 \quad \text{kHz}$$

7.5

Transient Inrush Current of a Transformer

Normally, transformers are so designed that the exciting current under steady-state operating conditions is a negligible fraction of the rated current. However, in the transient interval between the instant of switching the transformer onto a rated voltage supply and the eventual establishment of a steady-state condition, it is possible for the instantaneous value of the magnetizing current to be very high. In some instances, it may be even higher than the amplitude of the current that would flow if the transformer secondary were short circuited. A knowledge of the possible instantaneous magnitude of this transient current is of importance in determining the maximum mechanical stresses on the winding conductors. It is also important in designing the protection system for the transformer.

Consider a transformer with no secondary load and with a primary winding that may be switched on to a sinusoidal voltage supply. For simplicity, let the transformer be modeled by its winding resistance R and a nonlinear magnetizing branch L_m as shown in Fig. 7.15(a). Further, let us choose a simple approximation to the magnetizing characteristic as shown in Fig. 7.15(b) where negligible magnetizing current flows for a core flux linkage magnitude less than λ_k and where the relation is approximately linear with slope L_s for values of flux linkage magnitude in excess of λ_k.

When the transformer was last switched off, its flux linkage may have been left at some positive or negative value λ_r arising from hysteresis in the core material. Suppose, at time $t = 0$, a supply voltage is applied to the winding of

$$v = \hat{v} \sin \omega t \qquad \text{V} \tag{7.45}$$

Ignoring for the present the small effect of the winding resistance R, the flux linkage of the winding will be

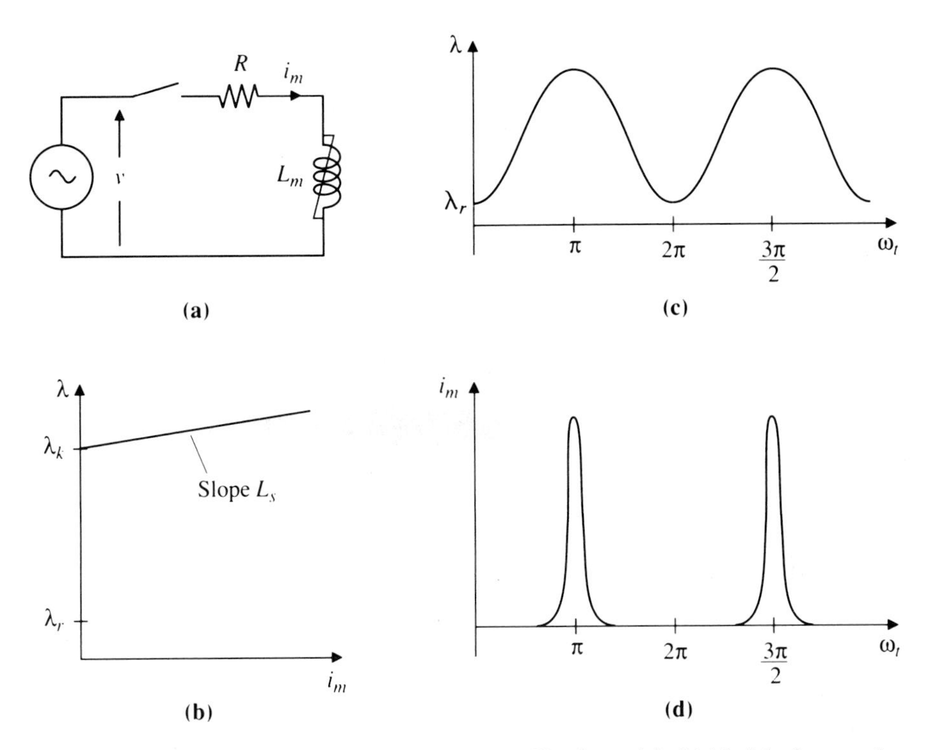

Figure 7.15 Transformer inrush current. (a) Circuit model. (b) Model of magnetizing branch. (c) Flux linkage waveform. (d) Magnetizing current waveform.

$$\begin{aligned} \lambda &= \int \hat{v} \sin \omega t \, dt \\ &= \lambda_r + \frac{\hat{v}}{\omega}(1 - \cos \omega t) \qquad \text{Wb} \end{aligned} \tag{7.46}$$

The peak value of this flux linkage will occur at $\omega t = \pi$ rad and will have a value

$$\hat{\lambda} = \lambda_r + 2\hat{v}/\omega \qquad \text{Wb} \tag{7.47}$$

The flux linkage over the first two cycles has the form shown in Fig. 7.15(c). From the magnetizing characteristic of Fig. 7.15(b), the magnetizing current will have the form shown in Fig. 7.15(d). For this simple approximate model, the peak current is given by

$$\hat{i}_m = \frac{\hat{\lambda} - \lambda_k}{L_s} \qquad \text{A} \tag{7.48}$$

It will be noted that the worst condition occurs when the sinusoidal voltage supply is switched on when the voltage is equal to zero and increasing as in Eq. (7.45), and when the residual flux linkage is at its maximum possible positive value. Cores with interleaved laminations usually have enough effective air gap to restrict the residual flux to a low value, but cores wound with a continuous strip may have a large residual flux density. Usually there is no control over the angle at which the switch may be effectively closed, applying voltage to the core.

The effect of the winding resistance R is normally negligible during the time period up to the initial peak current. However, the voltage drop across the resistance causes a reduction in successive flux-linkage peaks and a gradual reduction in the magnitude of the pulses of magnetizing current. Within a short time, which may be a few seconds in a large transformer, the magnetizing current will decrease to its small symmetrical steady-state value.

7.6

Transformers for Polyphase Systems

Most electrical energy is generated and transmitted using a three-phase system. A three-conductor transmission line can transmit essentially the same power as three single-phase lines which require a total of six conductors because, with a balanced load, the neutral or return current in the three-phase system is ideally zero. The three-phase power may be transformed either by use of suitably connected banks of single-phase transformers or by the use of special polyphase transformers. Some of the features of transformers in polyphase systems are discussed in the following sections.

7.6.1 • Connections of Transformers for Three-Phase Systems

Three similar single-phase transformers may be connected to give a three-phase transformation. Because the primary and secondary windings may be connected either in wye or in delta, there are four possible combinations of connections. The wye-wye connection allows the neutral of both primary and secondary to be grounded. It is also easy to analyze. However, it is rarely used in practice because of harmonic magnetizing currents in the ground circuits. A delta-delta connection has the advantage that, if one transformer breaks down, it may be removed, and reduced output of about $1/\sqrt{3}$ may be supplied by the remaining two units.

Many three-phase transformer banks use wye-delta connections. The wye connection permits grounding of the neutral point and is desirable for the high-voltage side because it limits the stress on the line-to-ground insulation. This connection is also used on the low-voltage side of many distribution transformers so that single-phase supplies with one side grounded can be made available. As will be seen in the following analysis, a delta connection on one side of a three-phase transformer is desirable to provide a path for the flow of harmonic magnetizing current components having frequencies that are multiples of three times the supply frequency, i.e., the so-called *triplen harmonics*.

Consider the delta-wye connection of three single-phase transformers shown in Fig. 7.16(a). The primary windings, designated as a, b, and c, are connected in delta to a three-phase supply. The secondary windings, designated as A, B, and C, are connected to a three-phase load. The turns ratio of each single-phase transformer is $a = N_1/N_2$. The voltage applied to each of the delta-connected windings is equal to a line-to-line voltage of the supply. If the supply voltages are sinusoidal, constant in magnitude and displaced uniformly in phase, and if the transformers are essentially ideal, the load voltages may be assumed to be sinusoidal, constant, and balanced, regardless of any unbalance in three-phase load.

If all components of the three-phase system are balanced, it is convenient to analyze the system by developing a single-phase equivalent circuit that is representative of each of the three phases, the only difference among them being the phase angle. Consider first the relation between the line-to-neutral voltages on the primary and secondary sides of the transformer bank. The line-to-neutral voltages V_{an}, V_{bn}, and V_{cn} of the supply with sequence a-b-c may be combined to give the line-to-line voltages V_{ab}, V_{bc}, and V_{ca}, as shown in the phasor diagram of Fig. 7.16(b). One of the line-to-neutral voltages on the wye-connected secondary side is then equal to V_{ab} divided by the turns ratio a, assuming ideal transformers. To achieve symmetry in notation, let us denote this secondary phase as C, i.e., the unique phase, a and b having been used as subscripts of the primary voltages. Thus,

$$V_{ab} = aV_{CN}; \quad V_{bc} = aV_{AN}; \quad V_{ca} = aV_{BN} \qquad \text{V} \tag{7.49}$$

Examination of the resultant phasors in Fig. 7.16(b) shows that the line-to-neutral voltage of the reference phase a of the primary is related to the line-to-neutral voltage of the reference phase A of the secondary by

$$\vec{V}_{an} = j\frac{a}{\sqrt{3}}\vec{V}_{AN} \quad (a\text{-}b\text{-}c) \qquad \text{V} \tag{7.50}$$

Thus, with a phase sequence a-b-c, the primary voltage leads the corresponding secondary voltage, both measured to neutral, by 90°. With another scheme of designation, this phase shift could have been 90 − 120° = −30° or 90 + 120° = 210°. Many power system publications use the −30° value. The 90° shift will be used in the following analysis because of the symmetry of the relations that it produces. If the phase sequence had been a-c-b, it can be readily shown that the voltage relationship would have been

$$\vec{V}_{an} = -j\frac{a}{\sqrt{3}}\vec{V}_{AN} \quad (a\text{-}c\text{-}b) \qquad \text{V} \tag{7.51}$$

The relations between the primary and secondary currents are shown in the phasor diagram of Fig. 7.16(c). With a balanced set of load currents $\vec{I}_A$, $\vec{I}_B$,

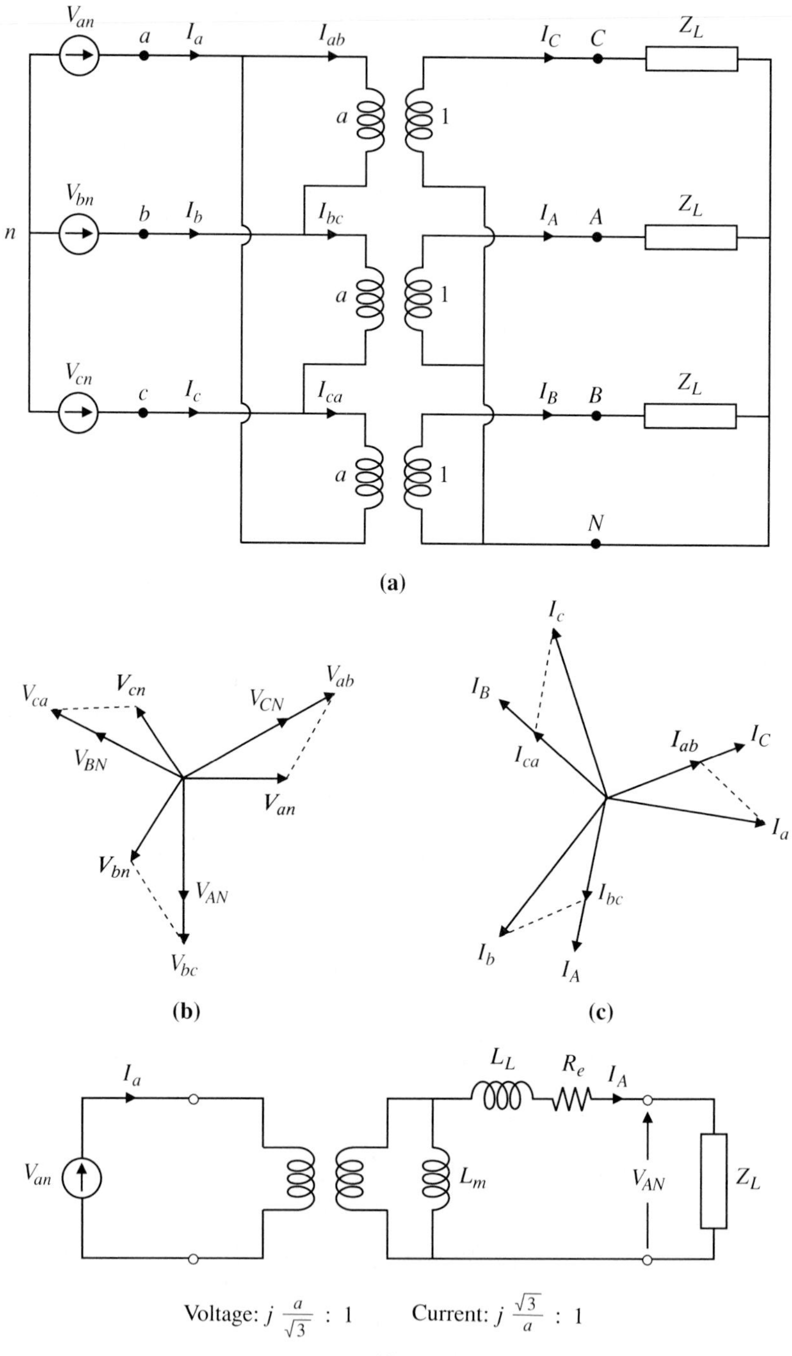

Figure 7.16 (a) Δ-Y connection of transformers. (b) Voltage phasor diagram. (c) Current phasor diagram. (d) Single-phase equivalent circuit.

and $\vec{I}_C$, and with sequence a-b-c, the currents in the primary windings of the transformers are

$$\vec{I}_{ab} = \frac{\vec{I}_C}{a}; \quad \vec{I}_{bc} = \frac{\vec{I}_A}{a}; \quad \vec{I}_{ca} = \frac{\vec{I}_B}{a} \qquad \text{A} \tag{7.52}$$

These currents now may be combined to give the supply currents. As seen in Fig. 7.16(c), the currents in the reference phases of primary and secondary are related by

$$\vec{I}_a = j\frac{\sqrt{3}}{a}\vec{I}_A \quad (a\text{-}b\text{-}c) \qquad \text{A} \tag{7.53}$$

Taken together, Eqs. (7.50) and (7.53) constitute a power invariant transformation, i.e., the complex power per phase is the same on each side (the asterisk denotes the conjugate of a phasor):

$$\begin{aligned} \vec{S}_{primary} &= \vec{I}_a\vec{V}^*_{an} \\ &= \left(j\frac{\sqrt{3}}{a}\vec{I}_A\right)\left(-j\frac{a}{\sqrt{3}}\vec{V}^*_{AN}\right) \\ &= \vec{I}_A\vec{V}^*_{AN} = \vec{S}_{secondary} \end{aligned} \tag{7.54}$$

An equivalent circuit for the delta-wye system is shown in Fig. 7.16(d). It includes the magnetizing inductance, the leakage inductance, and the equivalent winding resistances all referred to the secondary side. The magnitude of the voltage ratio of the ideal transformer in this circuit is $a/\sqrt{3}$ and the phase shift is 90°. In most analyses, this phase shift between the two three-phase systems connected by this transformer is not significant. However, where transformer banks are to be connected in parallel, similar connections are usually required in all banks.

If the analysis is focused on the supply side of the transformer bank of Fig. 7.16(a), the transformer parameters may be transferred from the secondary to the primary side. Note that, for any impedance Z_A on the secondary side, the equivalent impedance on the primary side is given by

$$\begin{aligned} Z_A &= \frac{\vec{V}_{AN}}{\vec{I}_A} = \frac{\left(j\dfrac{a}{\sqrt{3}}\vec{V}_{an}\right)}{\left(j\dfrac{\sqrt{3}}{a}\vec{I}_a\right)} = \frac{a^2}{3}\frac{\vec{V}_{an}}{\vec{I}_a} \\ &= \frac{a^2}{3}Z_a \qquad \Omega \end{aligned} \tag{7.55}$$

If each transformer in Fig. 7.16(a) has a sinusoidal voltage, its magnetizing current will contain a sequence of odd-order harmonics. The triple-frequency harmonics in the three phases will be equal in magnitude and displaced in phase by $3 \times 120° = 360°$, i.e., they will be in phase with each other. When the primary currents of two of the connected transformers in Fig. 7.16(a) are subtracted, it is seen that the resultant line current contains no third harmonic magnetizing current component. The same will be true for all harmonic orders that are multiples of three, i.e., the triplen harmonics. The triplen harmonic currents may be considered to circulate around the delta connection. Harmonics of 5th, 7th, 11th, etc. orders will flow in the supply lines, but each of these constitutes a balanced set of currents of sequence *a-c-b* or *a-b-c*.

In summary, the delta connection has the advantages of maintaining the balance and waveform of the voltages in spite of load unbalances and magnetizing current harmonics and of providing a path in which the triplen harmonics may flow without the need for a neutral return connection.

EXAMPLE 7.6 Three 100-kVA, 7600:460-V, 60-Hz transformers are connected in wye on the high-voltage side and delta on the low-voltage side to supply a heating load of 20 kW/phase and a three-phase, 460-V induction motor that requires a total of 210 kVA at a power factor of 0.8. The equivalent series impedance of each transformer referred to the low-voltage side is $0.05 + j0.095\ \Omega$.

(a) Using the line-to-neutral load voltage of phase a as a reference, determine the current phasors in the line connecting the transformer to the load and in the line connecting the supply to the transformer.

(b) Determine the line-to-neutral phasor voltage required at the supply to provide a load voltage of 460 V (L-L). Also determine the magnitude of the supply line-to-line voltage.

Solution

(a) Let us use the notation shown in Fig. 7.15(d) for the transformer. The line-to-neutral voltage on the load side can be expressed as

$$\vec{V}_{an} = \frac{460}{\sqrt{3}}\angle 0 = 266\angle 0 \qquad \text{V}$$

The phase angle of the induction motor load is

$$\theta_m = \cos^{-1} 0.8 = 36.9^\circ$$

The complex load power per phase is

$$S_a = 20\,000 + \frac{210\,000}{3}(0.8 + j\sin 36.9^\circ)$$

$$= 86\,800\angle -28.9^\circ \qquad \text{VA}$$

The load current is then given by

$$\vec{I}_a = \frac{S_a}{\vec{V}^*_{an}} = \frac{86\,800\angle -28.9^\circ}{266\angle 0} = 326\angle -28.9^\circ \qquad \text{A}$$

The turns ratio of one transformer is

$$a = \frac{460}{7600} = 0.0605$$

From the equivalent single-phase circuit of Fig. 7.15(d), the phasor current on the supply side can be represented as

$$\vec{I}_A = \frac{\vec{I}_a}{j\sqrt{3}/a} = -j\,\frac{0.0605}{\sqrt{3}}\,(326\angle -28.9^\circ) = 11.4\angle -118.9^\circ \qquad \text{A}$$

(b) It is convenient to transfer the series impedance of each transformer to the wye-connected high-voltage side:

$$Z = \frac{0.05 + j0.095}{0.0605^2} = 29.3\angle 62.2^\circ \qquad \Omega$$

The line-to-neutral phasor voltage at the supply can now be derived from the equivalent circuit of Fig. 7.15(d) as

$$\begin{aligned}\vec{V}_{AN} &= \frac{\vec{V}_{an}}{ja/\sqrt{3}} + Z\vec{I}_A \\ &= -j\,\frac{\sqrt{3}}{0.0605}\,(266.0 + j0) + (29.3\angle 62.2^\circ \times 11.4\angle -118.9^\circ) \\ &= 7880\angle -88.7^\circ \qquad \text{V}\end{aligned}$$

The required magnitude of the line-to-line supply voltage is

$$V = \sqrt{3} \times 7880 = 13\,650 \qquad \text{V}$$

In most situations, the absolute angles of the voltages and currents on the two sides of the transformer will not be relevant, and only the magnitudes will be required.

7.6.2 ▪ Polyphase Transformers

Instead of using a bank of three single-phase transformers, a polyphase transformer having three sets of primary and secondary windings on a common magnetic structure may be constructed. An advantage is that the weight and cost for a given rating is less than for a three-transformer bank. In addition, the external connections are reduced from 12 to 6; when these connections are brought out of the transformer tank through the elaborate bushings required in high-voltage transformers, this provides a considerable economy. A disadvantage is that if one phase breaks down, the whole transformer must be removed for repair. Provision of a spare three-phase stand-by transformer is more expensive than provision of one spare single-phase transformer.

To visualize the development of a three-phase, core-type transformer, consider the three single-phase core-type units shown in Fig. 7.17(a). Each of these has its two windings placed on one leg only; for simplicity, only the primary windings are shown. If the primary voltages are sinusoidal and balanced, the core fluxes Φ_a, Φ_b, and Φ_c also will be sinusoidal and balanced. Then, if the three legs carrying these fluxes are merged, the total flux in the combined leg will be zero, and this leg can be omitted, as shown in Fig. 7.17(b).

For cores made of stacked laminations, the in-line structure of Fig. 7.17(c) is more convenient. This structure can be evolved from Fig. 7.17(b) by eliminating the yokes of section *b* and fitting the remainder between *a* and *c*. The result does not have complete symmetry because the magnetic paths of *a* and *c* are somewhat longer than that of leg *b*; however, the resultant imbalance in the magnetizing currents is rarely significant.

The in-line configuration of Fig. 7.17(c) can be made of flat laminations and is used for most large core-type polyphase transformers. For smaller units, the configuration shown in Fig. 7.18 made of cores wound from a continuous strip may be used.

By using the methods developed in Section 7.1, an equivalent electric circuit can be produced for the three-phase, core-type transformer. The magnetic

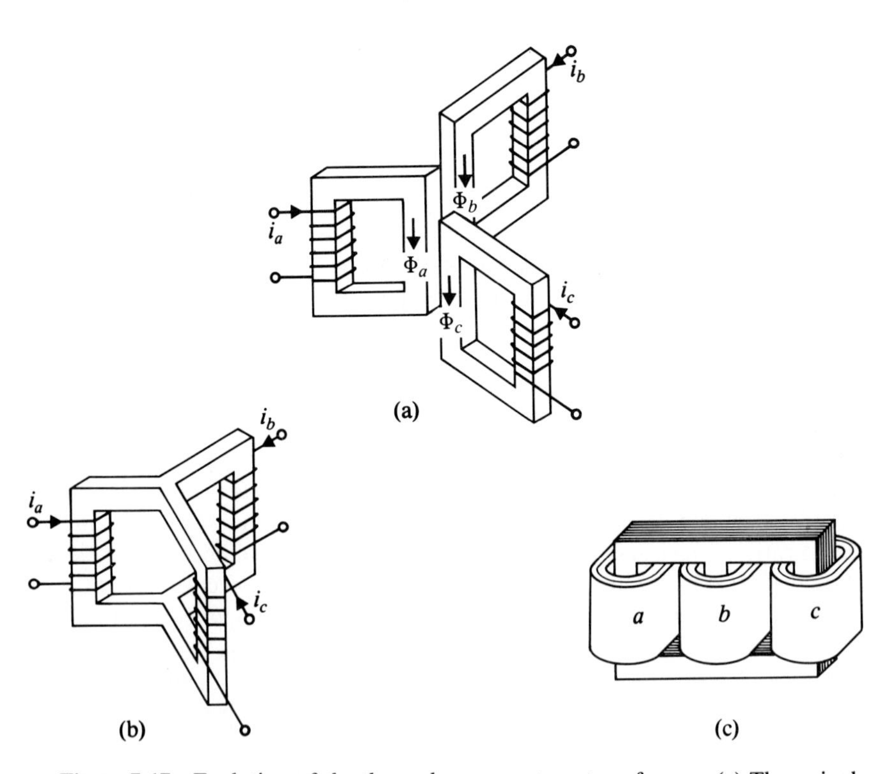

Figure 7.17 Evolution of the three-phase, core-type transformer. (a) Three single-phase transformers. (b) Return path for $\Phi_a + \Phi_b + \Phi_c$ removed. (c) In-line construction.

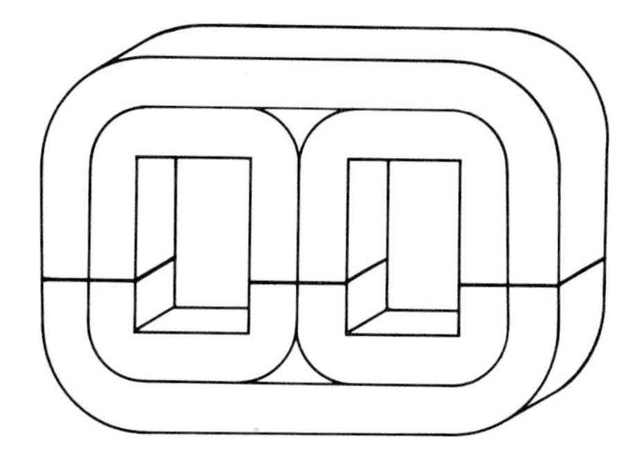

Figure 7.18 Wound core for three-phase transformer.

circuit consists mainly of three core sections in parallel. When the dual of this is formed, the resulting electric circuit has a form similar to that of three single-phase transformers connected in delta. Thus, a core-type polyphase transformer has inherently some of the features of a three-phase bank of individual transformers with one set of windings connected in delta. Because of the removal of the path for the combined return flux, the phase fluxes and also the phase voltages must sum to essentially zero, even with unbalanced loading. Also, the harmonic behavior of this type of transformer is similar to that of a delta-connected bank in that essentially no third-harmonic components of magnetizing current flow in the windings under balanced operation, irrespective of the winding connections. This can be appreciated by considering each phase leg as an equal in-phase source of harmonics. In the equivalent circuit, these harmonic sources are connected in delta and provide an internal short-circuited path independent of the winding connections.

An alternative form of polyphase transformer is the shell-type unit that is evolved by stacking three single-phase, shell-type units, as shown in Fig. 7.19(a). The winding direction of the center unit b is made opposite to that of units a and c. If the system is balanced with sequence a-b-c, the fluxes are also balanced, i.e.,

$$\vec{\Phi}_a = \alpha\vec{\Phi}_b = \alpha^2\vec{\Phi}_c \qquad \text{Wb} \tag{7.56}$$

where

$$\alpha = 1\angle 120°$$

The adjacent yoke sections of units a and b carry a combined flux of

$$\frac{\vec{\Phi}_a}{2} + \frac{\vec{\Phi}_b}{2} = \frac{\Phi_a}{2}(1\angle 0° + 1\angle -120°) = \frac{\Phi_a}{2}\angle -60° \qquad \text{Wb} \tag{7.57}$$

Thus, the magnitude of this combined flux is equal to the magnitude of each of its components. The cross-sectional area of the combined yoke section may be reduced to the same value as used in the outer legs and the top and bottom yokes, as shown in Fig. 7.19(b).

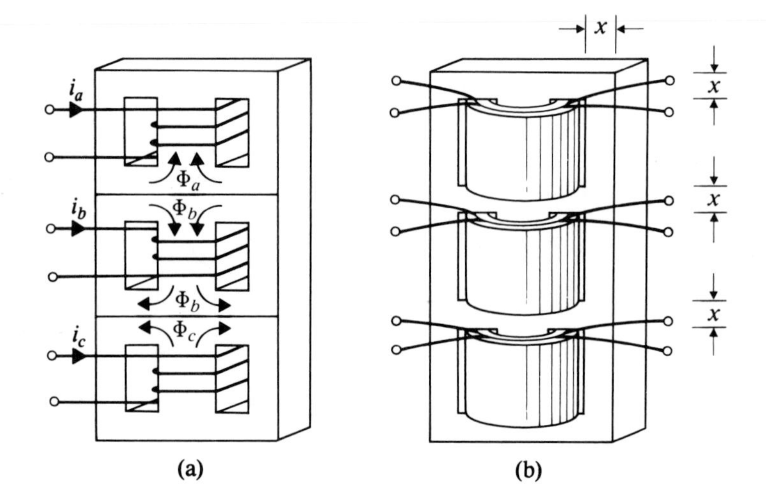

Figure 7.19 Evolution of the three-phase, shell-type transformer.

The slight imbalance in the magnetic paths among the three phases has very little effect on the performance of the three-phase shell-type transformer. Its behavior is essentially the same as that of a bank of three single-phase transformers. In contrast with the core-type unit, it normally must have one set of windings connected in delta to provide a path for triplen harmonic currents.

7.7

Some Special Transformation Devices

The analytical methods discussed earlier in this chapter are sufficient to determine the operating characteristics of most types of transformers. However, there are a few widely used devices that deserve further comment because of their special construction or application.

7.7.1 ▪ Autotransformers

A two-winding transformer provides for a change in voltage level and also provides for conductive isolation of its primary and secondary windings. If this isolation feature is not required (for example, when one side of both windings may be grounded), the voltage transformation can be achieved with a single tapped winding as shown in Fig. 7.20(a). If this autotransformer is regarded as ideal, its voltage ratio is

$$\frac{v_b}{v_a} = \frac{N_1 + N_2}{N_1} \tag{7.58}$$

As power is invariant in an ideal transformer, the current ratio is

$$\frac{i_b}{i_a} = \frac{N_1}{N_1 + N_2} \tag{7.59}$$

With an ideal core, the total mmf of the winding must be essentially zero. Thus, the current i_1, in the section of the winding having N_1 turns, is

$$i_1 = \frac{N_2}{N_1} i_b \quad \text{A} \tag{7.60}$$

An advantage of the autotransformer is that the amount of conductor required in its winding can be significantly less than required for a two-winding transformer of the same kVA rating. For the same core area, the copper ratio for the two types of transformer is approximately

$$\frac{\textit{Autotransformer}}{\textit{Two-winding transformer}} = \frac{N_1 i_1 + N_2 i_b}{N_1 i_a + (N_1 + N_2) i_b} = \frac{N_2}{N_1 + N_2} \tag{7.61}$$

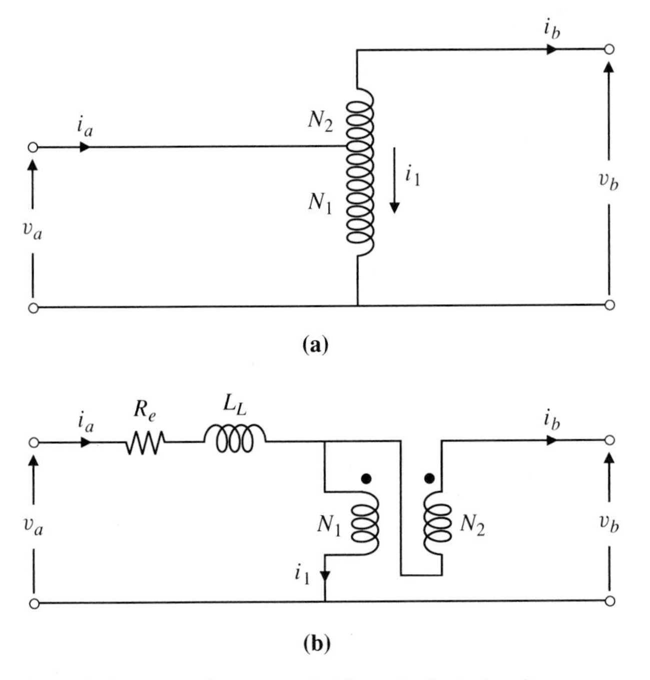

Figure 7.20 (a) Autotransformer and (b) equivalent circuit.

As compared with two-winding transformers of equivalent rating, autotransformers are smaller, are more efficient, and have lower internal impedance. They are used extensively in power systems where the voltages of the two systems coupled by the transformers do not differ by a factor greater than about three. For use in three-phase systems, an additional winding, usually denoted as a tertiary winding, may be added to each unit and connected in delta to provide a path for triplen harmonic currents, as discussed in Section 7.6.1.

Adjustable voltage output is obtainable from special autotransformers that have a single-layer winding around a toroidal core and a sliding brush that makes conductive contact with a bare sector of the winding turns.

For a more detailed analysis, it is preferable to consider an autotransformer as a two-winding transformer with its two windings connected in series. It then may be represented by the equivalent circuit shown in Fig. 7.20(b). In this, the magnetizing branch has been ignored. The leakage inductance and the equivalent winding resistance of the two-winding transformer have been referred to the N_1-turn side. Then, the series connections of the two windings section have been inserted.

EXAMPLE 7.7 A 10-kVA, 2300 : 230-V distribution transformer has its two windings connected in series to form an autotransformer providing a small increase in load voltage as compared with the available 2300-V supply.

(a) Determine the voltage transformation ratio on open circuit.

(b) Determine the permissible output of the transformer if the winding currents are not to exceed those for rated-load operation as a two-winding transformer.

(c) The efficiency of the two-winding transformer is 0.982 at rated load with a 0.9 power factor. Determine the efficiency of the autotransformer at its rated load at the same power factor.

Solution

(a) Using the notation of Fig. 7.20,

$$\frac{V_b}{V_a} = \frac{2300 + 230}{2300} = 1.1$$

(b) The current rating of the N_2 winding is

$$I_b = \frac{10\,000}{230} = 43.5 \quad \text{A}$$

For the N_1 winding, the rated current is

$$I_1 = \frac{10\,000}{2300} = 4.35 \quad \text{A}$$

As an autotransformer, the rating is

$$S = V_b I_b = 2530 \times 43.5 = 110.1 \qquad \text{kVA}$$

The same value may be obtained as

$$S = V_a I_a = 2300 \times (43.5 + 4.35) = 110.1 \qquad \text{kVA}$$

(c) The losses in the two-winding transformer are

$$P_L = (1 - 0.982)10\,000 \times 0.9 = 162 \qquad \text{W}$$

At 0.9 power factor, the rated power output as an autotransformer is

$$P_o = 0.9 \times 110\,100 = 99\,050 \qquad \text{W}$$

Then,

$$\eta = \frac{99\,050}{99\,050 + 162} = 0.9984$$

7.7.2 ▪ Instrument Transformers

Alternating current measuring instruments and protective relays are normally made to operate near earth potential and at currents not exceeding a few amperes. Where current must be measured in a high-potential conductor or where the current to be measured is large, it is usual to use a current transformer. The principles of a current transformer differ in no fundamental way from those of a power transformer; however, current transformers are so conservatively designed that they may be regarded as ideal with very small error.

A current transformer is usually constructed with a toroidal core wound from a continuous strip of low-loss, low-field-intensity magnetic material. Its secondary winding is wound around the toroidal core and usually has a current rating of 1 or 5 A, suitable for supply to an ammeter or wattmeter. A laboratory current transformer may have several sections of primary windings suitable for different current levels to be measured. For the largest current ratio, the current-carrying conductor is fed through the center of the current transformer core to provide a one-turn primary. The insulation around the secondary winding must be adequate for the voltage level of the current-carrying conductor.

Error in current measurement may arise from the exciting current of the transformer. To minimize this, the core is designed to operate at a low value of flux density. To achieve this in operation, the instrument load on the current transformer must be kept small. If the current transformer is used for current measurement, only the magnitude error is important. However, if it is used to supply the current coil of a wattmeter, the phase-angle error also is important.

In operating a current transformer, the secondary terminals should never be open circuited. If this were to happen, all the primary current would become magnetizing current driving the core rapidly between positive and negative saturation and producing very high-voltage pulses in the secondary winding. For

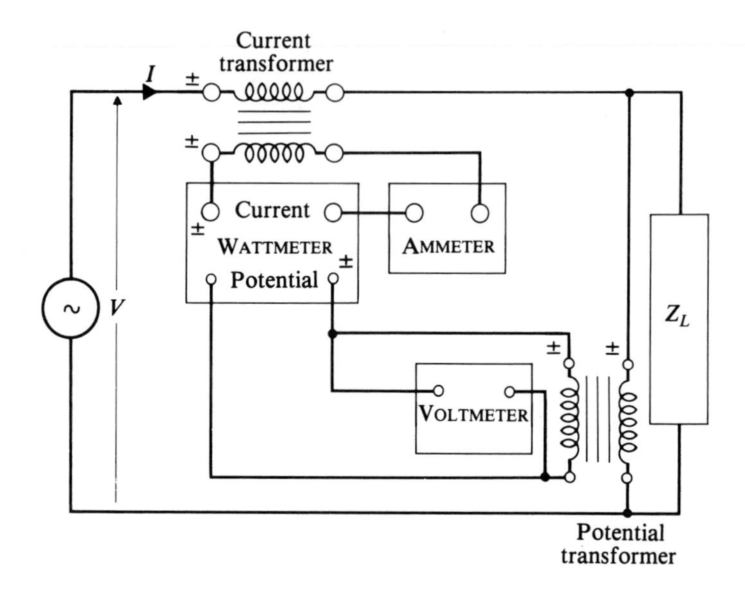

Figure 7.21 Connection of instrument transformers.

the protection of personnel and the transformer insulation, provision must be made to prevent any inadvertent break in the secondary circuit.

On all but low-voltage *a-c* systems, voltage is normally measured by use of a potential transformer, providing low voltage in the metering circuit and isolation of the metering systems from the high-voltage system. While the volt-ampere capacity of such potential transformers is very small, the transformers may be quite large because the primary must be fully insulated for the required voltage level.

Instrument transformers normally use ± markings to indicate terminal polarity. Corresponding markings are usually found on wattmeter terminals. Fig. 7.21 shows the required connections for measurement of voltage, current, and power in a single-phase *a-c* circuit using instrument transformers.

7.8

Forces in Saturable Magnetic Devices

The physical origin of forces between sections of ferromagnetic material carrying magnetic flux has been discussed in Section 3.2. While one can visualize these forces as arising from the attraction between aligned magnetic moments, this concept is not very useful for force calculation. An indirect approach involving the principle of energy conservation was used in Sections 3.6 and 3.7 to derive expressions relating the force and torque to the rate of change of inductance with displacement, or, alternatively, the rate of change of reluctance with displacement. For introductory simplicity, that development was restricted to magnetically linear systems where the iron section could be regarded as requiring negligible field intensity or having essentially constant permeability.

Practical devices based on the forces between ferromagnetic sections are normally operated well into the saturated regime to obtain the maximum force with minimum material. For these, the flux linkage becomes a nonlinear function of current, and additional concepts need to be introduced to provide means of force and torque prediction.

For example, consider the cylindrical actuator shown in cross section in Fig. 7.22. A current i in the coil produces flux across the air gap and around the closed ferromagnetic path. The force F tends to move the plunger or armature in such a direction as to close the air gap. The starting point for analysis of the force is the concept of conservation of energy. Consider a small displacement Δx of the armature in a time Δt. For simplicity, let us ignore hysteresis and eddy current losses in the core material. Further, let us associate the mechanical losses with the mechanical system and the winding losses with the electrical system. Then, from energy conservation, the electrical energy ΔW_e entering the actuator from the source, not including the resistance loss, must either be stored as energy ΔW_m in the magnetic field or it must appear as mechanical output ΔW_o, i.e.,

$$\Delta W_e = \Delta W_m + \Delta W_o \qquad \text{J} \tag{7.62}$$

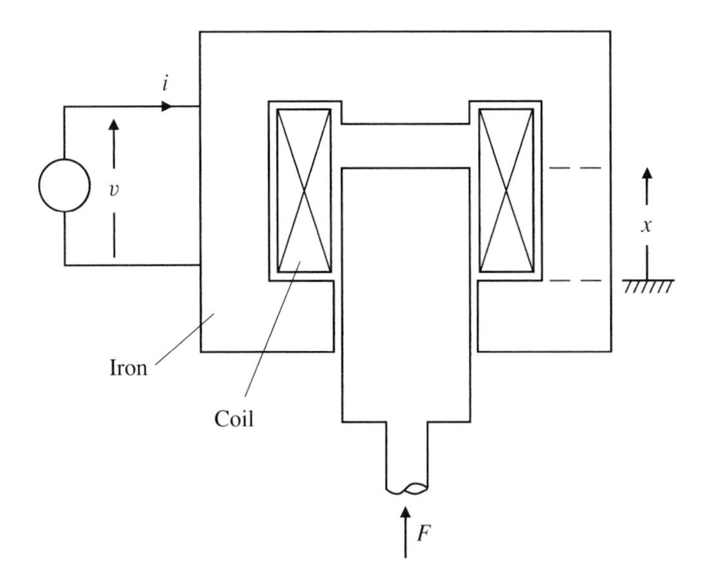

Figure 7.22 Cross section of cylindrical actuator.

Our approach will be to derive expressions for the energy input and the energy stored for a given displacement in a given time interval. Then, by process of elimination, we will have the mechanical output energy and thus the force.

A typical nonlinear relationship between coil flux linkage and coil current i is shown in Fig. 7.23 for a plunger position x_1. The energy stored in the field can be determined by considering the power input p from the source as the current i is increased from zero at position x_1. From Eq. (1.21),

$$p = vi = Ri^2 + i\frac{d\lambda}{dt} \quad \text{W} \tag{7.63}$$

Because the initial stored energy is zero, the energy stored in the field can be obtained by integrating the power in the second term of Eq. (7.63). Thus,

$$W_m = \int_o^{\hat{\lambda}} i\,d\lambda \quad \text{J} \tag{7.64}$$

As shown in Fig. 7.23, the stored energy is represented by the shaded area between the λ-i relation and the λ axis, measured in weber-amperes or joules.

Suppose the current is initially increased to a value i_1 with the plunger held at position $x = x_1$. Now suppose the gap length is decreased very slowly by a displacement Δx. With a slow movement, the rate of change of flux linkage with time may be ignored and the coil current may be considered to remain constant at i_1. Flux linkage-current curves for the initial position x_1 and the final position $x_2 = x_1 + \Delta x$ are shown in Fig. 7.24(a). During the displacement, the point of operation moves along the vertical locus from point 1 on the curve for $x = x_1$ to point 2 on the curve for $x = x_2$ at constant current i_1, increasing the flux linkage from λ_1 to λ_2. The electrical input energy to the system during the displacement is

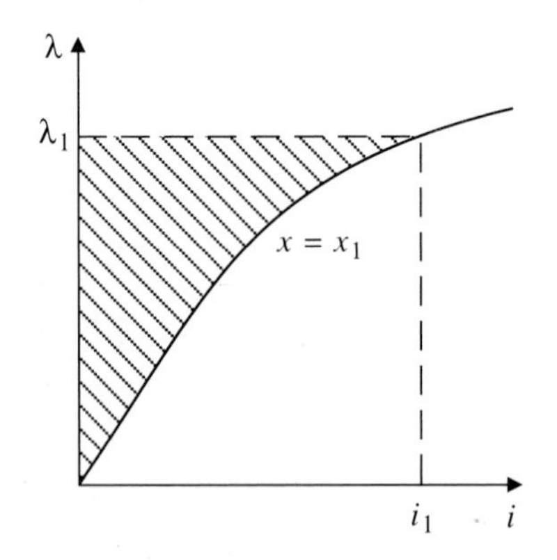

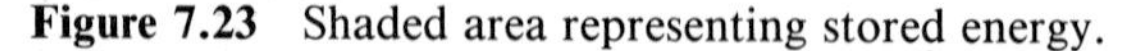
Figure 7.23 Shaded area representing stored energy.

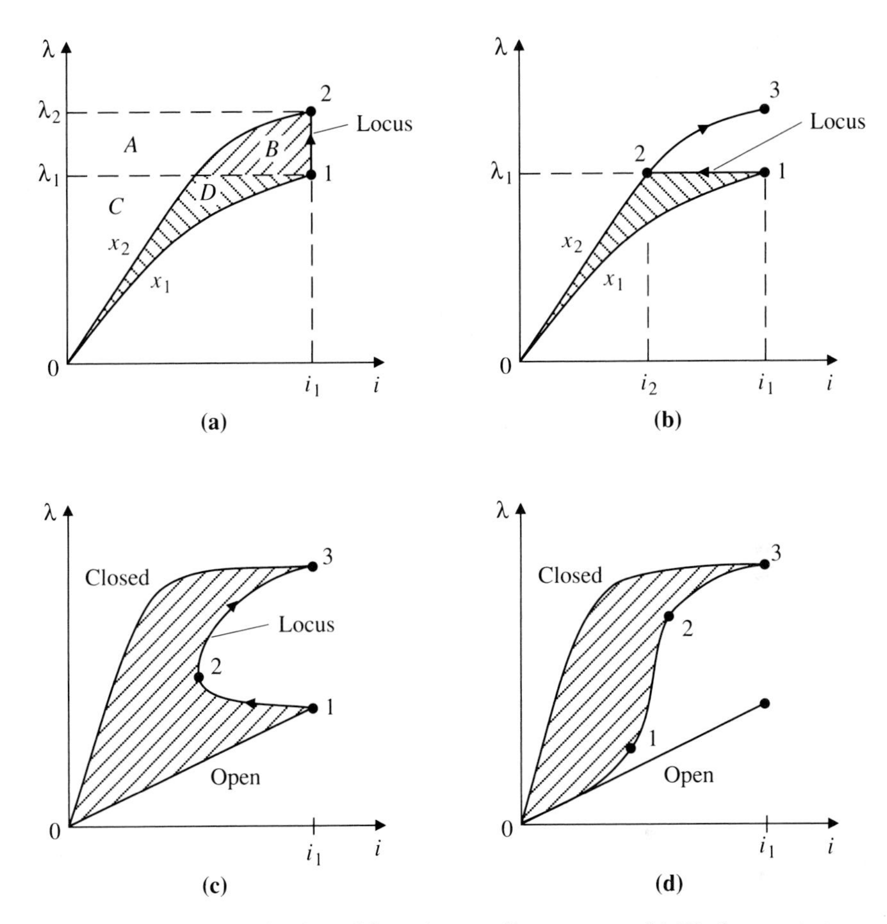

Figure 7.24 Determination of force in a nonlinear system. (a) Displacement at constant current. (b) Displacement with constant flux linkage. (c) General case of release at full current. (d) General case of applied voltage when open.

$$\begin{aligned} \Delta W_e &= \int_{\lambda_1}^{\lambda_2} i_1 \, d\lambda \\ &= i_1(\lambda_2 - \lambda_1) \\ &= \mathit{area}\ (A + B) \quad \text{in Figure 7.24(a)} \end{aligned} \tag{7.65}$$

The increase in stored energy during the displacement is equal to the stored energy W_2 at state 2 minus the initial stored energy W_1 at state 1. Using the curve for $x = x_1$,

$$W_1 = \int_o^{\lambda_1} i \, d\lambda = \mathit{area}\ (C + D) \qquad \text{J} \tag{7.66}$$

Similarly, using the curve for $x = x_2 = x_1 + \Delta x$,

$$W_2 = \int_o^{\lambda_2} i\,d\lambda = area\ (A + C) \quad \text{J} \tag{7.67}$$

Thus, the increase in stored energy during the displacement Δx is

$$\begin{aligned} \Delta W_m &= W_2 - W_1 \\ &= area\ (A + C) - area\ (C + D) \\ &= area\ (A - D) \quad \text{J} \end{aligned} \tag{7.68}$$

Substitution of the energy components of Eqs. (7.65) and (7.68) into Eq. (7.62) gives the mechanical output energy as

$$\begin{aligned} \Delta W_o &= \Delta W_e - \Delta W_m \\ &= area\ (A + B) - area\ (A - D) \\ &= area\ (B + D) \quad \text{J} \end{aligned} \tag{7.69}$$

The mechanical output energy is seen to be represented by the shaded area bounded by the locus 0-1-2-0. This is seen to be equal to the increase in the area under the λ-i curve caused by the displacement. This area has the dimension of energy and is complementary to the stored energy that has been shown in Fig. 7.23 to be the area between the λ-i curve and the λ axis. Because of its significance in force determination, it is given the name *coenergy* and is expressed as

$$W'_m = \int_o^{i} \lambda\,di \quad \text{J} \tag{7.70}$$

For the displacement Δx at constant current i_1, the mechanical energy output now can be expressed as the increase in coenergy, i.e.,

$$\Delta W_o = \Delta W'_m \quad \text{J} \tag{7.71}$$

For an infinitesimal displacement increment ∂x, recognizing that mechanical energy is the force times the displacement, the force F can be expressed as

$$F = \left. \frac{\partial W'}{\partial x} \right|_{i=constant} \quad \text{N} \tag{7.72}$$

i.e., the force acts in such a direction as to increase the coenergy. The expression of Eq. (7.72) is particularly useful when the flux linkage and/or the coenergy can be expressed as a function of current i and displacement x.

Let us now consider another limiting case in which, initially, the position is x_1 and the coil current is $i_1 = V/R$ where V is a constant-source voltage and R is the coil resistance. An incremental displacement Δx of the plunger is made so rapidly that the flux linkage remains substantially constant during the short time during which the displacement occurs. While the flux linkage may not change appreciably during this rapid displacement, its rate of change with time will be large, and the resultant induced voltage in the coil will cause the current to be reduced during the displacement as shown in the locus 1-2 of Fig. 7.24(b). After the displacement Δx is completed, the coil current will rise along the locus 2-3 toward its original value i_1, with all of the input electric energy going into field storage. As the flux linkage λ is constant at λ_1 during the displacement Δx, the electric energy input ΔW_e associated with the locus 1-2 is zero. Thus, from Eq. (7.62), the mechanical output energy must be obtained solely from the energy stored in the system, i.e.,

$$\Delta W_o = -\Delta W_m$$
$$= -\left[\int_o^{\lambda_1} i\,d\lambda_{along\ 0\text{-}2} - \int_o^{\lambda_1} i\,d\lambda_{along\ 0\text{-}1}\right] \quad \text{J} \qquad (7.73)$$

In Fig. 7.24(b), this mechanical output energy is represented by the shaded area bounded by 0-1-2-0.

To find the force at a particular value of displacement x, the increment Δx may be contracted to the infinitesimal ∂x, and the force can be expressed as

$$F = -\frac{\partial}{\partial x} W_m \bigg|_{\lambda = constant} \quad \text{N} \qquad (7.74)$$

This form of force expression is useful when the coil current and/or the stored energy is expressed as a function of the flux linkage λ and the displacement x.

Examination of two limiting cases has provided us with the expressions of Eqs. (7.72) and (7.74) which are useful in calculation of the force for a given position and for given values of current or flux linkage. In a more general situation, we may wish to find the total mechanical work done by the actuator as it moves from the open to the closed position. Typical λ-i curves for these two positions are shown in Fig. 7.24(c). First, consider a situation where a constant voltage V sets up a coil current of V/R while the actuator plunger is held in the open position for which the λ-i curve is 0-1. If the plunger then is released, the operating point may move along a locus such as 1-2-3 in Fig. 7.24(c). Eventually, when the gap is closed, the coil current i will return to the value V/R at point 3. If the plunger moves very slowly, this locus would be essentially a vertical line. If the plunger moves very rapidly, the initial locus 1-2 might be nearly at constant flux linkage. In any case, the mechanical work done would be equal to the enclosed shaded area. It will be noted that this mechanical work is highly dependent on the locus of the operating point. A complete analysis of this locus involves the simultaneous solution of three equations, the first being the electric system relation

$$V = Ri + \frac{d\lambda}{dt} \quad \text{V} \qquad (7.75)$$

the second being the differential equation for the mechanical system, and the third being the force expression of either Eq. (7.72) or Eq. (7.74), depending on whether λ is expressed as a function of i or i is expressed as a function of λ.

Figure 7.24(d) shows a somewhat more realistic situation in which a voltage V is applied to the coil when the plunger is in the open position. As the coil current i increases from its initial zero value, the operating locus will be nearly along the open λ-i locus until the force is sufficient to produce significant acceleration and displacement. Eventually the locus will arrive at the point 3 on the closed λ-i curve. The mechanical work done will be equal to the shaded area between the locus 0-1-2-3 and the λ-i curve 0-3 for the closed position.

EXAMPLE 7.8 The flux linkage-current relationship for an actuator can be expressed approximately as

$$\lambda = \frac{0.08}{g} i^{1/2} \qquad \text{Wb}$$

between the limits $0 < i < 5$ A and $0.02 \text{ m} < g < 0.10$ m where g is the air-gap length. If the current is maintained at 4 A, find the force on the armature for a gap of $g = 0.06$ m.

Solution For a nonlinear relationship, we might consider using either Eqs. (7.72) or (7.74) to derive the force. Because the flux linkage is given as a function of current, the former expression is more convenient. The coenergy expression is

$$W' = \int_o^i \frac{0.8}{g} i^{1/2}\, di = \frac{0.08}{g} \frac{2i^{3/2}}{3} \qquad \text{J}$$

Then,

$$F = \left.\frac{\partial W'}{\partial g}\right|_{i=4} = \frac{-0.08}{g^2} \frac{2i^{3/2}}{3} = -119 \qquad \text{N}$$

The negative sign indicates a force tending to reduce the gap g.

PROBLEMS

7.1 The magnetic system of Fig. 7.25 has a coil of 500 turns on its middle leg. The ferromagnetic material may be assumed to have a constant relative permeability of 4000. Leakage flux may be ignored.

(a) Derive a magnetic equivalent circuit for this system showing the values of all parameters.

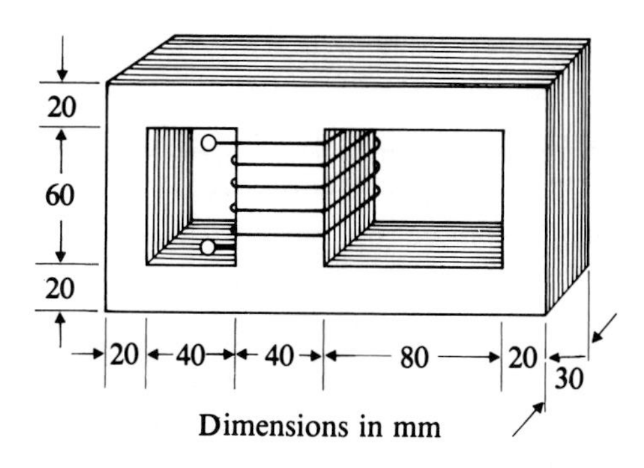

Figure 7.25 Magnetic system for Problem 7.1.

(b) Develop an electric equivalent circuit.

(c) Determine the total inductance of the 500-turn coil.

(d) Suppose a 100-turn coil is wound on the right-hand leg. Insert the appropriate ideal transformer into the electric equivalent circuit.

(e) If an alternating voltage of 100 V is applied to the 500-turn coil, what will be the open-circuit voltage on the 100-turn coil?

(Section 7.1.1)

7.2 In the two-coil magnetic device shown in Fig. 7.26, the magnetic material may be considered to be ideal. At each air gap, the magnetic field is not confined entirely to the gap. The fringing flux may be taken into account approximately by assuming that the region of uniform flux density extends outward from each of the gap edges by a distance equal to one-half the gap length.

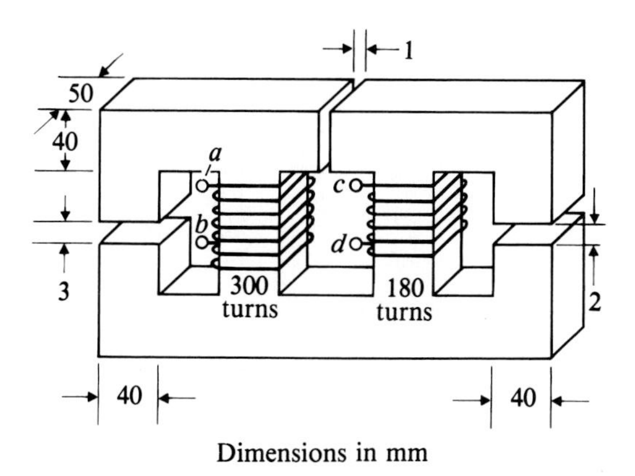

Figure 7.26 Magnetic device for Problem 7.2.

(a) Develop an electric equivalent circuit for this device showing the values of all parameters.

(b) If the frequency is 180 Hz, what is the reactance with the two coils connected in series with terminal b connected to terminal d?

(c) Repeat (b) with terminal b connected to terminal c.

(Section 7.1.1)

7.3 In electric welding, a voltage of 50 to 70 V is required to strike an arc. After the arc is struck, an essentially constant current supply to the arc is desired. Fig. 7.27 shows a two-winding transformer designed for use in a welder. To limit the load current, a low reluctance path for leakage flux has been provided between the primary and secondary windings.

(a) Derive an equivalent electric circuit for the transformer assuming ideal iron, no fringing flux around the air gap, no leakage flux around the coils, and no winding resistances.

(b) If the supply is 115 V (rms) at 60 Hz, determine the open-circuit secondary voltage and the short-circuit secondary current.

(c) If the arc can be regarded as a variable resistance with a value dependent on arc length, what is the maximum power that can be delivered to the arc?

(d) Would the power predicted in (c) have been increased if the effects of leakage flux around the windings had been included?

(Section 7.1.1)

7.4 A magnetic system consists of two toroidal ferromagnetic cores. Core A has a cross-sectional area of 300 mm^2 and an air gap of 1 mm length. Core B has an area of 500 mm^2 and an air gap of 0.5 mm length. A pri-

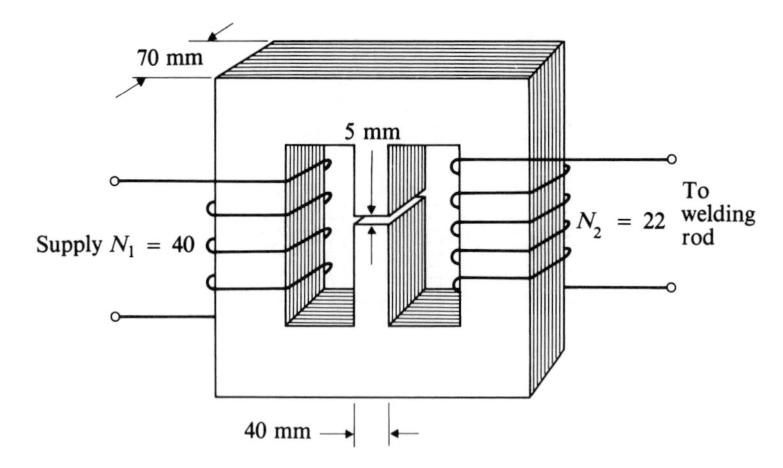

Figure 7.27 Welding transformer for Problem 7.3.

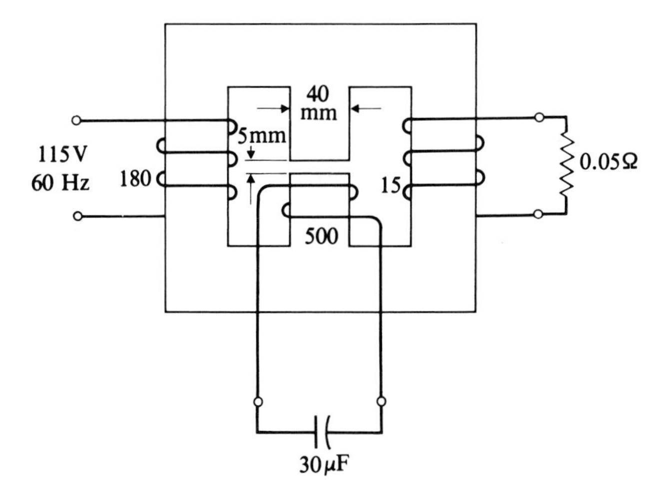

Figure 7.28 Device for Problem 7.5.

mary coil of 240 turns links both cores while a secondary coil of 400 turns links only core B. Each coil has a resistance of 0.04 Ω/turn. Assuming infinite permeability in the core material, develop an electric equivalent circuit for the system including the values of all parameters.

(Section 7.1.1)

7.5 The magnetic material in the device shown in Fig. 7.28 may be regarded as ideal. The core depth is 50 mm. Fringing around the gap and the coils may be ignored.

(a) Develop an electric equivalent circuit as seen from the resistive load.

(b) Determine the power delivered to the load.

(Section 7.1.1)

7.6 The transformer device shown in Fig. 7.29 is supposed to maintain a constant ratio of output voltage to input voltage regardless of the load while operating at a frequency of 400 Hz.

(a) Develop an electric equivalent circuit for the device. Winding resistances may be ignored and the magnetic material may be regarded as ideal. The leakage path around each of the three windings may be assigned a reluctance of 10^7 A/Wb.

(b) Determine the value of capacitance C required.

(c) Show that the performance of the device is similar to that of a two-winding transformer with a capacitor in series with one of the windings.

(Section 7.1.1)

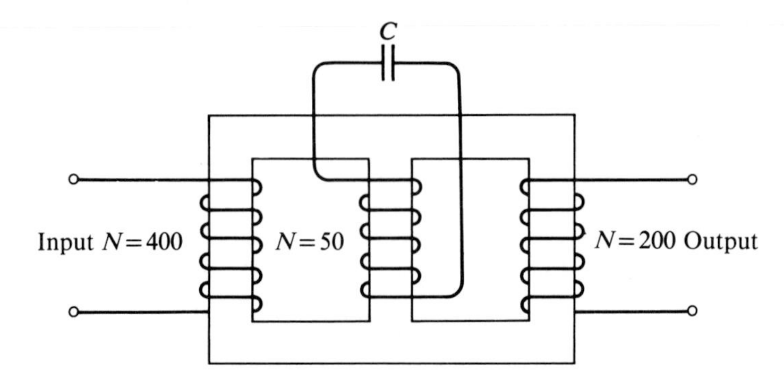

Figure 7.29 Transformer for Problem 7.6.

7.7 The system shown in Fig. 7.6 is to operate from a 230-V, 60-Hz source. Each section of the core has a cross section of 50 × 50 mm. The desired inductance in the equivalent circuit is 4 mH and the desired capacitance is 70 μF. The core material may be regarded as ideal. Fringing flux and winding resistances may be ignored.

(a) If the flux density in the core material is to be limited to a maximum value of 1.5 T, determine the appropriate number of turns in the source winding.

(b) Determine the appropriate length of the air gap.

(c) Suppose a 20 μF capacitor of suitable voltage rating is available. What should be the number of turns in the winding to which it is connected?

(Section 7.1.3)

7.8 It is desired to synthesize a system that will have the electric equivalent circuit of Fig. 7.30. A core with multiple windings is to be used, providing flexibility in the choice of source voltage, load voltage, and capacitor voltage. Develop an equivalent magnetic circuit for the appropriate magnetic device and sketch its cross section showing the windings and air gaps.

(Section 7.1.3)

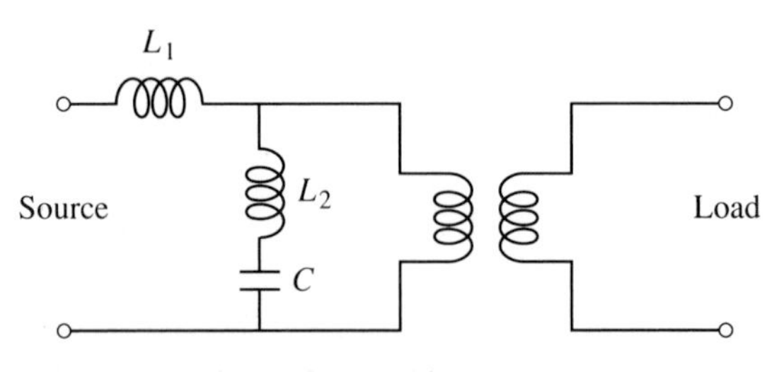

Figure 7.30 Equivalent circuit for Problem 7.8.

7.9 In the two-winding transformer shown in Fig. 2.7(a), both coils are nearly cylindrical in shape. The inside coil has an inner diameter of 80 mm. Each coil has a thickness of 15 mm and a height of 110 mm. The space provided for air circulation between the coils is 4 mm wide. Estimate the leakage inductance of this transformer referred to the 60-turn primary winding.

(Section 7.1.4)

7.10 Two single-phase transformers designated as A and B have the same physical proportions. The linear dimensions of all parts of transformer A are k times those of transformer B. The core laminations are of the same thickness in both transformers. The operating values of flux density, conductor current density, and frequency are the same for both. Determine the value of each of the following quantities for transformer A relative to the corresponding quality for transformer B: rated voltage, rated current, rated volt-amperes, mass, output per unit of mass, winding loss, winding resistances, core loss, heat loss per unit of surface area, total loss in per unit of rating, leakage inductance in henries, leakage reactance in ohms, per unit magnetizing reactance, per unit resistance, and per unit leakage inductance.

(Section 7.1.4)

7.11 Develop an approximate design for a 10-H inductor capable of carrying a continuous direct current of 0.6 A. A core of the shape shown in Fig. 7.9 may be used, assuming a square central leg and a stacking factor of 0.9. The core flux density should be limited to 1.5 T. The coil should be made of copper conductor with resistivity at 20°C of $1.72 \times 10^{-8}\ \Omega \cdot \text{m}$. The coil space factor may be assumed to be 0.45. Assume that heat is dissipated only from the outside vertical surface of the coil at a uniform rate of 30 $\text{W/m}^2 \cdot {}^\circ\text{C}$. The coil surface temperature is not to exceed 130°C, and the ambient temperature may be as high as 50°C. The temperature coefficient of resistivity of copper is 0.00393/°C. Determine appropriate values for (a) the dimension a, (b) the conductor current density, (c) the number of turns, and (d) the length of the air gap.

(Section 7.2.2)

7.12 The transformer shown in Fig. 7.31 is to operate with 115 V rms at 60 Hz on its primary winding. The secondary winding is to produce an output at 500 V rms. The core material is to operate at a peak flux density of 1.4 T, and its stacking factor is 0.95. The windings can be operated at a current density of 2 A rms/mm^2, and the winding space factor is 0.45.

(a) Determine the required number of primary and secondary winding turns.

(b) Determine the kVA rating of the transformer.

(c) At the operating temperature the copper resistivity is about 2×10^{-8} $\Omega \cdot \text{m}$. Assuming the projecting ends of the coils are semicircular in

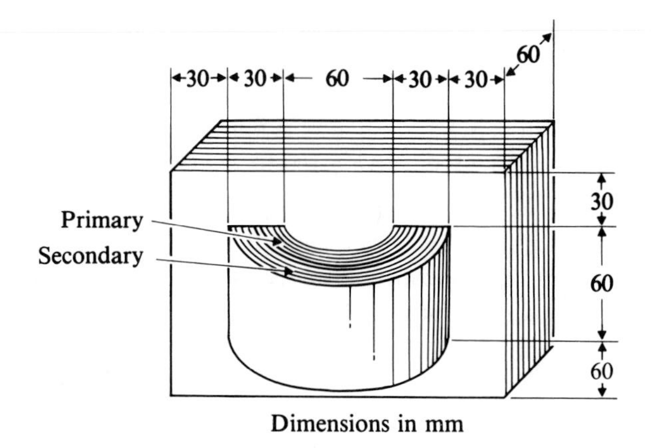

Figure 7.31 Transformer for Problem 7.12.

shape, estimate the power loss in the windings when operating at rated current.

(d) If the core laminations are made of M-36 sheet steel with the core-loss characteristics of Fig. 1.24 and a density of 7800 kg/m^3, find the power loss in the core.

(e) Find the efficiency of the transformer when operating with a rated resistive load.

(Section 7.2.3)

7.13 A 5-kVA, single-phase, 60-Hz, 4160 V : 230 V distribution transformer is to be made of steel laminations having the relative dimension ratios as shown in Fig. 7.9. The laminations are to be interleaved so that no effective air gap will occur in the core. The peak flux density in the core material can be set at 1.4 T. A stacking factor of 0.95 and a coil space factor of 0.5 can be assumed. The conductor current density can be taken as 3 A/mm^2. Assuming a square core cross section, find the appropriate value of the dimension a and the number of turns in the two windings.

(Section 7.2.3)

7.14 A magnetic field with a flux density of 1.5 T is required to be directed perpendicular to an area of 15 × 80 mm. The air-gap length of the field is to be 4 mm. An arrangement somewhat similar to that of Fig. 7.11(a) may be considered. Fringing flux may be ignored for a preliminary design, and the steel yoke may be considered to be magnetically ideal. A neodymium permanent-magnet material for which the B-H characteristic is given in Fig. 1.18 may be used. Its residual flux density is 1.2 T, and its relative per-

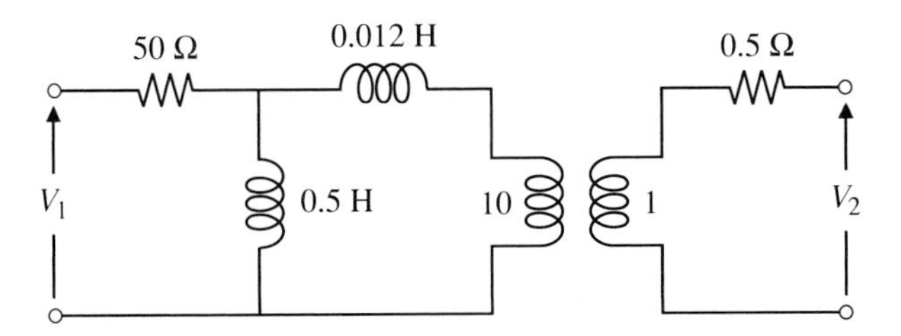

Figure 7.32 Equivalent circuit for Problem 7.15.

meability is 1.05. Design an appropriate device and determine the volume of magnet material required.

(Section 7.3)

7.15 A transformer for use at audio frequencies can be represented by the equivalent circuit of Fig. 7.32. A variable-frequency sinusoidal source provides a constant voltage V_1 to the transformer, and the voltage V_2 is applied to a resistive load of 10 Ω.

(a) Derive the ratio V_2/V_1 for the mid-frequency range.

(b) Determine the low and high frequencies at which the load power will be reduced to half that obtained in the mid-frequency range.

(Section 7.4)

7.16 A 5-kVA, 4160-V distribution transformer has 2500 turns in its primary winding. Its core has a cross-sectional area of 5000 mm^2 and a flux path length of 0.4 m. The core material can be modeled approximately by a *B-H* characteristic of the shape shown in Fig. 7.15(b) for which the residual flux density is 1.8 T and the relative permeability in the saturating region is 10. Initially, the core flux density can be assumed to be zero. A sinusoidal voltage having an rms value of 4160 V at 60 Hz is switched on to the primary winding at the instant that the voltage is zero. Estimate the peak value of the inrush current. Compare this with the peak rated current of the transformer.

(Section 7.5)

7.17 Three similar 2-kVA, 60-Hz, 220-V : 110-V, single-phase transformers are connected in turn in wye-wye, delta-wye, wye-delta, and delta-delta. The transformers may be considered to be ideal. Determine the rated values of primary and secondary line-to-line voltages, the rated primary and secondary line currents, the ratio of line-to-line voltages, and the ratio of line currents for each of the transformer connections.

(Section 7.6.1)

7.18 Three similar 10-kVA, 60-Hz, 2300-V : 115-V, single-phase transformers are connected in delta-wye to supply, on the low-voltage side, a balanced three-phase load of 24 kVA at 115 V, line to neutral. Determine the primary supply voltage (line to neutral) and the primary and secondary line currents. The transformers may be considered to be ideal.

(Section 7.6.1)

7.19 A three-phase, 60-Hz synchronous generator produces an output of 100 MVA at 15.8 kV (L-L). The output is to be fed into a transmission line at 735 kV (L-L). The required transformer bank is to consist of three two-winding transformers connected in delta on the generator side and in wye on the line side.

(a) Specify the required rated values of winding voltages and currents in each transformer.

(b) Suppose each transformer has a leakage reactance of 0.15 pu based on its rating. Determine the voltage (L-L) required at the generator terminals to provide 100 MVA to the transmission line at 0.8 power factor lagging and at 735 kV.

(Section 7.6.1)

7.20 Figure 7.33 shows an open-delta connection of two transformers that may be used to connect a three-phase load to a three-phase supply. If the transformers are similar and each is rated at 10 MVA, determine the balanced load that may be supplied by this arrangement without overloading either of the transformers. The transformers may be considered to be ideal.

(Section 7.6.1)

7.21 Figure 7.34 shows a transformer connection that can be used to produce a phase shift between the input and output three-phase systems.

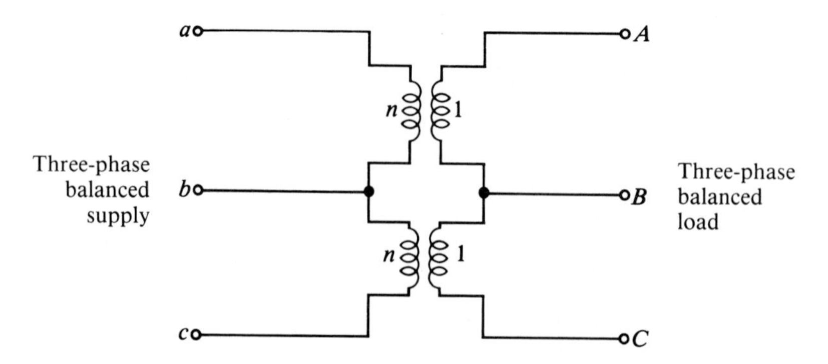

Figure 7.33 Open-delta transformer connection for Problem 7.20.

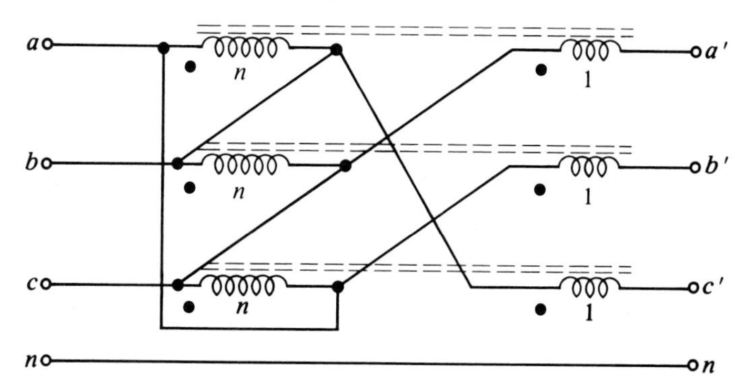

Figure 7.34 Transformer connection for Problem 7.21.

(a) Assuming ideal transformers, each with a turns ratio of $n:1$, show that the phase shift is given by

$$\theta = \tan^{-1}(\sqrt{3}/n)$$

(b) If the turns ratio is $5:1$, determine the ratio of the output to input voltage phasors.

(Section 7.6.1)

7.22 A three-phase core-type transformer of the form shown in Fig. 7.17(c) has windings of N_1 and N_2 turns on each leg, the N_1-turn winding being on the inside.

(a) Develop a magnetic equivalent circuit for the transformer assuming infinite permeability core material. Include the leakage paths between the pairs of windings and also a flux path from the top yoke to the bottom yoke outside the windings.

(b) Derive an electric equivalent circuit referring all inductances to the N_1 turn windings.

(c) Is it feasible to connect the three N_1-turn windings in parallel and use the resultant unit as a single-phase transformer?

(Section 7.6.2)

7.23 A three-phase, shell-type transformer is shown in Fig. 7.19(b). Assume windings of N_1 and N_2 turns in each window of the core. The relative permeability of the core material may be assumed to be infinite.

(a) Develop a magnetic equivalent circuit including the leakage paths between the pair of windings in each window.

(b) Derive an electric equivalent circuit.

(c) Is it feasible to connect the three N_1-turn windings in series and use the resultant unit as a single-phase transformer?

(Section 7.6.2)

7.24 An industrial process requires a single-phase supply of 500 kVA at 2400 V. The available utility supply is at 4000 V. An autotransformer is to be used.

(a) Determine the required current ratings of the two sections of the transformer winding.

(b) Suppose the two sections of the autotransformer are disconnected. Determine the kVA rating of the resultant two-winding transformer and the voltage rating of each of the windings.

(Section 7.7.1)

7.25 Figure 7.35 shows a variable-ratio transformer. A sliding brush makes contact with a bared part of the top of the single-layer winding. Determine the required number of turns and the dimension d for an autotransformer with an output voltage of 0 to 115 V rms at 60 Hz and an output current of 8.7 A rms. Square wire having a 0.25-mm insulation thickness may be used with a current density of 1 A rms/mm^2 in the copper. The toroidal core may be operated at a peak flux density of 1.5 T.

(Section 7.7.1)

7.26 A toroidal current transformer has a core with a mean diameter of 60 mm and a cross-sectional area of 100 mm^2. The core material has a relative permeability of 5000 for small values of flux density. The primary consists of a single bus bar passing through the aperture of the core and carrying a 60-Hz current.

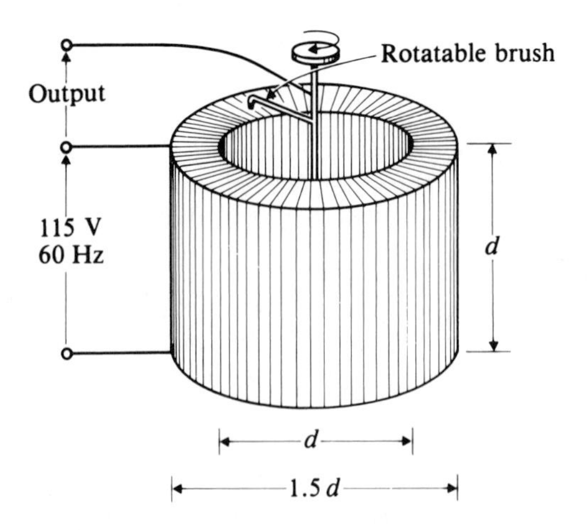

Figure 7.35 Variable-ratio transformer for Problem 7.25.

(a) How many turns are required on the core to give a nominal current ratio of 800 : 5?

(b) Determine the magnetizing reactance as seen by the secondary side.

(c) Suppose the secondary winding is connected by means of a two-conductor cable to a 5-A ammeter. Assume that the combined impedance of the meter, the cable, the winding resistance, and leakage reactance is $0.07 + j0.06\ \Omega$. Determine the magnitude error of the current measuring system.

(d) If the current transformer supplies the current coil of a wattmeter, the error in phase is significant. For the combined impedance given in (c), find the phase error.

(e) Would the phase error of (d) cause the wattmeter to read too high or too low if the short-circuit loss of a transformer were being measured?

(Section 7.7.2)

7.27 Over the intended operating range, the relationship among the flux linkage λ, the current i, and the displacement x of the moving member of an actuator are approximated by

$$\lambda = \frac{4 \times 10^{-4}}{x^2} i^{1/3}$$

Determine the force in the x direction when $i = 0.6$ A and $x = 5$ mm.

(Section 7.8)

7.28 A rotary actuator is illustrated in Fig. 7.36(a). When the coil is energized, the rotor rotates from its initial position at $\beta = 0$ to $\beta = \pi/2$, i.e., from the vertical to the horizontal orientation, the torque being absorbed by a mechanical load. The flux linkage-current relations for the coil are shown in idealized form in Fig. 7.36(b) for the initial and final rotor positions.

(a) Assuming the coil current is held constant at 3 A, determine the work done by the actuator in rotating from the initial to the final position.

(b) Determine the average torque during the motion of (a).

(c) Suppose the coil current is increased to 3 A while the rotor is held in the position $\beta = 0$. If the unloaded rotor then moves very rapidly to $\beta = \pi/2$, estimate the work done and the average torque.

(Section 7.8)

7.29 Consider the switched reluctance motor shown in Fig. 3.23(a) with the dimensions given in Example 3.4. In that example, the coil current was such that the magnetic system could be considered as magnetically ideal, i.e., with infinite permeability. The torque produced was 2.32 N·m. Let us now examine the more realistic operation in which the poles experience significant saturation. As an approximation, let us assume that no saturation occurs with the rotor poles aligned with the stator poles until the gap flux

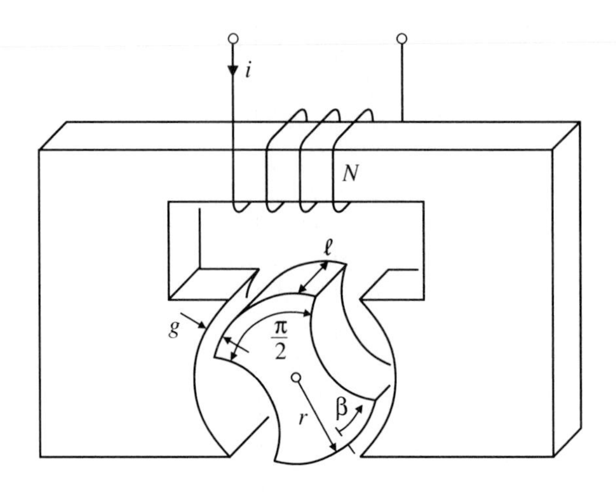

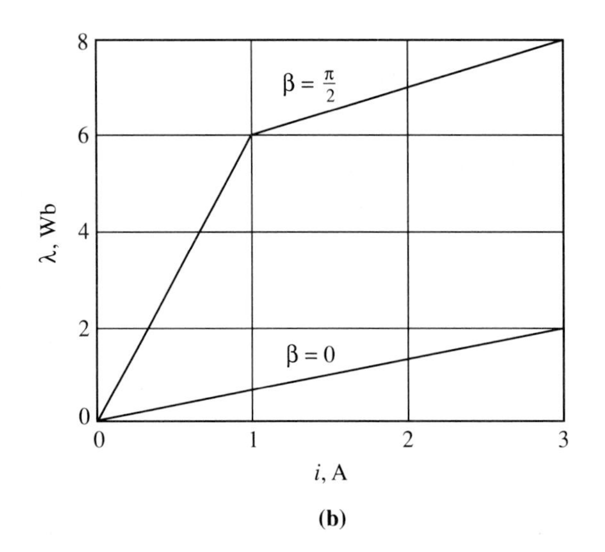

Figure 7.36 (a) Rotary actuator for Problem 7.28. (b) Flux linkage-current relations.

density reaches 1.5 T. For a higher current in the coils on the pair of stator poles, assume the slope of the flux linkage-current relation is 10% of the slope in the unsaturated region. When the rotor gaps are aligned with the energized stator poles, assume that the flux linkage-current relation is linear at 10% of the slope in the closed unsaturated condition. With a constant current of 10 A supplied to the phase coils, determine the average torque produced by this machine.

(Section 7.8)

GENERAL REFERENCES

Blume, L.F., *Transformer Engineering*, 2nd ed., John Wiley and Sons, New York, 1951.
Nasar, S.A., *Electric Machines and Transformers*, Macmillan, New York, 1984.
Say, M.G., *Alternating Current Machines*, 5th ed., Pitman, London, 1983.
Slemon, G.R., *Equivalent Circuits for Transformers and Machines, Including Nonlinear Effects*, *Proceedings*, Vol. 100, Part IV, Institution of Electrical Engineers, July 1953, 129.

CHAPTER

8

Power Semiconductor Converters

For many electric motor drives, it is adequate to have the ability to start and then run at essentially constant speed when connected to the constant-frequency a-c supply. However, there is a large and increasing range of applications for which adjustable speed, or precisely controllable speed, is required. One of the major factors influencing this trend to variable-speed drives is the concern for energy conservation which can be met by using the most efficient drive speed as the load demand changes.

To achieve variable speed, most types of electric motor require a source of controllable voltage or current. For commutator machines, the source is usually, but not always, direct voltage or current. For induction and synchronous machines, the source must have controllable voltage or current and, in addition, controllable frequency. Such supplies are best produced by the use of converters based on electronic switching devices which convert the available constant-voltage, constant-frequency utility sources to the required form.

In the following sections, some of the important types of power semiconductor converters are introduced. The approach is a simplified idealized one with the limited objective of presenting some of the major features that influence the drive performance. Many aspects of the internal design of converters will be ignored in this treatment. Those who are interested in these aspects should consult a text on power electronics.

8.1

Power Switching Devices

Power semiconductor converters are based on the use of solid-state semiconductor devices which generally act as switches that are either on or off. Ideally, these switches should have no voltage drop when on and carry no current when off, blocking any applied voltage. Actual devices have characteristics that approach this ideal closely enough that an appreciation of the main properties of converters can usually be achieved by assuming such ideal behavior. While there is a large and increasing catalogue of power semiconductor devices, they can be classified under only a few generic categories.

The first and simplest group of devices is *diode rectifiers*. These provide essentially unidirectional flow of current and have the idealized voltage-current characteristic of Fig. 8.1, which also contains the circuit symbol for the device. With a negative applied voltage v, the device current is near zero over its safe operating range up to a peak reverse voltage (PRV) that must not be exceeded. For positive applied voltage, the device current is established by the circuit in which the diode is connected. In most power diodes, the voltage drop in the forward direction is in the range 0.8 to 1.2 V. The permissible forward current i is limited by heating in the semiconductor material. A wide range of devices is available with current ratings available up to and beyond 5 kA and peak reverse voltages up to and in excess of 5 kV.

The next group of devices consists of *controlled rectifiers*, also known as *semiconductor* or *silicon controlled rectifiers* (SCR) or *thyristors*. These are three terminal devices that have properties similar to diode rectifiers when the applied

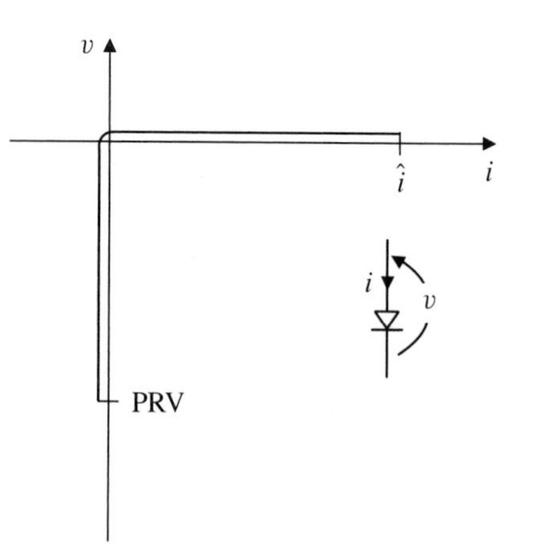

Figure 8.1 Idealized characteristic and symbol for a diode. PRV: peak reverse voltage.

voltage is negative. With positive applied voltage, the current is negligibly small until a positive low-power pulse is applied to the third terminal or gate, whereupon the device then acts much as a diode. To reestablish the off state, the device current must be reduced to zero for a short time. An alternative term used for this class of device is *naturally commutated semiconductor switches*, which reflects the property that they can be switched on but must be naturally turned off by the associated circuitry. The idealized voltage-current characteristics of a controlled rectifier are suggested by Fig. 8.2, which also contains the conventional circuit symbol. The forward voltage drop in thyristors when conducting is somewhat higher than that of diodes, being in the range 1 to 2.5 V depending on the voltage rating. The available limiting values of current and peak reverse voltage are comparable with those of diodes. The forward blocking voltage (FBV) is of the same order of magnitude as the peak reverse voltage. If this voltage is exceeded, the device is turned on.

The devices of the third category are denoted as *self-commutated semiconductor switches*. The general properties of devices in this class are the capability of being turned on or off by use of some form of low-power signal into a third or gate terminal and normally a capability of carrying current only in the forward direction. A number of physically distinctive types of switch are available in this category. Examples are gate turn-off switches (GTO), bipolar junc-

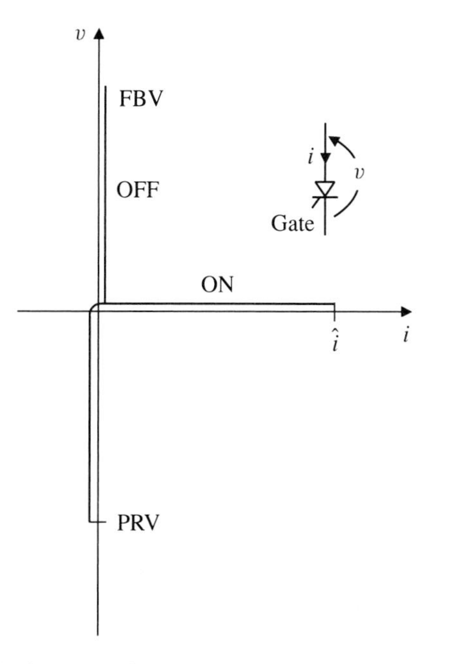

Figure 8.2 Idealized characteristic and symbol for a controlled rectifier. FBV: forward blocking voltage; PRV: peak reverse voltage.

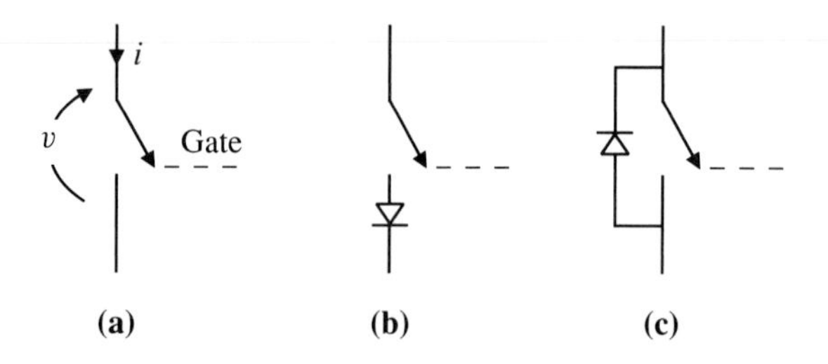

Figure 8.3 Self-commutated semiconductor switch. (a) Generic symbol. (b) With series diode for reverse blocking. (c) With parallel diode for reverse current.

tion transistors (BJT), insulated gate bipolar transistors (IGBT), static induction transistors (SIT), and metal-oxide semiconductor field-effect transistors (MOSFET). In electronic literature, each of these has a distinctive circuit symbol. In this text, we will represent all devices of this category by the generic symbol shown in Fig. 8.3(a), which is the classic symbol for a switch with the addition of an arrow to indicate the unidirectional-current property of these devices. The various types of self-commutated switches differ in their available current-carrying capacity, in their turn-on and turn-off times, and in their ability to withstand forward voltage in the off state. A majority of the switch types do not have significant reverse voltage blocking capability.

Where desirable, the generic symbols for the three categories of semiconductor devices can be combined. For example, where reverse voltage blocking is required, a diode may be placed in series with a self-commutated switch as shown in Fig. 8.3(b). Or, if current is to be permitted to flow freely in the reverse direction, a reverse diode can be placed in parallel with the self-commutated switch as shown in Fig. 8.3(c). None of the devices in the three categories has the simple properties of a mechanical switch, i.e., conduction in both directions when on and voltage-blocking ability in both directions when off.

8.2

Controlled Rectifier Systems

A *controlled rectifier system* is one that converts a-c power at constant voltage and frequency into controlled direct voltage or controlled direct current. The direct voltage or current output may be used to supply power to a commutator machine or may be used with an inverter to provide a-c power with controllable frequency, voltage, or current. Under some circumstances, provision can be made for power to be fed to the a-c source from the d-c side, i.e., for the rectifier to act as an inverter.

8.2.1 ▪ Single-Phase Controlled Rectifier

Consider a situation where we wish to supply a direct controlled voltage to a generalized load consisting of series resistance R, inductance L, and internal voltage e_d (such as might represent a commutator machine). Figure 8.4 shows a bridge connection of four thyristors between this load and a sinusoidal voltage source $v_p = \hat{v} \sin \omega t$. Thyristors S_1 and S_2 are turned on (or fired) simultaneously at the angle $\omega t = \alpha$ while S_3 and S_4 are turned on one-half cycle later. Figure 8.5 shows waveforms for an idealized situation with ideal switches, negligible series resistance R, and a large-enough series inductance L to keep the load current i_d continuously flowing. During the period $\alpha < \omega t < \alpha + \pi$ with S_1 and S_2 on, the average value of the load voltage shown in Fig. 8.5(b) is

$$\begin{aligned} \bar{v}_d &= \frac{1}{\pi} \int_{\alpha}^{\alpha+\pi} \hat{v} \sin \omega t \, d\omega t \\ &= \frac{2}{\pi} \hat{v} \cos \alpha \qquad \text{V} \end{aligned} \tag{8.1}$$

Thus, for this mode of operation, the average load voltage can be controlled by controlling only the firing angle α. A typical steady waveform of the steady-state load current i_d is shown in Fig. 8.5(c). If the resistance R is negligibly small, the average load voltage v_d is equal to the internal voltage e_d and

$$L \frac{di_d}{dt} = v_d - e_d \qquad \text{V} \tag{8.2}$$

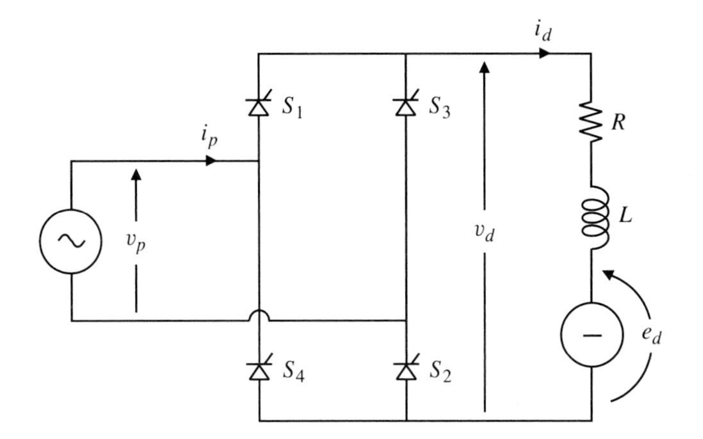

Figure 8.4 Single-phase controlled rectifier/inverter.

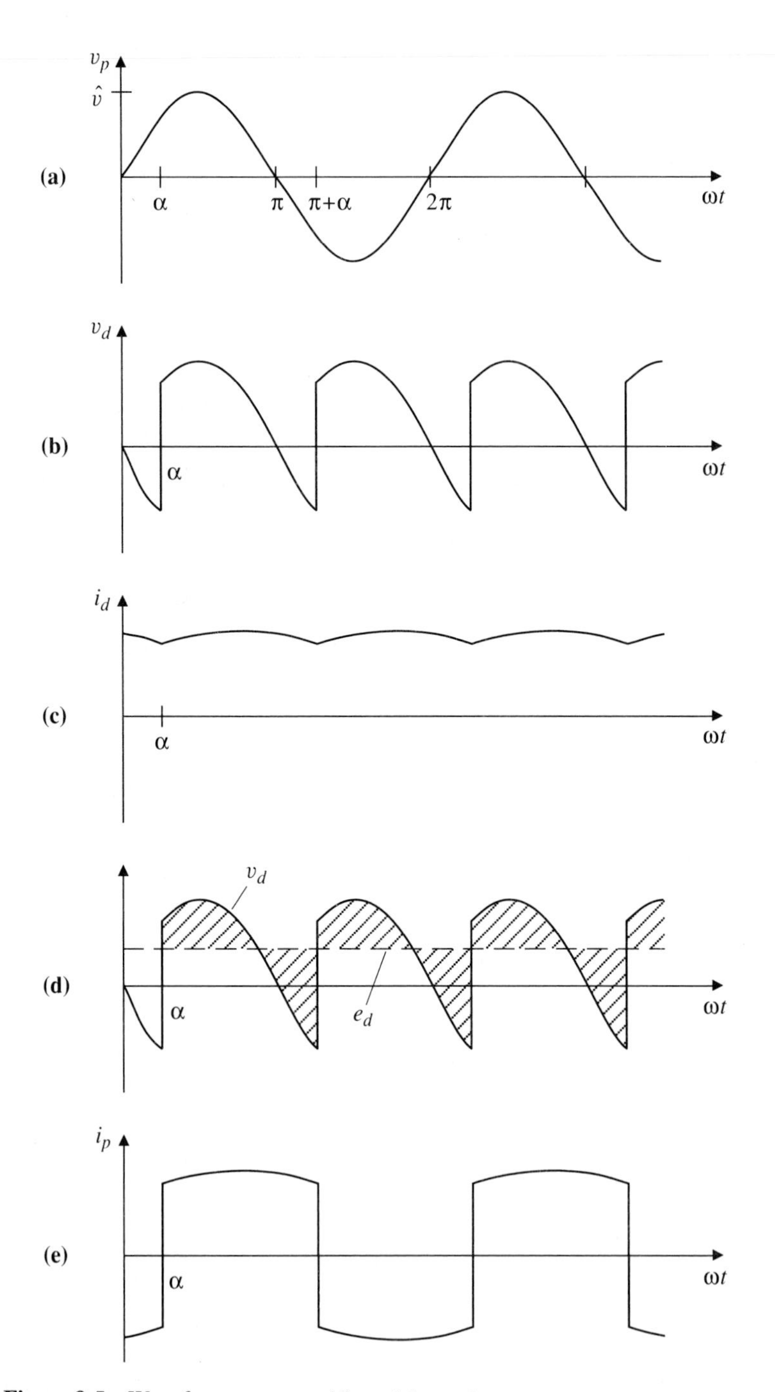

Figure 8.5 Waveforms as a rectifier with continuous current: (a) Supply voltage. (b) Load voltage. (c) Load current. (d) Shaded area representing inductor voltage. (e) Supply current.

As shown in Fig. 8.5(d), the load current increases during the upper shaded period and decreases by an equal amount during the lower shaded period. Because the average voltage across the inductance in the steady state is zero, the two shaded areas must be equal. The ripple of the load current, i.e., its departure from constancy, is inversely proportional to the inductance. The waveform of the supply current is shown in Fig. 8.5(e). With a large value of inductance, this approaches a square waveform.

The same system of Fig. 8.4 also may be operated in an inverter mode, feeding power from the d-c source e_d back to the a-c supply. Again, let us assume a negligible resistance and a large inductance in series with the d-c source e_d. If the delay angle α is made greater than 90°, the steady-state waveforms will be as shown in Fig. 8.6. The average voltage $\bar{v}_d$ is still given by Eq. (8.1) but this is now negative while the current i_d must be positive.

With a finite resistance R in the circuit, the average current $\bar{i}_d$ is

$$\bar{i}_d = \frac{\bar{v}_d - e_d}{R} \quad \text{A} \tag{8.3}$$

Assuming a large inductance L, the supply current will have an essentially square waveform of peak value $\bar{i}_d$. This will have a fundamental frequency component of peak value

$$\hat{i}_{1p} = \frac{4}{\pi}\,\bar{i}_d \quad \text{A} \tag{8.4}$$

and this fundamental component will lag the voltage by a phase angle α.

The square current wave also will have a set of odd-order harmonics with peak values given by

$$\hat{i}_{hp} = \frac{4}{h\pi}\,i_d \quad \text{for } h = 3, 5, 7, \ldots\ldots \quad \text{A} \tag{8.5}$$

These harmonic currents in the supply line may cause interference with other systems including possible resonance with line inductance and shunt capacitance. The harmonics also add to the rms value of the line current which, for a square wave, will be $I_p = \bar{i}_d$.

The apparent power factor at the supply terminals is defined as the ratio of real power to the product of rms supply voltage V_p and rms supply current I_p. Because the switches have been assumed to be perfect, the real power with a constant current is $\bar{v}_d\bar{i}_d$. Thus, using Eq. (8.1), the power factor is

$$pf = \frac{\bar{v}_d\bar{i}_d}{V_p I_p} = \frac{2\sqrt{2}}{\pi}\cos\alpha = 0.9\cos\alpha \tag{8.6}$$

Therefore, the supply power factor can be significantly lower than the ideal value of unity for a controlled rectifier, particularly when the output voltage is low.

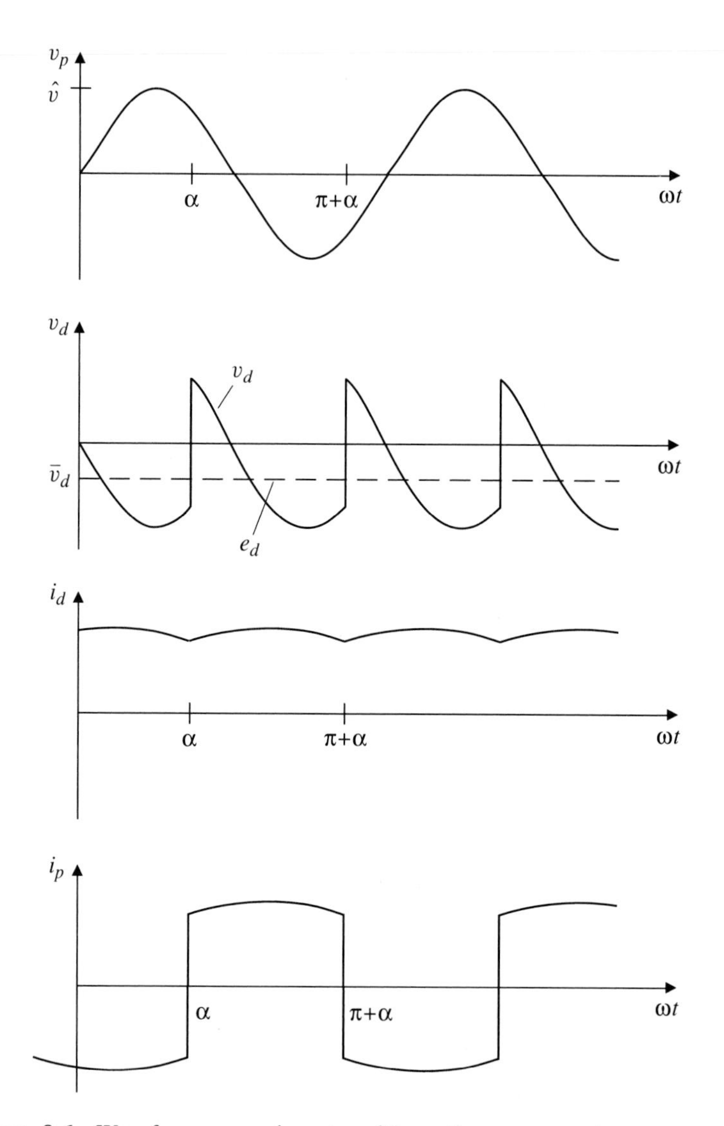

Figure 8.6 Waveforms as an inverter with continuous current.

So far we have considered only a load circuit containing a large value of inductance. The other extreme is one with negligible inductance, significant resistance R, and an internal direct voltage e_d. Typical waveforms for this mode of operation are shown in Fig. 8.7. Load current flows only as long as the instantaneous supply voltage v_p is larger than the internal voltage e_d, i.e.,

$$i_d = \frac{v_d - e_d}{R} \quad \text{for } i_d > 0 \qquad \text{A} \tag{8.7}$$

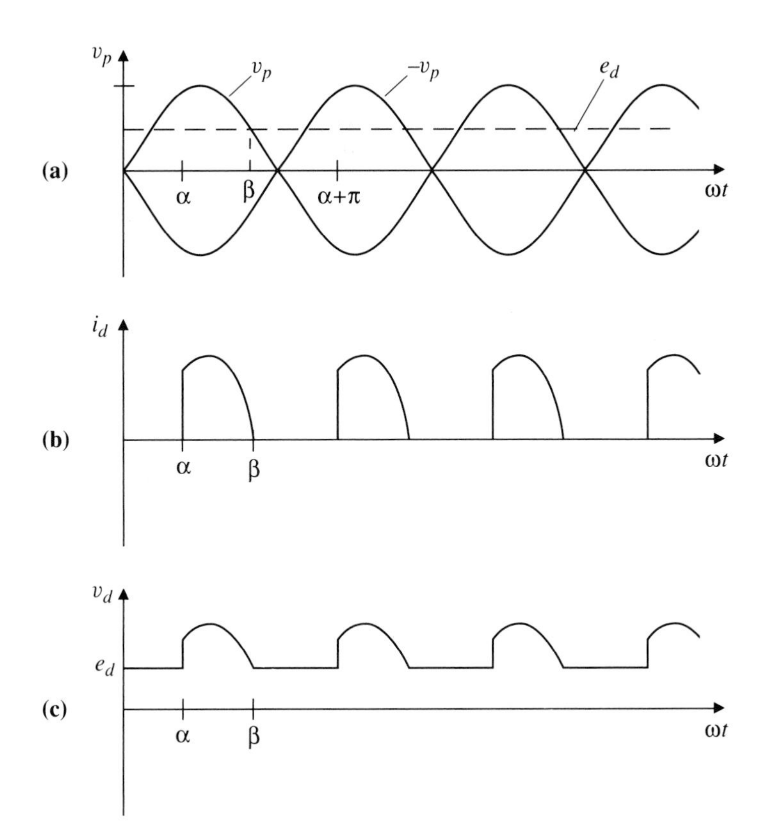

Figure 8.7 Waveforms as single-phase rectifier with discontinuous current.

The current is a series of discontinuous pulses as shown in Fig. 8.7(b). This is the usual condition of operation for small single-phase controlled rectifiers operating from a 50- to 60-Hz supply unless a large inductance is inserted into the load circuit. The load voltage waveform is shown in Fig. 8.7(c). The average value of the load voltage is no longer dependent only on the firing angle α as in Eq. (8.1) but also depends on the internal voltage e_d. As e_d approaches the peak value $\hat{v}$ of the supply, the average load current $\bar{i}_d$ goes to zero.

For a general condition with significant values of R, L, and e_d, the current and voltage waveforms can be determined by solving the first-order differential equation for the system of Fig. 8.4:

$$L \frac{di_d}{dt} + Ri_d + e_d = v_p \qquad \text{V} \tag{8.8}$$

over the interval from the firing angle α until the current returns to zero for discontinuous operation or until $\alpha + \pi$ for continuous-current operation.

An indication of the load current at which operation goes from discontinuous to continuous can be obtained by noting from Fig. 8.5(d) that the shaded areas representing the voltage difference $(v_d - e_d)$ are greatest at $\alpha = \pi/2$ rad, i.e., for $e_d = 0$. This voltage difference produces the rise and fall in the load current. The limiting condition for continuous-current operation at $e_d = 0$ occurs when the load current starts from zero at $\alpha = \pi/2$ and just reaches zero again at $\alpha = 3\pi/2$. For a system in which the resistance R can be ignored but which has significant inductance, the load current for this condition is given by

$$i_d = \frac{1}{L}\int \hat{v} \sin \omega t \, dt$$

$$= -\frac{\hat{v}}{\omega L} \cos \omega t \qquad \text{A} \tag{8.9}$$

The average value of this current pulse is

$$\bar{i}_d = \frac{2}{\pi}\,\hat{i}_d = \frac{2\hat{v}}{\pi \omega L} \qquad \text{A} \tag{8.10}$$

Operation at an average load current greater than this value at any voltage other than zero generally will be continuous because this current at $\alpha = \pi/2$ represents the condition with the highest ripple.

Where it is desired to have load current flow in both directions as well as reversible load voltage, two anti-parallel thyristor bridges, each of the form shown in Fig. 8.4, can be connected between the source and the load to form the system of Fig. 8.8, which is usually referred to as a *dual converter*.

If the thyristors in the system of Fig. 8.4 are replaced by diodes, the circuit is that of a full-wave diode bridge rectifier. Its output voltage is essentially constant for a constant a-c supply voltage at the value given in Eq. (8.1) with firing angle α set to zero.

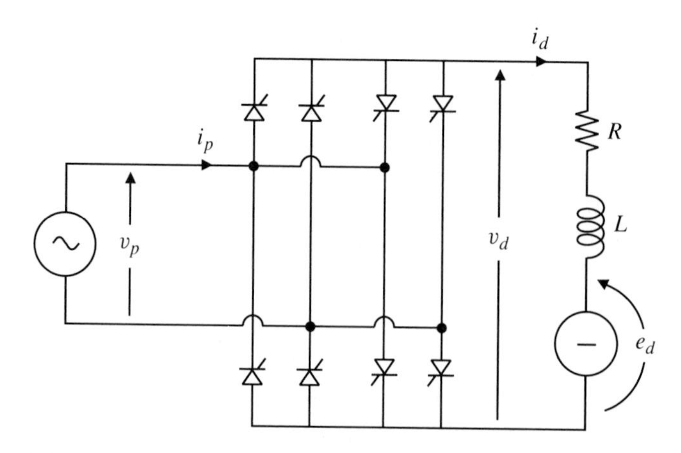

Figure 8.8 Dual converter for four-quadrant operation.

EXAMPLE 8.1 A particular load can be modeled as a direct-voltage source of 144 V in series with a resistance of 0.8 Ω and an inductance of 48 mH. It is desired to supply this load with an average current of 20 A using a 230-V, 60-Hz, single-phase supply and a single-phase controlled rectifier.

(a) Determine the required firing angle assuming continuous current.

(b) Estimate the average load current at which operation changes from the continuous to the discontinuous mode with the direct-voltage source set at zero, and use this answer to determine whether the condition in (a) is, in fact, continuous.

Solution

(a) The required average value of the load voltage is

$$\bar{v}_d = e_d + R\bar{i}_d = 144 + 0.8 \times 20 = 160 \qquad \text{V}$$

Then, from Eq. (8.1),

$$\alpha = \cos^{-1}\left[\frac{\pi}{2}\frac{\bar{v}_d}{\hat{v}}\right] = \cos^{-1}\left[\frac{\pi}{2}\frac{160}{\sqrt{2} \times 230}\right] = 39.4°$$

(b) Ignoring the resistance and setting $e_d = 0$, the average current at the boundary between continuous and discontinuous current is given approximately, from Eq. (8.10), by

$$\bar{i}_d = \frac{2\hat{v}}{\pi\omega L} = \frac{2\sqrt{2} \times 230}{\pi \times 377 \times 0.048} = 11.4 \qquad \text{A}$$

At larger values of the voltage e_d, the current ripple will be smaller. Thus, except for the small effect of the resistance, this rectifier will operate with continuous current for all values of direct-source voltage e_d up to 144 V.

8.2.2 ▪ Three-Phase Controlled Rectifier

When the required power is greater than a few kilowatts, the convenient source is usually a three-phase power system, and a three-phase bridge rectifier having a circuit such as that shown in Fig. 8.9 is frequently used. Six thyristors connect the three-phase voltage source to the generalized load circuit. The waveforms of the three line-to-neutral source voltages are shown in Fig. 8.10(a). The thyristors are turned on in the sequence in which they are numbered in Fig. 8.9 and at 60° intervals. At any instant in time, two thyristors are conducting. Suppose S_1 is fired at the angle α, as shown in Fig. 8.10(b), connecting source phase a to the positive load terminal. As a result of previous firing, S_6 is already connecting the negative load terminal to the source phase b which has the most negative voltage. As shown in Fig. 8.10(a), this connection will continue until S_2 is fired at $\alpha + 60°$. Because at this time v_{cn} is more negative than v_{bn}, this causes

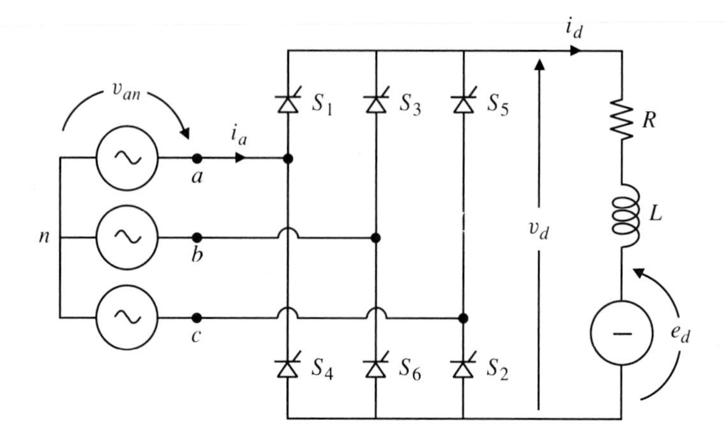

Figure 8.9 Three-phase controlled rectifier/inverter.

the current in S_6 to be extinguished, and S_6 is turned to its off condition. For the next 60° interval, the voltage v_{ac} will be applied to the load. This process is repeated in each 60° interval. The heavy lines in Fig. 8.10(a) indicate the voltages with respect to the source neutral of the two load terminals. The waveform of the voltage across the load is shown in Fig. 8.10(c).

It is convenient to choose the reference for measurement of firing angle α at the instant that gives the maximum value of average load voltage, i.e., where $v_{an} = v_{cn}$. The average load voltage is then given by

$$\begin{aligned}\bar{v}_d &= \frac{3}{\pi}\int_{\alpha+\pi/3}^{\alpha+2\pi/3} V_{ab}\, d\omega t \\ &= \frac{3}{\pi}\hat{v}_{L\text{-}L}\cos\alpha \\ &= \frac{3\sqrt{2}}{\pi}V_{L\text{-}L}\cos\alpha \\ &= 1.35\,V_{L\text{-}L}\cos\alpha \qquad \text{V}\end{aligned} \tag{8.11}$$

where $\hat{v}_{L\text{-}L}$ is the peak line-to-line source voltage and $V_{L\text{-}L}$ is its rms value. This expression applies only if the load current is continuous.

In the steady state, the average value of the load current will be

$$\bar{i}_d = \frac{\bar{v}_d - e_d}{R} \qquad \text{A} \tag{8.12}$$

If the load inductance L is large, the load current will be nearly constant at $i_d = \bar{i}_d$ and will have only a small ripple component. Under this condition, the current in a representative phase of the supply will have a waveform approximating the idealized form shown in Fig. 8.10(d). Each of the six thyristors conducts for an interval of $2\pi/3$ rad or 120°.

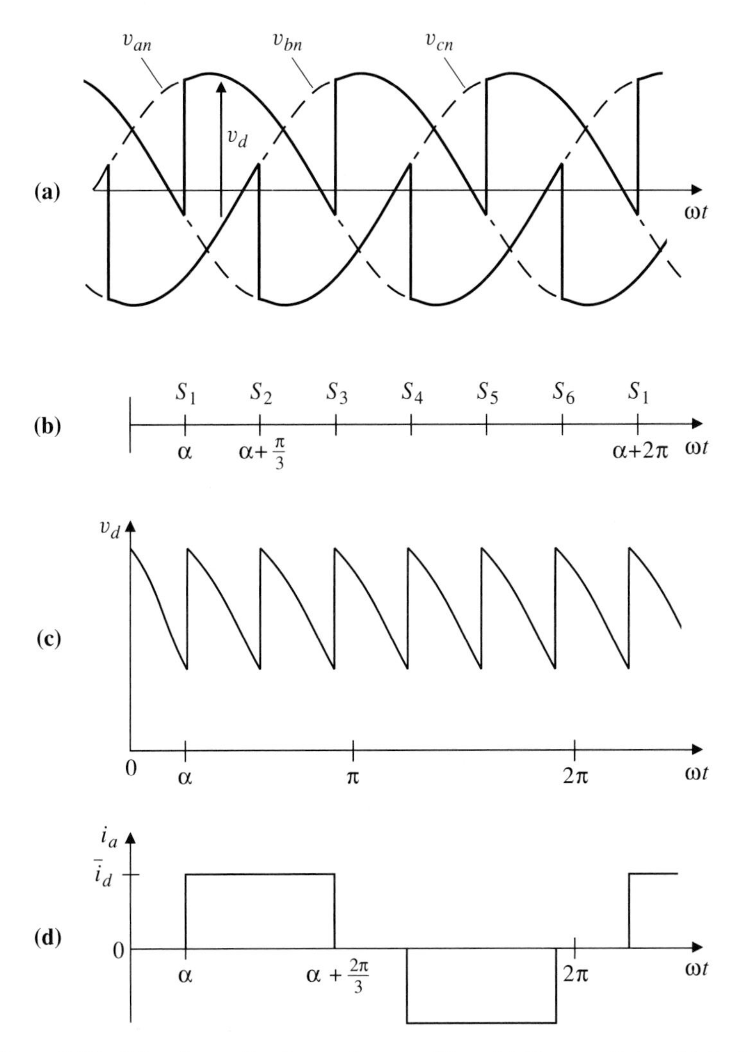

Figure 8.10 Waveforms for three-phase rectifier. (a) Phase voltages (dotted lines) with positive and negative load connections (heavy lines). (b) Firing sequence. (c) Load voltage. (d) Supply current.

The rms value of the fundamental frequency component of the source current can be obtained by Fourier analysis of the waveform of Fig. 8.10(d) as

$$
\begin{aligned}
I_{1a} &= \frac{1}{\sqrt{2}}\,\frac{4}{\pi}\,\sin\frac{\pi}{3}\,\bar{i}_d \\
&= \frac{\sqrt{6}}{\pi}\,\bar{i}_d \qquad (8.13)\\
&= 0.78\bar{i}_d \qquad \text{A}
\end{aligned}
$$

The phase angle of this fundamental-frequency current relative to the line-to-neutral voltage of phase a is the firing angle α. Using Eqs. (8.11) and (8.13), it can be shown readily that the source power arising from the fundamental component of source current is equal to the load power:

$$\begin{aligned} P &= \sqrt{3}\, V_{L\text{-}L} I_{1a} \cos\alpha \\ &= \sqrt{3}\left[\frac{\pi \bar{v}_d}{3\sqrt{2}}\right]\left[\frac{\sqrt{6}\,\bar{i}_d}{\pi}\right] \\ &= \bar{v}_d \bar{i}_d \qquad \text{W} \end{aligned} \tag{8.14}$$

The rms value of the source current including its odd-order harmonics terms is

$$I_{a(rms)} = \sqrt{\tfrac{2}{3}}\,\bar{i}_d = 0.82\bar{i}_d \qquad \text{A} \tag{8.15}$$

The source power factor then is given for this idealized current waveform by

$$\begin{aligned} pf &= \frac{\bar{v}_d \bar{i}_d}{\sqrt{3}\, V_{L\text{-}L} I_{a(rms)}} \\ &= \frac{3}{\pi}\cos\alpha \\ &= 0.955 \cos\alpha \end{aligned} \tag{8.16}$$

Thus, with a three-phase converter, the power factor is not reduced by harmonic currents as much as was derived for a single-phase converter in Eq. (8.6). However, the supply power factor can be much less than unity when the rectifier is operated with low output voltage.

Depending on the power rating and location of the converter, it may be necessary to reduce the current harmonics in the power supply lines. This can be achieved by use of an input filter consisting of series inductance in each phase and shunt capacitors connected between the phases on the rectifier side. With such a filter, the analysis of the rectifier system will become more complex, and reference should be made to a text on power electronics for further information.

The three-phase converter of Fig. 8.9 will operate as an inverter if the firing angle α is increased beyond 90°. The limit is a value of α just sufficiently less than 180° to ensure that each thyristor is turned on when its forward voltage is still positive. Also, for four-quadrant operation with reversible load current, a dual converter consisting of two anti-parallel, three-phase bridges can be used.

In some applications, a source of controlled direct current rather than a source of controlled voltage is required. This can be achieved by use of a controlled rectifier system including means of measuring the output direct current i_d and using the difference between this measured value and the desired value of current i_d^* to control the firing angle of the rectifier as shown in Fig. 8.11.

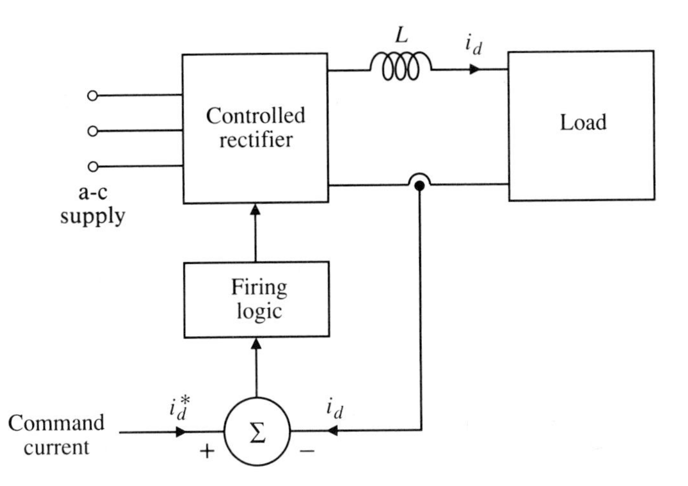

Figure 8.11 Controllable direct-current source.

EXAMPLE 8.2 A three-phase controlled rectifier is to be used to supply power to a commutator machine that can be modeled as a source voltage equal to 3.1 V per rad/s of speed in series with a resistance of 0.06 Ω and an inductance of 12.5 mH. The available supply is at 440 V (L-L) at 60 Hz.

(a) What should be the firing angle if the motor is to supply a mechanical power of 100 kW at a speed of 1700 r/min?

(b) Estimate the rms supply current and the supply power factor.

(c) Estimate the average load current at which the mode of operation would become discontinuous at zero speed.

Solution

(a) The speed is to be

$$\omega_0 = 1700 \times \frac{2\pi}{60} = 178 \qquad \text{rad/s}$$

The motor torque is then given by

$$T = \frac{P_0}{\omega_0} = \frac{100\,000}{178} = 562 \qquad \text{N}\cdot\text{m}$$

Using Eq. (4.16), the average load current is

$$\bar{i}_d = \frac{T}{k\Phi} = \frac{562}{3.1} = 181.2 \qquad \text{A}$$

The required average load voltage is then, from Eq. (4.22),

$$\bar{v}_d = k\Phi\omega_0 + R\bar{i}_d = 3.1 \times 178 + 0.06 \times 181.2 = 562.7 \qquad \text{V}$$

Then, from Eq. (8.11), the required firing angle is

$$\alpha = \cos^{-1} \frac{\bar{v}_d}{1.35 \times 440} = \cos^{-1} 0.947 = 18.7°$$

(b) Assuming that the series inductance is sufficient to maintain approximately constant load current, Eq. (8.15) can be used for an estimate of the rms line current.

$$I_{a(rms)} = 0.82\bar{i}_d = 148.6 \qquad \text{A}$$

and the power factor, from Eq. (8.16), will be

$$pf = 0.955 \cos \alpha = 0.905$$

(c) At zero speed, $e_d = 0$ and v_d can be assumed to be negligibly small. For this condition, the firing angle α is about 90°. The load voltage waveform will be

$$v_d = \hat{v}_{L\text{-}L} \sin \omega t \quad \text{for } \frac{5\pi}{6} < \alpha < \frac{7\pi}{6}$$

The load current for this interval will be, ignoring the resistance,

$$i_d = \frac{1}{L} \int v_d \, dt = \frac{\hat{v}_{L\text{-}L}}{\omega L} \left(-\cos \omega t - \frac{\sqrt{3}}{2} \right)$$

The average load current over this interval now can be found as

$$\begin{aligned} \bar{i}_d &= \frac{3}{\pi} \int_{5\pi/6}^{7\pi/6} i_d \, d\omega t \\ &= 0.089 \frac{\hat{v}_{L\text{-}L}}{\omega L} \\ &= \frac{0.089 \times \sqrt{2} \times 440}{377 \times 0.0125} \\ &= 11.75 \qquad \text{A} \end{aligned}$$

This is about 6.5% of the load current in (a). This rectifier can be expected to operate with continuous current over most of its potential load range.

8.3

Choppers

In some situations, a supply of constant direct voltage may be available, and the requirement may be for a source of adjustable direct voltage. An example would be the supply to the commutator motors on a subway train that receives power from contact with a trackside rail having a constant direct voltage which in turn might possibly be supplied from a rectifier. In this section, we will introduce some simple switching circuits that can convert power at a constant direct voltage to power at a variable direct voltage that is either less than or greater than the constant voltage, and can also accommodate a reversal of power flow. The general term *chopper* is used for these circuits because the approach is to use self-commutated semiconductor switches to chop sections out of the supply voltage and thus provide for control of the average value of the output voltage. The various types of choppers differ in the number of quadrants of the voltage-current diagram in which they are capable of operating.

8.3.1 ▪ Step-Down Choppers

Figure 8.12(a) shows the circuit diagram of a step-down chopper capable of feeding power at a controllable voltage v_d to a generalized load circuit from a direct-source voltage V, subject to the limitation that $\bar{v}_d \leq V$. S_1 is a self-commutated switch capable of being turned on or off by a timed gate signal; D_1 is a diode.

Assume an initial condition with zero load current and S_1 off. If e_d is positive, the initial current i_d is zero. When S_1 is turned on, the load current i_d will increase according to the expression

$$\frac{di_d}{dt} = \frac{V - e_d - Ri_d}{L} \quad \text{A/s} \tag{8.17}$$

as shown in Fig. 8.12(b). If the resistance drop is small, this rate of increase is approximately constant. If S_1 is turned off now, load current will continue to flow through diode D_1. The load current now will decrease at a rate

$$\frac{di_d}{dt} = \frac{-e_d - Ri_d}{L} \quad \text{A/s} \tag{8.18}$$

Figure 8.12(b) shows a sequence of switching actions that keep the average value of the load current nearly constant by appropriate choice of the relation between the time ON and the time OFF of the switch S_1.

Typical steady-state waveforms for this chopper are shown in Fig. 8.13. The switch S_1 is ON for a time t_{ON} and OFF for the remainder of the repetition period T. During the OFF period, the load voltage v_d is equal to zero as the load current flows through the diode that is assumed to be ideal. From Fig. 8.13(a), it is seen that the average load voltage is determined by the proportion of ON time in the period T, i.e.,

$$\bar{v}_d = \frac{t_{ON}}{T} V \quad \text{V} \tag{8.19}$$

The steady-state load current waveform is shown in Fig. 8.13(b). This waveform is made up of sections having the exponential form characteristic of the

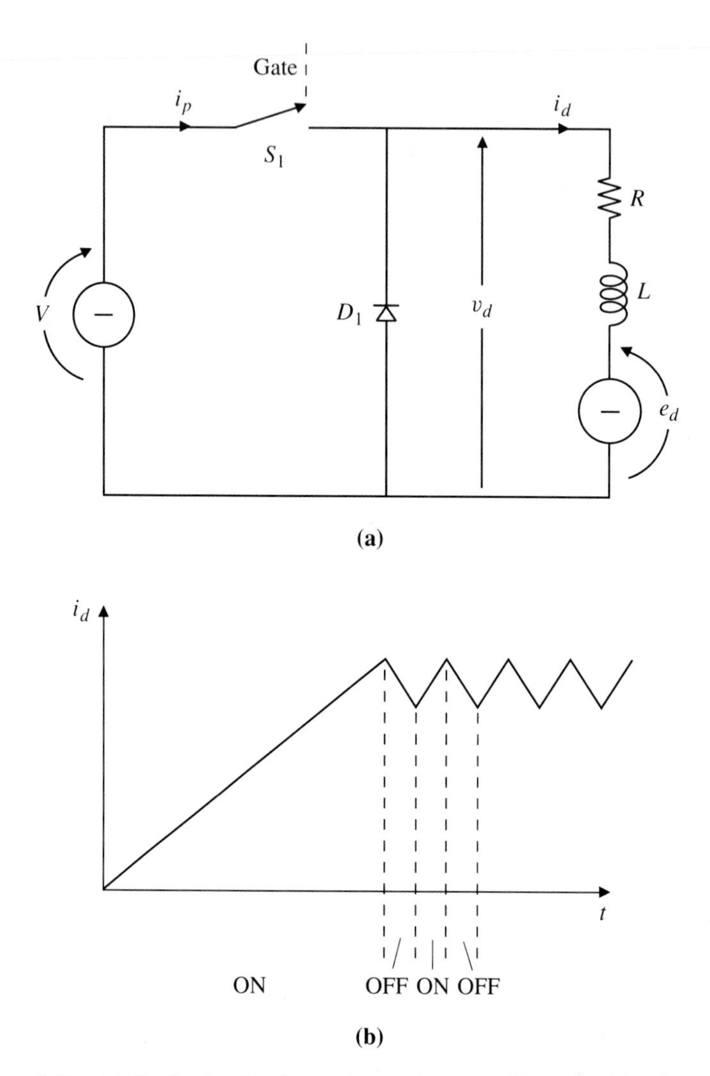

Figure 8.12 (a) Basic circuit of step-down chopper. (b) Typical load current on starting.

first-order differential Eqs. (8.17) and (8.18). However, if the time constant $\tau = L/R$ is considerably greater than the period T, these sections of exponentials may be assumed to be nearly straight lines. In any case, the average load current will be given by

$$\bar{i}_d = \frac{\bar{v}_d - e_d}{R} \qquad \text{A} \tag{8.20}$$

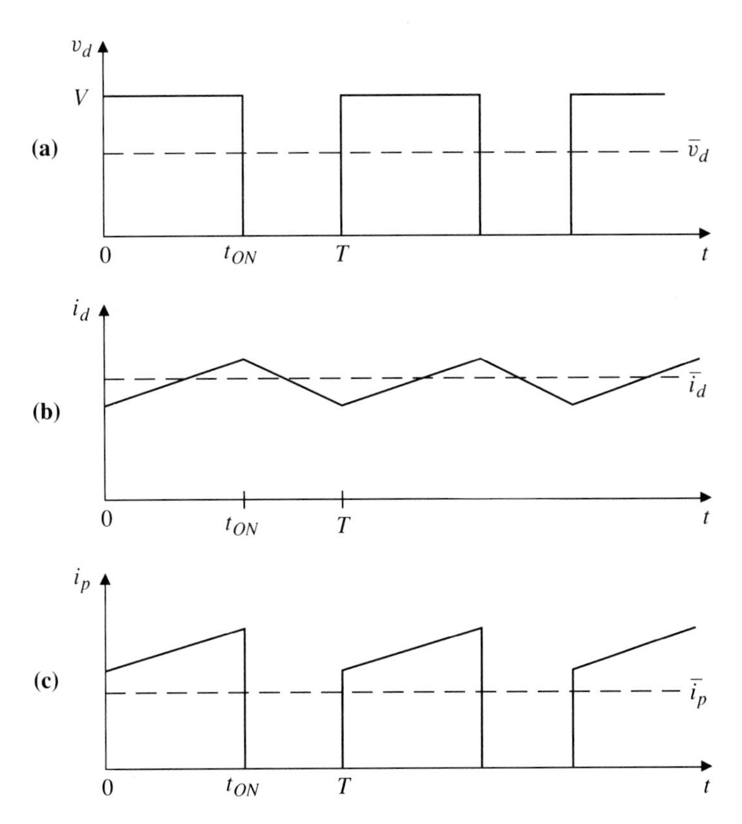

Figure 8.13 (a) Load voltage. (b) Load current. (c) Supply current.

The source current i_p is equal to the load current i_d during the ON period and is zero during the OFF period, as shown in the waveform of Fig. 8.13(c). Its average value is

$$\bar{i}_p = \frac{t_{ON}}{T}\,\bar{i}_d \qquad \text{A} \tag{8.21}$$

From Eqs. (8.19) and (8.21), it can be shown that the output power equals the input power in this idealized system.

The above relations apply as long as the load current is continuous. This is the normal operating condition of a chopper. In order to estimate the conditions under which the current can become discontinuous, note that, if the resistance R is small, the rates of rise and fall of the load current in Eqs. (8.17) and (8.18) are independent of the magnitude of the load current. Thus, with continuous current, the amplitude of the ripple component in the load current will remain the same as the average load current changes. The limit condition

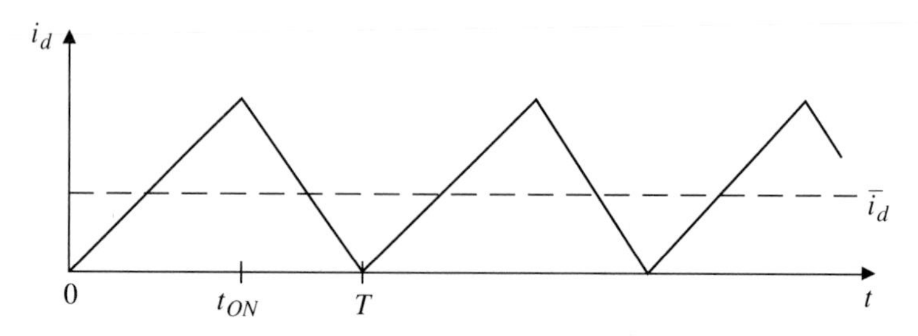

Figure 8.14 Limiting condition for continuous current.

for continuous load current is shown in Fig. 8.14. For the interval $0 < t < t_{ON}$, and ignoring resistance R,

$$i_d = \left(\frac{V - e_d}{L}\right)t \quad \text{A} \tag{8.22}$$

During the remainder of the period T, the load current decreases linearly to zero. Using Eq. (8.19) and noting that $e_d = \bar{v}_d$ when the effect of the resistance R is ignored, the average load current for this limit condition is

$$\bar{i}_d = \left(V - \frac{t_{ON}}{T}V\right)\frac{t_{ON}}{2L} \quad \text{A} \tag{8.23}$$

This can be shown to have a maximum value at $t_{ON} = T/2$. For this condition of half-voltage output, the maximum load current at the boundary of continuous current operation then is approximated by

$$\bar{i}_d = \frac{VT}{8L} = \frac{V}{8Lf} \quad \text{A} \tag{8.24}$$

where f is the frequency in cycles per second of the switch. Practical switching frequencies depend on the type and size of switch used. Typical values may be in the range 0.2 to 1 kHz for high current gate-turn-off devices or in the range 3 to 100 kHz for lower current transistors. In some devices, it is desirable to choose a switching frequency in excess of 20 kHz to avoid audible noise from oscillating magnetic forces in the system. However, there are power losses in the switch that increase with switching frequency.

The approximate expressions for load current in Eqs. (8.22) and (8.24) have been developed ignoring the circuit resistance. In cases where this approximation is not adequate, the straight line current segments of Fig. 8.14 can be replaced by the appropriate sections of the exponential solutions to the differential equations of Eqs. (8.17) and (8.18).

The basic chopper is a source of controlled direct voltage. It can be transformed into a source of controlled direct current using the same approach as shown for a controlled rectifier in Fig. 8.11. The difference between the measured output current and the desired or command current is used to control the ON-OFF switching of the chopper.

Where a controlled direct current is required and the original power source is an a-c system, an arrangement consisting of a diode rectifier to produce constant direct voltage followed by a chopper with control of its direct output current may be preferred to the system of Fig. 8.11. It overcomes the low power factor of the a-c supply which is characteristic of a controlled rectifier over a large part of its operating range.

EXAMPLE 8.3 A load consisting of a direct-voltage source of 72 V in series with a resistance of 0.015 Ω and an inductance of 0.85 mH is to be supplied with an average current of 420 A from a 120-V direct-voltage source. A chopper capable of a frequency of 1 kHz is to be used. Determine the required ON time of the chopper.

Solution To assess what assumptions can reasonably be made, let us first check the time constant of the load.

$$\tau = \frac{L}{R} = \frac{0.85 \times 10^{-3}}{0.015} = 57 \qquad \text{ms}$$

This is much larger than the period of $T = 1$ ms of the chopper; thus, the current waveform segments can be assumed to be at constant slope ignoring the resistance.

The required average load voltage is, from Eq. (8.20),

$$\bar{v}_d = e_d + R\bar{i}_d = 72 + 0.015 \times 420 = 78.3 \qquad \text{V}$$

Then, from Eq. (8.19), the time t_{ON} is

$$t_{ON} = \frac{\bar{v}_d}{V}\,T = \frac{78.3}{120} \times 10^{-3} = 0.65 \qquad \text{ms}$$

8.3.2 ▪ Step-Up Choppers

In some situations, the delivery of power from a low-voltage source to a higher-voltage system is desired. For example, it might be advisable to provide braking on a subway train by feeding power back to the direct-voltage supply from the commutator driving motors. Figure 8.15(a) shows a circuit that is capable of this mode of operation. The reference directions of the variables have been chosen to be the same as for the step-down chopper for reasons that will become evident in the next section. Because the internal voltage e_d is assumed to be less than the source voltage V, nothing happens until the self-commutated switch S_2 is turned on. Turning S_2 on causes the current i_d to increase in a negative direction. Now, when S_2 is turned off, the presence of the inductance L ensures that the load current will continue to flow through the diode D_2 into the source.

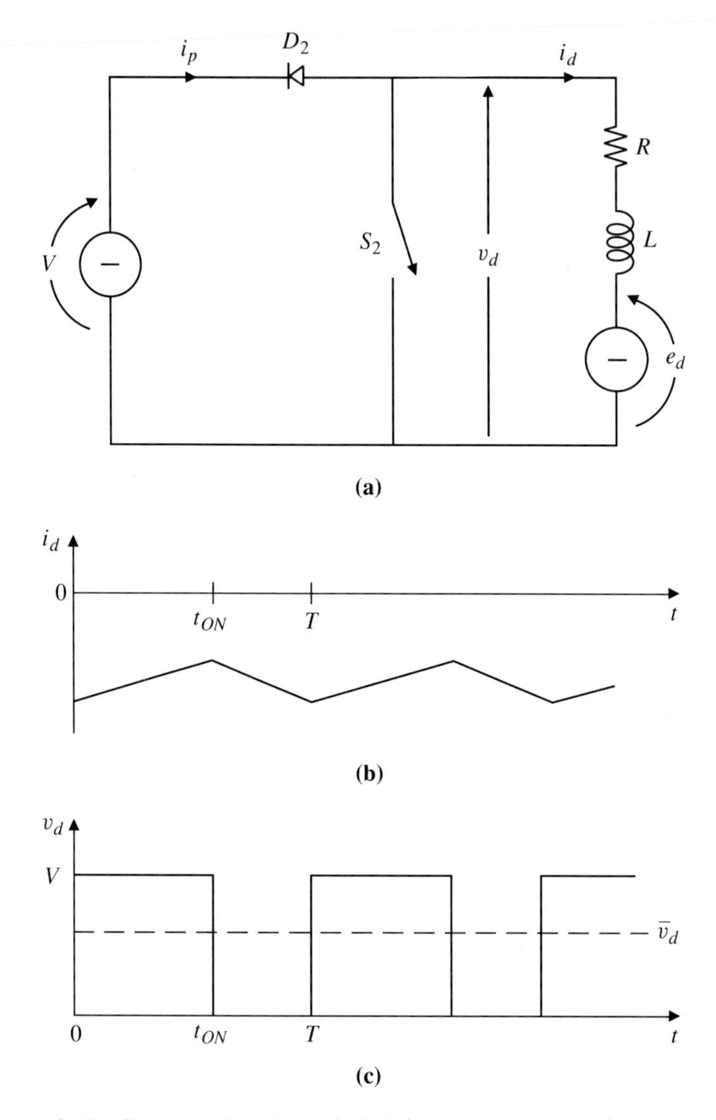

Figure 8.15 Step-up chopper: (a) circuit, (b) load current, (c) load voltage waveform.

During the time interval when S_2 is on,

$$\frac{di_d}{dt} = \frac{-e_d - Ri_d}{L} \qquad \text{A/s} \tag{8.25}$$

During the time when S_2 is off,

$$\frac{di_d}{dt} = \frac{V - e_d - Ri_d}{L} \quad \text{A/s} \tag{8.26}$$

Consider a starting point where load current i_d is already flowing back through D_2 to the source as shown in Fig. 8.15(b). In the next section, we will want to combine this chopper with a step-down unit to achieve a two-quadrant system. For this reason, we will designate t_{ON} as the time interval of conduction through the diode. During this interval, the rate of change of current is positive, i.e., from Eq. (8.26), the magnitude of the negative current i_d decreases. At t_{ON} the switch S_2 is closed and the rate of change of the current becomes negative, increasing the load current magnitude.

The load voltage is V when the diode is conducting and is zero with the switch S_2 closed, as shown in Fig. 8.15(c). Thus, the average load voltage is related to the supply voltage by

$$\bar{v}_d = \frac{t_{ON}}{T} V \quad \text{V} \tag{8.27}$$

The average load current is

$$\bar{i}_d = \frac{\bar{v}_d - e_d}{R} \quad \text{A} \tag{8.28}$$

Because of our choices of the current direction and the designation of t_{ON}, these expressions are the same as for the step-down chopper in Eqs. (8.19) and (8.20). The average source current is as given in Eq. (8.21).

8.3.3 ▪ Two- and Four-Quadrant Choppers

Frequently, a combination of the properties of the step-up and the step-down choppers is required. This can be provided by the circuit of Fig. 8.16(a). This circuit is capable of providing a controllable positive average load voltage v_d while accommodating load current in either direction with a smooth transition through zero current, i.e., it operates in the first and second quadrants of the v_d-i_d diagram shown.

In this chopper, the switches S_1 and S_2 are turned on alternately, as shown in Fig. 8.16(b), with a very short interval between to allow the switch being turned off to recover its forward blocking capability. For operation when i_d is always positive, switch S_1 and diode D_1 act as a step-down chopper. Neither S_2 or D_2 is operative with this current direction. Similarly, when i_d is always negative, switch S_2 and diode D_2 act as a step-up chopper, allowing S_1 and D_1 to be ignored. Figure 8.16(c) shows a typical current waveform during the transition from positive to negative current. When the current is positive, it flows through S_1 or D_1 while negative current flows through S_2 or D_2.

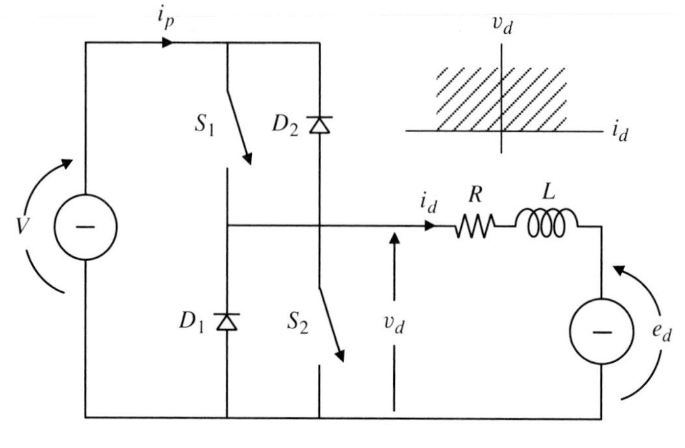

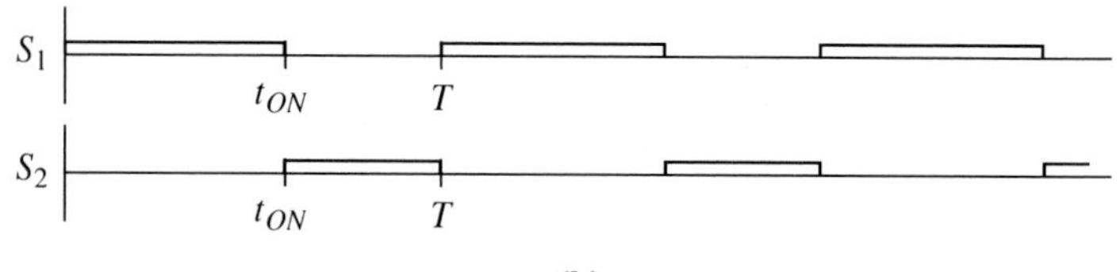

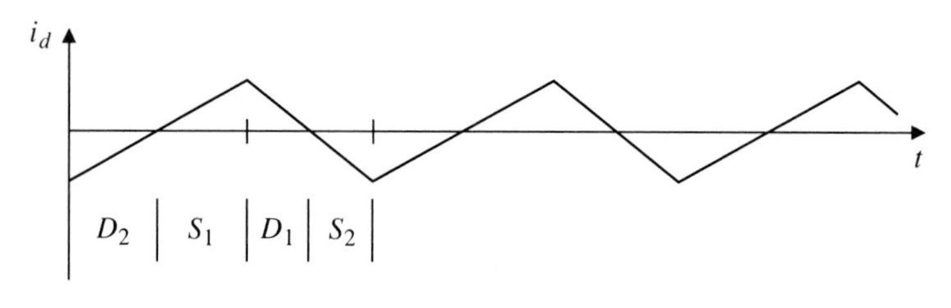

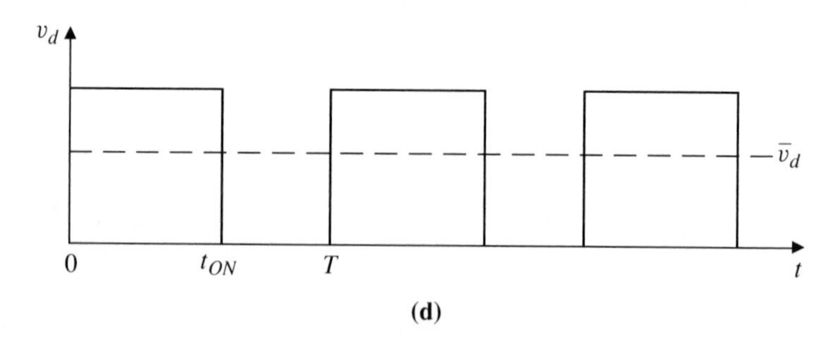

Figure 8.16 Two-quadrant chopper: (a) circuit, (b) switching sequence, (c) load current near zero value, (d) load voltage.

The voltage and current relations for this two-quadrant chopper are the same as for both step-down and step-up systems, i.e.,

$$\bar{v}_d = \frac{t_{ON}}{T} V \qquad \text{V} \tag{8.29}$$

and

$$\bar{i}_d = \frac{\bar{v}_d - e_d}{R} \qquad \text{A} \tag{8.30}$$

Some applications require a four-quadrant chopper, i.e., ability to provide voltage of reversible polarity as well as current of reversible polarity. An example would occur in a drive motor for a robot arm where positive current would be needed for acceleration and negative current for deceleration. Positive voltage would be needed for forward velocity and negative voltage for negative velocity. This four-quadrant operation can be achieved by use of two of the systems of Fig. 8.16(a) connected as shown in Fig. 8.17. For operation with positive voltage v_d, switch S_4 is held in the ON condition which, together with diode D_3, produces effectively the load connection of Fig. 8.16(a). For negative load voltage, switch S_2 is held in the ON condition, allowing devices S_3, D_3, S_4, and D_4 to act as a chopper in quadrants 3 and 4.

8.4 Inverters

The purpose of an *inverter* is to convert power from a direct voltage or current source to a controllable-frequency output with controllable voltage or current. In some inverters, the direction of power transfer also may be reversible. This

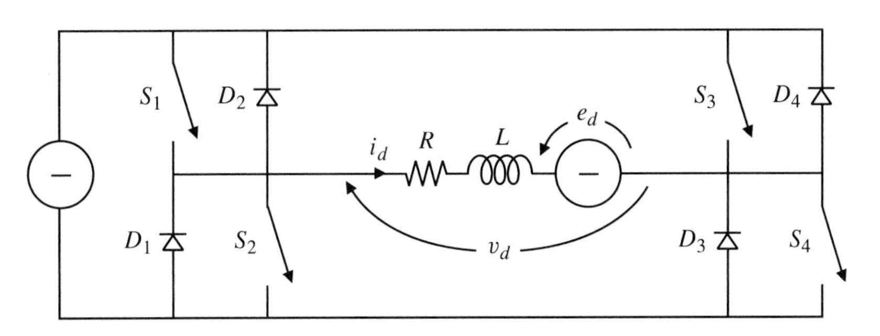

Figure 8.17 Four-quadrant chopper.

section introduces some of the simpler forms of inverter and discusses certain idealized properties. The reader who wishes a more detailed treatment should consult a power electronics text.

In most applications, the source of power is an a-c supply, either single-phase for relatively low power or three-phase for higher power loads. The direct voltage that is required as the input for most inverters can be produced using rectifiers as discussed in Section 8.3. In some instances, a direct-voltage supply may be available such as from batteries in an electric vehicle or from silicon cells in a solar-energy converter.

8.4.1 ▪ A Simple Voltage-Source Inverter

A major concern in this text is the provision of three-phase power at variable voltage and frequency to supply induction and synchronous motors. Figure 8.18(a) shows the circuit of a simple three-phase voltage-source inverter. This system uses six self-commutated switches, S_1 to S_6, to produce the variable-frequency output. It requires a controllable direct-voltage source to provide the variable alternating-voltage output. In Fig. 8.18(a), this controllable direct-voltage source is a controlled rectifier operating from the a-c supply. Typically, a large capacitor C is connected across the direct-voltage link between the rectifier and the inverter. This can be considered as the approximate equivalent of the internal source voltage represented in the load circuits of the rectifiers in Section 8.3.

In this inverter, the switches are turned on in the sequence in which they are numbered. The gate signals holding the switches in their ON condition are shown in Fig. 8.18(b). Each is held ON for just under one-half cycle and then is turned OFF. It will be seen that this normally causes three switches to be conducting at any time, two connected to one source terminal and one to the other. When any switch is ON, it and the diode connected in antiparallel with it constitute an effective short circuit. Thus, with two switches in the ON condition connected to one source terminal, there is a short circuit between the two corresponding output or load phases. The waveforms of the line-to-line output voltages are shown in Fig. 8.18(c). These consist in each half-cycle of a square wave of voltage equal in magnitude to the direct-link voltage v_d and having a duration of $2\pi/3$ rad or 120°. During the remaining 60°, the voltage is zero.

Normally, the three-phase load system will be balanced. With two phases short circuited together during each interval of 60°, the source voltage v_d will be distributed with two-thirds of its magnitude across one phase and the remaining third across the two parallelled phases. A typical six-step waveform for the voltage of reference phase a with respect to the load neutral is shown in Fig. 8.18(d). The waveforms for phases b and c are delayed by $2\pi/3$ and $4\pi/3$ rad, respectively.

The voltage waveforms of Fig. 8.18 can be represented as a series of sine-waves using Fourier analysis. For example, the series for the line-to-line voltage v_{ab} is given by

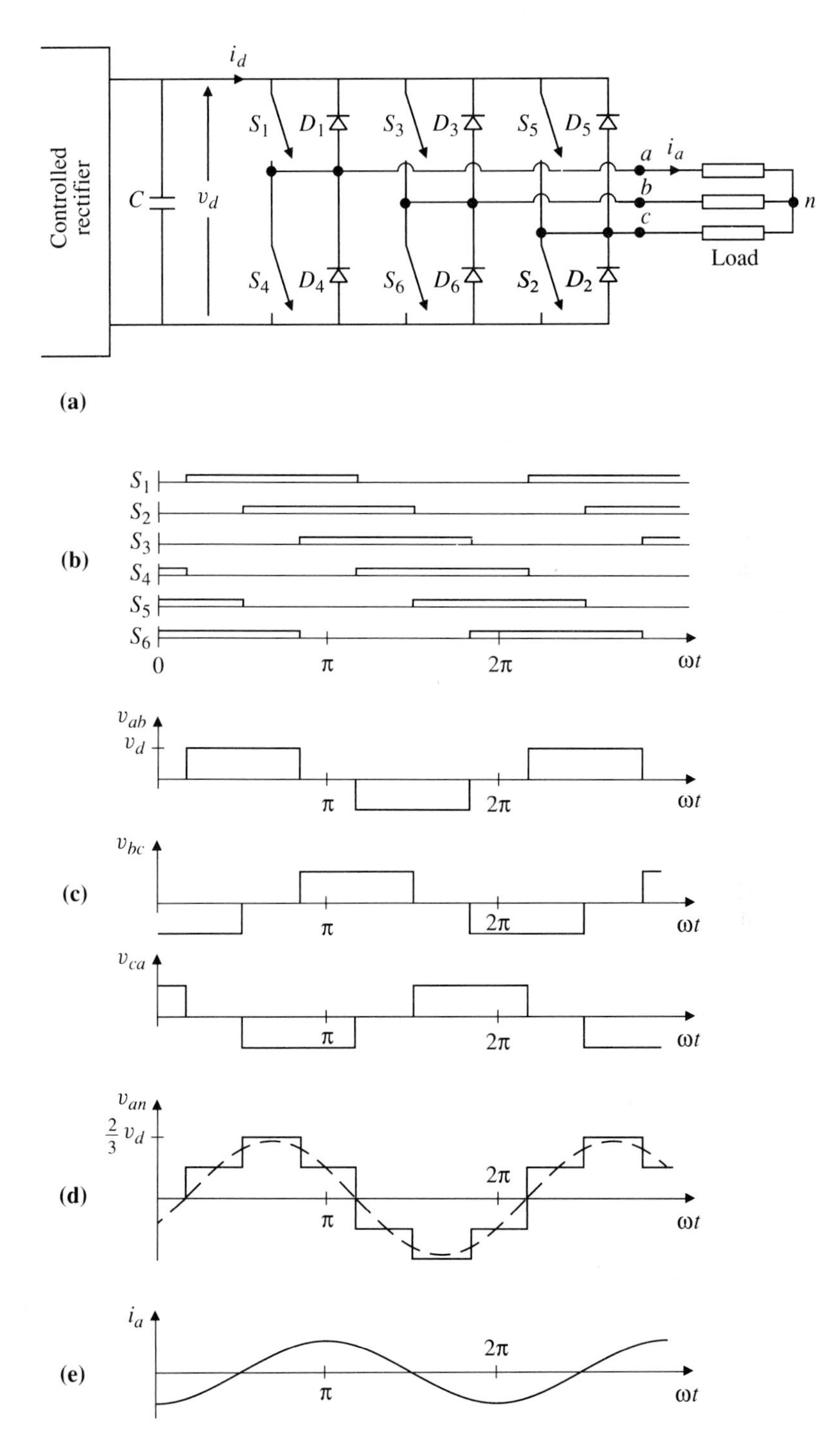

Figure 8.18 Voltage-source inverter with waveforms.

$$v_{ab} = \frac{4}{\pi}\cos(\pi/6)\left[\sin \omega t - \frac{1}{5}\sin 5\omega t + \frac{1}{7}\sin 7\omega t + (-1)^{(h+1)/2}\frac{1}{h}\sin h\omega t\right] \tag{8.31}$$

Note the absence of the triplen harmonics because these would be in phase with each other in the three line-to-line voltages and these three voltages must sum to zero. The remaining odd-order harmonics have magnitudes that are inversely proportional to the harmonic order h. For loads that are highly inductive, these harmonics may not be particularly significant because the inductive impedance is proportional to the harmonic order; thus, the harmonic current will be approximately inversely proportional to h^2. For such loads, the phase currents in the load may be nearly sinusoidal at the fundamental frequency. Therefore, it is convenient for many analyses to use only the fundamental component of the load voltage. The rms value of the line-to-line fundamental voltage is, from Eq. (8.31),

$$\begin{aligned} V_{1(L\text{-}L)} &= \frac{1}{\sqrt{2}}\frac{4}{\pi}\frac{\sqrt{3}}{2}v_d \\ &= \frac{\sqrt{6}}{\pi}v_d \\ &= 0.78v_d \qquad \text{V} \end{aligned} \tag{8.32}$$

and the line-to-neutral fundamental voltage is

$$V_{1(L\text{-}N)} = \frac{V_{1(L\text{-}L)}}{\sqrt{3}} = \frac{\sqrt{2}}{\pi}v_d = 0.45v_d \qquad \text{V} \tag{8.33}$$

Using these relations, phasor notation may be employed for analysis of the fundamental steady state of the load circuit.

The fundamental component of the line-to-neutral voltage is shown in dotted form in the phase-a waveform v_{an} of Fig. 8.18(d). If the load is somewhat inductive and the harmonic currents are ignored, the phase-a current will lag the phase voltage, as shown in Fig. 8.18(e). Note that the phase-a current is still negative when switch S_1 is turned ON. This negative current will flow back to the direct-voltage source through the anti-parallel diode D_1. The switch S_1 will actually conduct only when the phase current becomes positive. In this manner, the reactive energy of the inductive load is returned to the direct-voltage link.

The inverter of Fig. 8.18 also can act to feed power from the a-c side to the direct-voltage link. This can occur when an a-c drive motor is attached to the output terminals and it regenerates during braking action. For this condition, the current in phase a would lag the fundamental component of the line-to-neutral voltage on phase a by between 90 and 180° for a partially inductive a-c system, as shown in the waveforms of Fig. 8.19. Below these waveforms is a listing of the switches that are turned on in each time interval. It will be noted that, when the current $i_a = i_d$ is negative, it flows in diode D_1 and, when it is posi-

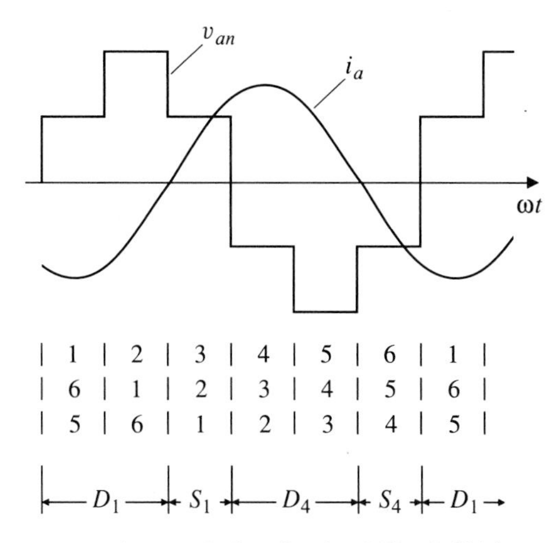

Figure 8.19 Regeneration mode for circuit of Fig. 8.18(a).

tive, it flows in switch S_1. For the next half-cycle with $i_a = -i_d$, diode D_4 and switch S_4 carry the phase current. When operating in this mode, the average link current i_d will be negative. The inverter is effectively acting as a rectifier. To accommodate the reversed-link current, the input rectifier must consist of two anti-parallel units, one of the form shown in Fig. 8.9 for positive-link current and another with reversed thyristors to act as an inverter to feed power back to the supply when the link current is negative.

For static a-c loads, the sequence of the three output phases is usually not significant. However, with electric motor loads, this sequence determines the direction of field rotation for the machine. The firing sequence *1-2-3-4-5-6* produces an output phase sequence *a-b-c*. The reversed phase sequence *c-b-a* can be readily produced by using the firing sequence *6-5-4-3-2-1*.

EXAMPLE 8.4 A three-phase supply is required having a frequency of 400 Hz with 208-V line to line or 120-V line to neutral. Using the simple voltage source inverter of Fig. 8.18(a), determine

(a) the required direct-link voltage

(b) the rms value of the line-to-line voltage.

Solution

(a) The rms value of the fundamental component of line-to-line voltage is to be 208 V. Thus, from Eq. (8.32), the required link voltage is

$$v_d = \frac{\pi}{\sqrt{6}}(208) = 267 \qquad \text{V}$$

(b) The actual line-to-line voltage wave is constant at 267 V over two-thirds of each half cycle. Thus,

$$V_{rms} = \sqrt{\tfrac{2}{3}}\,(267) = 218 \qquad \text{V}$$

8.4.2 ▪ Pulse-Width-Modulated Voltage-Source Inverters

One of the disadvantages of the simple voltage-source inverter of Section 8.4.1 is that control of the output voltage magnitude requires control of the firing angle α of the controlled rectifier. As was noted in Section 8.2.2, Eq. (8.16), this can lead to a low source power factor, particularly when the link voltage is low. An alternate, somewhat more sophisticated voltage-source inverter is shown in block diagram form in Fig. 8.20. This inverter draws power from a simple diode rectifier which produces an essentially constant link voltage V_d.

Control of the inverter output voltage is achieved by more frequent operation of the six self-commutated switches of Fig. 8.18(a) so that each half-wave of the line-to-line output voltages consists of a number of separate pulses of magnitude V_d and of variable width. Figure 8.21 shows line-to-line waveforms for a simple pulse-width modulated (PWM) inverter where there are only two voltage pulses for each half cycle. The fundamental component of this voltage, which can be found by Fourier analysis, is modified by varying the width of the pulses. It will be noted that there are intervals when all three line-to-line voltages are zero. This is achieved by turning on S_1, S_3, and S_5 simultaneously or, alternatively, S_2, S_4, and S_6, thus effectively connecting terminals a, b, and c together.

In most PWM inverters, a large number of pulses per cycle is used in order to synthesize a voltage wave that has a dominant fundamental sine-wave component and small harmonic components. A simple example is shown in the voltage waveform of Fig. 8.22 where the relative widths of the pulses are made approximately proportional to the desired sine wave and where the absolute widths of all the pulses are varied simultaneously to control the effective amplitude of the desired wave.

The larger the number of pulses per half cycle, the higher will be the order of the first significant harmonic component in the output voltage. In an induc-

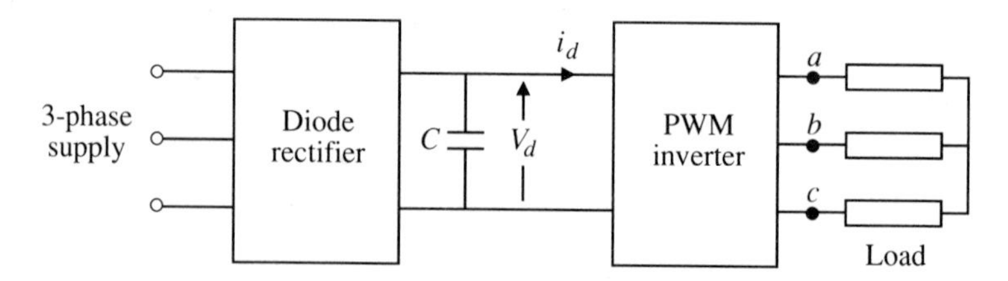

Figure 8.20 Block diagram of pulse-width-modulated inverter.

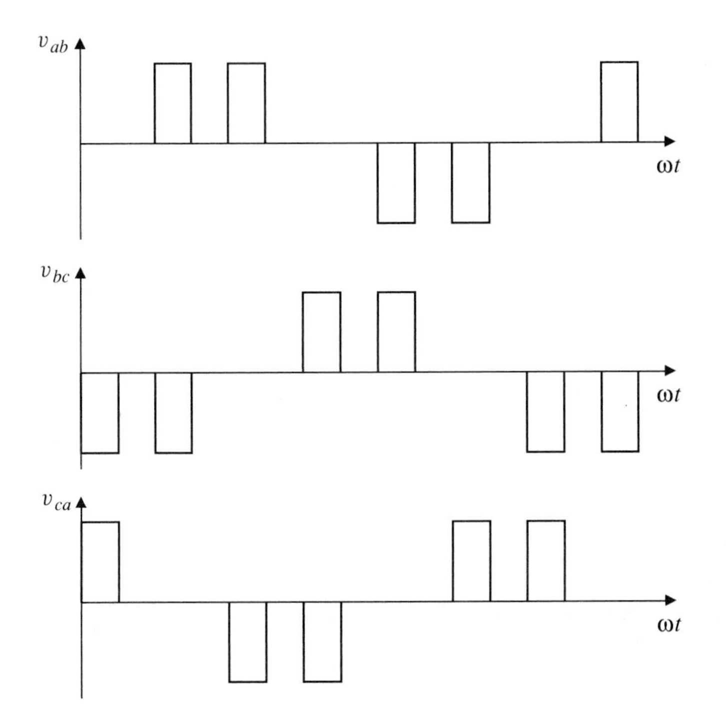

Figure 8.21 Simple pulse-width modulation of line-to-line voltages.

tive load, the currents due to these high-order harmonics are frequently negligible, leading to improved efficiency as compared with a simple voltage source inverter. On the other hand, the higher the switching frequency, the larger the power loss in the inverter because, practically, each turn-off operation involves

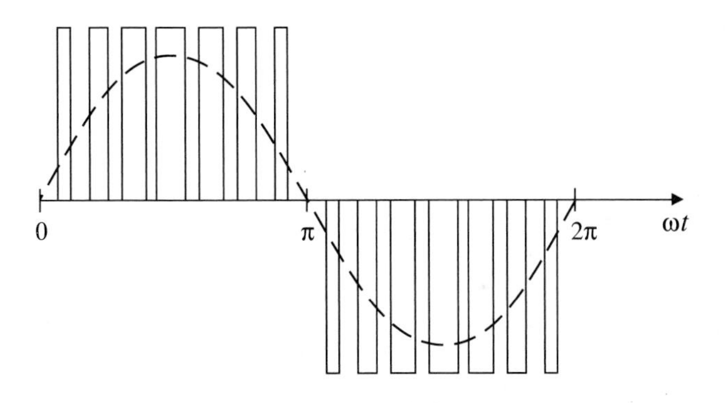

Figure 8.22 Sinusoidal pulse-width modulation.

some energy loss. In addition, each switch, depending on its type, requires a finite time to switch off, and this limits the maximum switching frequency.

8.4.3 ▪ Current-Source Inverters

In some applications, it is desirable to supply a three-phase load from a controlled alternating-current source rather than from a voltage source. An example might occur in the supply of a load such as an electric arc in which the impedance may vary over a wide range including a near short-circuit condition. A current-source inverter could ensure that the rated load current would not be exceeded.

A simple current-source inverter can be assembled using a source of direct current and a set of self-commutating switches. Normally, the original power source is an a-c power distribution system, either single or three phase, with essentially constant voltage and frequency. A direct-current source can be obtained by use of a controlled rectifier system with a feedback loop to maintain the desired value of direct current, as shown in the block diagram of Fig. 8.11. The inductor placed in the direct-current link to the inverter is made sufficiently large so that the current is nearly free of ripple.

A current source inverter with three-phase output can consist of six self-commutated switches, as shown in Fig. 8.23(a). Ideally, only two switches are on at a time, connecting two of the load phases in series to the direct-current supply. During each 60° interval, one load phase is open circuited. Idealized waveforms of the inverter output currents i_i are shown in Fig. 8.23(b) for the three phases. Each switch conducts for a 120° interval.

Most of the load systems of interest in our context are somewhat inductive. It is impossible to produce an instantaneous change in the current through an inductance. Accordingly, a suitable capacitor C may be connected across each phase of the load. With this capacitor in place, the phase currents in the load are more nearly sinusoidal.

The inverter output currents shown in Fig. 8.23(b) can be resolved into a set of fundamental frequency and harmonic components using Fourier analysis. For phase a,

$$i_{ia} = \frac{4}{\pi} \cos \frac{\pi}{6}\, i_d \left[\sin \omega t - \frac{1}{5} \sin 5\omega t + \frac{1}{7} \sin 7\omega t + (-1)^{(h+i)/2} \frac{1}{h} \sin h\omega t \right] \tag{8.34}$$

The rms value of the fundamental component of current is

$$I_i = \frac{1}{\sqrt{2}} \frac{4}{\pi} \frac{\sqrt{3}}{2}\, i_d = \frac{\sqrt{6}}{\pi}\, i_d = 0.78 i_d \qquad \text{A} \tag{8.35}$$

Because the impedance of the shunt capacitor to harmonics of order h is inversely proportional to h while the impedance of the inductive part of the load is proportional to h, much of the harmonic current flows in the capacitor.

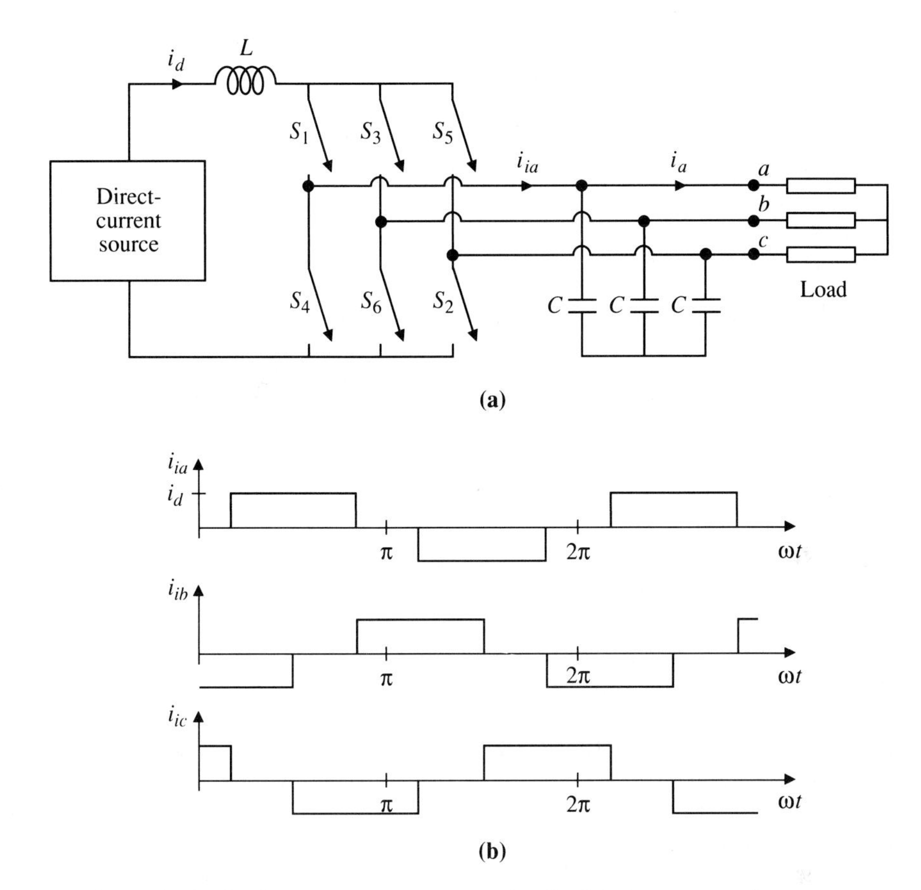

Figure 8.23 (a) Current source inverter circuit. (b) Current waveforms.

Some loads are particularly sensitive to harmonics in the phase currents. An example is an induction motor in which fifth and seventh harmonic currents produce an oscillatory torque at sixth harmonic frequency (see Section 10.2.2). Such troublesome harmonics can be effectively eliminated by use of pulse-width modulation of the current. A typical waveform is shown in Fig. 8.24. The positions of the pulses are constrained by the fact that the same pulse of current must occur in both of the phases that are connected in series. If there are N pulses in a half-cycle, it can be shown that $(N - 1)/2$ firing angles can be independently chosen so as to cause the effective elimination of the same number of current harmonics. For the five-pulse waveform of Fig. 8.24, the fifth and seventh harmonics which are the largest can be eliminated.

An alternate system for producing a near-sinusoidal current source makes use of the voltage source inverter circuit of Fig. 8.18(a). Signals proportional to the desired current waveform are generated and compared with signals from

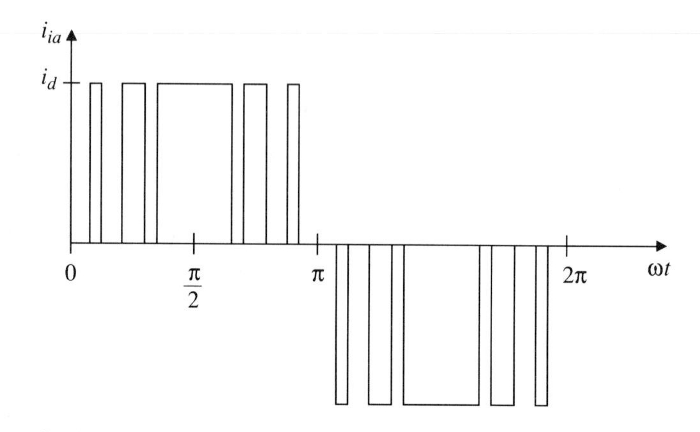

Figure 8.24 Pulse-width modulated current.

measurements of the actual phase currents. If a current is less than its desired value by a fixed amount δ, the terminal of that phase is connected to the positive side of the direct-voltage link. When the current is larger than the desired value by the increment δ, the terminal is switched to the negative side of the link. By making the increment δ small, any desired waveform can be approximated. However, decrease in δ requires an increase in the switching frequency. This system is usually known as a *hysteresis or bang-bang control.*

EXAMPLE 8.5 A particular static load takes a rated power of 25 kW at 0.9 lagging power factor from a 400-V line-to-line, three-phase, 50-Hz supply. This load is now to be supplied from a simple current-source inverter over the frequency range 0 to 50 Hz. In order to make best use of the inverter capacity, sufficient capacitance is connected in delta across the load to bring the net load to unity power factor at 50 Hz. The load can be modeled as a resistance in series with an inductance in each phase.

(a) Determine the required capacitance of each of the three capacitors.

(b) When the inverter is providing the above rated power at a frequency of 50 Hz, determine the fifth harmonic current per phase in the inverter and in the load.

(c) For the condition in (b), find the ratio of the fifth harmonic load voltage to the fundamental-frequency load voltage.

Solution

(a) Using the line-to-line neutral voltage as a reference,

$$\vec{V}_1 = \frac{400\angle 0}{\sqrt{3}} = 231\angle 0$$

Then, the load current per phase is

$$\vec{I}_{1L} = \frac{25\,000}{3 \times 231 \times 0.9} \angle \cos^{-1} 0.9$$
$$= 40.1 \angle -25.8$$
$$= 36.1 - j17.45 \qquad \text{A}$$

To correct the power factor to unity, the capacitive current per phase must be

$$\vec{I}_{1c} = j17.45 \qquad \text{A}$$

The admittance to neutral of the capacitance is

$$Y_{1c} = \frac{j17.45}{231} = j0.076 \qquad \mho$$

and the value of each capacitance connected in delta is

$$C = \frac{1}{3} \frac{Y_{1c}}{\omega} = \frac{0.076}{3 \times 314} = 80.7 \qquad \mu\text{F}$$

(b) When the power factor is corrected to unity, the fundamental component of the inverter current is

$$I_{1i} = 36.1 \qquad \text{A}$$

The fifth harmonic inverter current is then

$$I_{5i} = \frac{36.1}{5} = 7.22 \qquad \text{A}$$

The ratio of load current to inverter current at fifth harmonic is equal to the ratio of the admittance Y_{5L} of the load to the admittance Y_{5i} as seen by the inverter. At 50 Hz, the load impedance is

$$Z_{1L} = \frac{231 \angle 0}{40.1 \angle -25.8} = 5.18 + j2.51 \qquad \Omega$$

At fifth harmonic,

$$Z_{5L} = 5.18 + j(2.51 \times 5) = 13.58 \angle 67.6$$

and

$$Y_{5L} = 0.074 \angle -67.6 = 0.028 - j0.068 \qquad \mho$$

Using the fundamental admittance of the capacitor from (a), the total admittance is

$$Y_{5i} = Y_{5L} + Y_{5c}$$
$$= 0.028 - j0.068 - j(0.076 \times 5)$$
$$= 0.313 \angle 84.8 \qquad \mho$$

Then

$$I_{5L} = \frac{Y_{5L}}{Y_{5i}} I_{5i} = \frac{0.074}{0.313} (7.22) = 1.71 \qquad \text{A}$$

(c) The ratio of fifth to fundamental voltage is

$$\frac{I_{5i}}{Y_{5i} V_1} = \frac{7.22}{0.313 \times 231} = 0.10$$

8.5

Cycloconverters

A *cycloconverter* is a system of switches that is capable of converting a-c power from a constant-voltage, constant-frequency supply to a variable-voltage,

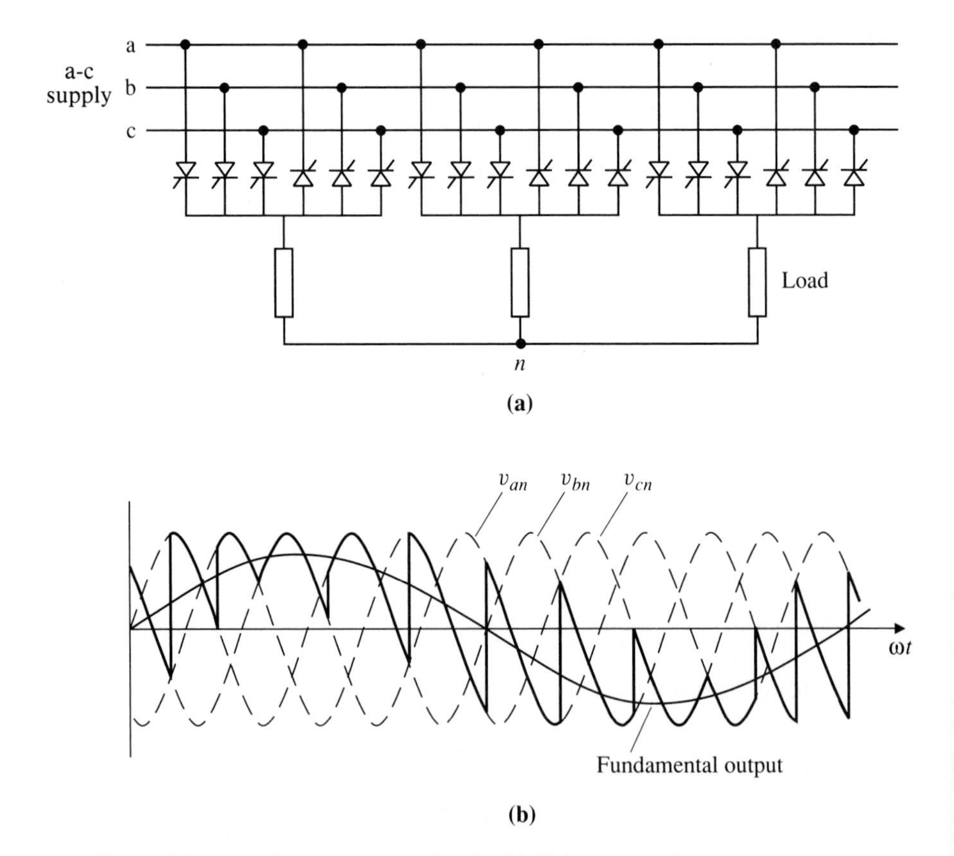

Figure 8.25 (a) Cycloconverter circuit. (b) Voltage waveform.

variable-frequency output. The basic circuit for a three-phase input, three-phase output system is shown in Fig. 8.25(a). Each output phase is supplied by two three-phase rectifiers, one to provide voltage during the positive half-cycle of the output and one to provide voltage during the negative half-cycle. The firing angles of the thyristors are so controlled that successive sectors from the input voltage waves produce an output waveform that has an approximately sine form plus a number of higher-frequency terms. A typical output voltage waveform is shown as the solid line in Fig. 8.25(b) together with its fundamental component.

Cycloconverters are useful for producing output at a low controllable frequency. Practically, the maximum useful output frequency is about one-third of the supply frequency for a three-phase system.

PROBLEMS

8.1 A single-phase controlled bridge rectifier connected to a 120-V, 60-Hz supply is used to charge a storage battery which may be modeled as a source of 96 V in series with a resistance of 0.6 Ω. Sufficient inductance is placed in series with the battery to make its current essentially ripple free.

(a) Determine the required firing angle for a charging current of 12 A.

(b) What is the supply power factor?

(Section 8.2.1)

8.2 A direct-current source is required to supply a controlled current of 20 A to a load in which the resistance varies from 0 to 12 Ω. The source consists of a 208-V (L-L), 60-Hz, three-phase supply, a controlled bridge rectifier, and a large series inductor.

(a) Determine the maximum and minimum values of the firing angle.

(b) Determine the maximum and minimum values of the source power factor.

(c) Determine the maximum and minimum values of the source kVA.

(Section 8.2.1)

8.3 Consider the system of Problem 8.2 with the series inductance eliminated. At what value should the firing angle be set to provide a power of 2.5 kW to the 12-Ω load?

(Section 8.2.1)

8.4 For the system described in Problem 8.2, estimate the value of series inductance required to keep the load current continuous at an average value of 20 A.

(Section 8.2.1)

8.5 A single-phase bridge rectifier/inverter is supplied from a 120-V, 60-Hz source. It is connected through a large inductance to a direct-voltage system that has an internal resistance of 0.8 Ω and a variable generated voltage. Determine the required firing angle when delivering power back to the alternating-voltage source with a direct current of 12 A and a direct generated voltage of 110 V.

(Section 8.2.1)

8.6 A three-phase controlled bridge rectifier is used to charge a battery with an internal voltage of 192 V and an internal resistance of 0.8 Ω. The alternating supply voltage is 230 V (L–L). Sufficient inductance is placed in series with the battery to maintain continuous current.

(a) Determine the maximum and minimum values of firing angle required if the maximum charging current is to be 15 A.

(b) Estimate the rms source current per phase for the condition of maximum charging current.

(c) Determine the power factor at the alternating source terminals for the condition of (b).

(Section 8.2.2)

8.7 A three-phase bridge rectifier/inverter couples a 460-V (L-L), 60-Hz, three-phase utility supply to a source of variable direct voltage with an internal resistance of 0.12 Ω. The direct-voltage source can vary from 500 to −500 V. What value of series inductance should be connected in series with the direct source to ensure continuous current operation if the minimum value of average direct current is to be 150 A?

(Section 8.2.2)

8.8 A step-down chopper is to be used to supply power from a 600-V overhead wire to the commutator motor on a trolley bus. The motor may be modeled as a generated direct voltage in series with a resistance of 0.2 Ω and an inductance of 25 mH. The frequency of the chopper is set at 500 Hz.

(a) When the bus is stopped, what should be the ON time of the chopper to provide a motor current of 150 A?

(b) What is the maximum possible generated voltage of the motor for which a current of 150 A can be supplied?

(c) What is the average value of the supply current for the conditions of (a) and (b)?

(d) If the motor current is to be continuous, what should be its minimum value?

(Section 8.3.1)

8.9 When the trolley bus in Problem 8.8 is going down a hill, its chopper is reconnected in the configuration shown in Fig. 8.15. For what length of time should the chopper switch be in the conducting condition if the generated voltage of the motor is 500 V and a power of 50 kW is to be fed back to the supply?

(Section 8.3.1)

8.10 A two-quadrant chopper of the form shown in Fig. 8.16 is used to control the speed of the commutator motor when driving and when braking. The motor has a resistance of 2.5 Ω and an inductance that is sufficiently large to make the motor current essentially free of ripple. The direct supply voltage is 48 V. The chopper frequency is 1 kHz.

(a) Determine the generated voltage of the motor when it is driving with a current of 4.5 A and the chopper is set at $t_{ON} = 200\ \mu s$.

(b) Determine the time t_{ON} when the machine is in the braking mode with a current of 4.5 A and a generated voltage of 30 V.

(Section 8.3.3)

8.11 In a simple, three-phase voltage-source inverter of the form shown in Fig. 8.18, the direct voltage v_d in the link is 550 V. The frequency of the inverter output is 200 Hz. Determine:

(a) the rms value of the fundamental component of the output voltage, line to line and line to neutral, and

(b) the rms value of the actual output voltage line to line and line to neutral.

(Section 8.4.1)

8.12 An emergency power supply is required to provide a three-phase sinusoidal voltage source of 208 V (L-L) at a frequency of 400 Hz. The power is to be obtained from a battery consisting of cells in series, each cell having an internal voltage of 2.2 V and an internal resistance of 0.006 mΩ. The load on the three-phase, 400-Hz system requires a fundamental frequency apparent power of 7.2 kVA at 0.9 power factor lagging. The inductance of the load is such that harmonic currents are negligible. How many cells should be connected in series to provide the required output? The switches may be regarded as ideal.

(Section 8.4.1)

8.13 A simple current-source inverter of the form shown in Fig. 8.23 supplies a three-phase load for which each phase can be modeled as a resistance of 1.8 Ω in series with an inductance of 1.2 mH. Three capacitors each of 150 μF are connected in delta in parallel with the load. If the direct-link current is 240 A and the frequency is 90 Hz,

(a) Determine the rms value of the fundamental component of the load voltage line to line.

(b) Determine the rms value of the fifth harmonic component of the load voltage line to line.

(c) Assuming an ideal inverter, estimate the voltage in the direct-current link.

(Section 8.4.3)

GENERAL REFERENCES

Baliga, B.J. and Chen, D.Y., *Power Transistors: Device Design and Application*, IEEE Press, New York, 1984.

Bose, B.K., *Power Electronics and AC Drives*, Prentice-Hall, Englewood Cliffs, New Jersey, 1986.

Brichart, F., *Forced Commutated Inverters—Design and Industrial Applications*, Macmillan, New York, 1984.

Dewan, S.B. and Straughen, A., *Power Semiconductor Circuits*, John Wiley and Sons, New York, 1975.

McMurray, W., *The Theory and Design of Cycloconverters*, Wiley Interscience, New York, 1971.

Pelly, B.R., *Thyristor Phase Controlled Converters and Cycloconverters*, Wiley Interscience, New York, 1971.

Sedra, A.S. and Smith, K.C., *Microelectronic Circuits*, 2nd ed., Holt, Rinehart and Winston, New York, 1987.

Shepherd, W., *Thyristor Control of AC Circuits*, Crosby Lockwood Staples, London, 1975.

CHAPTER

9

Commutator Motor Drives

Some commutator motors are used in constant-speed drives, operating from a constant direct-voltage supply. An example is the fan motor in an automobile. This mode of operation has been discussed in Section 4.6. This chapter examines the use of commutator motors in systems that require a wide range of accurate speed, position, or torque control. A typical example occurs in a paper mill which consists of a sequence of motor-driven pairs of rollers for which the roller speeds or angular positions must be accurately controlled so that the paper strip does not tear or droop between rollers. Other examples would be the drives that are required at each joint of an industrial robot to provide both rapid transfer of the robot head from place to place and precise positioning of the head relative to the work piece.

To appreciate some of the requirements a drive must meet, consider the elevator system sketched in Fig. 9.1(a). A cable wound around a grooved cylindrical sheave has the elevator cage of mass m_e on one end and a counterweight of mass m_c on the other. The sheave is driven by a motor through a reduction gear. One requirement might be the control of acceleration of the elevator during starting and stopping. For comfort and safety, this acceleration might be limited to about 1 to 1.5 m/s^2. To achieve this acceleration a, the net tangential force F on the sheave would have to be

$$F = m_e(g + a) - m_c(g - a) \qquad \text{N} \tag{9.1}$$

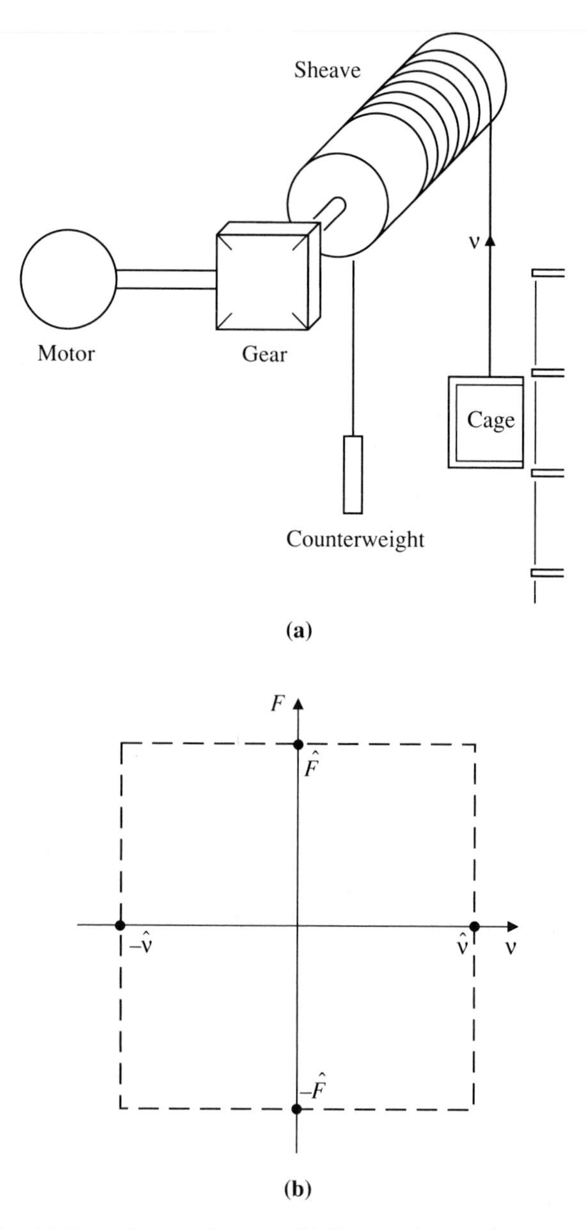

Figure 9.1 (a) Drive for an elevator. (b) Force-velocity plane.

where the gravitational constant $g = 9.81\ \text{m/s}^2$. This force may be either positive or negative depending on the relative values of the two masses as the elevator load changes and on the direction of the acceleration. Because the elevator must operate both up and down subject to a specified maximum velocity $\hat{v}$, the drive must be capable of operating in all four quadrants of the force-velocity plane of Fig. 9.1(b).

A further requirement at starting and stopping is usually the control of jerk j which is the rate of change of acceleration:

$$j = \frac{da}{dt} \qquad \text{m/s}^3 \tag{9.2}$$

For a public elevator, a typical value of maximum jerk might be 1 m/s^3 for 1 s until an acceleration of $a = 1$ m/s^2 is reached. The same constraint on jerk would apply on stopping. As the elevator comes to a stop, it will be required to be positioned to within an accuracy of 1 to 2 mm typically, regardless of its load. Other specifications might relate to freedom from vibration or noise, fail-safe operation in an emergency, low maintenance, and a long expected lifetime.

Each variable-speed drive is part of a system that consists of the mechanical system to be driven, the motor, the controllable electrical supply, and the measurement and control system, as shown in the block diagram of Fig. 9.2. Some knowledge of each of these aspects of the system is needed in order, firstly, to develop an understanding of the major drive properties, secondly, to be able to predict drive performance, and, thirdly, to design a system to meet the customer's specifications. In the following sections, we will attempt to draw together concepts and models from the study of commutator machines in Chapter 4 and the introduction of power semiconductor converters in Chapter 8, together with some mechanical system modeling to address these objectives. Some of the concepts discussed here also will be applicable in the study of other types of drives in later chapters.

9.1

Mechanical System Modeling

The mechanical system connected to the rotating shaft of the motor will usually have a significant polar moment of inertia J_L. The torque required to accelerate a pure inertia load as shown in Fig. 9.3 is

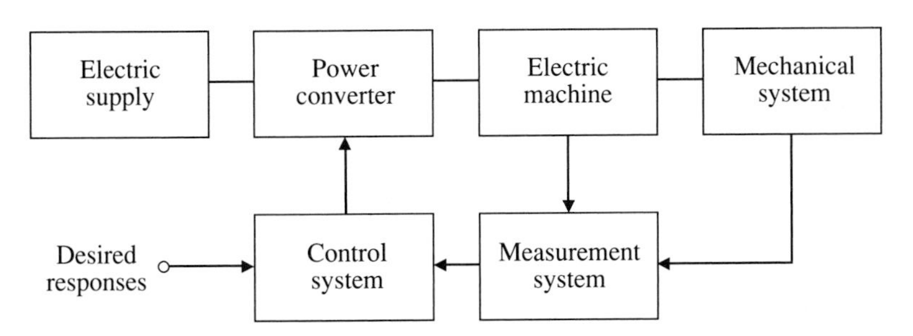

Figure 9.2 Block diagram of a drive system.

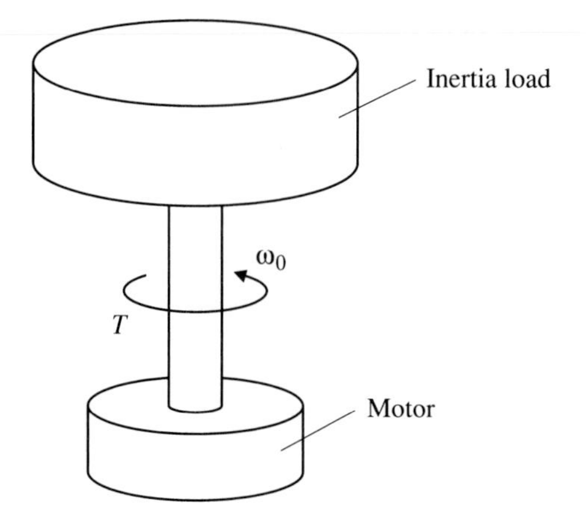

Figure 9.3 Motor driving pure inertia load.

$$T_J = J_L \frac{d\omega_0}{dt} \qquad \text{N}\cdot\text{m} \tag{9.3}$$

This torque will be zero for constant-speed operation. For purposes of estimating the inertia of a mechanical load, a thin, hollow cylinder of mass m and radius r rotating about its axis has a polar moment of inertia of

$$J = mr^2 \qquad \text{kg}\cdot\text{m}^2 \tag{9.4}$$

For a solid cylinder, integration with respect to r gives

$$J = \tfrac{1}{2}mr^2 \qquad \text{kg}\cdot\text{m}^2$$

If the solid cylinder is made of material of density ρ kg/m^3 and has an outer radius r and a length ℓ, its inertia can be expressed as

$$J = \frac{\pi}{2}\rho\ell r^4 \qquad \text{kg}\cdot\text{m}^2 \tag{9.5}$$

It is seen that inertia is very sensitive to the radius.

Many mechanical loads require most of their torque to overcome fluid or aerodynamic friction. Examples are fans, pumps, and high-speed vehicles. This type of friction load requires a torque that is approximately proportional to the square of the speed:

$$T = k\omega_0^2 \qquad \text{N}\cdot\text{m} \tag{9.6}$$

as shown in the graph of Fig. 9.4.

Some mechanical loads require a substantially constant torque when operating. An example is the ball mill of Fig. 9.5(a) used for grinding ore. The interior load is always unbalanced at an angle characteristic of the material and thus requires a torque that is nearly independent of speed over a wide range, as

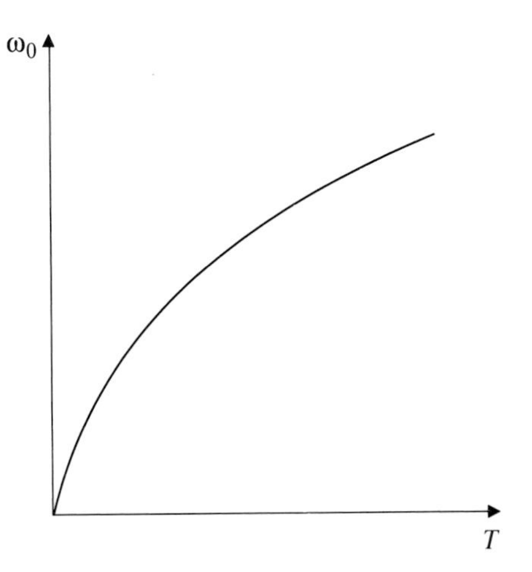

Figure 9.4 Speed-torque relation for fluid friction.

shown in Fig. 9.5(b). Another example of a load with near-constant torque is a compressor or pump operating against a nearly constant pressure. A further example is the elevator of Fig. 9.1 when operating at constant velocity.

Some mechanical loads are characterized by a requirement for a torque that is approximately proportional to speed. One example would be a coupled commutator generator supplying a constant resistance load. Because its generated

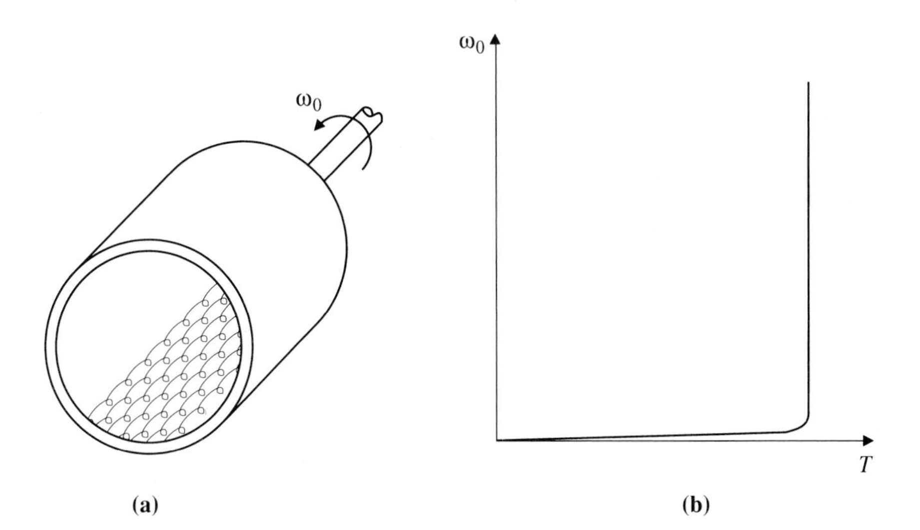

Figure 9.5 (a) A ball mill and (b) its speed-torque relation.

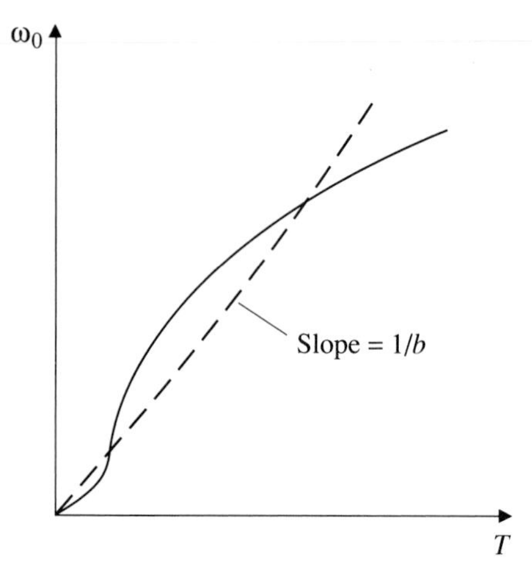

Figure 9.6 Typical composite load curve and its linear approximation.

voltage is proportional to speed, so also is its current and therefore its torque. Further examples are found in loads that have a dominantly viscous friction. The idealized relation for these loads is

$$T_b = b\omega_0 \qquad \text{N}\cdot\text{m} \tag{9.7}$$

Many loads will contain components of fluid friction, as, for example, from windage, plus viscous friction, as, for example, from bearings, plus some static or coulomb friction at a near-zero speed. The overall speed-torque relation may have a nonlinear form such as shown in Fig. 9.6. For purposes of analysis, such a relation frequently may be approximated by a straight line and represented by Eq. (9.7). However, the departures from linearity may be of critical significance in some instances.

In some situations, torsion may be important, particularly where mechanical resonance is being investigated. An example might occur in a long shaft from a motor to a rotating saw blade. The torque absorbed by the shaft in Fig. 9.7 is directly proportional to the twist angle as given by

$$T_K = T_0 - T_L = K(\theta_0 - \theta_L) \qquad \text{N}\cdot\text{m} \tag{9.8}$$

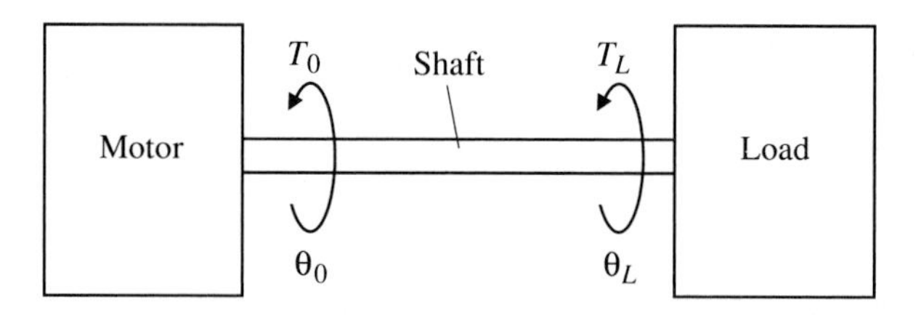

Figure 9.7 Torsion in a shaft.

The torque that can be produced by an electrical machine is limited by the flux density in its magnetic material and the current density in its conductors. To make efficient use of electric machines, it is usually necessary to operate them at relatively high rotational speeds. Such speeds may be much higher than optimum for the mechanical loads. Accordingly, the load is frequently coupled to the motor through a gear system. A simple pair of gears is shown in Fig. 9.8. Assuming no internal friction and ignoring the inertia of the gears themselves, the gear will be ideally efficient, i.e.,

$$\omega_0 T_0 = \omega_L T_L \qquad \text{W} \tag{9.9}$$

For gears with radii r_1 and r_2 and tooth numbers N_1 and N_2, respectively, the ideal gear relations are

$$\frac{\omega_L}{\omega_0} = \frac{T_0}{T_L} = \frac{N_1}{N_2} = \frac{r_1}{r_2} \tag{9.10}$$

The gear acts for a mechanical system as a transformer acts for an electrical system. Consider a load that can be represented by the relation

$$T_L = J_L \frac{d\omega_L}{dt} + b_L \omega_L \qquad \text{N·m} \tag{9.11}$$

If coupled to a motor through an ideal gear of ratio $N_1 : N_2$, the equivalent load as seen by the motor will be

$$\begin{aligned} T_0 &= \frac{N_1}{N_2} T_L \\ &= \left(\frac{N_1}{N_2}\right)^2 J_L \frac{d\omega_0}{dt} + \left(\frac{N_1}{N_2}\right)^2 b_L \omega_0 \qquad \text{N·m} \end{aligned} \tag{9.12}$$

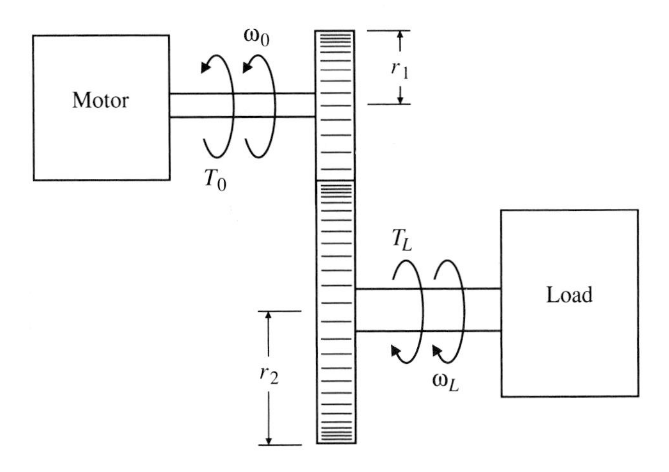

Figure 9.8 Gear coupling.

In modeling and analyzing a drive system, it is usually convenient to consider the motor inertia and its friction and windage as part of the total mechanical load to which the torque T produced at the motor air gap is applied.

9.2

Dynamic Relations for Drive Systems

A basic drive system using a commutator motor is shown in block form in Fig. 9.9. The motor is represented by its idealized equivalent circuit, from Fig. 4.12, consisting of armature circuit resistance R_a, armature circuit inductance L_a, and generated voltage e_a. A controllable armature terminal voltage v_t is supplied from a source such as a rectifier or a chopper. The motor field circuit is supplied from a suitable field voltage source v_f or, alternatively, the field may be provided by permanent magnets. In either case, a field flux Φ is produced in each pole. From Eq. (4.15), the generated voltage is related to the speed by

$$e_a = k\Phi\omega_0 \qquad \text{V} \tag{9.13}$$

and the torque T produced at the motor air gap is related to the armature current by

$$T = k\Phi i_a \qquad \text{N}\cdot\text{m} \tag{9.14}$$

where k is the constant for the motor armature winding given by Eq. (4.7). The armature circuit is described by the relation

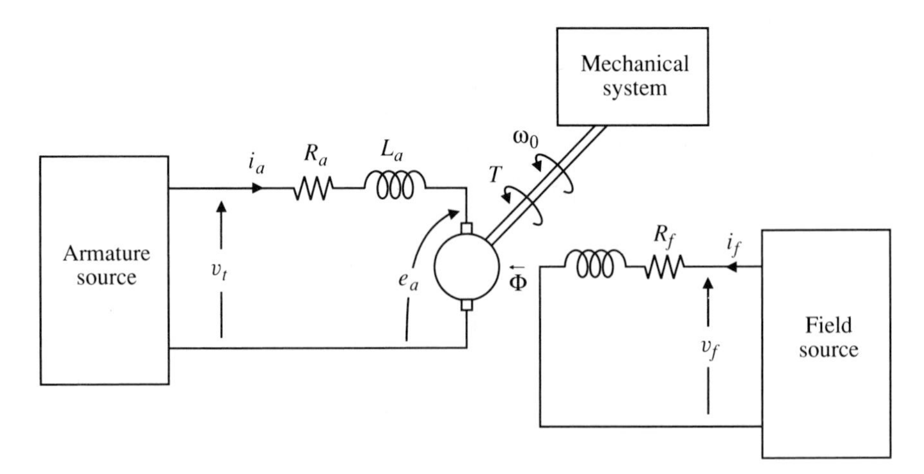

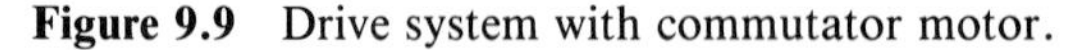
Figure 9.9 Drive system with commutator motor.

$$v_t = R_a i_a + L_a \frac{di_a}{dt} + k\Phi\omega_0 \qquad \text{V} \tag{9.15}$$

The mechanical load, including the mechanical components of the motor, can frequently be modeled by the total effective inertia J of the load and the motor as seen by the motor, plus a viscous friction coefficient b, plus, in some instances, a constant load torque T_c. Thus, the mechanical system may be represented by the idealized relation

$$T = J\frac{d\omega_0}{dt} + b\omega_0 + T_c \qquad \text{N}\cdot\text{m} \tag{9.16}$$

When there is a wound field connected to a source voltage v_f, the field circuit can be represented by the relation

$$v_f = R_f i_f + \frac{d\lambda_f}{dt} \qquad \text{V} \tag{9.17}$$

where the field flux linkage λ_f is a nonlinear function of the field current i_f. In many drives, the flux Φ is held constant at the maximum permissible value consistent with the limitation on field coil temperature. However, in some applications, slow reduction of the field current may be required to achieve higher speeds. It is infrequent that the transient behavior of the field circuit of a commutator machine is of importance, a fortunate circumstance in view of the nonlinear nature of Eq. (9.17).

In the above relations, it has been assumed that the flux interacting with the armature is not affected by the armature current. In some commutator machines, the flux Φ may be reduced somewhat when a large value of armature current i_a produces a distortion in the flux pattern. In machines used for control purposes, design features are introduced into the machine to avoid this interaction; these will be discussed in Section 9.8.

Because most readers of this text are familiar with the transient analysis of electric circuits, it is convenient to represent the complete drive system as an idealized electric equivalent circuit. Equations (9.13) and (9.14) can be used to replace the torque and speed variables in Eq. (9.16), providing a relation between the armature current i_a and the generated voltage e_a:

$$\begin{aligned} i_a &= \frac{T}{k\Phi} \\ &= \frac{J}{(k\Phi)^2}\frac{de_a}{dt} + \frac{b}{(k\Phi)^2}e_a + \frac{T_c}{k\Phi} \qquad \text{A} \end{aligned} \tag{9.18}$$

From the form of this equation, it will be seen that the mechanical system can be transformed into the electric circuit as a capacitance C representing the inertia, a resistance R_b representing the viscous friction load, and a current sink i_c representing any constant torque load, i.e.,

$$i_a = C\frac{de_a}{dt} + \frac{e_a}{R_b} + i_c \qquad \text{A} \tag{9.19}$$

where

$$C = \frac{J}{(k\Phi)^2} \quad \text{F}, \qquad R_b = \frac{(k\Phi)^2}{b} \quad \Omega, \qquad \text{and} \quad i_c = \frac{T_c}{k\Phi} \quad \text{A}$$

The system now may be represented by the equivalent electric circuit of Fig. 9.10.

Now that we have an idealized model for the motor and its mechanical load, let us consider how the speed of the drive responds to a change in the armature voltage v_t. Examination of the equivalent circuit of Fig. 9.10 indicates that there are two energy storage elements in the system and therefore two sources of delay in the response. In most commutator motor drives, the energy stored in the equivalent capacitance representing the inertia is much greater than that stored in the armature inductance. Therefore, we can obtain an approximate picture of the transient response by ignoring this inductance.

Let us consider first an unloaded motor. The transient behavior may be analyzed by use of the Laplace transforms of differential Eqs. (9.15) and (9.16). With zero initial conditions,

$$\begin{aligned} Js\omega_{(s)} &= T \\ &= k\Phi i_{a(s)} \\ &= \frac{k\Phi}{R_a}(v_{t(s)} - k\Phi\omega_{0(s)}) \qquad \text{N}\cdot\text{m} \end{aligned} \tag{9.20}$$

When rearranged, this becomes

$$\omega_{0(s)} = \frac{v_{t(s)}}{k\Phi(s\tau_e + 1)} \qquad \text{rad/s} \tag{9.21}$$

in which the effective time constant is

$$\tau_e = \frac{JR_a}{(k\Phi)^2} \qquad \text{s} \tag{9.22}$$

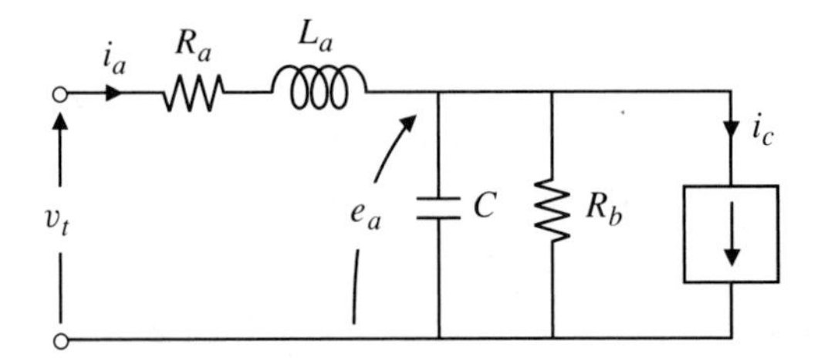

Figure 9.10 Equivalent electric circuit of an idealized drive system.

If v_t is a step voltage V applied at $t = 0$, the speed will be

$$\omega_0 = \frac{V}{(k\Phi)} (1 - \epsilon^{-t/\tau_e}) \qquad \text{rad/s} \tag{9.23}$$

and the armature current will be

$$i_a = \frac{V}{R_a} \epsilon^{-t/\tau_e} \qquad \text{A} \tag{9.24}$$

This current expression suggests one of the practical limitations that must be imposed. Usually, the maximum current that can be permitted in the armature circuit of a commutator machine is limited to about three to five times the continuous rated value. Thus, in all but very small machines, means must be introduced into the control of the armature supply voltage to limit the armature current during a transient. If the source is a controlled rectifier, it can be converted readily into a current source using the arrangement shown in Fig. 8.11.

If a mechanical load torque having the viscous friction term $b\omega_0$ is included in the analysis, the operational relation between the speed and the supply voltage can be derived from the equivalent circuit of Fig. 9.10 (still ignoring the inductance L_a) as

$$\omega_{0(s)} = \frac{e_{a(s)}}{k\Phi} = \left(\frac{R_b}{R_a + R_b}\right) \frac{v_{t(s)}}{k\Phi(s\tau_e + 1)} \qquad \text{rad/s} \tag{9.25}$$

where

$$\tau_e = C\frac{R_a R_b}{R_a + R_b} = \frac{JR_a}{(k\Phi)^2\left[1 + \dfrac{bR_a}{(k\Phi)^2}\right]} \qquad \text{s} \tag{9.26}$$

For a constant supply voltage, the steady-state speed is found by setting $s = 0$ in Eq. (9.25). This speed is seen to be dependent on the ratio $R_b/(R_a + R_b)$. This ratio is near unity for an efficient machine. For this type of mechanical load, the steady-state speed is proportional to the supply voltage. The transient time constant is seen to be equal to the capacitance C representing the total inertia multiplied by the resistance of the circuit as seen from that capacitance, i.e., R_a in parallel with R_b.

It may seem surprising that the motor with a friction load has a somewhat smaller time constant (Eq. [9.26]), and therefore reaches its final speed in a shorter time, than one without friction load. This can be explained by noting that the initial acceleration on application of a step of supply voltage is the same in both cases but the final speed with the friction load is somewhat lower.

In some drives, the effect of the armature inductance will be significant and should be included in the dynamic relations. For a system that can be represented by the equivalent circuit of Fig. 9.10 with $i_c = 0$, the armature current can be expressed as

$$i_{a(s)} = \frac{v_{t(s)} - e_{a(s)}}{R_a(s\tau_a + 1)} \qquad \text{A} \tag{9.27}$$

where $\tau_a = L_a/R_a$ s is the armature time constant. The generated voltage is

$$e_{a(s)} = \frac{R_b i_{a(s)}}{s\tau_L + 1} \qquad \text{V} \tag{9.28}$$

where the load time constant τ_L is given by

$$\tau_L = CR_b = \frac{J}{b} \qquad \text{s} \tag{9.29}$$

The operational relation between speed and armature voltage then is derived using Eqs. (9.27) and (9.28) as

$$\omega_{0(s)} = \frac{e_{a(s)}}{k\Phi} = \frac{R_b}{R_a k\Phi} \frac{(v_{t(s)} - k\Phi\omega_{0(s)})}{(s\tau_a + 1)(s\tau_L + 1)} \qquad \text{rad/s} \tag{9.30}$$

which, when rearranged, gives

$$\begin{aligned} \omega_{0(s)} &= \frac{K}{K+1} \frac{v_{t(s)}}{s^2 \dfrac{\tau_a \tau_L}{1+K} + s\dfrac{\tau_a + \tau_L}{1+K} + 1} \\ &= \frac{K}{K+1} \frac{v_{t(s)}}{(s\tau_1 + 1)(s\tau_2 + 1)} \qquad \text{rad/s} \end{aligned} \tag{9.31}$$

where $K = R_b/R_a$. The roots of the quadratic expression in s are normally real but in some systems with low inertia they may be complex, giving an oscillatory response. A typical speed response to a step in applied armature voltage is shown in Fig. 9.11. The armature current transient also is shown.

In a similar manner, the speed response to a transient torque disturbance may be found by letting i_c in Fig. 9.10 be a transient variable representing an applied mechanical torque T_c and setting v_t to zero. The result is

$$\omega_{0(s)} = \frac{(1/b)(s\tau_a + 1)T_{c(s)}}{(s\tau_a + 1)(s\tau_L + 1) + (k\Phi)^2/R_a B} \qquad \text{rad/s} \tag{9.32}$$

Readers who are familiar with Laplace transformer methods should now be able to produce predictions of transient behavior or frequency response for particular drive situations.

EXAMPLE 9.1 A commutator motor has a rated mechanical power of 15 kW at a speed of 3500 r/min. Its generated voltage at that speed is 230 V. Its armature resistance is 0.142 Ω, its armature inductance is 1.1 mH, and its polar moment of inertia is 0.088 kg·m². The motor is coupled to the mechanical load through a 3:1 speed reduction gear. The load power

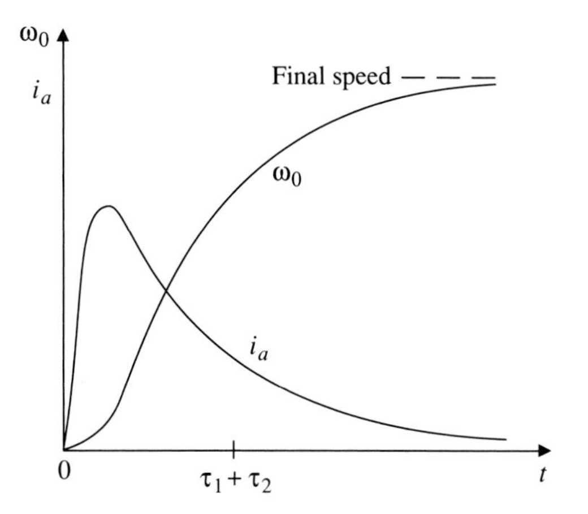

Figure 9.11 Speed response to a step in applied armature voltage together with the armature current transient.

may be considered to be dissipated in viscous friction, and the load inertia is 0.75 kg·m^2.

(a) Find the polar moment of inertia J as seen by the motor.

(b) Determine the capacitance C equivalent to this inertia.

(c) Find the friction coefficient b and its equivalent resistance R_b.

(d) Determine the armature time constant τ_a and the load time constant τ_L.

(e) Ignoring the armature time constant, determine the effective time constant of the system.

Solution The rated speed is

$$\omega_0 = \frac{3500 \times 2\pi}{60} = 366.5 \qquad \text{rad/s}$$

The flux constant for this machine is

$$k\Phi = \frac{230}{366.5} = 0.628 \qquad \text{V·s/rad}$$

(a) With a gear ratio $N_1/N_2 = 1/3$, from Eq. (9.12),

$$J = J_{motor} + \left(\frac{N_1}{N_2}\right)^2 J_{load} = 0.088 + \frac{0.75}{9} = 0.171 \qquad \text{kg·m}^2$$

(b) The equivalent capacitance is, from Eq. (9.19),

$$C = \frac{J}{(k\Phi)^2} = \frac{0.171}{0.628^2} = 0.434 \qquad \text{F}$$

(c) If 15 kW is taken as the mechanical power produced by the motor including its own friction and windage losses, the power can be expressed as

$$P = T\omega_0 = b\omega_0^2 \qquad \text{W}$$

from which

$$b = \frac{15\,000}{366.5^2} = 0.112 \qquad \text{N}\cdot\text{m}\cdot\text{s}$$

and

$$R_b = \frac{(k\Phi)^2}{b} = \frac{0.628^2}{0.112} = 3.52 \qquad \Omega$$

Note that the ratio R_a/R_b is 0.04, indicating that the loss in the armature resistance is about 4% of the load power at rated load.

(d) The armature time constant is

$$\tau_a = \frac{L_a}{R_a} = \frac{1.1 \times 10^{-3}}{0.142} = 7.75 \qquad \text{ms}$$

The load time constant is

$$\tau_L = CR_b = 0.434 \times 3.52 = 1.53 \qquad \text{s}$$

(e) Ignoring the armature inductance, the effective time constant, from Eq. (9.26), is

$$\tau_e = C\frac{R_a R_b}{R_a + R_b} = 0.434\frac{0.142 \times 3.52}{3.66} = 59.2 \qquad \text{ms}$$

9.3

Closed-Loop Speed Control

For many years, the commutator motor has been the preferred choice for many control applications because of the near constant and linear relation between speed and armature voltage. However, the simple drive arrangement of Fig. 9.9 may not be adequate for many drives either because the accuracy of maintenance of speed is not sufficient due to load or parameter changes or because the speed of response to a desired command is too slow. Such is likely to be the case with drives for machine tools, steel and paper mill drives, and robots.

Both the accuracy and the response time of speed control can be improved by the use of a feedback or closed-loop system such as that shown in Fig. 9.12. First, the actual speed of the drive is measured. This may be accomplished, for

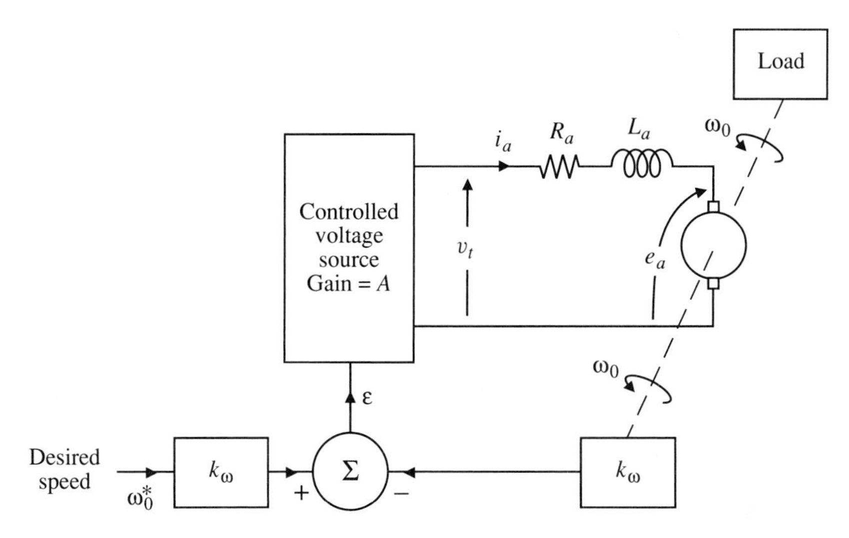

Figure 9.12 Closed-loop speed control system.

example, by a tachometer, which is a small commutator machine with a constant permanent-magnet field attached to the motor shaft. The measured speed signal is compared with the desired speed signal in an adder circuit, and the difference is used to control the armature voltage source. Optimization of the performance of such a system requires a number of concepts that are beyond the scope of this treatment and can be obtained from a standard control system text. In order to appreciate the main features of closed-loop control, we will examine only the simple situation in which the armature voltage is made directly proportional to the speed difference signal.

If the desired or command speed is ω_0^* and the tachometer constant is k_ω, the speed difference signal is

$$\epsilon = k_\omega(\omega_0^* - \omega_0) \qquad \text{V} \tag{9.33}$$

The output of the controlled voltage source to the motor is then

$$v_t = A\epsilon \qquad \text{V} \tag{9.34}$$

where A is the voltage amplification of the source. Suppose the motor has a viscous friction load and that the relation between speed and armature voltage is given by Eq. (9.25), i.e., ignoring armature inductance for simplicity, giving

$$\omega_{0(s)} = \frac{R_b}{R_a + R_b} \frac{v_{t(s)}}{k\Phi(s\tau_e + 1)} \qquad \text{rad/s} \tag{9.35}$$

If Eqs. (9.33), (9.34), and (9.35) are now combined, the drive speed will be related to the command speed by

$$\omega_{0(s)} = \left(\frac{G}{G+1}\right) \frac{\omega^*_{0(s)}}{s\dfrac{\tau_e}{G+1} + 1} \quad \text{rad/s} \tag{9.36}$$

where

$$G = A\frac{k_\omega}{k\Phi}\frac{R_b}{R_a + R_b} \tag{9.37}$$

The quantity G is known as the overall amplification factor of the closed loop. This factor is mainly dependent on the voltage gain A of the controlled voltage source. Examination of Eq. (9.36) shows that the accuracy with which the speed matches the command can be made very high in the steady state if the amplification factor G can be made large. Also, with $G \gg 1$, this accuracy becomes nearly independent of changes in parameters such as changes in load that would affect the parameter R_b or changes in flux. In addition, comparison of Eq. (9.36) with the relation for the open-loop system in Eq. (9.25) shows that the effective time constant is reduced by the factor $G + 1$. Thus, very rapid responses to small changes in command can be expected.

The idealized relation of Eq. (9.36) is potentially deceptive in that it gives the impression that the accuracy and speed of response can be improved indefinitely by increase of the gain A. However, if the effect of the armature inductance is included as in Eq. (9.31), the relation becomes

$$\omega_{0(s)} = \frac{G}{G+1} \frac{\omega^*_{0(s)}}{s^2\left(\dfrac{\tau_a\tau_L}{G+1}\right) + s\left(\dfrac{\tau_a + \tau_L}{G+1}\right) + 1} \quad \text{rad/s} \tag{9.38}$$

As the gain is increased, the speed response to a step in command speed becomes a damped sinusoidal oscillation with an increasing overshoot beyond the desired value, as shown in Fig. 9.13. The response is given by the expression

$$\omega_0 = \frac{G}{G+1}\omega_0^*\left[1 - \frac{\omega_n}{\omega_d}\epsilon^{-\alpha t}\cos(\omega_d t - \theta)\right] \quad \text{rad/s} \tag{9.39}$$

where

$$\omega_n = \left(\frac{G+1}{\tau_a\tau_L}\right)^{1/2} \quad \text{rad/s} \tag{9.40}$$

$$\alpha = \frac{1}{2}\left(\frac{1}{\tau_a} + \frac{1}{\tau_L}\right) \quad \text{s}^{-1} \tag{9.41}$$

$$\omega_d = (\omega_n^2 - \alpha^2)^{1/2} \quad \text{rad/s} \tag{9.42}$$

and

$$\theta = \cos^{-1}\frac{\omega_d}{\omega_n} \tag{9.43}$$

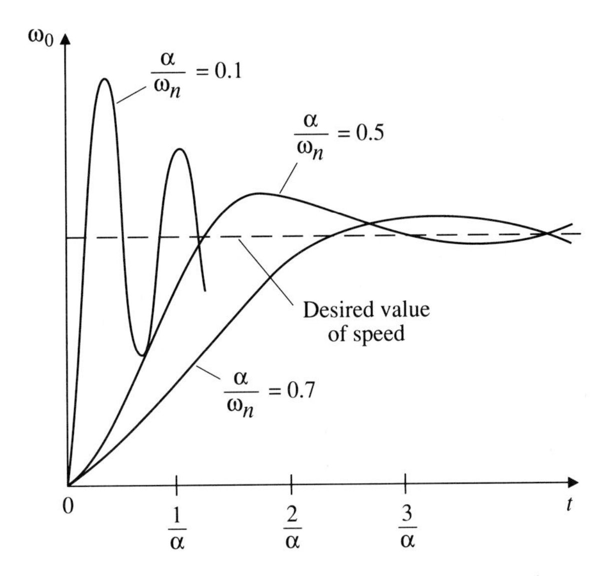

Figure 9.13 Transient response to a step in speed command.

The ratio of α to ω_n is known as the *relative damping ratio*. Transients for various values of this ratio are shown in Fig. 9.13. It is seen that the envelope within which the oscillation decays is independent of the gain. Only the frequency of oscillation increases as gain is increased. More sophisticated means of improving the speed response are discussed in control system texts.

EXAMPLE 9.2 In the speed control system shown in Fig. 9.12, the motor speed is measured by a tachometer that produces 0.2 V·s/rad. The controlled voltage source with its internal amplifier has an amplification A. The motor has an armature resistance of 0.4 Ω and an armature inductance of 0.02 H, and it produces a torque of 1.1 N·m/A of armature current. The polar moment of inertia of the motor and its load is 2 kg·m^2. Friction losses in both motor and load may be neglected.

(a) Develop a transfer function relating the load speed to the command speed.

(b) Determine the amplification A required to produce a relative damping ratio of 0.7.

(c) Suppose the system is originally at rest and a step of speed command signal is applied. Estimate the time required to reach the steady-state command speed for the first time.

Solution

(a) The armature current can be expressed in operational form, from Eq. (9.27), as

$$i_{a(s)} = \frac{v_{t(s)} - k\Phi\omega_{0(s)}}{R_a(s\tau_a + 1)} \qquad \text{A}$$

For this system with a lossless load, the torque is

$$T = k\Phi i_{a(s)} = Js\omega_{0(s)} \qquad \text{N}\cdot\text{m}$$

Rearranging these two expressions gives

$$\omega_{0(s)} = \frac{v_{t(s)}}{k\Phi[s\tau_e(s\tau_a + 1) + 1]} \qquad \text{rad/s}$$

where

$$\tau_e = \frac{JR_a}{(k\Phi)^2} \qquad \text{s}$$

With the loop closed,

$$v_{t(s)} = Ak_\omega(\omega^*_{0(s)} - \omega_{0(s)}) \qquad \text{V}$$

Thus,

$$\omega_{0(s)} = \frac{G}{G+1} \frac{\omega^*_{0(s)}}{s^2 \dfrac{\tau_e\tau_a}{G+1} + \dfrac{s\tau_e}{G+1} + 1} \qquad \text{rad/s}$$

where

$$G = Ak_\omega / k\Phi$$

(b) From the problem statement:

$$k_\omega = 0.2 \quad \text{V}\cdot\text{s}; \qquad R_a = 0.4 \quad \Omega$$

$$\tau_a = \frac{0.02}{0.4} = 0.05 \quad \text{s}$$

$$k\Phi = 1.1 \quad \text{V}\cdot\text{s/rad}$$

$$\tau_e = \frac{2 \times 0.4}{1.1^2} = 0.66 \quad \text{s}$$

$$G = A\,\frac{0.2}{1.1} = 0.182A$$

Thus, the transfer function is

$$\omega_{0(s)} = \frac{G}{G+1} \frac{\omega^*_{0(s)}}{s^2 \dfrac{0.66 \times 0.05}{G+1} + s \dfrac{0.66}{G+1} + 1}$$

The coefficient α is found as half the ratio of the coefficient of s to that of s^2:

$$\alpha = \frac{1}{2} \frac{1}{0.05} = 10 \qquad \mathrm{s}^{-1}$$

Then, for a damping ratio of 0.7, the natural frequency should be

$$\omega_n = \frac{\alpha}{0.7} = 14.3 \qquad \mathrm{rad/s}$$

Using the coefficient of s^2,

$$\omega_n^2 = \frac{G+1}{0.66 \times 0.05} = 204.1$$

Therefore,

$$G = 5.74 \qquad \text{and} \qquad A = \frac{5.74}{0.182} = 31.5$$

(c) From the typical responses shown in Fig. 9.13, the speed crosses the desired value for the first time for $\alpha/\omega_n = 0.7$ at about

$$t = \frac{2.35}{\alpha} = 0.235 \qquad \mathrm{s}$$

9.4

Position Control

Accurate position control is required in many applications such as in robots, machine tools, elevators, and disk drives. In the example shown in Fig. 9.14, the horizontal angle θ_L of a telescope is to be controlled in response to a command angle signal θ_L^*. The telescope platform is rotated by a commutator motor through a reduction gear of ratio n. The angle θ_L is measured using a position sensor such as a rotary potentiometer. The difference between the measured and desired values of angle is used to control the armature voltage.

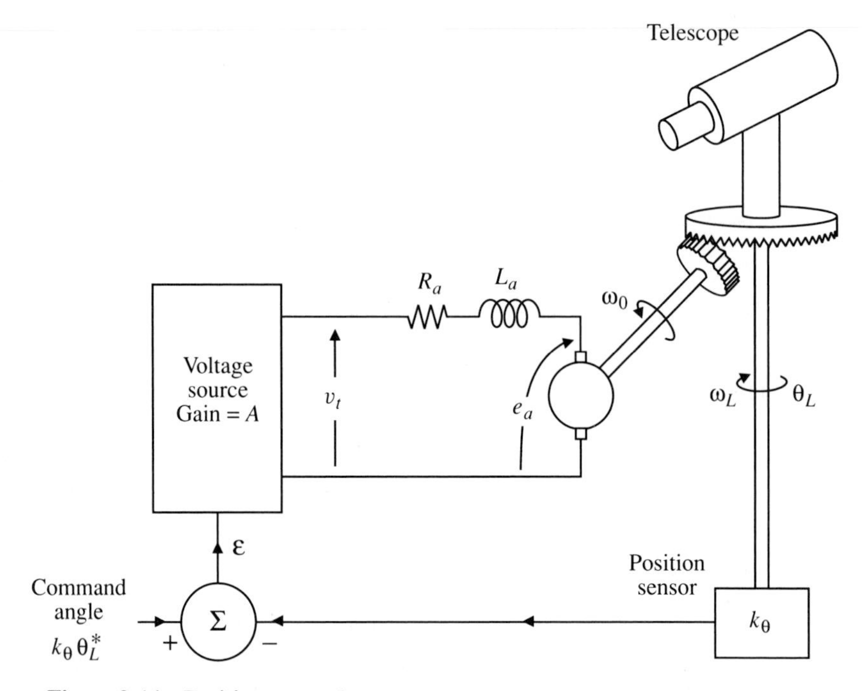

Figure 9.14 Position control system.

To appreciate the major features of this system, assume a lossless load consisting of the load inertia J_L and the motor inertia J_0. Then, from Eq. (9.12), the effective inertia as seen by the motor is

$$J = J_0 + J_L/n^2 \qquad \text{kg}\cdot\text{m}^2 \tag{9.44}$$

If the armature inductance is ignored, the open-loop speed response is, from Eq. (9.21),

$$\omega_{0(s)} = \frac{v_{t(s)}}{k\Phi(s\tau_e + 1)} \qquad \text{rad/s} \tag{9.45}$$

The feedback loop produces the relation between motor terminal voltage and the angle error,

$$v_{t(s)} = Ak_\theta(\theta_L^* - \theta_L) \qquad \text{V} \tag{9.46}$$

Because $\omega_L = d\theta_L/dt$, the load angle can be related to its command value by the relation

$$\theta_{L(s)} = \frac{\omega_{L(s)}}{s} = \frac{\omega_{0(s)}}{ns} = \frac{\theta_{L(s)}^*}{s^2\dfrac{\tau_e}{G} + \dfrac{s}{G} + 1} \qquad \text{rad} \tag{9.47}$$

where

$$G = \frac{Ak_\theta}{nk\Phi} \tag{9.48}$$

The result is a second-order expression from which the transient response to a step command will have the same general form as shown for speed in Fig. 9.13. If the effect of the armature inductance had been included, the result would have been a third-order relation. Setting $s = 0$ in Eq. (9.47) indicates that the steady-state error in matching the command angle setting is ideally zero. In practice, static friction in the driven system usually produces some steady-state error.

9.5

Chopper Drives

A set of chopper circuits capable of producing a controllable direct voltage in one, two, or four quadrants was introduced in Section 8.3. These choppers can be combined with a commutator motor to produce a variable-speed drive. This section examines some of the features of this type of drive.

A chopper drive system is shown in Fig. 9.15. A filter, to be discussed later, is inserted between the direct-voltage supply and the chopper. Let us assume that the voltage V at the chopper is constant. If the armature circuit inductance is sufficient to maintain continuous armature current, the average armature terminal voltage is, from Eq. (8.19),

$$\bar{v}_t = \frac{t_{ON}}{T} V \quad \text{V} \tag{9.49}$$

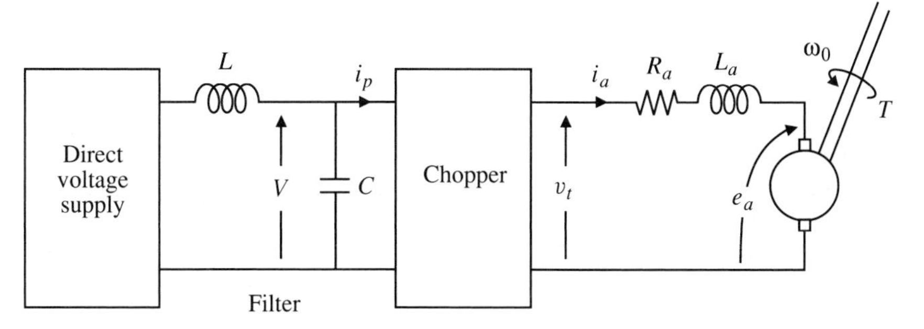

Figure 9.15 Commutator motor drive with chopper supply.

Thus, the steady-state behavior of the drive is given by

$$\omega_0 = \frac{V}{k\Phi}\frac{t_{ON}}{T} - \frac{R_a T}{(k\Phi)^2} \qquad \text{rad/s} \tag{9.50}$$

as developed in Section 4.5. The transient behavior has already been discussed in Sections 9.2 to 9.4.

However, there are a few additional considerations that are of significance with a chopper voltage source. One of these is the effect of harmonic components in the armature current. The Fourier series for the terminal voltage shown in Fig. 8.13 consists of the constant term of Eq. (9.49) plus a set of harmonics of order h having the peak magnitude

$$\hat{v}_h = \frac{2V}{h\pi}\sin\left(h\pi\frac{t_{ON}}{T}\right) \qquad \text{V} \tag{9.51}$$

for $h = 1,2,3\ldots$. The largest harmonic term is the first for $h = 1$ at frequency $f = 1/T$, and its maximum value occurs at $t_{ON} = T/2$, i.e., at half voltage output, where

$$\hat{v}_1 = \frac{2V}{\pi} \qquad \text{V} \tag{9.52}$$

The armature current will contain a component at this frequency. If the resistive part of the armature impedance is ignored, the rms value of this current component is given by

$$I_1 = \frac{\hat{v}_1}{\sqrt{2}}\frac{1}{\omega_1 L_a} = \frac{VT}{\sqrt{2}\pi^2 L_a} \qquad \text{A} \tag{9.53}$$

This current will add to the losses in the armature circuit resistance and therefore to heating of the machine. For a machine that has a continuous current rating of I_b, the maximum continuous average armature current $\bar{i}_a$ with a chopper supply should therefore be limited to

$$\bar{i}_a = [I_b^2 - I_1^2]^{1/2} \qquad \text{A} \tag{9.54}$$

Of course, there will be armature current components at higher harmonic frequencies. However, their effects on heating will be small because I_h is proportional to $1/h^2$ and therefore I_h^2 is proportional to $1/h^4$. Thus, Eq. (9.54) gives a reasonable indication of the amount by which the motor should be derated for use on a chopper supply.

Ripple in the armature current may also cause problems in the operation of the commutator; thus, motor manufacturers may specify the maximum permissible ripple current for the machine. Reduction of ripple current can be achieved either by increasing the chopper frequency or by adding external inductance in the armature circuit.

The current i_p that the chopper takes from its constant voltage supply V has a waveform of the type shown in Fig. 8.13(c). This waveform contains harmonics of frequencies hf in addition to its average value. In some installations, it may be necessary to restrict the harmonic currents in the source system to a small value to avoid interference with other systems. An example would be an electric trolley bus taking power at constant direct voltage from a pair of overhead conductors that are in turn supplied by a rectifier at a central station. Harmonic currents in these overhead conductors could cause serious interference with nearby communication circuits. Control of such harmonic currents frequently involves incorporation of a suitable filter.

If it is assumed that the armature current is essentially constant at its average value $\bar{i}_a$, the source current i_p in the system of Fig. 9.15 is a rectangular wave of peak value $\hat{i}_p = \bar{i}_a$ and an average value

$$\bar{i}_p = \frac{t_{ON}}{T} \bar{i}_a \quad \text{A} \tag{9.55}$$

The peak values of the harmonic components of order h in the source current then are given by

$$\hat{i}_{hp} = \frac{2\bar{i}_a}{h\pi} \sin\left(h\pi \frac{t_{ON}}{T}\right) \quad \text{A} \tag{9.56}$$

Again, it is seen that the largest harmonic occurs for the fundamental chopper frequency $f = 1/T$ and at the output voltage level $\bar{v}_t = V/2$. This usually is the most significant harmonic from the standpoint of filter design. Assuming that the filter consists of the ideal inductance L and capacitance C of Fig. 9.15, the peak value of the harmonic current in the line supplying the filter at harmonic order h is

$$\hat{i}_{hL} = \frac{\hat{i}_{hp}}{\omega_h^2 LC - 1} \quad \text{A} \tag{9.57}$$

To be effective at the lowest harmonic frequency, it is necessary that $LC \gg 1/\omega_1^2$. Thus, the rms harmonic line current components at frequencies hf can be expressed as

$$I_{hL} = \frac{\sqrt{2}\,\bar{i}_a}{h\pi(\omega_h^2 LC)} = \frac{\bar{i}_a}{2\sqrt{2}\pi^3 h^3 LCf^2} \quad \text{A} \tag{9.58}$$

Appropriate values of L and C can now be chosen to meet a specification on the permissible harmonic current.

EXAMPLE 9.3 A streetcar has a mass of 20 000 kg. It is to be accelerated at a constant rate of 1.1 m/s^2 on the level starting from standstill. A commutator motor is coupled to the drive wheel shaft through a reduction gear that gives a vehicle speed of 20 m/s for a motor speed of 420 rad/s. The motor is supplied by a chopper from a 600-V overhead conductor. The motor has an armature resistance of 0.04 Ω and provides a torque of 2.75 N·m per ampere of armature current. Mechanical losses in the motor, gearing, and vehicle friction as well as the motor and gear inertia may be ignored.

(a) Find the torque required to provide the desired acceleration and the corresponding value of armature current.

(b) What should be the switching ratio of the chopper to provide this acceleration at a vehicle speed of 6 m/s?

(c) Up to what vehicle speed can the full acceleration be maintained?

(d) For operation above the speed found in (c), the field flux is reduced, with the armature voltage and the input power held constant. To what value must the flux constant be reduced to produce a speed of 20 m/s and what will be the acceleration under this condition?

(e) Suppose the chopper frequency is 400 Hz and the armature inductance is 3 mH. By what amount should the motor be derated to operate on this chopper supply?

(f) If the maximum rms value of the line current at chopper frequency is to be 3 A, determine the required product LC of the line filter.

Solution

(a) For an acceleration $a = 1.1$ m/s^2, the required thrust on the vehicle is

$$F = ma = 20\,000 \times 1.1 = 22 \qquad \text{kN}$$

The motor speed ω_0 is related to the linear speed ν of the vehicle by

$$\omega_0 = \frac{420}{20}\nu = 21\nu$$

Then, for a lossless system, the required torque is

$$T = \frac{F\nu}{\omega_0} = \frac{22\,000}{21} = 1048 \qquad \text{N}\cdot\text{m}$$

The required armature current is

$$i_a = \frac{T}{k\Phi} = \frac{1048}{2.75} = 381 \qquad \text{A}$$

(b) For $\nu = 6$ m/s, $\omega_0 = 6 \times 21 = 126$ rad/s. At this motor speed, the required average armature voltage must be

$$\bar{v}_t = k\Phi\omega_0 + R_a\bar{i}_a = 2.75 \times 126 + 0.04 \times 381 = 361.7 \qquad \text{V}$$

Thus,

$$\frac{t_{ON}}{T} = \frac{361.7}{600} = 0.603$$

(c) With $t_{ON} = T$, $v_t = 600$ V. At this condition,

$$\omega_0 = \frac{600 - 0.04 \times 381}{2.75} = 212.6 \qquad \text{rad/s}$$

and

$$v = \frac{212.6}{21} = 10.13 \qquad \text{m/s}$$

(d) For the speed range $10.13 < v < 20$ m/s, the power input to the motor is to be constant at $P = 600 \times 381 = 228.6$ kW. At the maximum speed of $v = 20$ m/s or $\omega_0 = 420$ rad/s, the flux constant must be

$$k\Phi = \frac{600 - 0.04 \times 381}{420} = 1.39 \qquad \text{N·m/A}$$

For this flux, the motor torque is $T = 1.39 \times 381 = 530$ N·m and the thrust force on the vehicle will be $F = 530 \times 21 = 11\,130$ N. With this thrust, the vehicle acceleration will be

$$a = \frac{F}{m} = \frac{11\,130}{20\,000} = 0.56 \qquad \text{m/s}^2$$

(e) The chopper period is $T = 1/f = 2.5$ ms. From Eq. (9.53), the fundamental component of harmonic current in the armature under the worst condition is

$$I_1 = \frac{VT}{\sqrt{2}\pi^2 L_a} = \frac{600 \times 2.5 \times 10^{-3}}{\sqrt{2}\pi^2 \times 3 \times 10^{-3}} = 35.8 \qquad \text{A}$$

Thus, the rated or base current of the motor should be

$$i_{a(b)} = (381^2 + 35.8^2)^{1/2} = 382.7 \qquad \text{A}$$

Therefore, there is no significant derating of the motor required because of the harmonic current.

(f) From Eq. (9.57), at a frequency of 400 Hz, the LC product to restrict the fundamental ripple current to 3 A is

$$LC = \frac{\bar{i}_a}{2\sqrt{2}\pi^3 f^2 I_{1L}} = \frac{381}{2\sqrt{2}\pi^3 400^2 \times 3} = 9.1 \times 10^{-6} \qquad \text{s}^2$$

Appropriate values of L and C now may be chosen, usually on the basis of relative costs of the two elements.

9.6

Controlled Rectifier Drives

In industrial applications using commutator machines for drives, the available electric supply is usually three-phase, at a frequency of 50 or 60 Hz, and at reasonably constant and sinusoidal alternating voltage. For these situations, the controlled rectifier drive system such as that shown in Fig. 9.16 is widely employed.

The main characteristics of three-phase controlled rectifiers have been discussed in Section 8.2.2. In the normal continuous-current mode, the average armature voltage $\bar{v}_t$ is related to the line-to-line rms supply voltage by, from Eq. (8.11),

$$\bar{v}_t = 1.35 V_{L\text{-}L} \cos\alpha \tag{9.59}$$

where α is the firing angle. Consider the hoist drive of Fig. 9.16 in which the load is to be both raised and lowered but where the cable force F and thus the motor torque T are always positive. For this situation, the controlled rectifier of Fig. 8.9 can be used. The steady-state speed-torque relation is given by

$$\omega_0 = \frac{1.35 V_{L\text{-}L} \cos\alpha}{k\Phi} - \frac{R_a T}{(k\Phi)^2} \quad \text{rad/s} \tag{9.60}$$

A typical set of speed-torque relations for various values of α is shown in Fig. 9.17. Over most of the torque range, the relation is as given in Eq. (9.60).

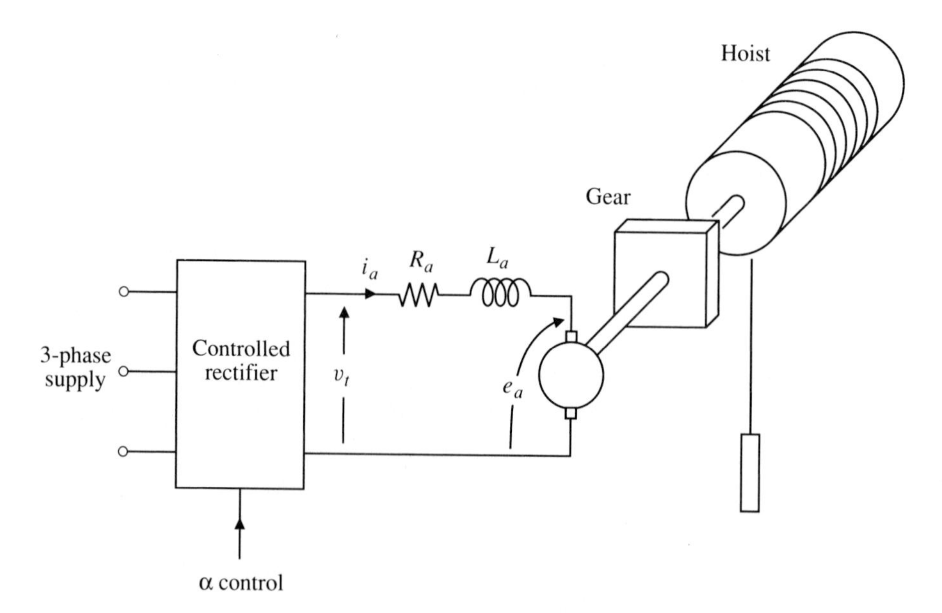

Figure 9.16 Hoist driven by commutator motor with controlled rectifier supply.

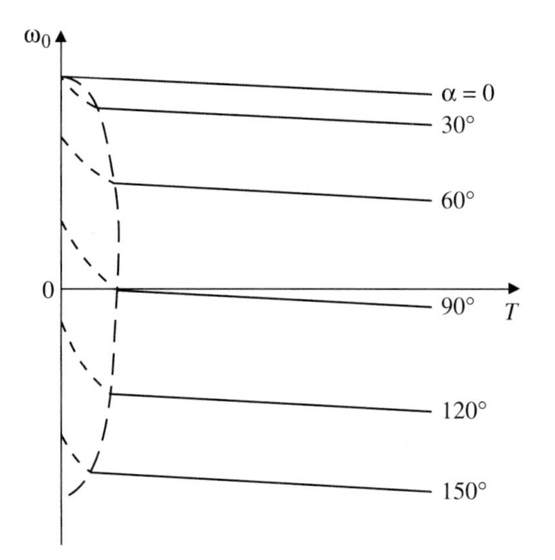

Figure 9.17 Speed-torque characteristics for a three-phase controlled rectifier drive.

However, there will be a region of low torque where the armature current will be discontinuous and in which the speed drops rapidly with increase in torque. With sufficient hoist load, this region may not be encountered. Lowering of the load is achieved with α equal to or greater than 90°. In situations where four-quadrant operation is required, a dual, three-phase controlled rectifier/inverter system can be employed. A single-phase version of a four-quadrant rectifier/inverter is shown in Fig. 8.8.

For more accurate and rapid speed control, the closed-loop system of Fig. 9.12 can be used. The analysis of this system has already been discussed in Section 9.3. One complication in using this analysis is the nonlinear relation between armature voltage and firing angle, as shown in Fig. 9.18. As a first approximation, this cosine curve can be approximated by the straight line shown which can be expressed as

$$\bar{v}_t \approx 0.9\hat{v}_t\left(\frac{\pi}{2} - \alpha\right) \qquad \text{V} \tag{9.61}$$

A typical output voltage waveform for a three-phase rectifier is shown in Fig. 8.10(c). The Fourier series for this waveform, for a supply frequency ω_s, is

$$v_{(t)} = \frac{3\sqrt{2}}{\pi} V_{L\text{-}L} \cos\alpha + \frac{6\sqrt{2}}{\pi} V_{L\text{-}L}\left[\left(\frac{\cos 7\alpha}{14} - \frac{\cos 5\alpha}{10}\right)\sin 6\omega_s t + \left(\frac{\sin 7\alpha}{14} - \frac{\sin 5\alpha}{10}\right)\cos 6\omega_s t\right] + 12\text{th} + 18\text{th harmonics} \qquad \text{V} \tag{9.62}$$

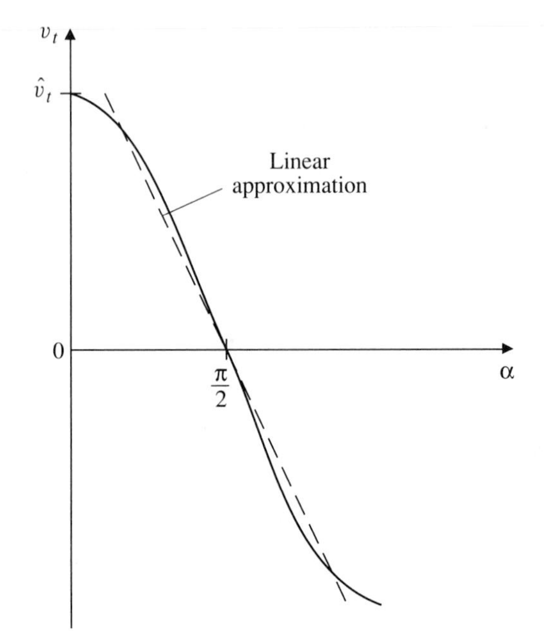

Figure 9.18 Linear approximation for rectifier control function.

The largest harmonic current is of order 6, and its maximum magnitude occurs at $\alpha = \pi/2$, i.e., at the condition of zero average output voltage. At this condition, the rms sixth harmonic voltage is

$$V_6 = \frac{6}{\pi} V_{L\text{-}L} \left(\frac{1}{14} + \frac{1}{10} \right) = 0.327 V_{L\text{-}L} \qquad \text{V} \tag{9.63}$$

The sixth harmonic current in the armature circuit now may be approximated by ignoring the armature resistance (erring on the high side) to give

$$I_6 = \frac{V_6}{6\omega_s L_a} \qquad \text{A} \tag{9.64}$$

Using this current, a check may be made to see the extent to which the motor should be derated for use in this system.

EXAMPLE 9.4 A commutator motor has an armature resistance of 0.045 Ω and an inductance of 0.73 mH. Its generated voltage is 230 V when driven at 3500 r/min. Its rated armature current is 89 A. The motor is to be supplied from a three-phase, 60-Hz source through a controlled rectifier.

(a) What should be the line-to-line supply voltage to provide a maximum direct-output voltage of 230 V?

(b) To what value of armature current should the armature be derated for this drive?

(c) What should the firing angle be to provide maximum output torque at a speed of 600 r/min? Mechanical losses may be ignored.

Solution

(a) To supply the motor with up to 230 V, the line-to-line supply voltage must be

$$V_{L\text{-}L} = \frac{230}{1.35} = 170 \qquad \text{V}$$

(b) The sixth harmonic voltage will be a maximum at $\alpha = \pi/2$, and will have the value

$$V_6 = 0.327 \times 170 = 55.6 \qquad \text{V}$$

The sixth harmonic current is

$$I_6 = \frac{55.6}{0.045 + j(12\pi \times 60 \times 0.73 \times 10^{-3})} = 33.7 \qquad \text{A}$$

For a rated armature current of 89 A, the maximum continuous current with controlled rectifier supply should be

$$i_a = [89^2 - 33.7^2]^{1/2} = 82.4 \qquad \text{A}$$

(c) The flux constant of the motor is

$$k\Phi = \frac{230}{3500 \times \dfrac{2\pi}{60}} = 0.628$$

The maximum torque with the allowed continuous current is

$$T = k\Phi i_a = 0.628 \times 82.4 = 51.7 \qquad \text{N}\cdot\text{m}$$

For a speed of $\omega_0 = 600$ r/min $= 62.8$ rad/s, the firing angle can be evaluated from Eq. (9.60) as

$$\theta = \cos^{-1}\frac{(k\Phi\omega_0 - R_a T/k\Phi)}{1.35\,V_{L\text{-}L}}$$

$$= \cos^{-1}\frac{0.628 \times 62.8 - 0.045 \times 51.7/0.628}{135 \times 170}$$

$$= \cos^{-1} 0.156 = 81^\circ$$

9.7

Series Commutator Motors

In a *series commutator motor*, the armature is connected in series with an appropriately designed series field winding. An equivalent circuit for the machine is shown in Fig. 9.19(a). It was noted in Section 4.5 that a wide speed range can be achieved by weakening the field flux, requiring the motor to run faster to generate a voltage comparable with the supply voltage. In a series motor, this field flux variation occurs inherently in response to load changes. An increase in load increases the armature current, which, being equal to the field current, in turn increases the field flux. The result is a motor that has a sleepy dropping speed characteristic with increased torque when operating from a constant voltage supply.

The basic steady-state speed-torque relation for a series commutator machine is, from Eq. (4.26),

$$\omega_0 = \frac{v_t}{k\Phi} - \frac{(R_a + R_s)T}{(k\Phi)^2} \qquad \text{rad/s} \tag{9.65}$$

where R_s is the resistance of the series field winding. The flux constant $k\Phi$ is related to the armature circuit current that flows in the field winding by a nonlinear curve of the form shown in Fig. 9.19(b). When the motor current is within the linear range of this magnetization curve, the flux constant can be expressed as

$$k\Phi = k_f i_a \tag{9.66}$$

For this condition, the torque produced in the motor is

$$T = k\Phi i_a = k_f i_a^2 \qquad \text{N}\cdot\text{m} \tag{9.67}$$

Inserting Eq. (9.67) into Eq. (9.65) gives the speed-torque relation in the linear flux regime as

$$\omega_0 = \frac{v_t}{(k_f T)^{1/2}} - \frac{R_s + R_a}{k_f} \qquad \text{rad/s} \tag{9.68}$$

A series motor can be operated as a rather inefficient adjustable speed drive by connecting it to a constant direct-voltage supply through a variable resistor R_d. A set of speed-torque relations for various values of R_d is shown in Fig. 9.19(c).

A significant drop in speed with increase in load torque is desirable for several types of mechanical load. For example, in a traction application, a low speed when climbing at high thrust and a high speed when running on the level at low thrust require more nearly constant power to be taken from the supply lines.

It is evident from the speed-torque curves of Fig. 9.19(c) that the series motor cannot be operated on a constant-voltage supply without mechanical load.

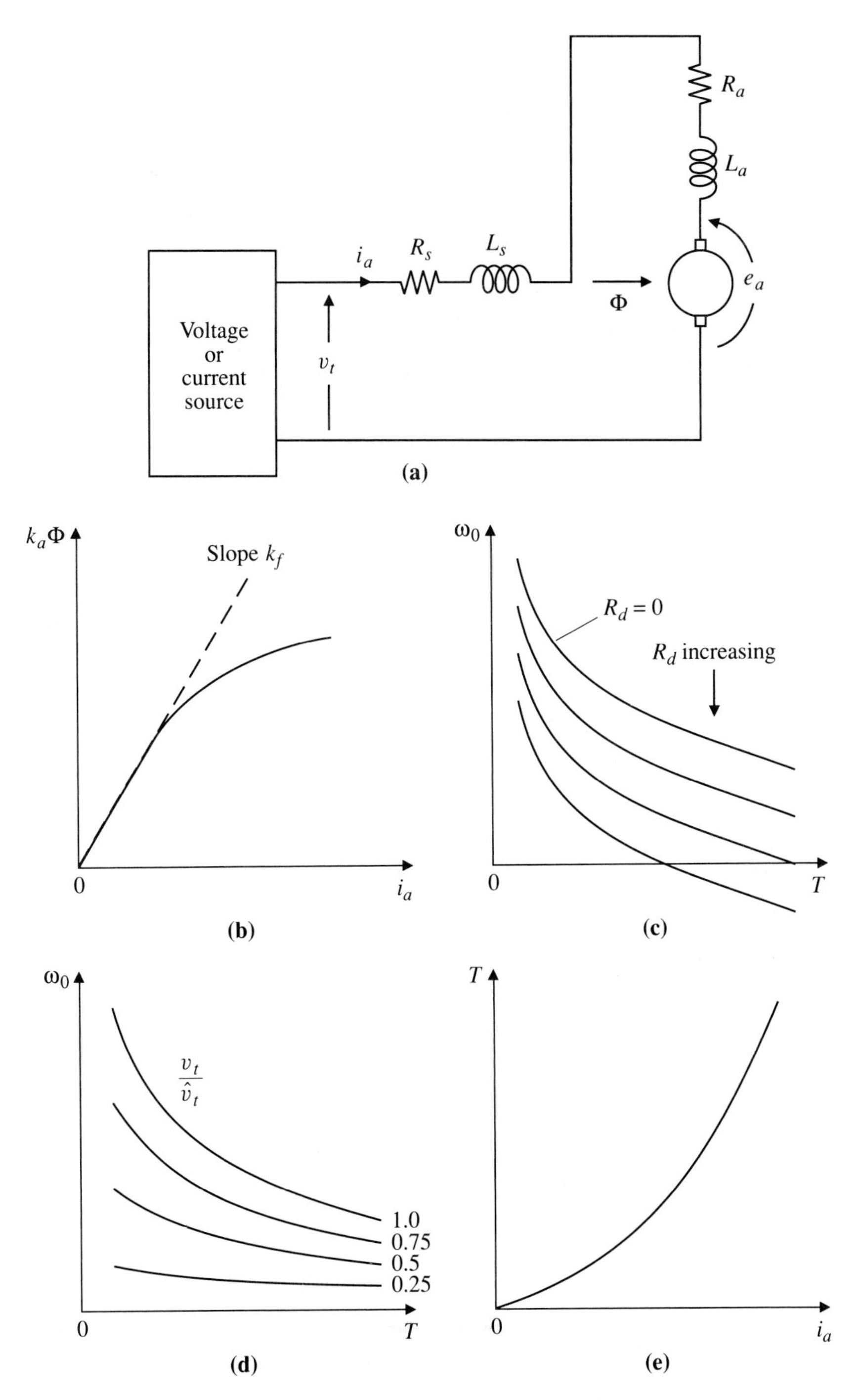

Figure 9.19 (a) Series motor. (b) Magnetization curve. (c) Speed-torque characteristic with series resistance Rd. (d) Speed-torque curves with controlled supply voltage. (e) Speed-torque relation with controlled supply current.

Removal of load causes an increase in speed, a reduction in current and flux, and, therefore, progressive increase in speed to a dangerously high level limited only by friction and windage.

Efficient variable-speed drives can be obtained using series motors with power semiconductor converter supplies. If the motor is supplied with controlled voltage from a controlled rectifier or a chopper, the speed-torque relations for various voltage settings will have the form shown in Fig. 9.19(d). If a variable-current source is used, the result is a controlled value of developed torque as given in Eq. (9.67). The torque varies as the square of the current for low values but becomes more nearly linearly proportional to current as the field system saturates. A typical curve is shown in Fig. 9.19(e).

Reversal of the motor current reverses both the flux direction and the direction of the armature current, leaving the torque direction unchanged, as seen from Eq. (9.67). Therefore, suitably designed series motors may be operated on an alternating voltage supply. With alternating current, the torque pulsates between zero and twice the average value, but the frequency of pulsation, being twice the line frequency, is usually high enough to avoid serious torsional vibration.

Alternating-current series motors require laminated poles and yokes to avoid excessive eddy-current losses. There are also special problems in commutation because of the voltage induced by transformer action in the coils which are short circuited during commutation. This factor restricts the torque rating of a-c series motors.

For a series motor connected to a supply of alternating voltage of rms value V_t and frequency ω, the armature current phasor can be derived from the equivalent circuit of Fig. 9.19(a) in the magnetically linear regime as

$$\vec{I}_a = \frac{\vec{V}_t}{(k_f\omega_0 + R_a + R_s) + j\omega(L_a + L_s)} \quad \text{A} \qquad (9.69)$$

where L_s is the inductance of the series field circuit. For starting, the current is limited mainly by the inductive impedance, and direct-line starting is permissible.

Small series motors that can be used with either direct or alternating voltage supplies are known as *universal motors*. They are used extensively in hand tools and domestic appliances. Their power ratings are generally less than 1 kW, and their full-load speeds may be in excess of 10 000 r/min.

EXAMPLE 9.5 A universal series motor produces a standstill torque of 2 N·m when carrying a direct current of 3 A. The resistance between its terminals is 2.5 Ω and the inductance is 0.04 H. Magnetic linearity may be assumed and rotational losses may be ignored. Suppose the motor is connected to a 115-V, 60-Hz, alternating-voltage supply.

(a) Determine the average starting torque of the motor.

(b) Determine the speed, torque, and mechanical power produced when the motor current is 3 A.

(c) Determine the power factor for the condition of (b).

Solution From the torque measurement with direct current

$$T = k_f i_a^2; \qquad k_f = \frac{2.0}{3^2} = 0.222$$

(a) At standstill

$$\vec{I}_a = \frac{115\angle 0}{2.5 + j(2\pi \times 60 \times 0.04)} = \frac{115\angle 0}{15.29\angle 80.6} = 7.52\angle -80.6 \quad \text{A}$$

$$T = k_f I_a^2 = 12.6 \quad \text{N}\cdot\text{m}$$

(b) The generated voltage in the motor is

$$E_a = k_f I_a \omega_0$$

Thus,

$$|V_s| = |E_a + (R + j\omega L)I_a|$$

$$\frac{115}{3} = |0.222\omega_0 + 2.5 + j15.1| = 38.3 \quad \Omega$$

and

$$\omega_0 = \frac{1}{0.222}[(38.3^2 - 15.1^2)^{1/2} - 2.5] = 147.2 \quad \text{rad/s}$$

$$T = 0.222 \times 3^2 = 2 \quad \text{N}\cdot\text{m}$$

$$P_0 = T\omega_0 = 2 \times 147.2 = 294 \quad \text{W}$$

(c) The motor impedance at $I_a = 3$ A is

$$Z = 35.2 + j15.1 \quad \Omega$$

$$\text{pf} = \frac{35.2}{38.3} = 0.92$$

9.8

Armature Reaction and Commutation

It has been assumed so far that the magnetic flux crossing the air gap and interacting with the armature current is dependent only on the mmf of the field coils or of the field magnets. However, the armature winding also produces an mmf. Under high armature current conditions, this armature mmf may have a significant effect on the useful flux.

An understanding of this effect can be obtained from an examination of the two-pole cross section shown in Fig. 9.20(a). With a uniform air gap and no armature current, the flux density in the air gap under the pole is essentially con-

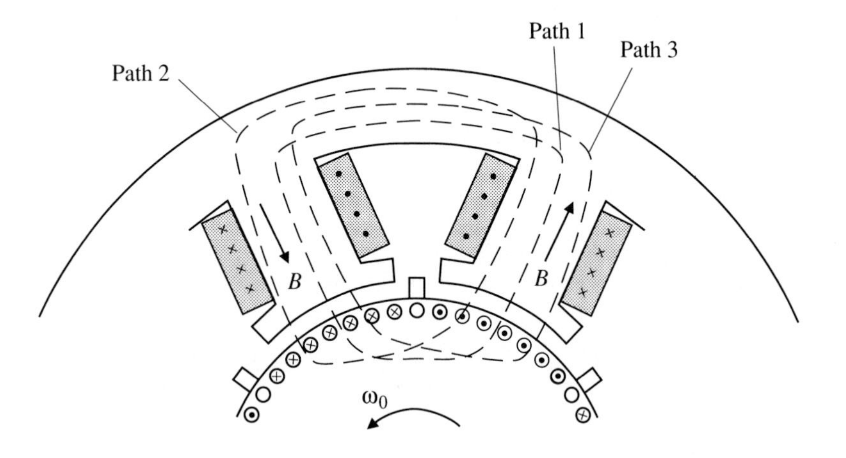

(a)

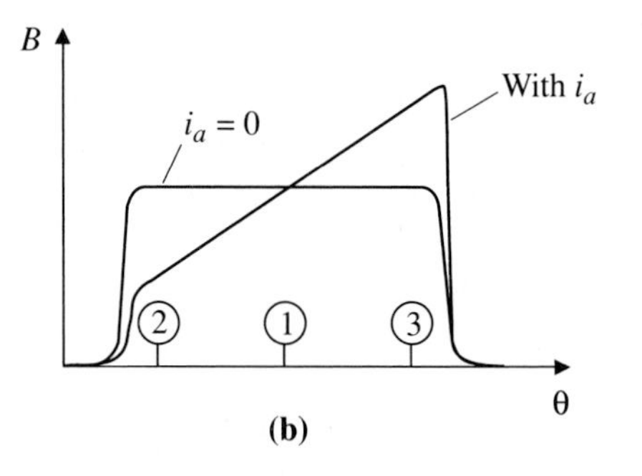

(b)

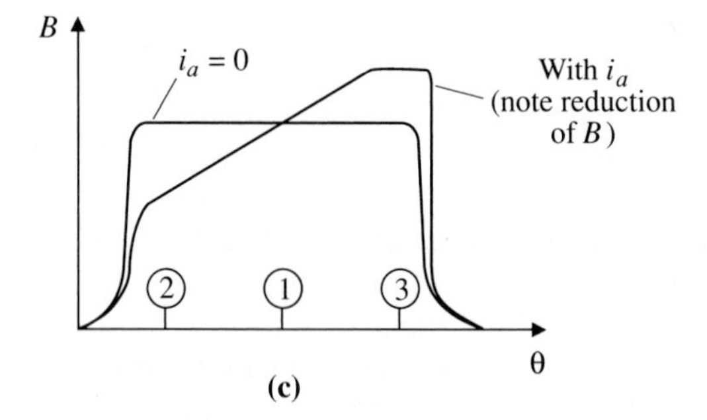

(c)

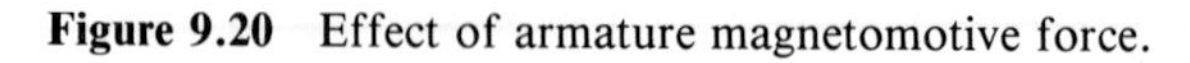

Figure 9.20 Effect of armature magnetomotive force.

stant over the pole face, as shown in Fig. 9.20(b). Armature coils undergo commutation when they are midway between poles. Thus, all armature conductors under a pole carry current in one direction while, under an adjacent pole, all currents are in the opposite direction. For the path 1 of Fig. 9.20(a), the net armature current enclosed is zero, and there is no effect on gap flux density at the center of the pole. For path 2, the net armature mmf is in such a direction as to oppose the field mmf, thus reducing the flux density on this side of the pole. For path 3, the armature mmf adds to the field mmf, increasing the flux density.

If the magnetic system of the machine were truly linear, the reduction in flux on one side of the pole would be exactly balanced by the increase in flux on the other side, as shown in Fig. 9.20(b). The shape of the voltage waveform generated in each armature coil would be changed but its average value would be unaffected. However, if the armature mmf is increased to a value such that saturation occurs in the armature teeth and/or the pole iron near one pole tip, the flux density distribution may have the form shown in Fig. 9.20(c). Therefore, the total pole flux may be reduced by the effect of a large armature mmf. This effect is commonly known as *armature reaction.*

Usually, commutator machines are designed so that the effect of armature mmf on pole flux is negligible up to rated load. Even at the highest transient value of armature current, the reduction in flux is limited to a relatively small percentage. In many machines, this is accomplished by choosing the air-gap length so that the field mmf per pole is significantly greater than the armature mmf at rated current.

In a commutator generator, a reduction of flux due to armature current would cause a reduction in generated voltage and thus cause the terminal voltage to decrease more than the amount due to the voltage drop across the armature resistance. In a motor, the effect of a reduction in flux on the steady-state speed can be appreciated by examining the expression

$$\omega_0 = \frac{v_t - R_a i_a}{k\Phi} \quad \text{rad/s} \tag{9.70}$$

If the flux reduction is large enough, the speed may actually increase as the load torque and armature current are increased. This condition may lead to mechanical instability and is to be avoided.

The flux distortion caused by armature mmf may be effectively eliminated by feeding the armature current through a set of compensating coils placed in slots or holes in the pole faces, as shown in Fig. 9.21. The number of conductors in this compensating winding is so chosen that its mmf is equal and opposite to that produced by the armature. Such compensating windings are frequently incorporated into control motors which may experience large transient values of armature current.

In a permanent-magnet commutator motor such as that shown in Fig. 4.11, the armature mmf can have the same effect of distorting the flux-density distribution across the magnet arc. It may cause saturation in the teeth under one

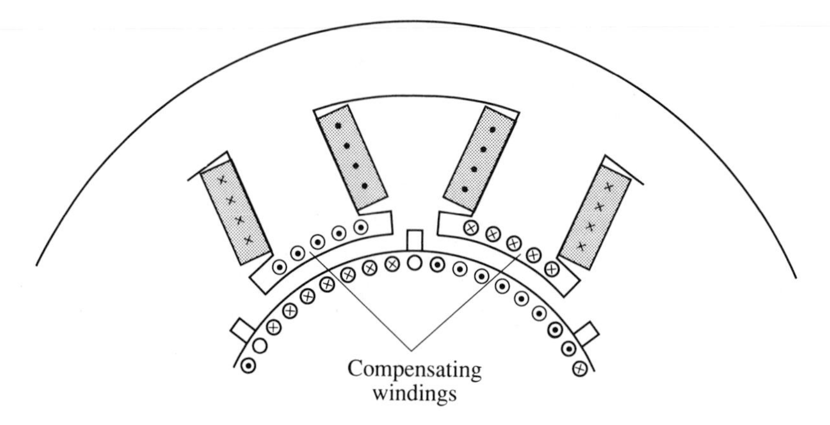

Figure 9.21 Use of compensating windings.

pole edge and thus reduce the net flux per pole. An additional consideration for a permanent-magnet machine is that the flux density at the other edge of the magnet not be reduced to a value that would result in permanent demagnetization, as would occur if the flux density were reduced below the critical value B_D in the B-H loop of Fig. 1.18.

Commutation is the process of switching armature conductors from one path to another as the machine rotates. A particular coil in the machine of Fig. 9.22 has a current i_a/a when its sides are under a pair of poles, where a is the number of parallel paths in the armature winding circuit. When the coil sides are midway between poles, the coil is short circuited for a short time t_c by the stationary brush on the commutator. As soon as the coil passes the brush, its current becomes $-i_a/a$. The reversal of this coil current should be such that no significant sparking occurs at the point of contact between the edge of the brush and the commutator bar.

It would be ideal if the coil current changed at a linear rate over the commutation period t_c, arriving at the correct value at the instant that it is open circuited. This can be accomplished approximately by generating an appropriate voltage e_c in the coil while it is undergoing commutation. This coil has an inductance L_c which depends to some extent on the shape of the armature slot in which it resides. If the resistances of the coil and the brush contacts are ignored, the value of this generated voltage should be

$$e_c = L_c \frac{di}{dt} = L_c \frac{2i_a}{at_c} \qquad \text{V} \tag{9.71}$$

Figure 9.22 shows how this voltage may be produced. A narrow commutating pole, or *interpole*, with a winding carrying the armature current produces flux density in the region of the coil undergoing commutation. The voltage generated in the coil will thus be proportional to the armature current and also will

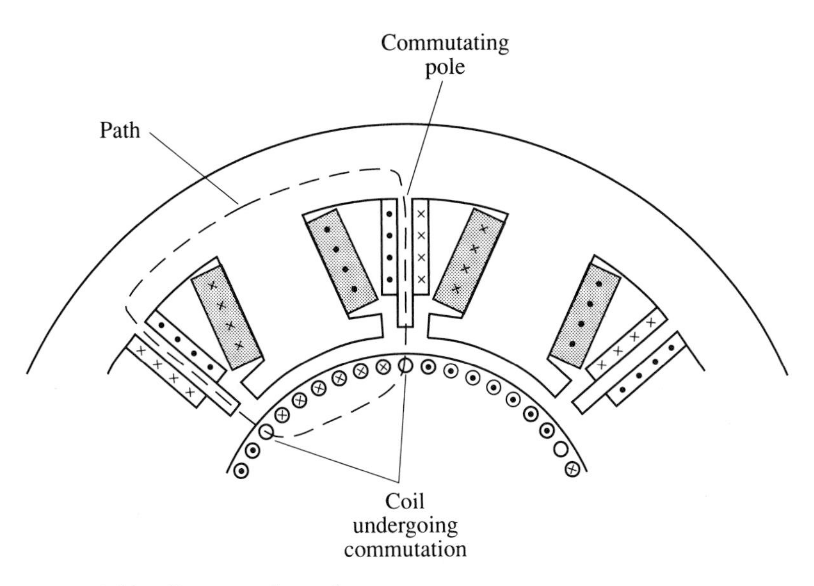

Figure 9.22 Commutating poles.

be proportional to the speed. This satisfies Eq. (9.71) because the time t_c is inversely proportional to speed.

In practice, it is difficult to achieve good spark-free commutation over a wide range of loading conditions, particularly in machines that experience flux weakening at high speed. Commutators require periodic maintenance, and brushes require periodic replacement.

PROBLEMS

9.1 A mechanical load having a polar moment of inertia of 170 $kg \cdot m^2$ and requiring a torque of 50 $N \cdot m$ per rad/s of speed is to be driven at a maximum rotational speed of 400 r/min. The maximum speed of the commutator drive motor is to be 3000 r/min. Its polar moment of inertia is 1.8 $kg \cdot m^2$.

(a) Choose an appropriate gear ratio.

(b) Ignoring all losses and the inertia of the gears, determine the equivalent total polar moment of inertia as seen from the air gap of the motor and the constant b of the equivalent friction load.

(c) Ignoring losses, find the motor power at maximum speed.

(Section 9.1)

9.2 An elevator system of the type shown in Fig. 9.1(a) has a commutator motor coupled through a 12:1 near-ideal reduction gear to a drum of 0.3 m radius around which the elevator cable is coiled. The empty cage has a mass of 800 kg, and the counterweight has a mass of 1200 kg. For the following conditions, find the upward force on the cage, the angular velocity of the drum, the angular velocity of the motor, the torque of the motor, and the motor power, ignoring all losses,

(a) with the elevator moving upward with twelve 75-kg passengers at a constant velocity of 10 m/s.

(b) with the elevator moving upward with its twelve-passenger load accelerating at 1.5 m/s^2 at the instant it reaches a velocity of 10 m/s.

(c) with the empty elevator moving down at a velocity of 4 m/s and decelerating at 1.5 m/s^2.

(Section 9.1)

9.3 A rapid transit car has a total mass of 20 000 kg. It is driven by two motors each coupled through a 7:1 reduction gear to an axle that has a pair of steel wheels of radius 0.4 m. Each wheel set has a polar moment of inertia of 30 $kg \cdot m^2$. Each motor has a polar moment of inertia of 0.5 $kg \cdot m^2$. Ignoring all losses, what total force must be exerted at the wheel rims to accelerate the car at 1.2 m/s^2 on a level track, and what is the corresponding torque required of each motor?

(Section 9.1)

9.4 A permanent-magnet commutator motor has the following name plate data: 230 V, 40 A, 7.4 kW (mech), 1000 r/min, $R_a = 0.5\ \Omega$.

(a) Find the value of the quantity $k\Phi$ for the motor.

(b) Assume that the loss torque in the machine is proportional to speed. Find the constant of proportionality.

(c) The motor is coupled through an 8:1 reduction gear to a load requiring a constant torque of 100 $N \cdot m$. What armature terminal voltage is required to drive the load at a speed of 62.5 r/min?

(d) For the condition of (c), what resistance would be required in series with the armature if a 230-V direct-voltage supply is to be used?

(Section 9.2)

9.5 A commutator machine is driven at a speed of 1170 r/min. The open circuit armature voltage v_t is measured for a set of values of field current i_f giving the following data:

i_f (A)	0.1	0.2	0.3	0.4	0.5	0.6	0.7	0.8
v_t (V)	55	110	161	218	265	287	303	312

(a) Determine the quantity of $k\Phi$ for the machine when its field current is 0.7 A.

(b) What torque will the machine produce with a field current of 0.7 A and an armature current of 25 A?

(c) When the machine is operating in the magnetically linear range, estimate the value of the quantity k_f which is the torque per unit of the product of the field and armature currents.

(Section 9.2)

9.6 The field circuit of a commutator machine has an inductance of 18.4 H and a resistance of 9 Ω.

(a) Suppose the field circuit is supplied from a 125-V source in series with a variable resistor that is adjusted to give a steady-state field current of 10 A. At what time after this source is connected will the field current reach 9 A?

(b) A resistor is frequently connected across the field circuit to protect the insulation of the field coils from the excessive voltage that might otherwise occur when a switch connecting the field circuit to the source is opened. Suppose the insulation will safely withstand a maximum voltage of 2000 V. What value of protective resistance is required to limit the voltage at the field circuit terminals to this value if the initial field current is at the maximum value obtainable from the 125-V source?

(Section 9.2)

9.7 Suppose the commutator drive motor in Problem 9.1 has a generated voltage of 230 V at a speed of 4000 r/min.

(a) Determine the capacitance which is equivalent to the total inertia of the system as seen from the motor.

(b) Determine the resistance which is equivalent to the friction load as seen from the motor.

(c) If the motor has an armature circuit resistance of 0.045 Ω and the armature inductance is assumed to be negligible, find the time constant at which the load speed responds to a change in supply voltage.

(Section 9.2)

9.8 A commutator motor is used to rotate a load having a polar moment of inertia of 0.8 $kg \cdot m^2$ and requiring negligible torque for continuous rotation. The motor has a polar moment of inertia of 0.05 $kg \cdot m^2$, an armature resistance of 0.5 Ω, negligible armature inductance, and negligible rotational loss. With an armature voltage of 60 V, the motor has a steady-state speed of 120 rad/s.

(a) At a certain time, the armature voltage is suddenly increased from 60 to 80 V. Determine the speed of the motor 1 s after the change is made.

(b) When the motor is operating steadily with an armature voltage of 60 V, a mechanical load requiring a constant torque of 5 N·m is applied to its shaft. Determine the speed of the motor 1 s after this change is made.

(Section 9.2)

9.9 A 220-V, 1750-r/min, 89-A commutator motor has an armature resistance of 0.086 Ω, an armature circuit inductance of 2.2 mH, and a polar moment of inertia of 0.30 kg·m^2. Its rotational losses may be ignored. It is driving a load that may be regarded as a pure inertia of polar moment 120 kg·m^2 through a lossless reduction gear of ratio 9:1.

(a) Ignoring the armature inductance, develop the transfer function relating the motor speed to the armature voltage. Evaluate the time constant for this transfer function.

(b) With the system initially at standstill, a direct voltage of 20 V is applied to the armature terminals. Derive the speed and the armature current as functions of time and plot them.

(c) Repeat (a) and (b) including the armature inductance. Compare the results to test the validity of ignoring armature inductance.

(Section 9.2)

9.10 A single-phase source of low and variable frequency up to 5 Hz is required to supply a sinusoidal voltage of 115 V rms to a 50-Ω resistive load. An available commutator machine has a rating of 200 V, 500 W, and 1200 r/min. This machine requires a field current of 0.2 A to provide rated voltage on no load at rated speed. The machine may be regarded as magnetically linear. Other parameters of the machine are armature resistance 5 Ω, armature inductance 0.11 H, field resistance 200 Ω, and field inductance 10 H. A low-frequency electric oscillator having negligible internal impedance is connected to the field. The machine is driven at rated speed.

(a) Determine the rms value of the voltage that must be applied to the field terminals at a frequency of 5 Hz.

(b) Determine the phase shift between the load voltage and the field voltage for the condition of (a).

(c) Determine the power amplification of the machine when operating at 5 Hz with the above load.

(Section 9.2)

9.11 The starter of a permanent-magnet commutator motor consists of a single resistor, the value of which is adjusted to limit the maximum armature current during starting to twice the rated value. The generated voltage is 200 V at a speed of 1800 r/min. Rated armature current is 18 A and the armature resistance is 1.2 Ω. The polar moment of inertia of the motor and its connected load is 2.5 kg·m^2. Rotational losses and armature circuit inductance may be ignored. The supply voltage is 200 V. For how long a time should the starting resistor be left in the circuit if the armature current is to rise to just twice its rated value when the starting resistor is shorted out?

(Section 9.2)

9.12 The permanent-magnet starting motor for a jet turbine engine is geared to the engine rotor through a 1:5 speed increaser. The polar moment of inertia of the engine rotor is 0.4 $kg \cdot m^2$. The windage and friction torque of the engine rotor is approximately 0.02 N·m per rad/s. The starting motor produces a torque of 0.8 N·m per ampere of armature current. Its armature resistance is 0.8 Ω and its polar moment of inertia is 0.01 $kg \cdot m^2$. If a 100-V direct-voltage supply is used, how long a time is required for the engine to reach its starting speed of 250 rad/s?

(Section 9.2)

9.13 A speed control system for a turntable is shown in Fig. 9.23. A permanent-magnet commutator motor is coupled through a 4:1 reduction gear to a turntable which has a polar moment of inertia of 2 $kg \cdot m^2$. The speed of the turntable is measured by a permanent magnet tachometer which has a generated voltage of 0.25 V per rad/s of speed. The difference between the tachometer voltage and a command voltage v^* is used as the input to a voltage amplifier of gain A. The motor parameters are $R_a = 0.3\ \Omega$, $L_a = 2.3$ mH, $J = 0.07\ kg \cdot m^2$, $k\Phi = 0.9$. The voltage amplification A of the source is set at 135. A step of direct-command voltage is applied at time $t = 0$. Derive and sketch the speed-time relation for the system.

(Section 9.3)

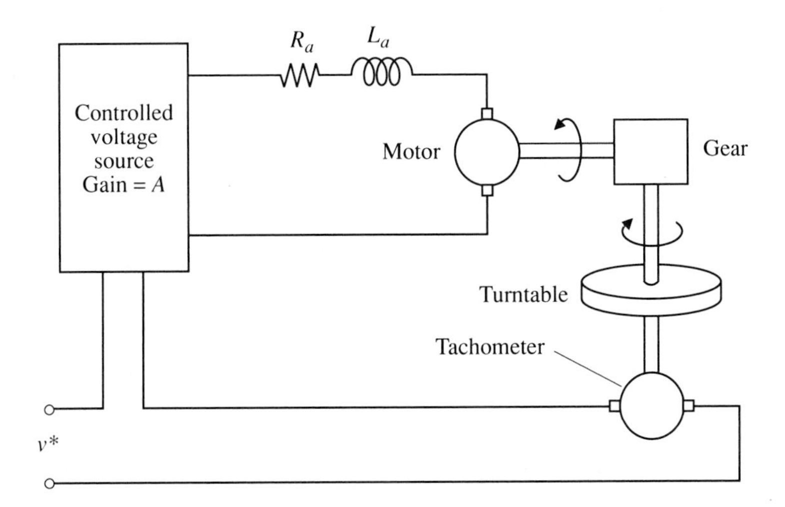

Figure 9.23 Speed-control system for Problem 9.13.

9.14 A system for positioning a machine tool is shown in Fig. 9.24. The tool is moved 2.5 mm for each revolution of the motor shaft. The position of the tool is measured by a linear potentiometer which gives an output voltage of 0.1 V/mm of displacement. The permanent-magnet motor has an armature resistance of 12 Ω, negligible armature inductance, and negligible friction losses. The polar moment of inertia of the motor and its load is 10^{-5} kg·m^2. The motor produces a torque of 0.1 N·m per ampere of armature current. The amplifier has an amplification A and an internal resistance of 10 Ω.

(a) Develop a transfer function relating the position x of the tool to the command voltage v^*.

(b) Determine the amplification required to produce a relative damping ratio of 0.5.

(c) A change in tool position is initiated by a step change in the command voltage. Estimate the time required for the tool to remain within 5% of the total change in position.

(Section 9.4)

9.15 A 230-V, 100-A, 1200-r/min commutator motor is supplied from a two-quadrant chopper with an input voltage of 300 V and a chopping frequency of 300 Hz. The armature resistance of the motor is 0.12 Ω and its inductance is 2 mH. Rotational losses may be ignored.

(a) Determine the rated torque of the motor with continuous direct current.

(b) Determine the switching ratio of the chopper to operate the machine in the regenerating mode with 90% of the torque found in (a) and at a speed of 500 r/min.

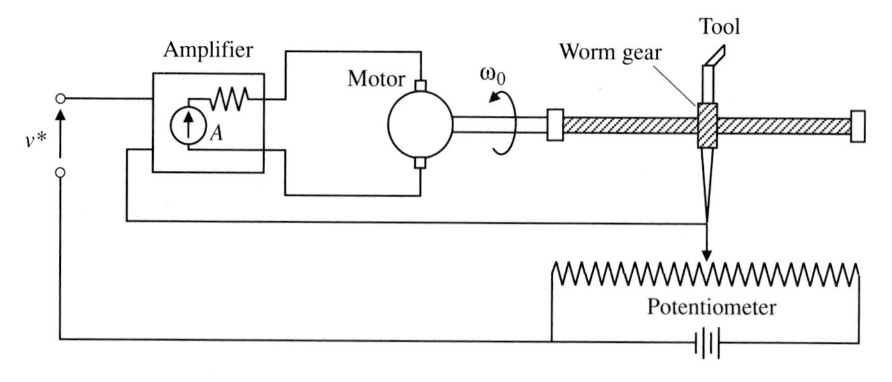

Figure 9.24 Position control system for Problem 9.14.

(c) Estimate the maximum torque that this motor can provide continuously without overheating the armature at any value of speed. Express the torque as a ratio of the rated value found in (a).

(Section 9.5)

9.16 A 230-V, 10-kW, 50-A, 1500-r/min, $R_a = 0.3\ \Omega$, $L_a = 3$ mH commutator motor is supplied from a four-quadrant chopper and a 250-V direct-voltage source. Rotational losses may be considered to be proportional to speed. The chopper frequency is 400 Hz.

(a) Find the switching ratio required for each switch to drive a load with a shaft torque of 50 N·m at 1000 r/min.

(b) Repeat (a) for a shaft torque of 50 N·m at −1000 r/min.

(Section 9.5)

9.17 An electric trolley bus having a loaded mass of 20 000 kg is to be capable of driving up or down a 6% grade continuously at any speed up to a maximum of 60 km/h. The direct-voltage electric supply on the trolley wires is 600 V. Supply to the commutator motor is through a 500-Hz chopper. All losses in the motor, the chopper, the gears, and the wheels may be ignored.

(a) The bus is equipped with a contactor for use on the occasions that it is to operate in reverse. Sketch the circuit of an appropriate chopper.

(b) Estimate the voltage, current, and power ratings of the required commutator motor.

(c) Determine the switching ratio of the chopper when the bus is going down a 4% grade at a constant speed of 30 km/h.

(d) Determine the average motor current and average supply current for the condition of (c).

(Section 9.5)

9.18 In the electric trolley bus of Problem 9.17, suppose the maximum permissible alternating current in the overhead trolley wires at the lowest harmonic frequency is to be limited to 1.0 A rms. An inductance-capacitance filter is to be used. Determine appropriate values for the inductance and the capacitance on the assumption that the cost of storing energy in an inductor is five times that of storing energy in a capacitor.

(Section 9.5)

9.19 A commutator motor has the following ratings on constant current: 110 V, 2000 r/min, 12 A, $R_a = 0.5\ \Omega$, $L_a = 2.5$ mH. The motor is supplied from a step-down chopper connected to a 96-V battery supply. The chopper operates at a frequency of 500 Hz. Rotational losses may be ignored.

(a) What is the maximum torque that this motor can produce continuously without overheating at any set value of speed? Assume no loss in cooling capacity with reduction in speed.

(b) At what speed will the limiting condition of (a) occur?

(Section 9.5)

9.20 Each vehicle on a personal rapid transit system has a capacity for six persons. Its fully loaded mass is 1500 kg. It is to be capable of an acceleration of 1.5 m/s^2 while travelling up a 6% grade until it has reached a velocity of 5.5 m/s. In the speed range from 5.5 to 11 m/s, the field current of the commutator motor is to be reduced. The commutator motor is connected to the 0.5-m-diameter driving wheels through a gear box. The motor is supplied from a chopper which is connected to a 600-V supply through a contact with a trackside rail.

(a) Choose an appropriate type of chopper for this application.

(b) Ignoring all losses, determine the maximum power required from the motor and also its maximum armature current.

(c) If the maximum speed of the motor is to be 4500 r/min, determine an appropriate gear ratio.

(c) Determine the maximum and minimum values for the flux constant $k\Phi$ of the motor.

(Section 9.5)

9.21 A single-phase bridge rectifier is used to control the speed of a 3.75-kW, 115-V, 30-A, 1200-r/min commutator motor. The armature resistance is 0.4 Ω. The available utility supply is at 120 V, 60 Hz.

(a) Determine the firing angle to give a motor speed of 1000 r/min when the motor is operating with rated armature current.

(b) Determine the supply power factor for the condition of (a).

(c) If the air-gap torque is 16 N·m and the firing angle is 60°, what is the motor speed?

(d) What is the rms current in each switch for the condition of (c)?

(Section 9.6)

9.22 A permanent magnet commutator motor rated at 230 V, 50 A, 1500 r/min is supplied from a three-phase, 230-V (L-L), 60-Hz controlled rectifier/inverter. The armature resistance is 0.5 Ω and the armature inductance is high enough that the armature current can be considered to be ripple free. Rotational losses may be ignored.

(a) Find the firing angle necessary to drive a load requiring a constant torque of 55 N·m at a speed of 1200 r/min.

(b) Find the firing angle for the same torque at a negative speed of 600 r/min.

(c) Find the supply power factor for the load condition of (a).

(d) The maximum speed is not to exceed 2250 r/min. Can the firing angle be reduced to zero?

(Section 9.6)

9.23 A winch is required to raise and lower a mass of 8000 kg at a maximum speed of 1.2 m/s. The winch cable is wound around a drum with a radius of 0.2 m. The drum is driven by a commutator motor through a gear box. The motor is to be supplied from a three-phase controlled rectifier/inverter connected to a 230-V (L-L), 60-Hz supply. The maximum motor speed is to be 3500 r/min.

(a) Determine an appropriate gear ratio.

(b) Choose an appropriate value for the rated voltage of the motor.

(c) Assume that the winch is raising or lowering the load 40% of the time and is idle for the remaining time. Choose an appropriate value for the rated current of the motor.

(d) Estimate the rms current in each rectifier thyristor when raising the load.

(Section 9.6)

9.24 A series commutator motor is to be used to drive a load requiring a torque of 4 N·m at 25 r/s. A 200-V direct-voltage supply is available. A test on the motor at standstill shows that a terminal current of 5 A produces a shaft torque of 4 N·m with a terminal voltage of 10 V. The rotational losses of the machine are considered to be negligible.

(a) What resistance is required in series with the 200-V supply to meet the required load condition?

(b) Assuming that the motor is magnetically linear, repeat (a) for a torque of 2 N·m at 25 r/s.

(Section 9.7)

9.25 A four-pole series-wound fan motor rotates at 100 rad/s and takes 20 A from its 200-V direct-voltage supply when its field coils are connected in series. In order to increase the speed, it is decided to reconnect the field coils in two parallel groups, each group having two coils in series. The fan load requires a power that varies as the cube of the speed. The motor may be considered magnetically linear and all motor losses may be ignored. Determine the new value for the motor speed and motor current.

(Section 9.7)

9.26 A 115-V, 60-Hz universal series motor has a total resistance between terminals of 1.3 Ω and a total inductance of 0.025 H. It takes a current of 8 A rms from the supply when driving a load with a constant torque of 2.5 N·m. Magnetic linearity may be assumed and rotational losses may be ignored.

(a) Determine the speed of the motor.

(b) Determine the input power factor.

(c) If the motor is now connected to a 115-V direct-voltage supply, at what speed would it drive the 2.5-N·m torque load?

(Section 9.7)

9.27 The fan motor of an automobile has a ferrite permanent-magnet field. The motor is supplied from a 11.2-V battery. At a temperature of 20°C, the motor parameters are $R_a = 0.55\ \Omega$, $k\Phi = 0.05$. The combined torque of the fan and the motor mechanical losses is 0.24 N·m. As the temperature of the motor rises from 20 to 120°C, will its speed increase or decrease? The temperature coefficient of the flux density in the magnet material is −0.2%/°C, and the temperature coefficient of the copper in the winding is 0.393%/°C.

GENERAL REFERENCES

Dewan, S.B., Slemon, G.R., and Straughen, A., *Power Semiconductor Drives*, John Wiley and Sons, New York, 1984.

Dubey, G.K., *Power Semiconductor Controlled Drives*, Prentice-Hall, Englewood Cliffs, New Jersey, 1989.

Kusko, A., *Solid-State DC Motor Drives*, M.I.T. Press, Cambridge, Mass., 1969.

Leonhard, W., *Control of Electric Drives*, Springer-Verlag, New York, 1985.

Lightband, D.A. and Bicknell, D.A., *The Direct Current Traction Motor—Its Design and Characteristics*, Business Books, London, 1970.

Nasar, S.A., *Handbook of Electric Machines*, McGraw-Hill, New York, 1987.

Nene, V.D., *Advanced Propulsion Systems for Urban Rail Vehicles*, Prentice-Hall, Englewood Cliffs, New Jersey, 1985.

Sen, P.C., *Thyristor DC Drives*, Wiley Interscience, New York, 1981.

CHAPTER

10

Induction Motor Drives

As compared with commutator motors, induction motors are robust, easily maintained, and more reliable. They have lower cost, weight, and inertia. They can operate in dirty and explosive environments. As a result, a substantial majority of all electric motors are of this type. Most of these are installed in applications for which an essentially constant speed is acceptable.

As the capabilities of semiconductor switches have improved and as their cost has decreased, it has become both possible and economical to produce converters providing variable-frequency voltage and/or current supplies. These can be used with induction motors to produce drives that can compete with commutator machine drives in both performance and cost. Many applications that have conventionally used constant-speed induction motors are now being converted to adjustable-speed drives so that the optimum speed can be chosen for each operating condition, thus achieving higher system efficiency and energy conservation. Also, induction motors with semiconductor converters are being used extensively for applications requiring precise and rapid speed and position control.

The basic structure and the steady-state properties of induction motors have been introduced in Chapter 5. This chapter begins with the development of concepts and equivalent circuits that will allow prediction of both transient and steady-state behavior of induction machines. Several useful drive configurations are then introduced and discussed. Finally, some special induction machine applications are described.

Throughout this chapter and the next, we will normally denote the rotational speed of a machine by the variable ω_0 in electrical radians per second unless we are specifically discussing its torque, in which case the term mechanical speed, ω_{mech}, in mechanical radians per second will be used.

10.1

A Transient Model of the Induction Machine

A three-phase induction machine has three current variables and three voltage variables for its stator. If it is of wound rotor construction, it has a similar number of rotor variables. If it has a squirrel-cage rotor, it could have as many current variables as it has rotor bars. At first glance, the development of a transient model for such a machine would appear to be very complex. However, a relatively simple transient model can be produced because of the symmetrical arrangement of the phase windings and because of the near-sinusoidal distribution of these windings.

The induction machine is based on the concept of using variable currents in stator windings to produce a revolving magnetic field. To achieve this, the stator slots are fitted with windings that are arranged in a nearly sinusoidal angular distribution around the periphery. This sinusoidal distribution is chosen so that the total effect of a set of angularly displaced windings will always be sinusoidally distributed in angular space for any set of instantaneous winding currents.

Consider the machine stator shown in cross section in Fig. 10.1. The winding for phase a shown has its effective number of turns N_{se} distributed with a conductor density of

$$n_a = \frac{N_{se}}{2} \sin \theta_s \tag{10.1}$$

For simplicity, a two-pole machine will be analyzed first and the effect of multiple-pole pairs will be derived later. Windings b and c are similar to winding a except that their axes are located $2\pi/3$ and $4\pi/3$ rad ahead of winding a. A current i_a produces an mmf across the stator-to-rotor air gap (assuming ideal iron) of

$$\mathfrak{F}_{a\theta} = \frac{N_{se} i_a}{2} \cos \theta_s \quad \text{A} \tag{10.2}$$

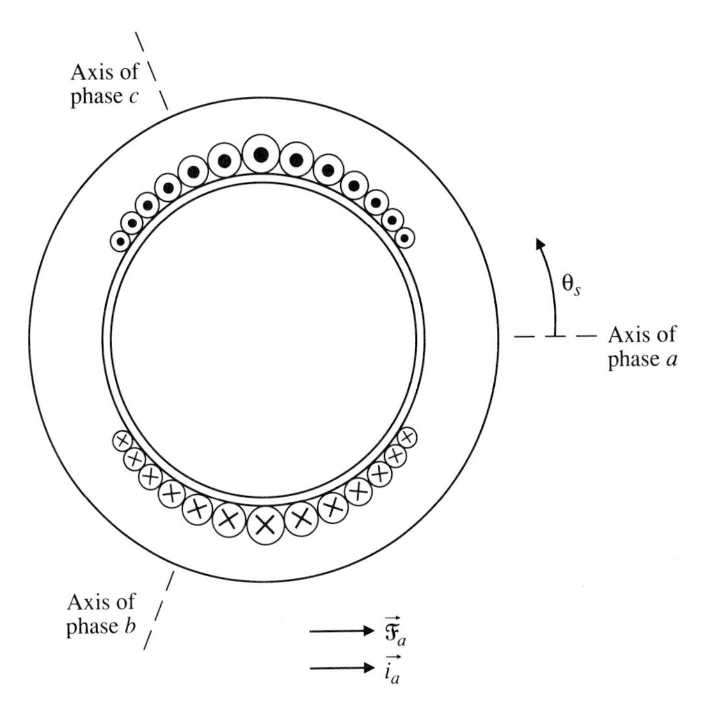

Figure 10.1 Sinusoidal distribution of conductors of phase a, together with the space vectors of magnetomotive force and current.

as derived in Section 5.1. Because this and other variables will be similarly distributed around the air-gap periphery, it is convenient to represent them by space vectors. Each of these space vectors is represented by a complex number having a magnitude equal to the peak amplitude of the variable and an angle equal to the angular position of this positive peak value. When appropriate, each space vector is considered to have a cosinusoidal distribution around the machine periphery. Thus, the mmf due to current i_a in winding a as expressed in Eq. (10.2) is represented by the symbol $\vec{\mathfrak{F}}_a$ and is denoted by the space vector along with the "a" axis, as shown in Fig. 10.1.

Because the mmf distribution $\mathfrak{F}_{a\theta}$ is fixed in space by the axis of the a winding and is directly proportional to i_a, it is convenient to represent the winding current also as a space vector $\vec{i}_a$ so that

$$\vec{\mathfrak{F}}_a = \frac{N_{se}}{2}\vec{i}_a \quad \text{A} \tag{10.3}$$

where $\vec{i}_a$ lies along the axis of winding a as shown in Fig. 10.1 and has a magnitude equal to the instantaneous value of the winding current i_a.

In a similar manner, the effects of the phase currents i_b and i_c in windings b and c may be represented by space vectors $\vec{i}_b$ and $\vec{i}_c$ along the axes of these windings. If a positive rotation of $2\pi/3$ rad or 120° is represented by

$$\alpha = \epsilon^{j(2\pi/3)} \tag{10.4}$$

then

$$\vec{i}_b = \epsilon^{j(2\pi/3)} i_b = \alpha i_b \qquad \text{A} \tag{10.5}$$

and

$$\vec{i}_c = \epsilon^{j(4\pi/3)} i_c = \alpha^2 i_c \qquad \text{A} \tag{10.6}$$

The total effect of a set of instantaneous currents in all three phase windings is then found by simple vector addition as

$$\vec{i} = \vec{i}_a + \vec{i}_b + \vec{i}_c \qquad \text{A} \tag{10.7}$$

Figure 10.2(b) shows this vector addition for the particular instant in the balanced set of sinusoidal phase currents shown in Fig. 10.2(a).

Most induction machine stator windings have only three terminals, and no neutral connection is available. Thus, for this condition

$$i_a + i_b + i_c = 0 \tag{10.8}$$

Then, the current vector $\vec{i}$ can be expressed as

$$\begin{aligned} \vec{i} &= i_a + \left(-\frac{1}{2} + j\frac{\sqrt{3}}{2}\right) i_b + \left(-\frac{1}{2} - j\frac{\sqrt{3}}{2}\right) i_c \\ &= \frac{3}{2} i_a + j\frac{\sqrt{3}}{2}(i_b - i_c) \qquad \text{A} \end{aligned} \tag{10.9}$$

While we could use the space vector $\vec{i}$ to represent the combined effect of the stator currents, it is more convenient to use a magnitude for the current space vector that is equal to the peak value of one of the phase currents when operating in the sinusoidal steady state. If this stator current vector is denoted as $\vec{i}_s$, then

$$\begin{aligned} \vec{i}_s &= \tfrac{2}{3}\vec{i} \\ &= \tfrac{2}{3}(\vec{i}_a + \vec{i}_b + \vec{i}_c) \\ &= i_a + j\frac{i_b - i_c}{\sqrt{3}} \\ &= i_a + j\frac{i_a + 2i_b}{\sqrt{3}} \qquad \text{A} \end{aligned} \tag{10.10}$$

Having concentrated the effect of the three phase currents into a single complex variable, let us now consider the converse process. If, at any instant in time,

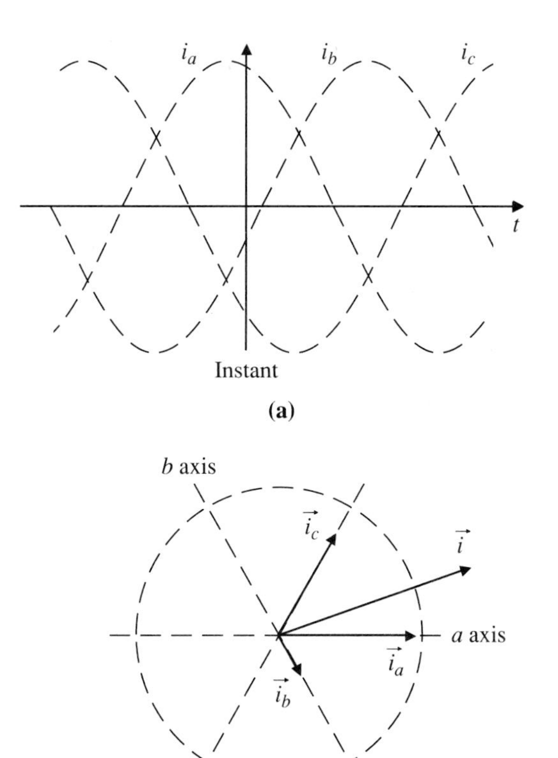

Figure 10.2 (a) A particular instant in a set of phase currents. (b) Summation of phase current space vectors to give the total effect.

we know the space vector $\vec{i}_s$, we can find the instantaneous values of the individual phase currents by projecting the vector $\vec{i}_s$ on the a, b, and c axes as shown graphically in Fig. 10.3. The analytical equivalent of this projection is

$$i_a = \Re\vec{i}_s; \qquad i_b = \Re\alpha^2\vec{i}_s; \qquad i_c = \Re\alpha\vec{i}_s \qquad \text{A} \tag{10.11}$$

where $\Re$ represents the "real part of" the complex number.

Next, let us consider a wound rotor that also has a set of three phase windings A, B, and C where the axis of phase A is at an angle β to the axis of stator phase a at the instant of interest, as shown in Fig. 10.4. Let us assume sinusoidally distributed windings, each of N_{re} effective turns, so that the conductor distribution of phase A is

$$n_A = \frac{N_{re}}{2} \sin\theta_r \tag{10.12}$$

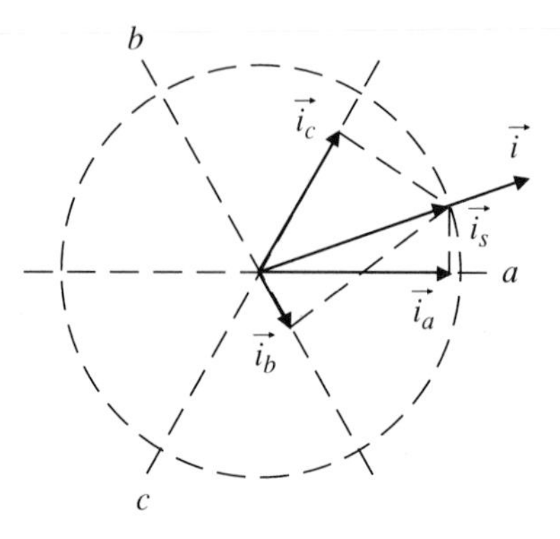

Figure 10.3 Derivation of phase current vectors from the stator current space vector $\vec{i}_s$.

As was shown in Chapter 5, it is usually convenient to refer all rotor variables to the equivalent number of stator turns. Thus, the actual winding of n_A turns per radian carrying current i_A is replaced for analytical purposes by a winding of n'_A turns per radian carrying current i'_A where

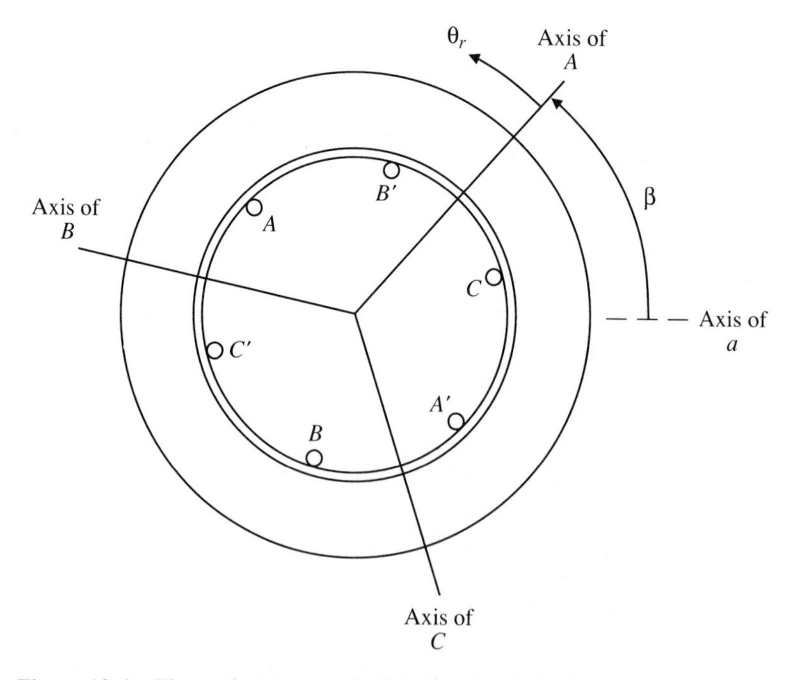

Figure 10.4 Three-phase rotor winding showing only the center conductors of sinusoidally distributed windings.

$$n'_A = \frac{N_{se}}{N_{re}} n_A \qquad \text{and} \qquad i'_A = \frac{N_{re}}{N_{se}} i_A \qquad \text{A} \tag{10.13}$$

The total effect of a set of instantaneous phase currents in the rotor windings now can be expressed as

$$\vec{i}'_r = \tfrac{2}{3}(\vec{i}'_A + \vec{i}'_B + \vec{i}'_C) = \tfrac{2}{3}(i'_A + \alpha i'_B + \alpha^2 i'_C) \qquad \text{A} \tag{10.14}$$

This is a space vector whose angle is measured relative to the axis of rotor phase A. If this vector is known, it can be projected on the A, B, and C axes to obtain the instantaneous rotor phase currents as was done for stator currents in Eq. (10.11).

The effect of the rotor currents as seen by an observer on the phase a axis of the stator can be represented by a space vector $\vec{i}_r$ at angle β ahead of $\vec{i}'_r$ such that

$$\vec{i}_r = \vec{i}'_r \epsilon^{j\beta} \qquad \text{A} \tag{10.15}$$

It is the combined effect of the stator current $\vec{i}_s$ and this transformed rotor current $\vec{i}_r$ that produces the magnetic field in the air gap of the machine. Their sum can be denoted as the magnetizing current space vector $\vec{i}_m$ given by

$$\vec{i}_m = \vec{i}_s + \vec{i}_r \qquad \text{A} \tag{10.16}$$

This magnetizing current produces an air gap mmf $\vec{\mathfrak{F}}_m$. By analogy with Eq. (10.3) and recalling the 2/3 factor of Eq. (10.10), this mmf can be represented as

$$\vec{\mathfrak{F}}_m = \tfrac{3}{4} N_{se} \vec{i}_m \qquad \text{A} \tag{10.17}$$

If the effective air-gap length is g_e and the stator and rotor iron can be regarded as ideal, the air-gap flux density can be represented by a space vector

$$\vec{B}_g = \frac{\mu_0}{g_e} \vec{\mathfrak{F}}_m \qquad \text{T} \tag{10.18}$$

i.e., a cosinusoidal distribution with maximum positive value along the same axis as that of the magnetizing mmf vector.

In Section 5.3, it was shown that an air-gap flux density $\vec{B}_g$ due to a magnetizing current $\vec{i}_m$ produced a flux linkage $\vec{\lambda}_m$ in a winding of N_{se} distributed turns of value

$$\vec{\lambda}_m = L_m \vec{i}_m \qquad \text{Wb} \tag{10.19}$$

where, from Eq. (5.50), the magnetizing inductance of a p-pole machine is related to the air-gap radius r and axial length ℓ by the expression

$$L_m = \frac{3\pi}{2p^2} N_{se}^2 \frac{\mu_0 \ell r}{g_e} \qquad \text{H} \tag{10.20}$$

Typical relationships among the space vectors of Eqs. (10.16) to (10.19) are shown in Fig. 10.5.

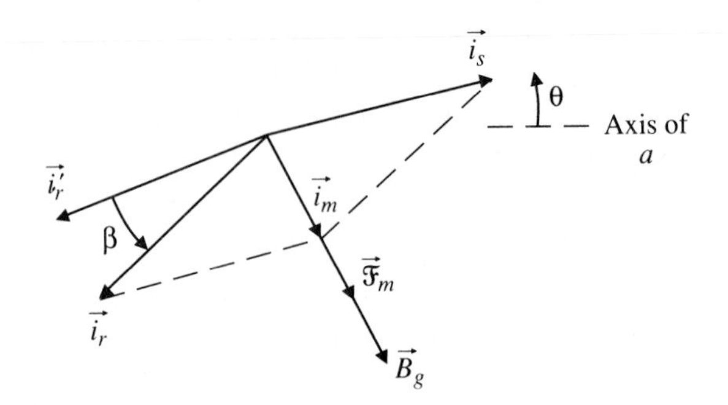

Figure 10.5 Current, mmf, and flux-density space vectors.

The flux linkage of the stator winding arising from flux that crosses the air gap to the rotor is λ_m. In addition, there is a component of stator flux that is established along the path from stator tooth to stator tooth around the stator periphery and also around the stator end windings. This leakage flux linkage is directly proportional to the stator current assuming ideal iron. The proportionality constant is the stator leakage inductance $L_{\ell s}$. Similarly, of all the flux that crosses the air gap, a part will link the rotor windings, but another part, proportional to the rotor current, will take a leakage path around the rotor periphery, tooth to tooth, and also around its end connections. This part is represented as the flux linkage $L_{\ell r} i_r$, referred to the N_{se} equivalent stator turns, where $L_{\ell r}$ is the rotor leakage inductance. However, the separation of the total leakage into stator and rotor components is somewhat artificial. Models will be developed later in this chapter that use only the total leakage inductance.

The flux-linkage space vectors of the stator winding and of the referred rotor winding now can be related to the winding currents by the set of equations

$$\begin{aligned} \vec{\lambda}_s &= (L_{\ell s} + L_m)\vec{i}_s + L_m \vec{i}_r \\ \vec{\lambda}_r &= L_m \vec{i}_s + (L_{\ell r} + L_m)\vec{i}_r \qquad \text{Wb} \end{aligned} \tag{10.21}$$

The machine now can be represented by the flux linkage-current equivalent circuit of Fig. 10.6. The final step in developing an analytical model for the ma-

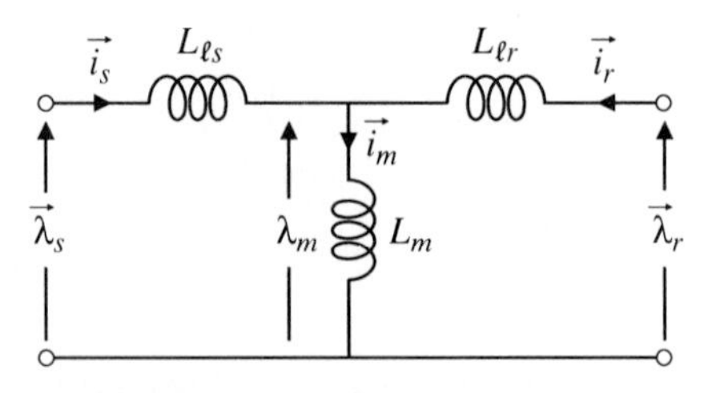

Figure 10.6 Flux linkage-current model.

chine is the derivation of the stator and rotor voltages. For the stator, the induced voltage is the rate of change with time of the stator flux linkage. If the stator winding resistance per phase is R_s, the space vector representing the stator voltages is given by

$$\vec{v}_s = R_s\vec{i}_s + p\vec{\lambda}_s \qquad \text{V} \tag{10.22}$$

where $p = d/dt$.

If the resistance of winding A on the rotor is R_A, the equivalent resistance, referred to the stator turns, is

$$R_r = \left(\frac{N_{se}}{N_{re}}\right)^2 R_A \qquad \Omega \tag{10.23}$$

The voltages and currents of the rotor windings can be expressed by the space vectors $\vec{v}_r'$ and $\vec{i}_r'$, which are the actual terminal values in space vector form referred through the turns ratio. The rotor terminal voltage vector is related to the rotor flux linkage and the rotor current by

$$\vec{v}_r' = R_r\vec{i}_r' + p\vec{\lambda}_r' \qquad \text{V} \tag{10.24}$$

In addition to referring all rotor quantities to the equivalent number of stator turns, we have also chosen to develop a model in which the stator phase a is used as angle reference for all space variables. The current $\vec{i}_r'$ in the rotating rotor has its angle related to the axis of rotor phase A. This current has already been transformed into an equivalent current i_r in a stationary or stator reference frame using Eq. (10.15). Similar transformations through the angle $\epsilon^{j\beta}$ will now produce the new set of variables $\vec{\lambda}_r$ and $\vec{v}_r$ from the rotor variables $\vec{\lambda}_r'$ and $\vec{v}_r'$. Care must now be taken in transforming Eq. (10.24) because, for a moving rotor, the quantity $\epsilon^{j\beta}$ is also a function of time because the rotor position β varies with time. Thus, using Eq. (10.15),

$$\begin{aligned}\vec{v}_r &= \vec{v}_r'\epsilon^{j\beta} \\ &= R_r\vec{i}_r'\epsilon^{j\beta} + (p\vec{\lambda}_r')\epsilon^{j\beta} \\ &= R_r\vec{i}_r + p(\vec{\lambda}_r\epsilon^{-j\beta})\epsilon^{j\beta} \\ &= R_r\vec{i}_r + p\vec{\lambda}_r - j\omega_0\vec{\lambda}_r \qquad \text{V}\end{aligned} \tag{10.25}$$

because, for a two-pole machine, the instantaneous rotor speed ω_0 is

$$\omega_0 = p\beta \qquad \text{rad/s} \tag{10.26}$$

An equivalent circuit incorporating these instantaneous space vector quantities can now be constructed using Eqs. (10.22) and (10.25) together with Eq. (10.21). The result is shown in Fig. 10.7 in which the source voltage $\vec{e}_0$ is given by

$$\vec{e}_0 = -j\omega_0\vec{\lambda}_r \qquad \text{V} \tag{10.27}$$

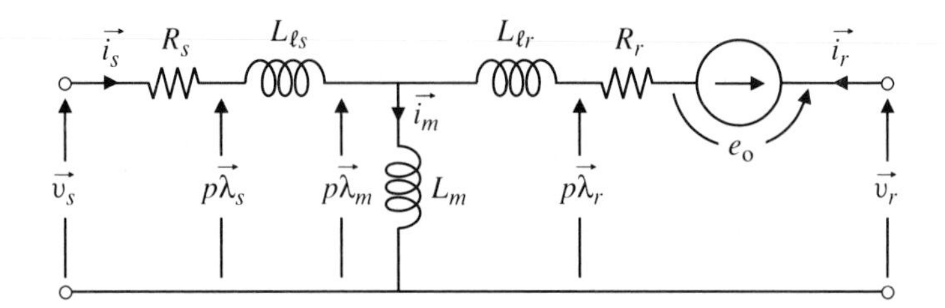

Figure 10.7 Transient equivalent circuit for induction machine.

In a squirrel-cage machine, the rotor bars are shorted. Thus, the rotor terminal voltages and their space vector equivalent $\vec{v}_r'$ referred to the stator turns are all equal to zero.

10.1.1 ▪ An Alternate Form of Transient Model

For an induction machine that can be assumed to be magnetically linear, the equivalent circuit of Fig. 10.7 is actually more complex than is necessary. It contains three inductance parameters as compared with the two that were used in the elementary introduction to the induction machine in Chapter 5. By use of a simple change of variables, a mathematically equivalent circuit with only two inductances can be produced.

A set of flux linkage-current relations are stated in Eq. (10.21). Suppose we introduce a new set of rotor variables related to the original set by

$$\vec{\lambda}_R = \gamma \vec{\lambda}_r \qquad \text{Wb} \tag{10.28}$$

and

$$\vec{i}_R = \frac{\vec{i}_r}{\gamma} \qquad \text{A} \tag{10.29}$$

Substitution of these new variables into the flux-linkage relations of Eq. (10.21) gives

$$\begin{aligned} \vec{\lambda}_s &= (L_{\ell s} + L_m)\vec{i}_s + \gamma L_m \vec{i}_R \\ \vec{\lambda}_R &= \gamma L_m \vec{i}_s + \gamma^2 (L_{\ell r} + L_m)\vec{i}_R \end{aligned} \tag{10.30}$$

The choice of the value of γ can be made arbitrarily without changing the validity of Eq. (10.30). Suppose we choose the value

$$\gamma = \frac{L_{\ell s} + L_m}{L_m} = \frac{L_M}{L_m} \tag{10.31}$$

Then, Eq. (10.30) can be rewritten as

$$\begin{aligned}\vec{\lambda}_s &= L_M\vec{i}_s + L_M\vec{i}_R \\ \vec{\lambda}_R &= L_M\vec{i}_s + (L_L + L_M)\vec{I}_R \qquad \text{Wb}\end{aligned} \tag{10.32}$$

in which the total effective leakage inductance is

$$L_L = \gamma L_{\ell s} + \gamma^2 L_{\ell r} \qquad \text{H} \tag{10.33}$$

The transient relations of Eqs. (10.22) and (10.25) now can be rewritten as

$$\begin{aligned}\vec{v}_s &= R_s\vec{i}_s + p\vec{\lambda}_s \\ &= R_s\vec{i}_s + pL_M\vec{i}_s + pL_M\vec{i}_r\end{aligned} \tag{10.34}$$

$$\begin{aligned}\vec{v}_R &= R_R\vec{i}_R + p\vec{\lambda}_R - j\omega_0\vec{\lambda}_R \\ &= R_R\vec{i}_R + (p - j\omega_0)\,[L_M\vec{i}_s + (L_L + L_M)\vec{i}_R]\end{aligned} \tag{10.35}$$

The effective rotor resistance R_R in this expresion is related to the resistance R_r of Eq. (10.23) by

$$R_R = \gamma^2 R_r \qquad \Omega \tag{10.36}$$

With this change of variables and the revised notation for reflected rotor quantities, the relations of Eqs. (10.34) and (10.35) can be incorporated into the equivalent circuit of Fig. 10.8. This circuit has the same inductance terms as derived in Chapter 5 and as shown in Fig. 5.16. The circuit of Fig. 10.8 has been designated as the Γ model because of the similarity of this symbol to the connection of the inductance elements, in contrast to the T model of Fig. 10.7.

The voltage $\vec{e}_O$ is related to the voltage variable $\vec{e}_0$ by

$$\vec{e}_O = -j\omega_0\vec{\lambda}_R = \gamma\vec{e}_0 \qquad \text{V} \tag{10.37}$$

The change of variables stated in Eqs. (10.28) and (10.29) can be considered as equivalent to transforming across an ideal transformer of ratio γ. Thus, for a

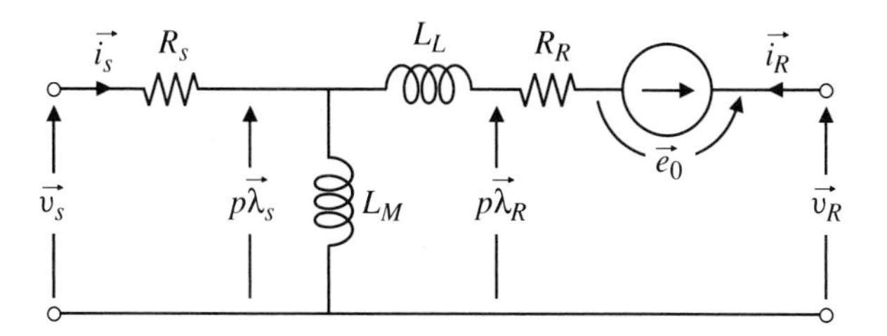

Figure 10.8 Alternate Γ-type equivalent circuit.

wound-rotor machine, the rotor terminal voltage $\vec{v}_r$ of Fig. 10.7 can be transformed to

$$\vec{v}_R = \gamma \vec{v}_r \qquad \text{V} \tag{10.38}$$

in the revised equivalent circuit. Also, any circuit elements connected to the slip rings of a wound-rotor machine can be similarly transformed to be incorporated into the circuit of Fig. 10.8.

10.1.2 • Instantaneous Power and Torque Relations

The power flow through the machine can be followed using the equivalent circuit model of Fig. 10.8. The instantaneous power entering the stator windings can be found from the space vectors of stator current and stator voltage using the expression

$$p_s = \tfrac{3}{2} \Re (\vec{v}_s^* \vec{i}_s) \qquad \text{W} \tag{10.39}$$

The product of the current and the conjugate of the voltage is analogous to the product of their corresponding phasors in steady-state analysis. The factor of 2 arises from the fact that $\vec{v}_s$ and $\vec{i}_s$ are peak rather than rms quantities, and the factor of 3 accounts for the three phases. Proof of the validity of Eq. (10.38) is given by the following, noting that the sum of the stator currents is zero:

$$\begin{aligned} p_s &= \frac{3}{2} \Re \left[\frac{2}{3} (v_a + \alpha^2 v_b + \alpha v_c) \frac{2}{3} (i_a + \alpha i_b + \alpha i_c) \right] \\ &= \frac{2}{3} \Re \left\{ v_a \left[i_a - 0.5(i_b + i_c) + j \frac{\sqrt{3}}{2} (i_b - i_c) \right] \right. \\ &\quad + v_b \left[i_b - 0.5(i_c + i_a) + j \frac{\sqrt{3}}{2} (i_c - i_a) \right] \\ &\quad \left. + v_c \left[i_c - 0.5(i_a + i_b) + j \frac{\sqrt{3}}{2} (i_a - i_b) \right] \right\} \\ &= v_a i_a + v_b i_b + v_c i_c \qquad \text{W} \end{aligned} \tag{10.40}$$

The instantaneous power loss in the stator windings is

$$p_{R_s} = \tfrac{3}{2} \Re (R_s \vec{i}_s \vec{i}_s^*) = \tfrac{3}{2} R_s \hat{i}_s^2 \qquad \text{W} \tag{10.41}$$

An expression similar to Eq. (10.41) gives the power dissipation in the rotor. The instantaneous flow of power into energy storage in each of the two inductances in Fig. 10.8 can be found similarly using the appropriate instantaneous space vectors of current and induced voltage. If the machine has its rotor windings connected through slip rings to an external system, the power output to that system would be given by

$$p_r = -\tfrac{3}{2} \Re (\vec{v}_R^* \vec{i}_R) \qquad \text{W} \tag{10.42}$$

Having now accounted instantaneously for the electrical input and output power, the loss power and the power into magnetic field storage, it follows from the law of conservation of energy that the mechanical power output is

$$\begin{aligned} p_0 &= \tfrac{3}{2}\,\mathcal{R}\,(\vec{e}_O^{\,*}\vec{i}_R) \\ &= \tfrac{3}{2}\,\mathcal{R}\,(-j\omega_0\vec{\lambda}_R^{\,*}\vec{i}_R) \\ &= -\tfrac{3}{2}\,\omega_0\,\mathcal{I}(\vec{\lambda}_R^{\,*}\vec{i}_R) \qquad \text{W} \end{aligned} \tag{10.43}$$

The instantaneous torque produced by a p-pole machine is then given by

$$\begin{aligned} T &= \frac{p_0}{\omega_{mech}} \\ &= \frac{p}{2}\frac{p_0}{\omega_0} \\ &= -\frac{3p}{4}\,\mathcal{I}(\vec{\lambda}_R^{\,*}\vec{i}_R) \qquad \text{N}\cdot\text{m} \end{aligned} \tag{10.44}$$

It is hoped that the use of the standard symbol p for instantaneous power, for the number of poles, and for time differentiation has not become too confusing.

Alternate expressions for the torque can be derived by noting that all components of flux linkage in inductance elements are colinear with their corresponding currents and thus do not contribute to the imaginary component in the torque expression. Thus, with the appropriate substitutions from Eq. (10.21) into Eq. (10.44), it can be shown that

$$\begin{aligned} T &= -\frac{3p}{4}\,L_M\mathcal{I}(\vec{i}_s^{\,*}\vec{i}_R) \\ &= \frac{3p}{4}\,\mathcal{I}(\vec{\lambda}_s^{\,*}\vec{i}_s) \qquad \text{N}\cdot\text{m} \end{aligned} \tag{10.45}$$

This torque is applied to the mechanical system including the mechanical load plus the inertia and loss torque of the motor itself. It can be described by a relation of the type discussed in Section 9.2, such as

$$T = J\frac{d\omega_{mech}}{dt} + T_L \qquad \text{N}\cdot\text{m} \tag{10.46}$$

where J is the polar moment of inertia and T_L is the load torque, which may be a function of speed and/or time.

The transient performance of the machine may now be predicted using the three differential Eqs. (10.34), (10.35), and (10.46). If, for example, the phase voltages applied to the stator are known as functions of time, they can be expressed as a time-varying space vector using a relation analogous to Eq. (10.10), i.e.,

$$\vec{v}_s = v_a + j\,\frac{v_b - v_c}{\sqrt{3}} \qquad \text{V} \tag{10.47}$$

With known initial conditions for the stator and rotor currents (and, therefore, for their space vectors) plus the initial value of the rotor speed, the three differential equations can be solved using complex numbers to provide the variables as a function of time. Closed solution will be possible in only a limited number of situations because of the nonlinear nature of the torque expression, involving as it does a product of two variables. For general analysis, the equations can be evaluated by numerical computation. For this process it is frequently convenient to rearrange the equations as a set of first-order relations as follows. Assuming a shorted rotor,

$$\begin{aligned} \frac{d\vec{\lambda}_s}{dt} &= \vec{v}_s - R_s\vec{i}_s \\ \frac{d\vec{\lambda}_R}{dt} &= j\omega_0\vec{\lambda}_R - R_R\vec{i}_R \\ \frac{d\omega_{mech}}{dt} &= -\frac{1}{J}\left[-\frac{3p}{4}\,\mathscr{I}(\vec{\lambda}_R^{\,*}\vec{i}_R) - T_L\right] \end{aligned} \tag{10.48}$$

At each step in the numerical integration, the currents can be evaluated from the known flux linkages using the expressions

$$\begin{aligned} \vec{i}_R &= \frac{\vec{\lambda}_R - \vec{\lambda}_s}{L_L} \\ \vec{i}_s &= \frac{\vec{\lambda}_s}{L_M} - \vec{i}_R \end{aligned} \tag{10.49}$$

Alternatively, the equations can be programmed in a digital simulator.

EXAMPLE 10.1 A four-pole squirrel-cage induction motor has a leakage inductance of 5 mH, a stator resistance of 0.1 Ω, and a rotor resistance as seen from the stator of 0.13 Ω. It is to be started on a three-phase, 60-Hz supply with 208-V rms (L-L). The magnetizing inductance is sufficiently high that the magnetizing current can be ignored during starting conditions.

(a) Determine and plot the phase currents for the first two cycles after the motor is switched on to the supply.

(b) Determine and plot the torque produced for the first two cycles after switching the motor on to the supply.

Solution

(a) Because the rotor of a squirrel-cage induction machine is shorted, the rotor voltage $\vec{v}_R$ is equal to zero in the equivalent Γ circuit of Fig. 10.8.

Suppose the motor is connected to the supply at time $t = 0$ when the voltage on stator phase a is a maximum. Then, after $t = 0$,

$$\vec{v}_s = \hat{v}_s \epsilon^{j\omega_s t} \qquad \text{V}$$

where

$$\hat{v}_s = \frac{208}{\sqrt{3}} \sqrt{2} = 169.8 \qquad \text{V}$$

and

$$\omega_s = 2\pi(60) = 377 \qquad \text{rad/s}$$

Let us assume that the speed remains at essentially zero during the first few cycles of the supply current. Thus, $\vec{e}_O = 0$. The stator current is then given by

$$\vec{i}_s = \frac{\vec{v}_s}{R_s + R_R + pL_L} = \frac{169.8\epsilon^{j377t}}{0.23 + p(5 \times 10^{-3})} \qquad \text{A}$$

The solution of this first-order differential equation consists of a particular integral for which $p = j\omega_s$ and a simple exponential complementary function with time constant

$$\tau = \frac{5 \times 10^{-3}}{0.23} = 21.7 \qquad \text{ms}$$

To meet the initial condition that $\vec{i}_s = 0$ at $t = 0$, the stator current expression, with all angles in radians, is

$$\vec{i}_s = \frac{169.8\epsilon^{j377t}}{0.23 + j(5 \times 10^{-3} \times 377)} + (const)\epsilon^{-t/\tau}$$

$$= 89.2\epsilon^{j(377t - 1.45)} - 89.2\epsilon^{-j1.45}\epsilon^{-t/0.0217}$$

Now that the current has been found in space-vector form, the individual phase currents can be evaluated using Eq. (10.11):

$$i_a = \Re\vec{i}_s = 89.2\cos(377t - 1.45) - 10.8\epsilon^{-t/0.0217}$$

$$i_b = \Re\alpha^2\vec{i}_s = 89.2\cos(377t + 2.74) + 82.1\epsilon^{-t/0.0217}$$

$$i_c = \Re\alpha\vec{i}_s = 89.2\cos(377t + 0.64) - 71.3\epsilon^{-t/0.0217}$$

These phase currents are shown in Fig. 10.9(a). In this example, the peak phase current occurs in phase b and is about 60% greater than the peak value of the sinusoidal component due to the unidirectional component which has a time constant somewhat greater than a period of the sine term. If this time constant had been large with respect to a period, the peak currents could have approached twice the peak value of the sinusoidal component.

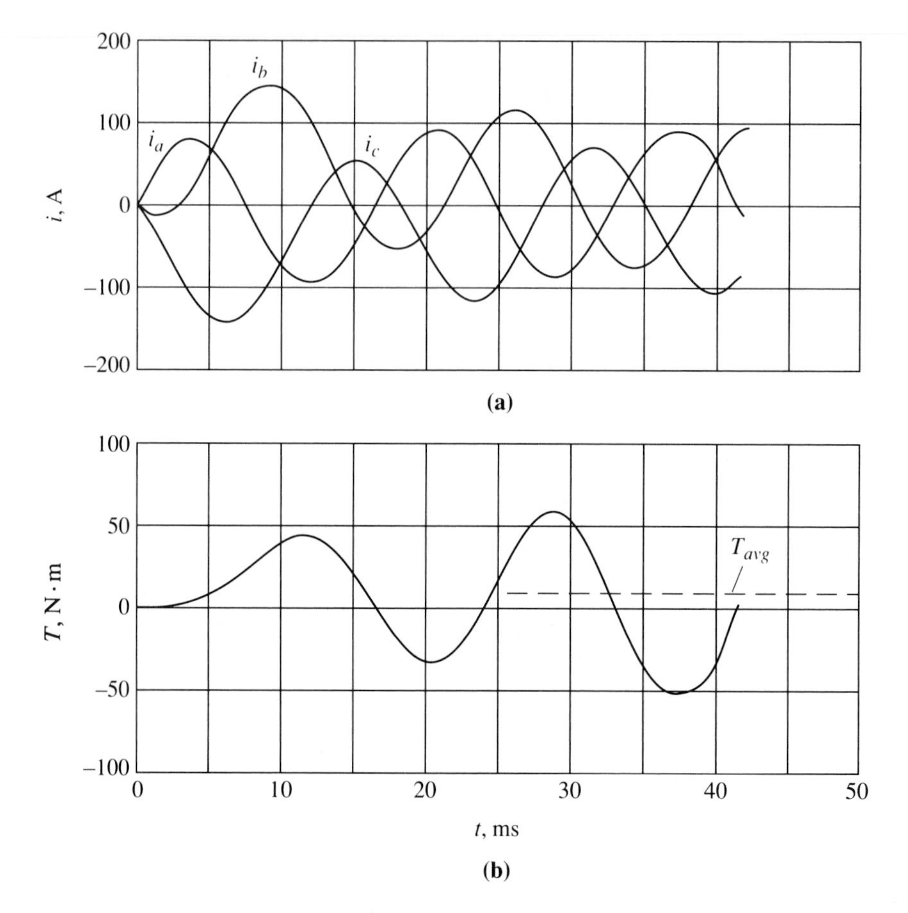

Figure 10.9 (a) Phase currents for Example 10.1. (b) Transient torque.

(b) Expressions for torque have been given in Eqs. (10.44) and (10.45). The one involving L_M is inappropriate here because this inductance has been assumed to be large enough to be ignored. The stator flux linkage could be found by subtracting the resistance drop from the stator voltage and integrating the result. A more direct approach appears to be an evaluation of the rotor flux linkage $\vec{\lambda}_R$ as the integral of the voltage across the rotor resistance because $\vec{v}_R = 0$ and $\vec{e}_O = 0$ at standstill. Let

$$\vec{i}_s = \hat{i}_s \epsilon^{-j\alpha}(\epsilon^{j\omega t} - \epsilon^{-t/\tau}) = -\vec{i}_R$$

The rotor flux linkage is equal to zero at time zero and is given by

$$\hat{\lambda}_R = \int R_R \vec{i}_s \, dt$$

$$= R_R \vec{i}_s \epsilon^{-j\alpha}\left[\frac{\epsilon^{j\omega t} - 1}{j\omega} + \tau(\epsilon^{-t/\tau} - 1)\right]$$

From Eq. (10.44), the torque is

$$T = \frac{3p}{4}\, \mathcal{I}[\vec{\lambda}_R^{*}\vec{i}_S]$$

$$= \frac{3p}{4}\, R_R i_S^2 \mathcal{I}\left\{\left[\frac{\epsilon^{-j\omega t} - 1}{-j\omega} + \tau(\epsilon^{-t/\tau} - 1)\right](\epsilon^{j\omega t} - \epsilon^{-t/\tau})\right\}$$

$$= \frac{3 \times 4}{4}\, \frac{0.13 \times 89.2^2}{377}$$

$$\times\, \mathcal{I}[(\epsilon^{-t/\tau} + 1)(1 - \cos \omega t) + \tau\omega_0(\epsilon^{-t/\tau} - 1) \sin \omega t]$$

$$= 8.23\,[1 + \epsilon^{-t/0.0217}(1 - \cos 377t)$$

$$+ 8.18(\epsilon^{-t/0.0217} - 1) \sin 377t] \qquad \text{N}\cdot\text{m}$$

This torque waveform is plotted in Fig. 10.9(b). The starting torque is seen to be highly oscillatory. The average torque of 8.23 N·m is produced by the rotating components of flux linkage and current. The large oscillatory components of torque arise from the interaction of rotating components of flux or current with stationary decaying flux and current components.

These expressions for transient currents and torque ignore the effects of the increase in speed and of the magnetizing current. When these are included in a numerical solution of the three equations of Eq. (10.48), the oscillatory torque is found to decay toward its average value derived from a steady-state model after a number of supply cycles.

10.1.3 ▪ Steady-State Analysis

In many applications, the induction machine is supplied with three-phase sinusoidal voltage of constant peak amplitude, and the speed is either constant or varying slowly with time. For these situations, there is an interest in predicting the steady-state relationships among speed, torque, input current, power factor, and efficiency. Even when the supply is not sinusoidal, as obtains with inverter supplies, the response arising from the fundamental-frequency component is of predominant interest.

Suppose the supply-voltage space vector is

$$\vec{v}_s = \hat{v}_s \epsilon^{j\omega t} \qquad \text{V} \tag{10.50}$$

and the stator-current space vector is

$$\vec{i}_s = \hat{i}_s \epsilon^{j\omega t + \phi_s} \qquad \text{A} \tag{10.51}$$

In the steady sinusoidal state, $p\vec{i}_s = j\omega_s \vec{i}_s$, and it is convenient to use phasor notation where, by definition, the phasor magnitude is $1/\sqrt{2}$ of the vector magnitude and the phasor angle is the angle of the vector at time zero:

$$\vec{V}_s = \frac{\vec{v}_s}{\sqrt{2}} \epsilon^{-j\omega t} = \frac{\hat{v}_s}{\sqrt{2}} \epsilon^{j0} \quad \text{V} \tag{10.52}$$

and

$$\vec{I}_s = \frac{\hat{i}_s}{\sqrt{2}} \epsilon^{j\phi_s} \quad \text{A} \tag{10.53}$$

The expressions developed in Eqs. (10.35) and (10.45) for general transient analysis may now be written for the steady state:

$$\vec{V}_s = (R_s + j\omega_s L_M)\vec{I}_s + j\omega_s L_M \vec{I}_R \tag{10.54}$$

$$\vec{V}_R = j(\omega_s - \omega_0) L_M \vec{I}_s + [R_R + j(\omega_s - \omega_0)(L_L + L_M)]\vec{I}_R \tag{10.55}$$

To develop an equivalent electric circuit, we need a set of equations with bilateral mutual impedances. To transform Eq. (10.55) into a form in which it has this property together with Eq. (10.54), the expression for $\vec{V}_R$ can be multiplied by the factor

$$\frac{\omega_s}{\omega_s - \omega_0} = \frac{\omega_s}{\omega_r} \tag{10.56}$$

where ω_r is the actual frequency of the rotor currents. The transformed rotor equation then becomes

$$\frac{\omega_s}{\omega_r} \vec{V}_R = j\omega_s L_M \vec{I}_S + \left[R_R \frac{\omega_s}{\omega_r} + j\omega_s(L_L + L_M)\right]\vec{I}_R \tag{10.57}$$

For a squirrel-cage machine, the rotor voltage $\vec{V}_R$ is zero. Equations (10.55) and (10.57) now represent the steady-state equivalent circuit of Fig. 10.10(a). This is the same circuit as was derived for the steady state from a more physical point of view in Chapter 5 and shown in Fig. 5.16. As discussed there, the resistance $R_R\omega_s/\omega_0$ contains a component R_R in which the rotor losses per phase occur plus a component $R_R\omega_0/\omega_r$ representing the effect per phase of the mechanical load. Because the mechanical power is the torque multiplied by the mechanical angular velocity, the steady-state torque for the squirrel-cage machine is

$$\begin{aligned} T &= \frac{3p}{2\omega_0}\left(\frac{R_R\omega_0}{\omega_r} . I_R^2\right) \\ &= \frac{3p}{2}\frac{R_R}{\omega_r} I_R^2 \quad \text{N}\cdot\text{m} \end{aligned} \tag{10.58}$$

The quantity ω_r/ω_s is commonly known as the *slip s*. This is a per-unit quantity that is useful for the many situations for which the stator frequency is constant. Because much of our emphasis in this chapter is on variable-frequency operation, the identity of the stator and rotor frequencies will be retained in the following treatment.

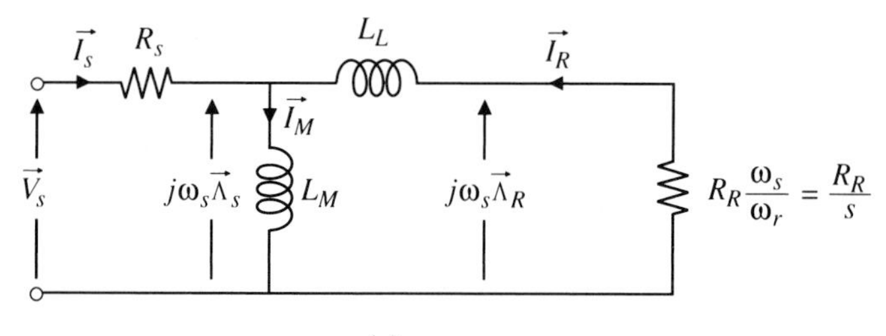

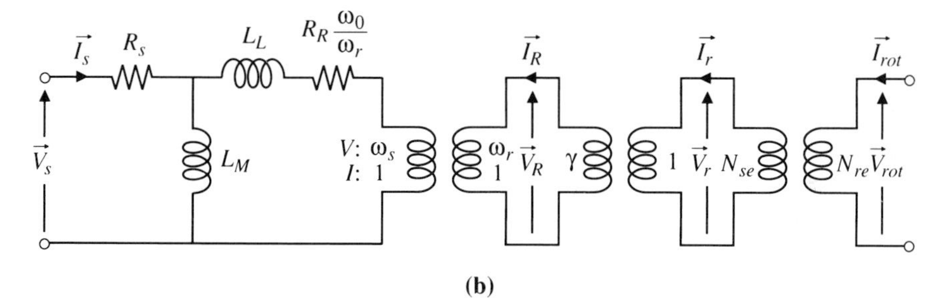

Figure 10.10 Steady-state equivalent circuits for induction machine. (a) Squirrel cage. (b) Wound rotor.

For a wound-rotor induction machine, it is important that the equivalent circuit model include means of relating the actual rotor voltage and current phasors, $\vec{V}_{rot}$ and $\vec{I}_{rot}$, to the internal phasors $\vec{V}_R$ and $\vec{I}_R$, so that various parameters may be transferred from one side to the other. These relations are shown in step form in the equivalent circuit of Fig. 10.10(b). One ideal transformer represents the turns ratio $N_{se} : N_{re}$. A second ideal transformer accommodates the quantity γ of Eq. (10.31). Finally, an ideal induction machine is introduced to represent the voltage ratio $\omega_s : \omega_r$ while retaining a current ratio of $1 : 1$. All three of these transformations may be combined for convenience when referring quantities from rotor to stator or vice versa.

For any polyphase induction machine operating in the balanced steady state, the torque may be found as the total power crossing the air gap to the rotor divided by the synchronous mechanical speed. From Eqs. (10.44) and (10.45), appropriate expressions for torque are

$$\begin{aligned} T &= \frac{3p}{2\omega_s}\,\Re\left[(j\omega_s\vec{\Lambda}_s)^*\vec{I}_s\right] \\ &= \frac{3p}{2}\,\Im(\vec{\Lambda}_s^*\vec{I}_s) \\ &= -\frac{3p}{2}\,\Im(\vec{\Lambda}_R^*\vec{I}_R) \qquad \text{N}\cdot\text{m} \end{aligned} \tag{10.59}$$

The second of these expressions is particularly useful in those frequent situations where the stator flux $\vec{\Lambda}_s$ is maintained essentially constant by the application of stator voltage of constant magnitude and frequency.

10.1.4 ▪ Modeling of Saturation Effects

Thus far in the modeling of the induction machine, the iron of the rotor and stator has been assumed to be magnetically ideal, requiring no significant magnetic field intensity to establish the magnetic flux in the iron parts of the flux path. However, it is normal to operate machines at a flux level where iron saturation is relevant because this allows a higher voltage level for a given iron and winding structure and therefore a smaller machine for a given rating. This section examines appropriate means by which the nonlinear effects of the iron sections can be introduced with adequate accuracy into the equivalent circuit models.

Consider the machine shown in the cross section of Fig. 10.11(a). The circles represent a sinusoidal distribution of magnetizing current in the stator wind-

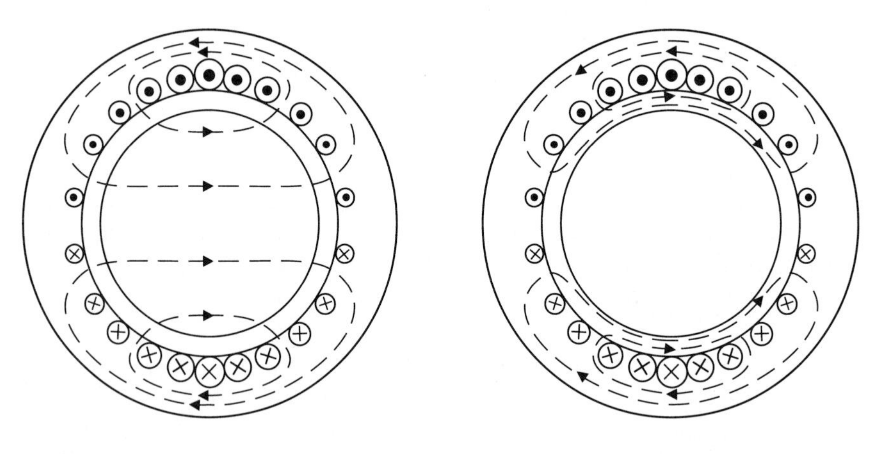

(a) (b)

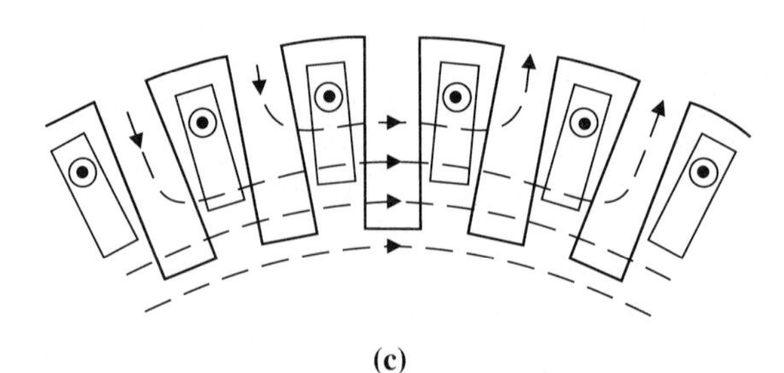

(c)

Figure 10.11 (a) Mutual flux. (b) Stator leakage flux. (c) Paths of stator leakage flux.

ings. The dotted lines suggest the pattern of the mutual flux that crosses the air gap from stator to rotor. In the stator, this flux is carried by the stator teeth and the stator yoke. If the stator current is represented as a space vector $\vec{i}_s$, which in this case is the magnetizing current $\vec{i}_m$, the stator flux linkage will be a space vector $\vec{\lambda}_m$, colinear with $\vec{i}_m$ as shown in the equivalent circuit of Fig. 10.7. Next, consider the leakage flux linkage set up in the stator winding due to a stator current $\vec{i}_s$. The pattern of this flux is suggested by Fig. 10.11(b). While some of this flux follows a path along the air gap, most of it is in a path from toothtip to toothtip, as suggested by Fig. 10.11(c). This component of flux linkage is proportional to and colinear with the stator current. It has been denoted as $L_{\ell s}\vec{i}_s$, as shown in Fig. 10.7. Adding these two components together gives the total stator flux linkage as

$$\vec{\lambda}_s = L_{\ell s}\vec{i}_s + \vec{\lambda}_m \qquad \text{Wb} \tag{10.60}$$

With ideally sinusoidally distributed windings and no magnetic saturation, both these components of flux would set up sinusoidally distributed flux densities in the stator teeth. The stator-tooth flux density space vector $\vec{B}_{ts}$ then would be the vector sum of the flux density components due to the magnetizing and the stator currents. Establishing this flux density in the stator teeth will require an additional component of magnetizing current, which will be a nonlinear function of the stator flux linkage $\vec{\lambda}_s$. In addition, it is noted from Fig. 10.11 that the stator yoke carries the total flux that links the stator. Thus, the component of magnetizing current for the yoke also will be dependent nonlinearly on the stator flux linkage. Therefore, the combined effect of the mmf required by the stator teeth and yoke may be incorporated into our equivalent circuit models by adding an appropriate nonlinear inductive element across the flux linkage $\vec{\lambda}_s$. In the equivalent circuit of Fig. 10.12(a), this has been combined with the previously linear inductance L_M to give the single nonlinear $\vec{\lambda}_s - \vec{i}_M$ relation of Fig. 10.12(b), which for convenience will still be denoted as L_M.

In the rotor, the flux in the rotor teeth and in the rotor core is equal to the flux crossing the air gap minus the rotor leakage flux that bypasses the rotor winding by taking a path around the rotor periphery, mainly from tooth to tooth along the rotor surface. Therefore, an additional magnetizing current component may be required to represent the additional mmf required by the rotor teeth and core. This component will be a nonlinear function of the net rotor flux linkage $\vec{\lambda}_r$ shown in the equivalent circuit of Fig. 10.7. It could be represented in that circuit by a nonlinear inductance across $\vec{\lambda}_r$. It could be represented equally well by a nonlinear inductance element across $\vec{\lambda}_R$ in the Γ form of the equivalent circuit of Fig. 10.12(a) using the transformations of Eqs. (10.28) and (10.29).

In many analyses of induction machines, the effect of magnetic nonlinearity in the rotor is small in comparison with that of the stator. An extreme example would be the starting performances of an induction motor where the stator flux linkage $\vec{\lambda}_s$ is determined by the constant applied stator voltage $\vec{v}_s$, but the rotor flux linkage $\vec{\lambda}_R$ is relatively small, most of the stator flux linkage

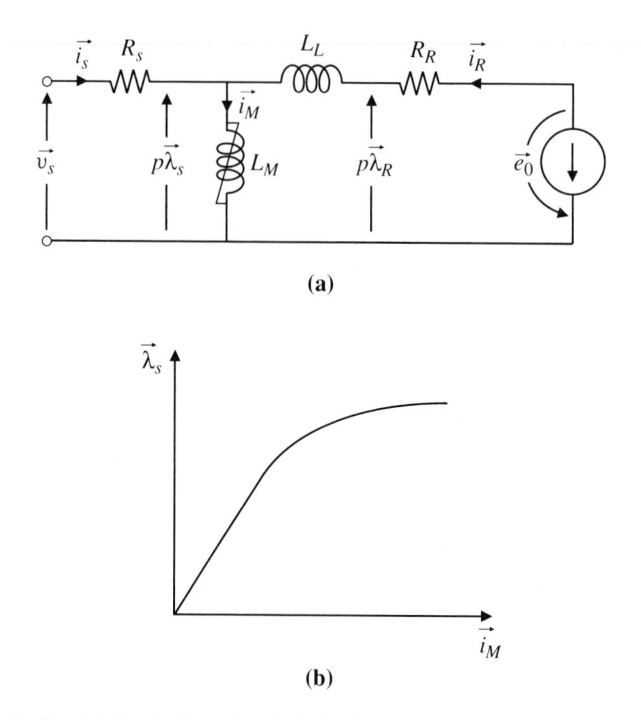

Figure 10.12 (a) Equivalent circuit including stator iron saturation. (b) Graph of equivalent magnetizing inductance L_M.

being absorbed in the leakage inductance L_L. Even in a normal load condition, the magnitude of the rotor flux linkage will be reduced below that of the stator flux linkage by the vector addition of the drop across the leakage inductance. Thus, in these situations, the rotor nonlinearity frequently may be ignored. The circuit of Fig. 10.12(a) with all the magnetizing nonlinearity associated with the stator thus may be used as an adequate approximation.

The effect of eddy current and hysteresis losses in the iron also may be incorporated into the equivalent circuit if desired. In the stator, these losses are dependent on the flux densities in the stator teeth and yoke. Thus, following the approach developed in Section 1.10, these core losses can be represented by a nonlinear resistance R_c in parallel with the magnetizing inductance L_M in Fig. 10.12(a). In analyses that are intended to predict the relations among voltage and current vectors or phasors, the core loss current usually can be ignored because it is small. However, core losses are significant in efficiency prediction.

Iron losses also occur in the rotor iron but usually can be ignored. Under normal operating conditions, the rotor frequency ω_r is very small; thus, the loss is small. Under both starting and running conditions, both eddy and hysteresis currents in the rotor iron actually contribute to the developed torque of the machine and should not be classed as losses. In an extreme case, a rotor made of a solid iron cylinder would have large eddy currents and would produce a useful torque. Such rotors are sometimes used in very high-speed induction machines.

Other refinements may be added to the simple equivalent circuit model of Fig. 10.12(a) if required. For example, some of the stator leakage flux linkage arises from flux around the stator end windings. For greater precision, this could be represented by a separate linear inductance in series with the resistance R_s. Such refinements are seldom justified in practice and present the difficulty that the additional circuit elements cannot readily be determined from measurements made at the machine terminals.

10.1.5 ▪ Modeling of Deep Rotor Bar Effects

In the previous development of Section 10.1.4, the rotor has been represented as a single circuit with leakage inductance and resistance. However, as was discussed in Section 5.8, most induction motors are designed with either a double cage as shown in Fig. 5.19(a) or with a deep, solid rotor bar. The objective of this design is to provide higher effective rotor-circuit resistance during on-line starting, giving a high starting torque when the rotor frequency is high, combined with a low effective rotor circuit resistance when the rotor frequency is low under running conditions.

This effect can be readily modeled into the transient and steady-state equivalent circuits, as shown in Fig. 10.13 for a simple double-cage rotor. The symbols are analogous to those of Fig. 5.19(b). In some situations, a more complex

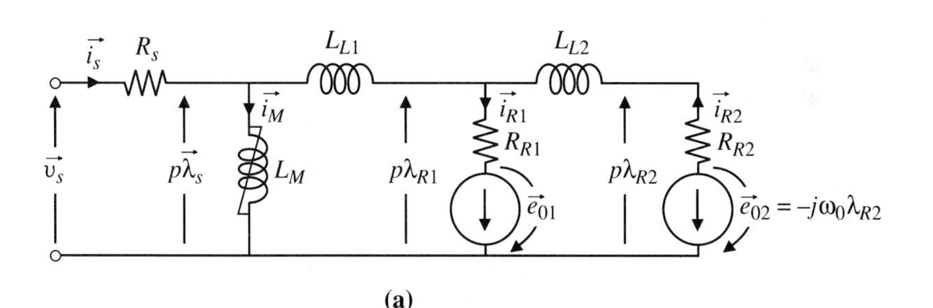

(a)

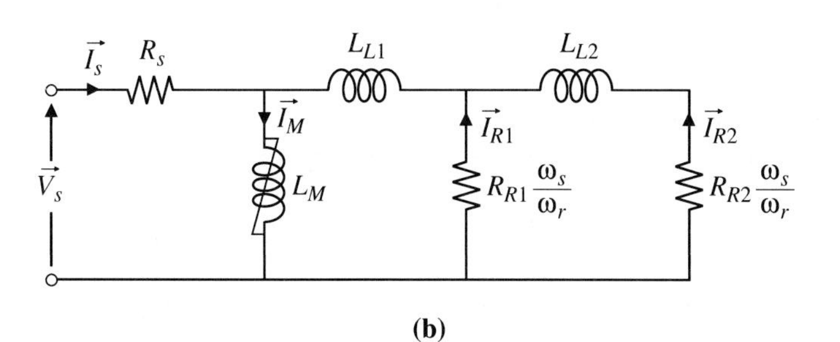

(b)

Figure 10.13 Equivalent circuits for induction machine with double-cage rotor. (a) Transient. (b) Steady state.

ladder of leakage inductances and rotor resistances may be desired to represent a deep-bar rotor. This network may be developed using the magnetic-to-electric circuit transformation techniques developed in Section 7.1.1.

10.2

Variable-Voltage, Variable-Frequency Operation

An induction motor supplied from a variable-voltage, variable-frequency inverter provides a drive that is capable of both high efficiency and good transient performance. A block diagram for such a drive system using a simple voltage-source inverter is shown in Fig. 10.14(a). This inverter has been discussed in Section 8.4.1, and its circuit is shown in Fig. 8.18.

10.2.1 ▪ Fundamental-Frequency Performance

Let us first review the basic properties of this drive considering only the fundamental-frequency component of the applied stator voltage. Given a direct-voltage source v_d of controllable magnitude, the inverter produces a 120°

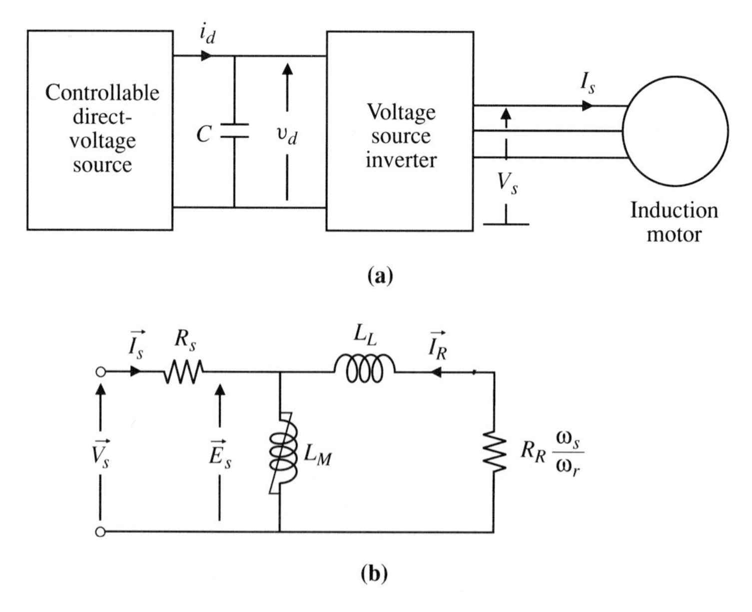

Figure 10.14 Induction motor drive with voltage-source inverter. (a) System. (b) Motor equivalent circuit.

rectangular voltage waveform for each phase, as shown in Fig. 8.17. From Eq. (8.33), the rms fundamental-frequency component of line-to-neutral or phase voltage is

$$V_s = \frac{\sqrt{2}}{\pi} v_d = 0.45 v_d \qquad \text{V} \tag{10.61}$$

An induction motor drive with variable-frequency supply always operates with a small value of rotor frequency to obtain its torque with the least loss and thus the highest efficiency. Thus, the equivalent circuit of Fig. 10.14(b) may be used for steady-state analysis even with a double-cage or deep-bar rotor. Moreover, the basic properties of the drive usually can be approximated adequately by ignoring both the stator resistance and the leakage inductance. To obtain maximum torque with minimum current and therefore minimum winding loss, the machine is normally operated at or near its rated value of stator flux linkage of rms value Λ_s. Then, to a first approximation ignoring stator resistance drop, the fundamental frequency stator voltage for any operating frequency ω_s should be adjusted so that

$$\vec{V}_s = j\omega_s \vec{\Lambda}_s \qquad \text{V} \tag{10.62}$$

The required direct-link voltage then is given by

$$v_d = \frac{\omega_s \Lambda_s}{0.45} \qquad \text{V} \tag{10.63}$$

Because of this relationship, this mode of operation is frequently denoted as "constant volts per hertz" control, although some adjustment to compensate for the stator resistance drop is usually required, particularly at low speed.

The steady-state torque has been derived in Eq. (10.58) as

$$T = \frac{3p}{2} \frac{R_R}{\omega_r} I_R^2 \qquad \text{N}\cdot\text{m} \tag{10.64}$$

If the leakage inductance effect is ignored, the rotor current can be approximated by

$$\begin{aligned} \vec{I}_R &\approx -\frac{\vec{E}_s \omega_r}{R_R \omega_s} \\ &= -\frac{j\omega_s \vec{\Lambda}_s \omega_r}{R_R \omega_s} \\ &= -j\vec{\Lambda}_s \frac{\omega_r}{R_R} \qquad \text{A} \end{aligned} \tag{10.65}$$

The torque now may be expressed as a function of the stator flux linkage using Eqs. (10.58) and (10.65):

$$T \approx \frac{3p}{2} \Lambda_s^2 \frac{\omega_r}{R_R} \qquad \text{N}\cdot\text{m} \tag{10.66}$$

The approximate speed-torque relationship for the machine is therefore

$$\begin{aligned} \omega_{mech} &= \frac{2}{p}(\omega_s - \omega_r) \\ &= \frac{2}{p}\omega_s - \left(\frac{2}{p}\right)^2 \frac{R_R T}{3\Lambda_s^2} \qquad \text{rad/s} \end{aligned} \tag{10.67}$$

This relation is shown in Fig. 10.15 for various values of stator frequency. Negative values of the stator frequency ω_s are obtained by use of the reversed phase sequence *a-c-b*. In order to operate in quadrants 2 and 4, the direct-voltage source must be capable of absorbing the power fed back by the machine acting as a generator, i.e., the direct current i_d must be reversible. Examples of such sources are the four-quadrant chopper of Fig. 8.17 and the dual rectifier/inverter of Fig. 8.8.

The total input power to the stator of the motor is

$$P_s = 3\Re(\vec{V}_s^*\vec{I}_s) \qquad \text{W} \tag{10.68}$$

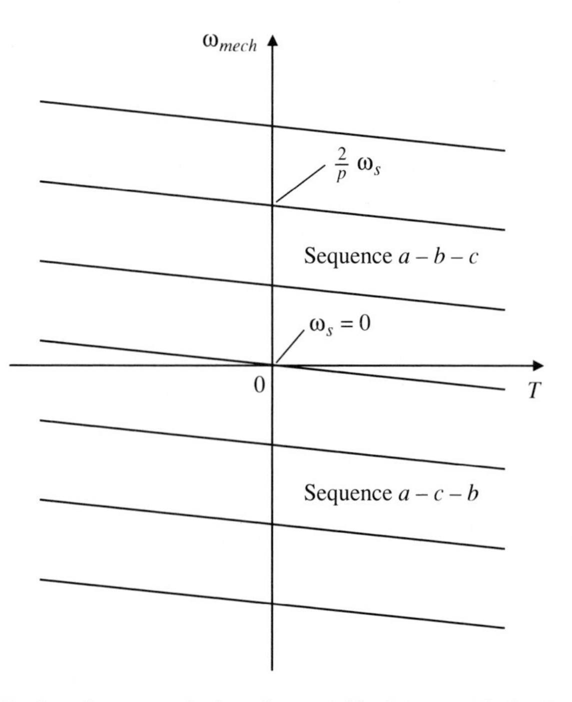

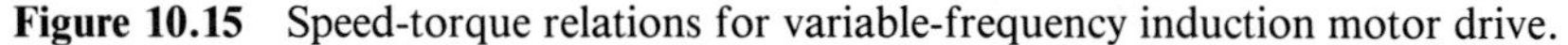

Figure 10.15 Speed-torque relations for variable-frequency induction motor drive.

Assuming an ideal lossless inverter, the average direct current in the link will be

$$\bar{i}_d = \frac{P_s}{v_d} \qquad \text{A} \tag{10.69}$$

The instantaneous current i_d also will contain components at integral multiples of frequency $6\omega_s$. The capacitor C is connected across link voltage partially to absorb these components.

A practical direct-voltage source has a maximum available value of the link voltage $\hat{v}_d$. Therefore, there is a maximum stator frequency ω_b at which the full or rated stator flux linkage can be maintained. Up to this frequency and its corresponding speed, full-rated torque can be produced. In some drives, it is desired to operate above this value of speed at reduced torque, approximating a constant power characteristic.

For operation at $\omega_s > \omega_b$, with constant link voltage v_d and therefore constant stator voltage V_s, the rms stator flux linkage is given approximately by

$$\Lambda_s = \frac{V_s}{\omega_s} = \frac{\omega_b}{\omega_s}\Lambda_{s(\text{max})} \qquad \text{Wb} \tag{10.70}$$

Reduction of the stator flux linkage causes an increase in the downward slope of the speed-torque curve in this speed range, as seen from Eq. (10.67). Also, as frequency increases, the effect of the leakage inductance becomes more significant. From Section 5.7, the maximum torque that can be produced for a given value of Λ_s is

$$\hat{T} = \frac{3p}{4}\frac{\Lambda_s^2}{L_L} = \left(\frac{\omega_b}{\omega_s}\right)^2 \hat{T}_{\text{at }\omega_b} \qquad \text{N}\cdot\text{m} \tag{10.71}$$

A set of speed-torque curves for various stator frequencies at and above ω_b is shown in Fig. 10.16. The maximum available torque $\hat{T}$ decreases rapidly with increased frequency. A near-constant output power characteristic thus can be maintained for only a limited speed range.

EXAMPLE 10.2 A 440-V, 60-Hz, six-pole induction motor is to be incorporated into a variable-speed drive where the supply is obtained from a 460-V, three-phase, 60-Hz source, a controlled rectifier, and a simple voltage-source inverter. The motor has a leakage inductance of 8 mH and a rotor resistance of 0.3 Ω.

(a) What is the maximum value of fundamental-frequency voltage that the inverter can supply?

(b) Up to what stator frequency can the motor be operated at its rated value of flux linkage?

(c) What is the maximum torque that can be produced at the frequency of (b) and at what value of rotor speed will it occur?

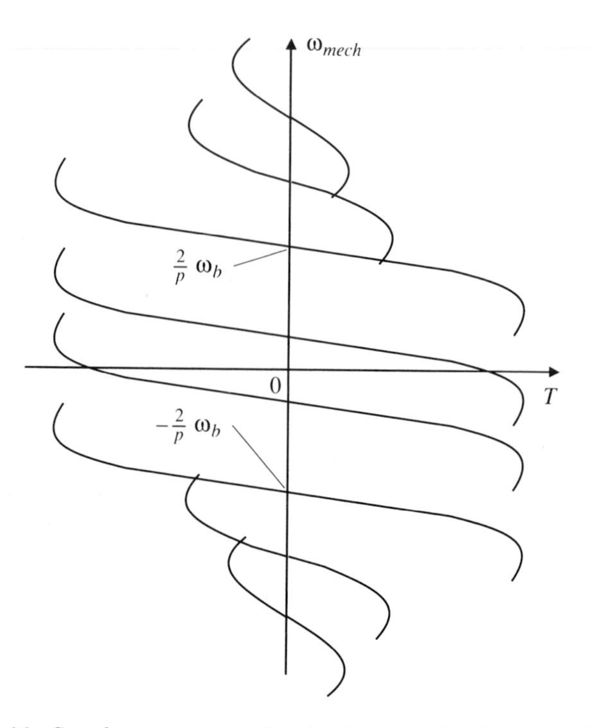

Figure 10.16 Speed-torque curves showing the reduction in torque above base frequency ω_b.

(d) If the stator frequency is increased to 1.6 times the value obtained in (b), what maximum torque can be obtained?

(e) Typically, this motor might have a rated torque of about 40% of the peak value found in (c). Up to what approximate speed would this drive be capable of maintaining a constant power characteristic starting at this rated torque?

Solution

(a) With the rectifier set at $\alpha = 0$, the maximum value of the link voltage is, from Eq. (8.11),

$$\hat{v}_d = 1.35 V_{L\text{-}L} = 1.35(460) = 621 \qquad \text{V}$$

At this link voltage, the fundamental phase voltage in the motor stator will be, from Eq. (10.61),

$$V_s = 0.45\,\hat{v}_d = 0.45(621) = 279.5 \qquad \text{V}$$

(b) Because the drop across the stator resistance can be ignored, the rated stator flux linkage of the motor is

$$\Lambda_b = \frac{440}{\sqrt{3} \times 2\pi \times 60} = 0.67 \qquad \text{Wb}$$

The breakpoint frequency up to which this flux linkage can be maintained is then

$$\omega_b = \frac{V_s}{\Lambda_b} = \frac{279.5}{0.67} = 415 \qquad \text{rad/s}$$

(c) The maximum torque is, from Eq. (10.71),

$$\hat{T} = \frac{3p}{4} \frac{\Lambda_s^2}{L_L} = \frac{3 \times 6}{4} \frac{0.67^2}{0.008} = 256 \qquad \text{N·m}$$

This maximum torque occurs when the rotor frequency is

$$\omega_r = \frac{R_R}{L_L} = \frac{0.3}{0.008} = 37.5 \qquad \text{rad/s}$$

Then, the motor speed is

$$\omega_{mech} = \frac{2}{p}(\omega_s - \omega_r) = \frac{2}{6}(415 - 37.5) = 125.8 \qquad \text{rad/s}$$

(d) The flux linkage will be reduced by the factor 1.6. Therefore, the maximum torque will be

$$T = \frac{256}{1.6^2} = 100 \qquad \text{N·m}$$

(e) For the constant power characteristic, the rated or base torque is

$$T_b = 0.4(256) = 102.4 \qquad \text{N·m}$$

The base speed is $\omega_{b(mech)} = 125.8$ rad/s. For a higher speed, the torque on the constant power curve is

$$T = \frac{T_b \omega_{b(mech)}}{\omega_{mech}} \qquad \text{N·m}$$

The peak torque available at frequency ω_s is

$$\hat{T} = 256\left(\frac{415}{\omega_s}\right)^2 \qquad \text{N·m}$$

and this occurs at a mechanical speed of

$$\omega_{mech} = \frac{2}{p}(\omega_s - \omega_r) = \frac{\omega_s - 37.5}{3} \qquad \text{rad/s}$$

Thus, the maximum frequency that can be achieved on the constant power curve occurs when $T = \hat{T}$, i.e.,

$$\frac{(102.4 \times 125.8)3}{\omega_s - 37.5} = 256\left(\frac{415}{\omega_s}\right)^2$$

or

$$\omega_s^2 - 1140\omega_s + 42\,800 = 0$$

from which $\omega_s = 1100$ rad/s and

$$\omega_{mech} = \frac{1100 - 37.5}{3} = 356 \qquad \text{rad/s}$$

10.2.2 • Effects of Voltage Harmonics

The stepped voltage waveform produced by a simple voltage-source inverter contains a set of harmonics of all odd orders h except for multiples of three, i.e.,

$$h = 1 + 6k \qquad \text{where } k = 0, \pm 1, \pm 2 \ldots. \tag{10.72}$$

From Eq. (8.31), the rms values of the harmonic voltages are inversely proportional to the harmonic order:

$$V_{hs} = \frac{1}{h} V_{1s} \qquad \text{V} \tag{10.73}$$

If the phase sequence of the fundamental is *a-b-c*, the phase sequence of the fifth harmonic can be seen to be *a-c-b* because its phase *b* lags phase *a* by $120 \times 5 = 600° = 240°$. Harmonic orders with positive values of k in Eq. (10.72) have positive sequence while those with negative values of k have negative sequence.

The effect of these harmonic voltages on the stator current can be approximated by examination of the harmonic equivalent circuit of Fig. 10.17(a). Under normal operating conditions, the equivalent, two-pole rotor speed ω_0 is approximately equal to the supply frequency ω_s. The harmonic rotor frequency is therefore

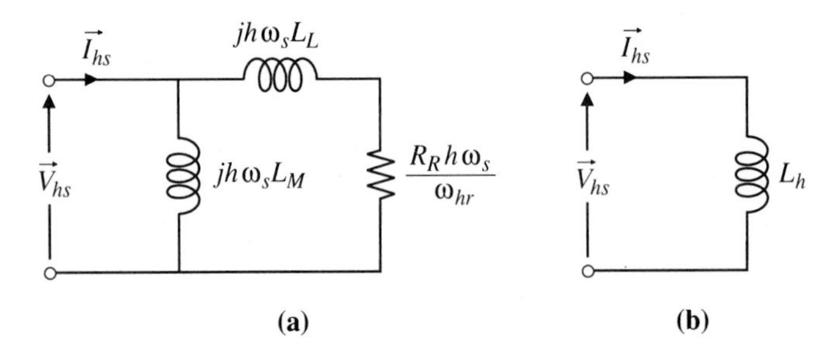

Figure 10.17 (a) Harmonic equivalent circuit. (b) Approximate circuit.

$$\begin{aligned} \omega_{hr} &= h\omega_s - \omega_0 \\ &\approx -6\omega_s \qquad \text{for } h = -5 \\ &\approx 6\omega_s \qquad \text{for } h = 7 \end{aligned} \tag{10.74}$$

The effective resistance in the rotor circuit of Fig. 10.17(a) thus is approximately equal to the rotor resistance R_R and is much less than the leakage reactance $h\omega_s L_L$. Thus, for harmonic frequencies, the rotor can be represented approximately as an inductance L_h, as shown in Fig. 10.17(b) where

$$L_h = \frac{L_M L_L}{L_M + L_L} \approx L_L \qquad \text{H} \tag{10.75}$$

ignoring the resistance of both stator and rotor. The harmonic phase current in the motor then is approximated by

$$I_{hs} = \frac{V_{hs}}{h\omega_s L_h} = \frac{V_{1s}}{h^2\omega_s L_h} = \frac{\Lambda_s}{h^2 L_h} \qquad \text{A} \tag{10.76}$$

The effect of these harmonic currents in heating the windings can be assessed by noting that typically an induction motor may have a leakage inductance in the range 0.12 to 0.2 pu. Thus, with 1 pu stator flux linkage, the most significant harmonic currents are

$$I_{5s} \approx 0.2 \text{ to } 0.33 \qquad \text{pu}$$

and

$$I_{7s} \approx 0.1 \text{ to } 0.17 \qquad \text{pu}$$

Note that these values are independent of the torque and the speed if the fundamental flux linkage is kept constant. Because rated stator current normally is based on the fundamental frequency current alone, the operating value of stator current should be derated below this value. The amount of derating for a machine with 0.12 pu leakage inductance can be estimated from the first four harmonics as

$$1 = (I_{1s}^2 + 0.33^2 + 0.17^2 + 0.07^2 + 0.05^2)^{1/2} \tag{10.77}$$

from which

$$I_{1s} \approx 0.925 \qquad \text{pu}$$

The actual heating of the machine may be significantly more than indicated by this expression because the effective rotor resistance increases significantly with rotor frequency, as discussed for the deep-bar or squirrel-cage motor. The effective stator resistance also may increase somewhat with frequency. There also is an increase in the eddy current losses in the stator iron due to the flux harmonics. Thus, derating factors frequently are derived from operating experience.

The average torque produced by these harmonic voltages is usually negligible. From the equivalent harmonic circuit of Fig. 10.17(a), the harmonic rotor current is approximately Λ_{hs}/L_L. Then the torque at harmonic h is approximately

$$\begin{aligned} T_h &= \frac{3p}{2}\frac{R_R}{\omega_{hr}}\left(\frac{\Lambda_{hs}}{L_L}\right)^2 \\ &\approx \frac{3p}{2}\frac{R_R}{(h-1)\omega_s}\left(\frac{\Lambda_{1s}}{hL_L}\right)^2 \quad \text{N·m} \end{aligned} \tag{10.78}$$

This torque can be compared with the starting torque of the machine:

$$T_{start} \approx \frac{3p}{2}\frac{R_R}{\omega_s}\left(\frac{\Lambda_{1s}}{L_L}\right)^2 \quad \text{N·m} \tag{10.79}$$

Comparison of these expressions shows that $T_h \approx T_{start}/h^3$ and is thus of negligible significance in most situations. It may be noted that the torque is negative for the largest component with $h = 5$.

The most significant effect of the harmonics is the production of harmonic torques which can cause serious vibrations and noise. Suppose we represent the motor by the harmonic equivalent circuit of Fig. 10.18(a). From Eq. (10.45), the fundamental torque can be expressed as

$$T_1 = \frac{3p}{4}\,\mathcal{I}(\vec{\lambda}_{1s}^*\vec{i}_{1s}) \approx \frac{3p}{4}\,\hat{\lambda}_{1s}\hat{i}_{1s} \quad \text{N·m} \tag{10.80}$$

because, having ignored the magnetizing current and the leakage inductance,

$$\vec{i}_{1s} \approx \frac{\vec{v}_{1s}\omega_{1r}}{R_R\omega_s} = j\,\frac{\vec{\lambda}_{1s}\omega_r}{R_R} \quad \text{A} \tag{10.81}$$

The fundamental torque is produced by the interaction of the sinusoidally distributed flux density B_1 and the sinusoidally distributed rotor current density J_1, as shown in Fig. 10.18(b), both rotating forward or counterclockwise at angular velocity ω_s.

A fifth harmonic current, represented by its space vector $\vec{i}_{5s}$, produces a sinusoidally distributed current density in the stator rotating backward or anticlockwise at angular velocity $-5\omega_s$, as shown in Fig. 10.18(c). The fifth harmonic rotor current will be approximately equal and opposite. This rotor current density interacts with the forward rotating fundamental flux linkage $\vec{\lambda}_{1s}$ to produce an oscillatory torque T_{15} at the sixth harmonic frequency, the magnitude given by

$$T_{15} \approx \frac{3p}{4}\,\mathcal{I}(\vec{\lambda}_{1s}^*\vec{i}_{5s}) \quad \text{N·m} \tag{10.82}$$

Similarly, the seventh harmonic current $\vec{i}_{7s}$ produces a sinusoidally distributed current density, rotating forward at an angular velocity $7\omega_s$, which interacts

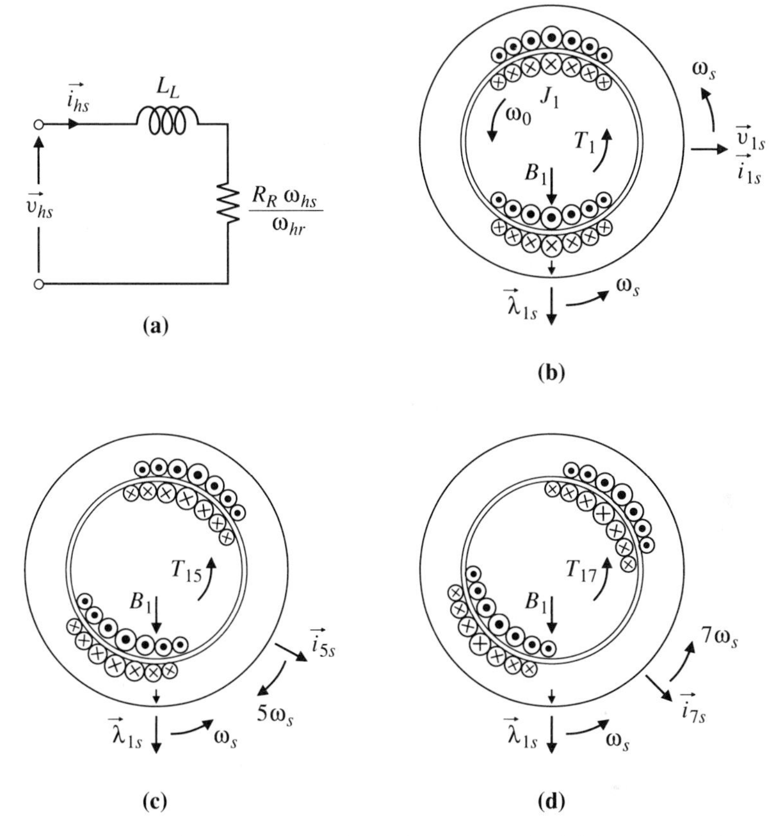

Figure 10.18 (a) Simple harmonic equivalent circuit. (b) Fundamental space vectors and torque. (c) Torque due to 5th harmonic current and fundamental flux. (d) Torque due to 7th harmonic current and fundamental flux.

with the fundamental flux density, represented by $\vec{\lambda}_{1s}$, which is rotating forward at velocity ω_s, as shown in Fig. 10.18(d) to produce a sixth harmonic torque

$$T_{17} \approx \frac{3p}{4}\,\mathcal{J}(\vec{\lambda}^*_{1s}\vec{i}_{7s}) \qquad \text{N}\cdot\text{m} \tag{10.83}$$

There are also harmonic torque components that are produced by the harmonic flux linkages $\vec{\lambda}_{5s}$ and $\vec{\lambda}_{7s}$ interacting with the fundamental current $\vec{i}_{1s}$ to produce sixth harmonic torques:

$$T_{51} = \frac{3p}{4}\,\mathcal{J}(\vec{\lambda}^*_{5s}\vec{i}_{1s}) \qquad \text{N}\cdot\text{m} \tag{10.84}$$

$$T_{71} = \frac{3p}{4} \mathcal{I}(\vec{\lambda}_{5s}\vec{i}_{7s}) \qquad \text{N}\cdot\text{m} \tag{10.85}$$

The total sixth harmonic torque is the vector sum of the four components in Eqs. (10.82) to (10.85). For a simple voltage source inverter, the result is a sixth harmonic peak torque which is usually 10 to 20% of the fundamental torque. Torque components at frequencies $12\omega_s$ and $18\omega_s$ are also produced by pairs of higher-frequency harmonics.

Many variable-speed drives have a limited operating speed range. For example, a fan drive with torque proportional to speed squared has its mechanical power reduced to about 34% at 70% speed. For these drives, the torque harmonics may not be significant because they are absorbed by the motor and load inertia with little speed fluctuation. However, drives that must operate at low values of speed will be very jerky due to torque harmonics if the motor is supplied by the waveform of the simple voltage inverter. For these drives, a more nearly sinusoidal supply voltage can be provided using a pulse-width modulated voltage source inverter of the type discussed in Section 8.4.2.

10.2.3 ▪ Closed-Loop Speed Control

When precise control of speed is desired, a drive system such as that shown in Fig. 10.19 may be used. The motor speed converted to its two-pole value ω_0 is measured and compared with the desired speed ω_0^*. The difference can be envisaged as a demand value of rotor frequency ω_r^* which, from Eq. (10.66), will produce a proportional torque tending to reduce the speed error. This rotor frequency is now added to the speed ω_0 to produce the required stator frequency ω_s. This frequency signal is used to control the firing of the inverter. It also is used to establish the desired level of link voltage proportional to frequency as required by Eq. (10.62).

A variety of refinements can be added to the basic system of Fig. 10.19. For example, a limit normally is placed on the maximum value of rotor frequency $\hat{\omega}_r^*$. Thus, a large step of demand speed ω_0^* produces an essentially constant rotor frequency ω_r^* and a constant torque until the actual speed is within close range of its demand value.

10.3

Variable-Current, Variable-Frequency Operation

Inverter systems for producing a controllable-current source with variable frequency were discussed in Section 8.4.3. The performance of an induction motor on a current source differs markedly from that with a voltage source. Our

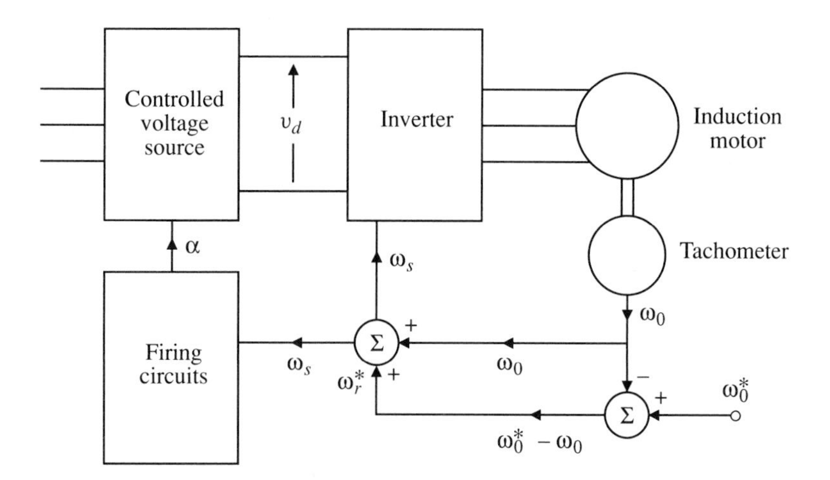

Figure 10.19 Closed-loop speed control system.

analysis of both transient and steady-state performance of a current-driven motor will be facilitated by development of yet another form of equivalent circuit.

In Section 10.1.1, it was shown that the flux linkage-current equations of the T form of equivalent circuit as shown in Fig. 10.6 could be transformed into the Γ form of circuit of Fig. 10.7 by appropriate choice of ratio γ. Now consider an alternate transformation of the relations of Eq. (10.21) using

$$\gamma = \frac{L_m}{L_m + L_{\ell r}} \tag{10.86}$$

to give a new set of rotor variables

$$\vec{\lambda}'_R = \gamma \vec{\lambda}_r \qquad \text{Wb} \tag{10.87}$$

and

$$\vec{i}'_R = \vec{i}_r / \gamma \qquad \text{A} \tag{10.88}$$

The resulting relations are

$$\begin{aligned} \vec{\lambda}_s &= (L'_L + L'_M)\vec{i}_s + L'_M \vec{i}'_R \\ \vec{\lambda}'_R &= L'_M \vec{i}_s + L'_M \vec{i}'_R \end{aligned} \tag{10.89}$$

where

$$L'_L = L_{\ell s} + \gamma L_{\ell r} \qquad \text{and} \qquad L'_M = \gamma L_m \tag{10.90}$$

These can be incorporated into the transient equivalent circuit of Fig. 10.20(a). This circuit is denoted as the inverse Γ form, reflecting the structure of its inductances. The remainder of its parameters and variables are transformed in the same way as in Eqs. (10.36) to (10.38). The corresponding steady-state equivalent circuit is shown in Fig. 10.20(b).

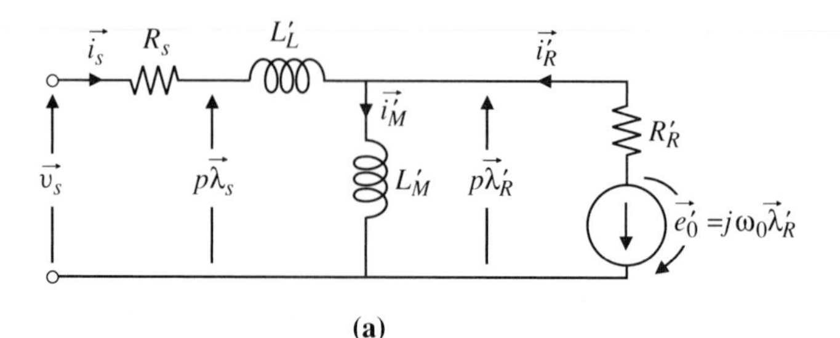

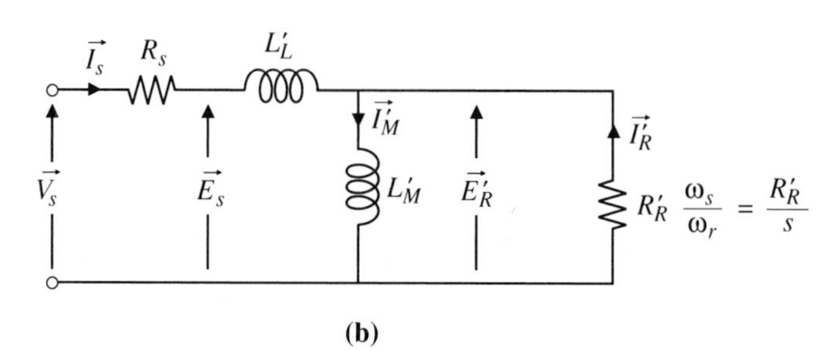

Figure 10.20 Inverse-Γ form of equivalent circuit. (a) Transient. (b) Steady state.

Consider first the steady-state behavior when a sinusoidal current phasor $\vec{I}_s$ at frequency ω_s is injected into the circuit of Fig. 10.20(b). The rotor current will be

$$\begin{aligned}\vec{I}'_R &= \frac{j\omega_s L'_M}{R'_R \dfrac{\omega_s}{\omega_r} + j\omega_s L'_M}\vec{I}_s \\ &= \frac{\vec{I}_s}{1 - j\dfrac{R'_R}{\omega_r L'_M}} \quad \text{A}\end{aligned} \tag{10.91}$$

The torque then is given by

$$\begin{aligned}T &= \frac{3p}{2\omega_s} R'_R \frac{\omega_s}{\omega_r} I'^2_R \\ &= \frac{3pR'_R}{2\omega_r} \frac{I_s^2}{1 + \left(\dfrac{R'_R}{\omega_r L'_M}\right)^2} \quad \text{N}\cdot\text{m}\end{aligned} \tag{10.92}$$

Over much of the useful operating range, $\omega_r \gg R'_R/L'_M$, and the torque can be approximated by

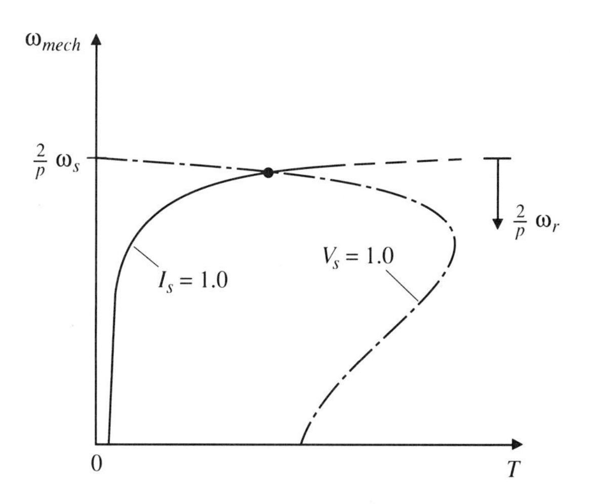

Figure 10.21 Speed-torque relation with current source (curve with voltage source shown for comparison).

$$T \approx \frac{3pR'_R}{2\omega_r} I_s^2 \qquad \text{N}\cdot\text{m} \tag{10.93}$$

The torque relation of Eq. (10.92) is shown in Fig. 10.21 for operation with rated stator current of 1 pu at a constant stator frequency and variable rotor frequency. The corresponding speed-torque relation for operation at rated stator voltage also is shown for comparison.

Theoretically, a peak torque occurs in this constant current relation when $\omega_r = R'_R/L'_M$. At this condition,

$$I'_M = I'_R = \frac{I_s}{\sqrt{2}} \qquad \text{A} \tag{10.94}$$

This condition would be valid only for very low values of stator current because a high value of magnetizing current would otherwise drive the machine far into saturation. The region where $\omega_r < R'_R/L'_M$ normally is not used because most of the stator current would be routed through the magnetizing branch.

An important feature of a current-driven induction motor is that operation at any point on the solid part of its speed-torque curve of Fig. 10.21 is unstable unless special control means are employed. Consider the application of a load torque when the stator frequency is held constant. This will decrease the speed, thus increasing the rotor frequency. From Eq. (10.93), the effect will be to decrease the developed torque, causing further deceleration and decrease in speed, i.e., instability. Therefore, closed-loop speed control is required.

A speed control system for a current-driven induction motor is shown in Fig. 10.22. The motor speed ω_0 is measured. To this is added a constant de-

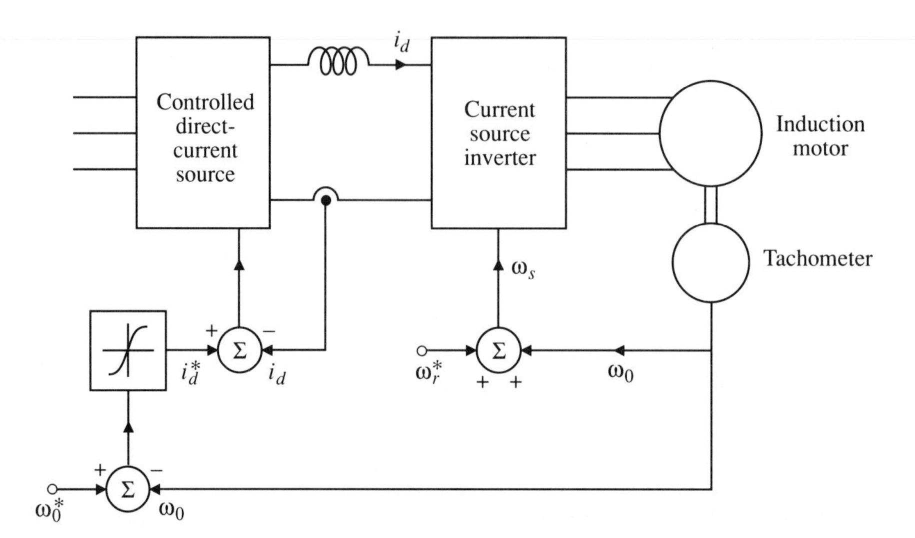

Figure 10.22 Current-driven induction motor with constant rotor frequency.

mand value of rotor frequency ω_r^* to give the required stator frequency ω_s at which the current source inverter will be fired. An appropriate choice of rotor frequency might be the rated value for the motor, i.e., the value at the intersection of the two speed-torque curves of Fig. 10.21. To control the speed, the error between the desired speed ω_0^* and the actual speed ω_0 is processed to give a demand value for the link current i_d^*. Because the stator current is proportional to i_d and the torque is proportional to I_s^2, a torque proportional to speed error would require that the link current i_d be made proportional to the square root of speed error. With positive rotor frequency, this drive is capable of positive torque only. If negative torque is required, a negative value of ω_r^* should be used when the speed error is negative.

A drive based on a current-driven induction motor can be made to have superior transient performance by using an approach known as *vector control.* Suppose it were possible to measure the angular position of the rotor flux linkage space vector $\vec{\lambda}_R'$. Examination of the equivalent circuit of Fig. 10.20(a) suggests that the stator current should have a magnetizing component of magnitude $\vec{i}_M'$, which is in space phase with $\vec{\lambda}_R'$. If this current magnitude is held constant, the flux linkage will be constant in magnitude at the desired value. The stator current now can be forced to have a further component which is 90° in angular space behind the rotor flux linkage $\vec{\lambda}_R'$. The magnitude $\hat{i}_R'$ of this component can be made proportional to the demand torque. From Eq. (10.44), the torque can then be expressed as

$$T = -\frac{3p}{4}\,\mathcal{I}(\vec{\lambda}_R'\vec{i}_R') = \frac{3p}{4}\,\hat{\lambda}_R'\hat{i}_R' \qquad \text{N}\cdot\text{m} \tag{10.95}$$

The required stator current is the vector sum of these two components. If this stator current could be supplied from an essentially ideal three-phase current source in which both the magnitude $\hat{i}_s$ and the space angle can be instantaneously established, the drive theoretically would be capable of producing instantaneous torque response to a command signal. Such a controlled stator current can be produced approximately by use of the hysteresis control scheme described in Section 8.4.3.

Direct measurement of the angular position of the net rotor flux is not normally feasible because the path of the desired flux is deep in the rotor winding and the rotor is rotating. However, indirect methods may be employed to evaluate the angular position θ_R' of the rotor flux linkage space vector $\vec{\lambda}_R'$. One approach is shown in the system of Fig. 10.23. Two stator voltages and two stator currents are measured and manipulated to produce the space vectors $\vec{v}_s$ and $\vec{i}_s$, each as a complex number. The appropriate expressions for $\vec{i}_s$ and $\vec{v}_s$ are given in Eqs. (10.10) and (10.47). The resistance drop $R_s\vec{i}_s$ is subtracted from the stator voltage $\vec{v}_s$ to give the induced voltage vector $\vec{e}_s$. This is then integrated to obtain the stator flux linkage $\vec{\lambda}_s$ from which the leakage flux linkage $L_L'\vec{i}_s$ may be subtracted to derive the rotor flux linkage space vector $\vec{\lambda}_R' = \hat{\lambda}_R' \epsilon^{j\theta_R'}$. The angle of this space vector is now multiplied by the desired magnitude $\hat{i}_M'^*$ to obtain the space vector $\vec{i}_M'^*$ of the required magnetizing current. The same angle de-

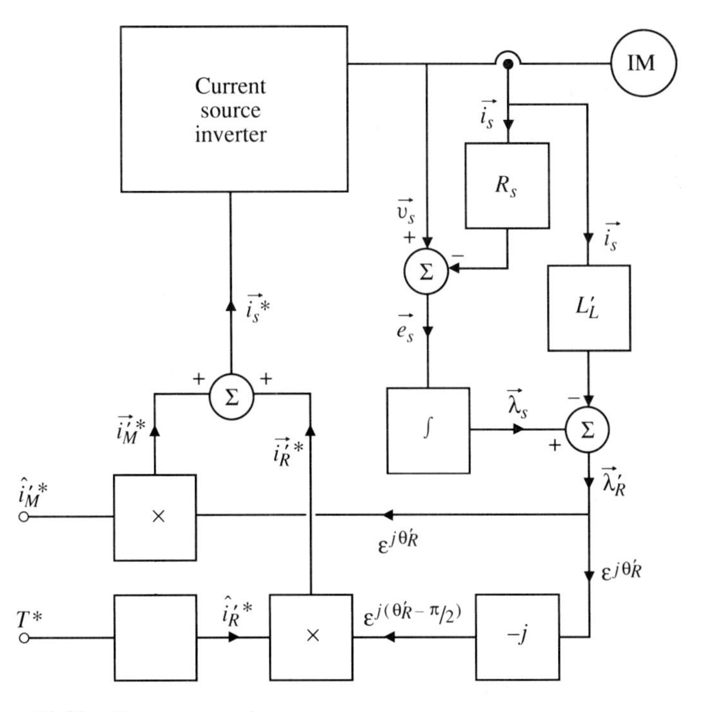

Figure 10.23 Vector control system.

layed by $\pi/2$ rad is then multiplied by the desired rotor current magnitude $\vec{i}_R'^*$ which is proportional to the demand torque T^*. The two vector components are now added to produce the required space vector of the stator current $\vec{i}_S^*$. The hysteresis controller of the current-source inverter then proceeds to produce the required instantaneous phase currents to inject into the stator windings. Normally, most if not all of the operations in the system of Fig. 10.23 are performed digitally.

The vector control system shown in Fig. 10.23 cannot operate at very low values of speed because the induced voltages in the phase windings become too small to obtain an adequate estimate of the flux angle. Other versions of vector control use a shaft position sensor to obtain the angular position of the rotor and then add to this a computed angle derived from integration of the predicted rotor frequency ω_r to obtain a prediction of the angle of the rotor flux linkage.

10.4

Variable-Voltage, Constant-Frequency Operation

Some drives require only a small operating speed range in addition to an ability to start. An example might be a fan drive where the fan power can be reduced to about 50% by reducing the speed to about 80%. A drive with this limited capability can be achieved using the simple system shown in Fig. 10.24(a). Back-to-back thyristors are connected in series with each stator phase. If each thyristor is turned on at the beginning of the corresponding sine wave of supply voltage, full voltage will be supplied to the stator, as shown in the waveforms of Fig. 10.24(b). In fact, the gate signal of the thyristor must be held on until the phase current goes positive after a delay of the power factor angle ϕ.

If the firing angle of each thyristor is delayed after the voltage zero by more than the power factor angle ϕ, a chopped sine wave of voltage will be applied to the stator winding. This voltage waveform will continue until the discontinuous pulse of stator current goes to zero, turning off the thyristor as shown in Fig. 10.24(c). Thus, the effective or fundamental-frequency component of the applied stator voltage can be reduced by delay of the firing angles of the thyristors. Ignoring the stator resistance drop, the fundamental stator flux linkage Λ_s at the constant frequency ω_s thus can be controlled. Using the simplified equivalent circuit of Fig. 5.13 in which the leakage inductance is ignored, the torque of the motor is given, from Eq. (10.66), by

$$T = \frac{3p}{2} \frac{\Lambda_S^2 \omega_r}{R_R} \quad \text{N}\cdot\text{m} \tag{10.96}$$

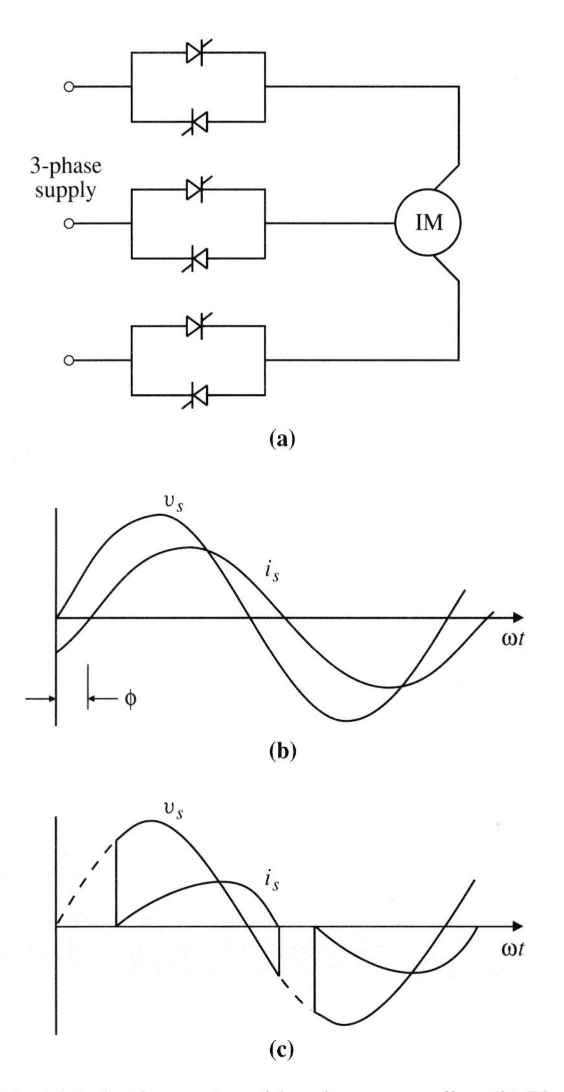

Figure 10.24 (a) Induction motor with voltage controller. (b) Phase voltage and current at full voltage. (c) Phase voltage and current with delayed firing of thyristors.

Because the supply frequency remains constant, speed reduction is achieved only by increase in the rotor frequency ω_r. Figure 10.25 shows a set of torque speed curves for various values of effective stator voltage or flux linkage. It also shows a speed-torque relation for a fan load of the form

$$T = k\omega_{mech}^2 \qquad \text{N}\cdot\text{m} \tag{10.97}$$

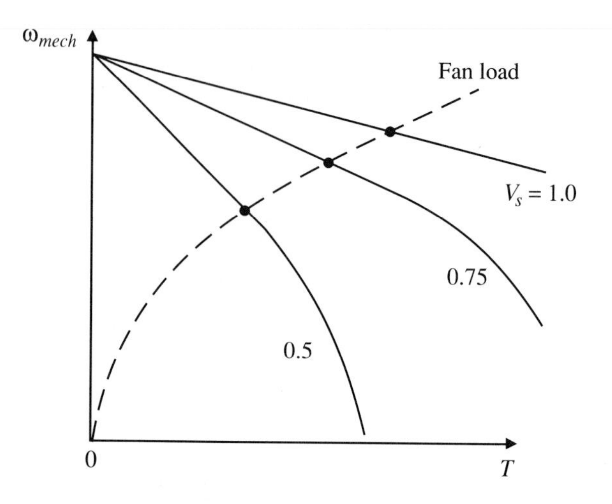

Figure 10.25 Control of fan speed by adjustment of stator voltage (V_s in per unit of rated value).

The intersections of this curve with the motor curves gives possible values of steady-state speed.

Induction motors for use in this system usually have a high value of rotor resistance and are of the Class-D design shown in Fig. 5.19(c). Their efficiency is relatively low when operating at rated voltage and decreases as the stator voltage is reduced because, even ideally, only the fraction ω_0/ω_s of the air-gap power is converted into mechanical power. There are also losses due to the harmonics in the supply voltage and current. However, a low efficiency at a reduced power output may be acceptable in some instances.

The rotor current and rotor losses in a motor driving a fan load will not necessarily be a maximum at the full-voltage, maximum-speed condition. For a two-pole motor, the developed torque can be expressed as

$$T = \frac{3R_R}{\omega_r} I_R^2 \qquad \text{N·m} \tag{10.98}$$

Combining this with Eq. (10.97), the rotor current is related to the rotor frequency by

$$I_R = \left[\frac{k(\omega_s - \omega_r)^2 \omega_r}{3R_R}\right]^{1/2} \qquad \text{A} \tag{10.99}$$

Differentiating I_R with respect to ω_r and setting the result equal to zero gives the condition for maximum rotor current as

$$\omega_r = \omega_s/3 \qquad \text{rad/s} \tag{10.100}$$

i.e., at a slip of one-third. The rotor frequency of a typical Class-D motor at rated load is usually substantially less than one-third of rated frequency. Thus, the motor current will tend to rise as its speed is reduced by voltage reduction. Therefore, the motor should be appropriately derated so that it will not overheat at any expected load condition.

The arrangement of Fig. 10.24 is sometimes used to provide a "soft start" for an induction motor. If an induction motor is switched directly on the supply line, the initial starting current will frequently be in the range of 5 to 8 times rated. With reduced stator voltage, the starting current and thus the influence on other equipment attached to the same supply is proportionally reduced at the expense of reduced starting torque.

EXAMPLE 10.3 A Class-D, four-pole, 60-Hz induction motor has a rated full-load speed of 1530 r/min. By what factor should this motor be derated if it is to drive a fan load with variable stator voltage over a wide speed range?

Solution The rotor frequency of this motor on rated load is, using subscript b for base or rated,

$$\omega_{rb} = \left[\frac{1800 - 1530}{1800}\right] 60 = 9 \qquad \text{Hz}$$

Let the rated rotor current be I_{Rb}. Using an idealized model for the motor in which leakage inductance is ignored, the rotor current will be directly proportional to the rotor frequency. From Eq. (10.100), the maximum value of rotor current $\hat{I}_R$ will occur at $\omega_r = \omega_s/3$ or 20 Hz. Thus, from Eq. (10.99),

$$\frac{\hat{I}_R}{I_{Rb}} = \left[\left(\frac{\omega_s - \hat{\omega}_r}{\omega_s - \omega_{rb}}\right)^2 \frac{\hat{\omega}_r}{\omega_{rb}}\right]^{1/2} = \left[\left(\frac{40}{51}\right)^2 \frac{20}{9}\right]^{1/2} = 1.17$$

The current rating and therefore the power rating of the motor should thus be reduced by a factor $1/1.17 = 0.86$. Alternatively, a motor having a rating at least 17% larger than needed for full-load power should be chosen.

10.5

Speed Control by Rotor-Power Recovery

A wound-rotor induction motor has a conventional three-phase winding in slots on its rotor. These windings are brought out to three slip rings on the rotor shaft. In Section 10.1.3 it was shown that, of the power crossing the air gap, a fraction ω_0/ω_s is converted to mechanical power while the remaining fraction ω_r/ω_s is transferred to the rotor circuit. In the squirrel-cage motor, this latter fraction is power loss. With a wound-rotor motor, it is possible to recover most of this power transferred to the rotor circuit by converting it back to supply voltage and frequency using the system of Fig. 10.26. In this system, the brushes on the rotor slip rings are connected to a three-phase rectifier of the form shown in Fig. 8.9 using diodes instead of thyristors. The power in the direct-voltage link then is supplied back to the supply lines using a three-phase inverter with six thyristors of the same form shown in Fig. 8.9.

The basic properties of this system can be derived using the steady-state equivalent circuit of Fig. 10.10(b) together with the rectifier-inverter relations from Section 8.2.2. Suppose the ratio of rotor voltage to stator voltage with the rotor open circuited at standstill is denoted as n. Ignoring the stator resistance,

$$n = \gamma \frac{N_{se}}{N_{re}} \tag{10.101}$$

where N_{se} and N_{re} are the effective numbers of sinusoidally distributed turns per phase on stator and rotor, and γ is the near-unity factor introduced in Eq. (10.31) in deriving the Γ model equivalent circuit. When the motor is operating at a speed $\omega_{mech} = \omega_0(2/p)$, the rotor frequency is

$$\omega_r = \omega_s - \omega_0 \qquad \text{rad/s} \tag{10.102}$$

A first approximation to the system behavior can be obtained by considering the motor to be ideal, i.e., neglecting both stator and rotor resistances and the leakage inductance. The rms-induced voltage per phase in the rotor winding then is given by

$$V_R = \frac{V_s}{n}\frac{\omega_r}{\omega_s} \qquad \text{V} \tag{10.103}$$

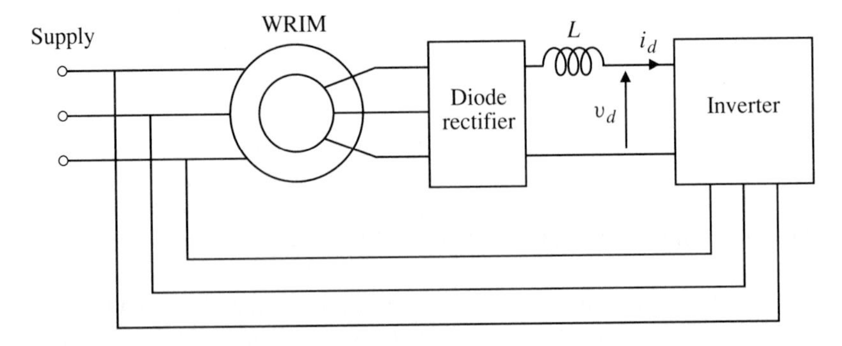

Figure 10.26 Speed control using rotor-power recovery.

where V_s is the rms supply voltage per phase. When this rotor voltage is rectified in the diode rectifier, the link voltage will be, from Eq. (8.11),

$$\bar{v}_d = \frac{3\sqrt{6}}{\pi} V_R = 2.34 V_R \qquad \text{V} \tag{10.104}$$

The link voltage is now coupled to the supply voltage using the inverter relation of Eq. (8.11), i.e.,

$$\bar{v}_d = -2.34 V_s \cos\alpha \qquad \text{V} \tag{10.105}$$

The minus sign in this expression arises from the fact that, when a rectifier is operated in the inverter mode with delay angle $\alpha > \pi/2$, its direct voltage is reversed. A combination of Eqs. (10.103) and (10.105) then gives the idealized relations

$$\omega_r = -\omega_s n \cos\alpha \tag{10.106}$$

$$\omega_{mech} = \frac{2}{p}(\omega_s - \omega_r) = \frac{2}{p}\omega_s(1 + n\cos\alpha) \tag{10.107}$$

Therefore, the speed of the drive may be adjusted from its synchronous value of $\omega_s(2/p)$ with $\alpha = 90°$ down to the fraction $(1 - n)$ of that value as α approaches $180°$.

The torque for the idealized motor can be derived using Eq. (8.13) and noting that

$$P_{mech} = T\omega_{mech} = \frac{\omega_0}{\omega_s} P_{air\ gap} \qquad \text{W} \tag{10.108}$$

Thus,

$$\begin{aligned} T &= \frac{3p}{2}\frac{V_s}{\omega_s}(0.78 i_d) \\ &= 1.17 p \Lambda_s i_d \qquad \text{N}\cdot\text{m} \end{aligned} \tag{10.109}$$

In a practical drive, the speed-torque relation will have a negative slope due to the rotor resistance and the losses in both rectifier and inverter, as shown in Fig. 10.27, for $n = 0.5$. In many applications, the required speed range is small, typically less than a 25% reduction. Thus, the ratio n is small and the rating of the rectifier and the inverter is only approximately the fraction n of the motor power rating. For systems with such a limited speed range, the motor may be started with resistances connected to the rotor terminals. The rectifier-inverter system can then be connected when the motor is up to speed.

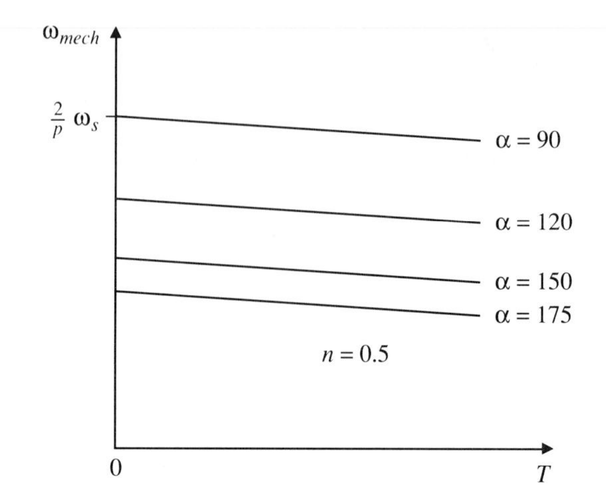

Figure 10.27 Speed-torque relations for drive with rotor-power recovery.

10.6

Unbalanced Operation

Throughout this chapter, it has been assumed that the supply voltages have been balanced, i.e., equal in magnitude and symmetrically arranged with $2\pi/3$ rad phase displacement. While the practical situation is usually close enough to this ideal to justify the assumption, there are situations in which machines operate, either temporarily or permanently, under unbalanced voltage conditions. For example, an induction motor might be operated temporarily with one phase open circuited, or its supply might be unbalanced due to the effect of a large single-phase load on its three-phase supply. Therefore, an ability to predict torque capability and stator currents during an unbalanced condition is important.

The steady-state analysis of unbalanced three-phase systems is facilitated by the use of a technique, based on the principle of superposition, known as *symmetrical components*. In this, any unbalanced set of phase voltages is represented as the superposition of three sets, each of three voltages. The first is a balanced set $\vec{V}_{1a}$, $\vec{V}_{1b}$, and $\vec{V}_{1c}$ with a phase sequence *a-b-c* which is denoted as a positive sequence. The second is a balanced set $\vec{V}_{2a}$, $\vec{V}_{2b}$, and $\vec{V}_{2c}$ with a negative sequence *a-c-b*. The third set, $\vec{V}_{0a}$, $\vec{V}_{0b}$, and $\vec{V}_{0c}$ consists of equal voltage phasors in all three phases and is denoted as the zero-sequence set. At first glance, this expansion from three voltages to nine might appear to complicate the issue. However, we have developed simple steady-state models for three-phase machines when operating with balanced sets of voltages and can use these for ready analysis.

The voltage phasor $\vec{V}_a$ is given by

$$\vec{V}_a = \vec{V}_{1a} + \vec{V}_{2a} + \vec{V}_{0a} \quad \text{V} \tag{10.110}$$

Because we normally use phase a as representative of all phases in a balanced system, it is convenient to state the other phase voltages in terms of the same variables as in Eq. (10.110). Thus,

$$\begin{aligned} \vec{V}_b &= \vec{V}_{1b} + \vec{V}_{2b} + \vec{V}_{0b} \\ &= \alpha^2\vec{V}_{1a} + \alpha\vec{V}_{2a} + \vec{V}_{0a} \quad \text{V} \end{aligned} \tag{10.111}$$

and

$$\vec{V}_c = \alpha\vec{V}_{1a} + \alpha^2\vec{V}_{2a} + \vec{V}_{0a} \quad \text{V} \tag{10.112}$$

where

$$\alpha = 1\angle 2\pi/3 = 1\angle 120^\circ$$

and

$$\alpha^2 = 1\angle 4\pi/3 = 1\angle 240^\circ$$

If it is understood that all quantities are to be expressed in terms of the phase-a components, the subscript a may be dropped. The synthesis of sequence voltages to produce any unbalanced set can now be expressed in matrix form as

$$\begin{vmatrix} \vec{V}_a \\ \vec{V}_b \\ \vec{V}_c \end{vmatrix} = \begin{vmatrix} 1 & 1 & 1 \\ \alpha^2 & \alpha & 1 \\ \alpha & \alpha^2 & 1 \end{vmatrix} \begin{vmatrix} \vec{V}_1 \\ \vec{V}_2 \\ \vec{V}_0 \end{vmatrix} \quad \text{V} \tag{10.113}$$

Solution of the three relations in Eq. (10.113) gives the analysis of three unbalanced voltage phasors into their symmetrical components:

$$\begin{vmatrix} \vec{V}_1 \\ \vec{V}_2 \\ \vec{V}_0 \end{vmatrix} = \frac{1}{3} \begin{vmatrix} 1 & \alpha & \alpha^2 \\ 1 & \alpha^2 & \alpha \\ 1 & 1 & 1 \end{vmatrix} \begin{vmatrix} \vec{V}_a \\ \vec{V}_b \\ \vec{V}_c \end{vmatrix} \quad \text{V} \tag{10.114}$$

Figure 10.28 shows a typical set of unbalanced voltage phasors together with its sets of symmetrical components.

With a balanced set of supply voltages, the steady-state equivalent circuit for an induction machine derived in Section 10.1.3 and shown in Fig. 10.29 may be used for analysis. For use with positive-sequence voltages, the stator frequency ω_s is considered to be positive. The rotor frequency, denoted as ω_{1r}, being equal to ω_s minus the speed ω_0, is also positive unless the speed is greater than the synchronous value, as would be the case for generating action. With negative-sequence voltages, the field in the machine rotates in the opposite direction. For this sequence, it is convenient to denote the stator frequency ω_{2s} as

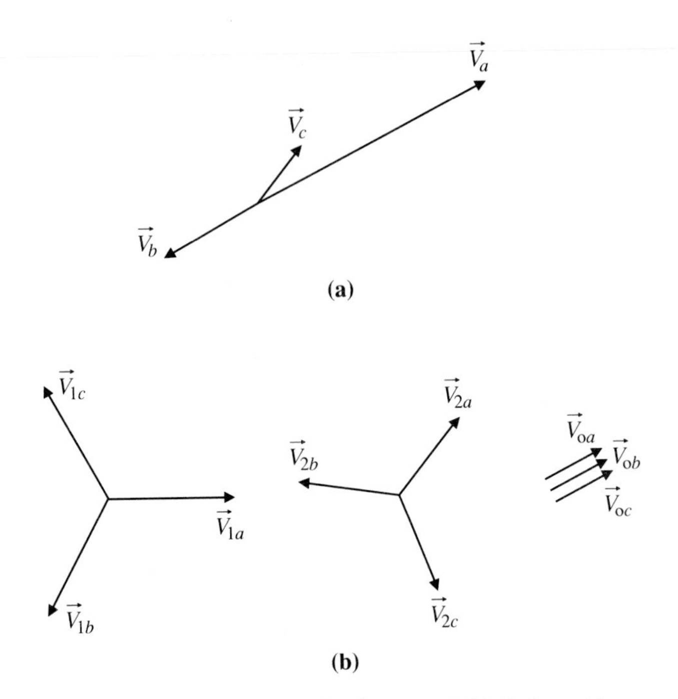

Figure 10.28 (a) A set of unbalanced voltages and (b) their positive, negative, and zero sequence component sets.

negative, i.e., equal to $-\omega_{1s}$. The corresponding rotor frequency for the negative sequence is

$$\omega_{2r} = \omega_{2s} - \omega_0 = -(\omega_{1s} + \omega_0) \qquad \text{rad/s} \tag{10.115}$$

When an induction machine is rotating near synchronous speed in its forward direction, its positive-sequence rotor frequency is small but its negative-sequence rotor frequency is nearly twice the supply frequency. The effective resistance of the rotor branch of Fig. 10.29 for negative-sequence operation near positive synchronous speed is approximately $R_R/2$. Therefore, the impedance

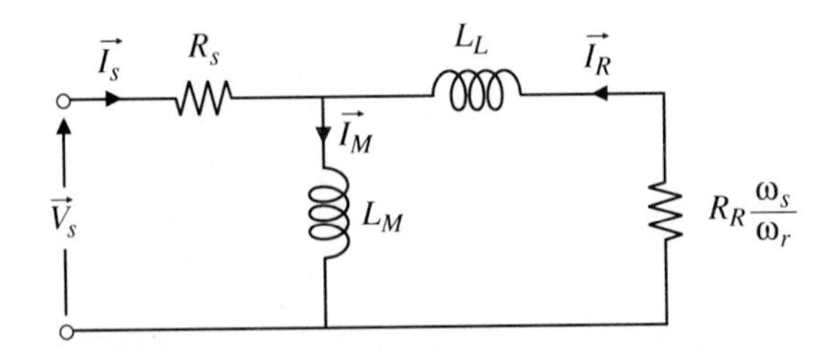

Figure 10.29 Induction motor equivalent circuit.

of the machine for negative sequence is mainly inductive and comparable in magnitude to the impedance for positive sequence during starting or at standstill. The effective resistance for many machines increases with rotor frequency as discussed for deep-bar machines in Section 5.8. For accurate analysis, this effect should be included.

From the above discussion, it can be appreciated that an induction machine is very sensitive to the application of voltages that have a negative-sequence component because even a small negative-sequence voltage can produce a large negative-sequence current due to the low value of the impedance to negative-sequence currents.

Many induction machines have only three stator terminals, so the sum of the three-phase currents must be zero. It follows that there can be no zero sequence or equal components of current in the three phases. Therefore, the zero-sequence impedance of the machine with a delta-connected stator winding or with a star-connection with open neutral is infinite. If there is a neutral connection, typically to ground, equal currents could flow in the three phases. If each of the windings were ideally distributed in angular space around the machine periphery, the net mmf produced across the air gap would be zero. Moreover, the net current density at any point around the machine would theoretically be zero. Ideally, the impedance to zero-sequence currents would be only the winding resistance R_s.

In practice, windings are not ideally sinusoidally distributed. Most stator slots will have significant net zero-sequence current and will thus have local leakage flux. Even the slots with no net zero-sequence current will have flux around and in each conductor. In addition, there will be leakage flux around the stator end windings. As a result, there will be a small but finite inductance L_0 for zero-sequence currents.

The positive-, negative-, and zero-sequence currents $\vec{I}_1$, $\vec{I}_2$, and $\vec{I}_0$ may now be evaluated by applying the positive-, negative-, and zero-sequence voltages $\vec{V}_1$, $\vec{V}_2$, and $\vec{V}_0$ to the appropriate steady-state equivalent circuits. The final step is the determination of the actual phase currents. By analogy with Eq. (10.113), these are given by

$$\begin{vmatrix} \vec{I}_a \\ \vec{I}_b \\ \vec{I}_c \end{vmatrix} = \begin{vmatrix} 1 & 1 & 1 \\ \alpha^2 & \alpha & 1 \\ \alpha & \alpha^2 & 1 \end{vmatrix} \begin{vmatrix} \vec{I}_1 \\ \vec{I}_2 \\ \vec{I}_0 \end{vmatrix} \quad \text{A} \tag{10.116}$$

EXAMPLE 10.4 A 460-V, 60-Hz, four-pole induction motor is operating at a speed of 1770 r/min when two of its phases become shorted together. The result is a voltage of 440 V from one phase to the other two. The stator resistance is 0.1 Ω, the rotor resistance 0.13 Ω, and the leakage inductance 5 mH. Magnetizing current can be ignored. The stator has only three terminals.

(a) Assuming no speed change, determine the steady-state phase currents following the short circuit.

(b) Compare the maximum phase current in (a) with the maximum phase current that would be expected during a normal start at rated voltage.

Solution

(a) For symmetry about phase a, assume that phases b and c are shorted together. Because the zero-sequence impedance is infinite, the zero-sequence current is zero and the value of the zero-sequence voltage is immaterial. Therefore, the neutral point can be freely chosen. Let

$$\vec{V}_a = 220\angle 0°; \qquad \vec{V}_b = \vec{V}_c = 220\angle 180°$$

Using Eq. (10.114),

$$\begin{vmatrix} \vec{V}_1 \\ \vec{V}_2 \end{vmatrix} = \frac{1}{3}\begin{vmatrix} 1 & \alpha & \alpha^2 \\ 1 & \alpha^2 & \alpha \end{vmatrix}\begin{vmatrix} \vec{V}_a \\ \vec{V}_b \\ \vec{V}_c \end{vmatrix} = \begin{vmatrix} 146.7\angle 0° \\ 146.7\angle 0° \end{vmatrix}$$

The impedance of the machine is

$$Z = 0.1 + 0.13\,\frac{\omega_s}{\omega_r} + j(0.005 \times 377)$$

For positive sequence,

$$\frac{\omega_{1s}}{\omega_{1r}} = \frac{1800}{30} = 60$$

For negative sequence,

$$\frac{\omega_{2s}}{\omega_{2r}} = \frac{-1800}{-3570} = 0.504$$

The sequence currents are now given by

$$\begin{vmatrix} \vec{I}_1 \\ \vec{I}_2 \end{vmatrix} = \begin{vmatrix} \vec{V}_1/Z_1 \\ \vec{V}_2/Z_2 \end{vmatrix} = \begin{vmatrix} 17.6 - j4.2 \\ 6.7 - j77.8 \end{vmatrix}$$

The phase currents can now be evaluated using Eq. (10.116) as

$$\begin{vmatrix} \vec{I}_a \\ \vec{I}_b \\ \vec{I}_c \end{vmatrix} = \begin{vmatrix} 85.6\angle -73.4° \\ 60.4\angle 31.4° \\ 91.2\angle 146.4° \end{vmatrix}$$

(b) For starting with rated voltage,

$$\vec{V}_1 = \frac{460}{\sqrt{3}} \angle 0°$$

$$Z_1 = 0.23 + j1.89$$

$$\vec{I}_1 = 139.5 \angle -83°$$

Thus, the phase c current in (a) is about 65% of the design starting current and about four times the stator current prior to the short.

10.6.1 ▪ Operation with One Phase Open

Symmetrical components also may be used to analyze situations where the phase voltages of the supply are balanced but there is an unbalance in the connections to the motor or in line impedances. Consider, for example, the situation shown in Fig. 10.30(a) where one phase connection to an induction motor is open. From the observation that $\vec{I}_a = 0$, it follows that

$$\vec{I}_1 + \vec{I}_2 + \vec{I}_0 = 0 \tag{10.117}$$

Next, consider the set of voltages between the source and the motor. Note that

$$\vec{V}_{b'b} = \vec{V}_{c'c} = 0$$

From Eq. (10.113), it follows that the symmetrical components of these source-to-motor voltages must have the relations

$$\alpha^2\vec{V}_1 + \alpha\vec{V}_2 + \vec{V}_0 = 0 \tag{10.118}$$

$$\alpha\vec{V}_1 + \alpha^2\vec{V}_2 + \vec{V}_0 = 0 \tag{10.119}$$

Subtracting these two equations from each other gives

$$(\alpha^2 - \alpha)\vec{V}_1 + (\alpha - \alpha^2)\vec{V}_2 = 0 \tag{10.120}$$

from which

$$\vec{V}_1 = \vec{V}_2$$

Inserting this into either equation gives

$$\vec{V}_0 = \vec{V}_1 = \vec{V}_2 \tag{10.121}$$

The system can now be represented by the equivalent circuit connection of Fig. 10.30(b). Because the source has been considered ideal, it has only the positive-sequence voltage $\vec{V}$ and has no impedance to negative- and zero-sequence currents. A more elaborate representation of the source system could have been used if required. The network of Fig. 10.30(b) now can be analyzed to find the sequence currents $\vec{I}_1$, $\vec{I}_2$, and $\vec{I}_0$ into the machine. From these, the phase currents can be evaluated.

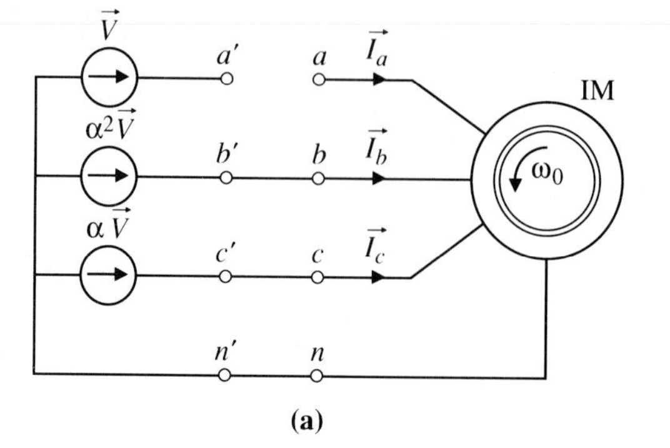

(a)

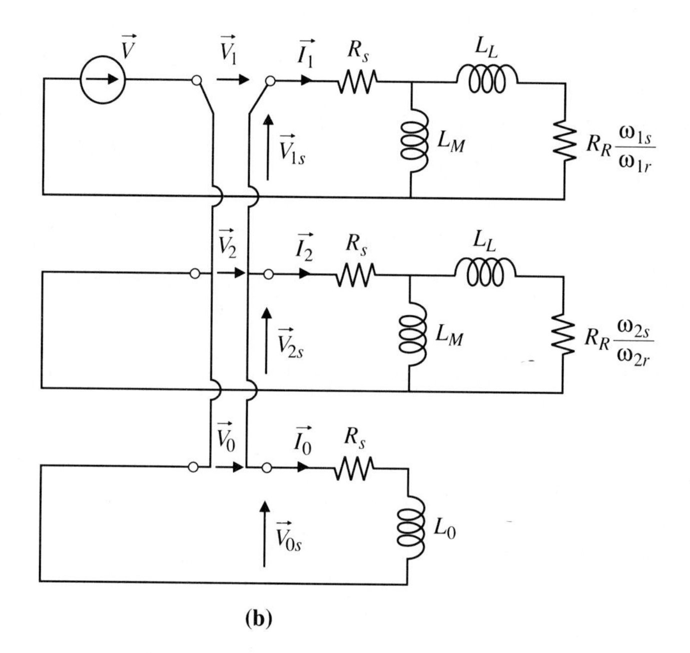

(b)

Figure 10.30 (a) Induction machine with one phase open. (b) Interconnection of sequence equivalent circuits to represent condition of (a).

In the previous section it was noted that the zero-sequence impedance is generally small. Examination of Fig. 10.30(b) shows that the zero-sequence impedance Z_0 is connected in parallel with the negative-sequence impedance Z_2 and the combination is connected in series with the positive-sequence circuit. Because Z_0 is small, the voltages $\vec{V}_1$, $\vec{V}_2$, and $\vec{V}_0$ are also small at any value of speed. Thus, the positive-sequence current in the stator will be little changed because of the opening of one phase; however, the phase currents will be distinctively different after the phase is opened.

Following the reasoning developed earlier for steady-state operation, the torque is equal to the power crossing the air gap divided by the synchronous mechanical velocity. Under unbalanced conditions, there is a power P_{1M} per phase at the air gap of the positive-sequence circuit and a power P_{2M} at the air gap of the negative-sequence circuit. The net torque in the positive direction of rotation is then

$$T = \frac{3p}{2\omega_s}(P_{1M} - P_{2M}) \qquad \text{N}\cdot\text{m} \tag{10.122}$$

If the motor in Fig. 10.29 has no neutral, the connection to the zero-sequence circuit in Fig. 10.30(b) can be omitted. It can be seen that, at zero speed, the positive- and negative-sequence circuits are now effectively in series opposition with $\vec{I}_2 = -\vec{I}_1$. Under starting conditions, the positive- and negative-sequence air-gap powers will be equal, resulting in no net starting torque. However, if the machine is given a starting thrust in either direction of rotation, the effective rotor circuit resistance in that sequence will increase while that in the other sequence will decrease. The result will be a torque in the direction of rotation. Near synchronous speed, the torque will be only slightly less than it would have been for phase a connected because the power in the sequence opposite to rotation will be relatively small. This provides an alternate explanation for the behavior of the single-phase induction motors which were introduced in Section 5.9.

The technique demonstrated in this example of an open phase can be applied in a wide variety of unbalanced operating conditions. The general approach is to focus on the point of unbalance and first state the three constraints on the phase voltages and/or currents that occur at that point. These constraints can next be converted into relations among the symmetrical components of voltage and current at that point. Usually, these relations then can be represented as connections among the sequence equivalent circuits of the balanced parts of the system.

10.7

Practical Windings and Space Harmonics

The analysis of this chapter has so far been based on an assumption of sinusoidally distributed windings. Even at best, a compromise must be made with this idealization because the number of slots is limited. One approximation to a sinusoidal winding distribution was shown in Fig. 5.3(a). This concentric winding is particularly useful in small machines with relatively few slots. With larger machines, the production and installation of coils is greatly simplified if all the coils are identical in shape and in number of turns. Although this might appear as a severe restriction, we shall find that approximately sinusoidal fields can be produced by identical-coil phase windings that are far from sinusoidal in angular distribution of conductors.

Consider first the extreme case of a three-phase winding in which each phase has only one concentrated coil of N turns. This is shown in Fig. 10.31(a) as a rotor winding for convenience in drawing. The mmf distribution across the air gap for phase a is shown in Fig. 10.31(b). This square wave can be expressed as a Fourier series of angular-space harmonics:

$$\mathfrak{F}_{a\theta} = \frac{2}{\pi} N i_a \left[\sin\theta + \frac{\sin 3\theta}{3} + \frac{\sin 5\theta}{5} + \ldots \right] \quad \text{A} \tag{10.123}$$

If $i_a = \hat{i}\sin\omega t$, the mmf of phase a can be expressed as

$$\mathfrak{F}_{a\theta} = \frac{2}{\pi} N i \sum_{h=1}^{h=\infty} \sin\omega t \, \frac{\sin h\theta}{h} \quad \text{A} \tag{10.124}$$

where h is a series of odd numbers indicating the order of the space harmonic. If phases b and c are displaced in angular space by $2\pi/3$ and $4\pi/3$ rad with respect to phase a and the currents in phases b and c are delayed in phase angle by similar angles, the total mmf will be

$$\mathfrak{F}_{\theta} = \frac{2N\hat{i}}{\pi h} \sum_{h=1}^{h=\infty} \left[\sin\omega t \, \frac{\sin h\theta}{h} + \sin\left(\omega t - \frac{2\pi}{3}\right) \sin h\left(\theta - \frac{2\pi}{3}\right) \right.$$
$$\left. + \sin\left(\omega t - \frac{4\pi}{3}\right) \sin h\left(\theta - \frac{4\pi}{3}\right) \right] \quad \text{A} \tag{10.125}$$

After some manipulation, the fundamental space distribution for $h = 1$ is found to be

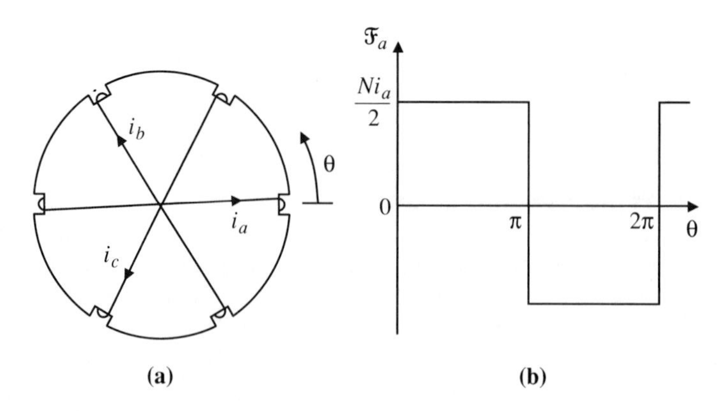

Figure 10.31 (a) Concentrated winding of N turns per phase. (b) Magnetomotive force distribution for phase a.

$$\mathfrak{F}_{1\theta} = \frac{3N\hat{i}}{\pi} \cos(\omega t - \theta) \qquad \text{A} \tag{10.126}$$

Comparison of this with Eq. (5.9) shows that the concentrated windings of N turns produce the same fundamental rotating field as would $(4/\pi)N$ sinusoidally distributed turns. This represents a significant saving in the required number of turns and a corresponding reduction in winding resistance.

Of the various space harmonic terms, all those in which h is divisible by three sum to zero. For the others,

$$\begin{aligned} \mathfrak{F}_{h\theta} &= \frac{3N\hat{i}}{\pi h} \cos(\omega t - h\theta) \qquad \text{for } \frac{h-1}{3} = \text{integer} \\ &= \frac{3N\hat{i}}{\pi h} \cos(\omega t + h\theta) \qquad \text{for } \frac{h+1}{3} + \text{integer} \end{aligned} \tag{10.127}$$

The magnitude of the harmonic mmf is seen to be inversely proportional to the harmonic order h. These harmonics, particularly those of lower order, would have a significant effect on motor performance if only concentrated windings were used.

In practice, it is usual to arrange each phase winding of N turns in a number c of coils distributed in slots over a 60° band on each side of the machine, as shown in Fig. 10.32(a) for one phase of a two-pole machine. Each of these coils produces an mmf with a rectangular form that can be expressed as a Fourier series of odd harmonics as in Eq. (10.123). The space phase displacement between successive coils is $\alpha = 2\pi/s = \pi/3c$, where s is the total number of slots in the two-pole machine. From the geometrical construction of Fig. 10.32(b), note that the fundamental space vector of each coil forms a chord of a circle over an angle α while the total mmf vector is a chord over an angle $c\alpha$. The result is that the fundamental mmf field produced by the winding is reduced with respect to a concentrated winding by a distribution factor k_{1d} where

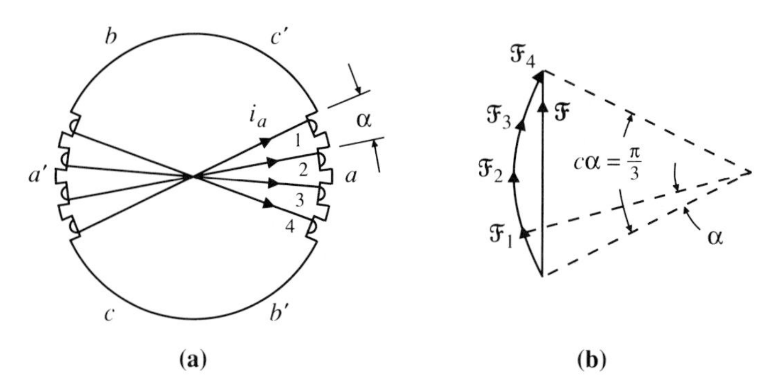

Figure 10.32 (a) Distributed phase coils ($c = 4$). (b) Summation of fundamental space vectors of coil magnetomotive forces.

$$k_{1d} = \frac{\sin(c\alpha/2)}{c\sin(\alpha/2)} = \frac{0.5}{c\sin(\pi/6c)} \tag{10.128}$$

A distributed winding of N turns therefore produces the same fundamental space wave as a concentrated winding of $k_{1d}N$ turns or a sinusoidally distributed winding of N_e turns where

$$N_e = \frac{4}{\pi} k_{1d} N \tag{10.129}$$

For a three-phase winding, k_{1d} is between 0.966 and 0.955 for $c > 1$.

The space vectors of the unwanted space harmonics are displaced from each other by an angle $h\alpha$ for the hth space harmonic. Thus, the distribution factor for the hth harmonic is

$$k_{hd} = \frac{\sin(hc\alpha/2)}{c\sin(h\alpha/2)} = \frac{\sin(h\pi/6)}{c\sin(h\pi/6c)} \tag{10.130}$$

This harmonic distribution factor is always less than unity and may be made small for certain space harmonics by appropriate choice of the number c.

An effective means of minimizing one or more space harmonics is the use of coils that span less than π rad for a two-pole machine (or $2\pi/p$ rad for a p-pole machine). For an s-slot, two-pole machine, the span of a coil may be made $\pi - \beta$ where β is a multiple of $2\pi/s$ rad. This is known as *short pitching*. Normally, one side of each coil is placed in the bottom half of a slot while the other side is placed in the top half of a slot. For example, the coil span can be shortened by one slot by shifting all the top coil sides by one slot. One effect is to reduce the mmf for the hth space vector by a pitch factor k_{hp} where

$$k_{hp} = \cos\frac{h\beta}{2} \tag{10.131}$$

Another effect is to reduce the length of the end connections and therefore the mass and loss of the winding. By proper choice of the angle β, the pitch factor for a particular harmonic can be made small or even equal to zero. For example, all fifth harmonic mmf can be eliminated by making $\beta = \pi/5$ or 30°. In a two-pole, 36-slot machine, this would involve short pitching each coil by 3 slots. The effect of this on the fundamental mmf would be relatively small because $\cos(\pi/12) = 0.966$.

The effects of coil distribution and short pitching can be combined into a single winding factor k_{hw} where

$$k_{hw} = k_{hd}k_{hp} \tag{10.132}$$

for harmonic order h. A phase winding of N distributed and short-pitched turns then has an equivalent number of fundamental sinusoidally distributed turns of

$$N_{se} = \frac{4}{\pi} k_{1w} N \tag{10.133}$$

In a p-pole machine, all the coils of a particular phase are normally connected in series. Both ends of each phase winding may be brought out to terminals. More commonly for an induction machine, the winding is internally connected in star or delta and only three terminals are brought out.

Space harmonics of mmf can produce air-gap flux that can link with a squirrel-cage winding to produce torque. Generally, these torque components are undesirable and should be minimized. The nature of these components can be visualized by examination of the mmf expressions of Eqs. (10.126) and (10.127). For a two-pole machine, the maximum positive value of the fundamental mmf component occurs at an angle

$$\theta_1 = \omega_s t \tag{10.134}$$

where ω_s is the stator frequency. The hth space harmonic in a two-pole machine produces $2h$ poles. From Eq. (10.127), a maximum positive value for one of these harmonic poles occurs at an angle

$$\theta_h = \pm \frac{\omega_s t}{h} \tag{10.135}$$

Thus, this harmonic field rotates at $1/h$ of the angular velocity of the fundamental field. Its synchronous speed is $1/h$ of the fundamental synchronous speed. The fifth-space harmonic field rotates in a direction opposite to that of the fundamental field while the seventh harmonic rotates in the same direction as the fundamental. The effect on the speed-torque relation for an induction machine is shown in Fig. 10.33 in somewhat exaggerated form for a poorly designed winding. One effect of the seventh-harmonic torque is that a mechanical load that would have been expected to be accelerated from standstill might lock in at a speed of about one-seventh of synchronous speed.

A further effect of space harmonics of mmf is that they may produce flux that links with the stator winding but not with the rotor winding. This is, by definition, a leakage flux and it is proportional to the stator current. It contributes to the total leakage inductance of the motor, reducing its maximum torque capability and reducing its stator power factor.

10.8

Linear Induction Motors

In some applications, *linear* or *translational motion* is desired in contrast to the rotating motion normally associated with electric motors. An example is the propulsion of a vehicle such as an electric bus or train. While the propulsion can be provided by a rotating machine coupled through gears to the drive wheels, there are significant advantages to a drive system that does not depend on the

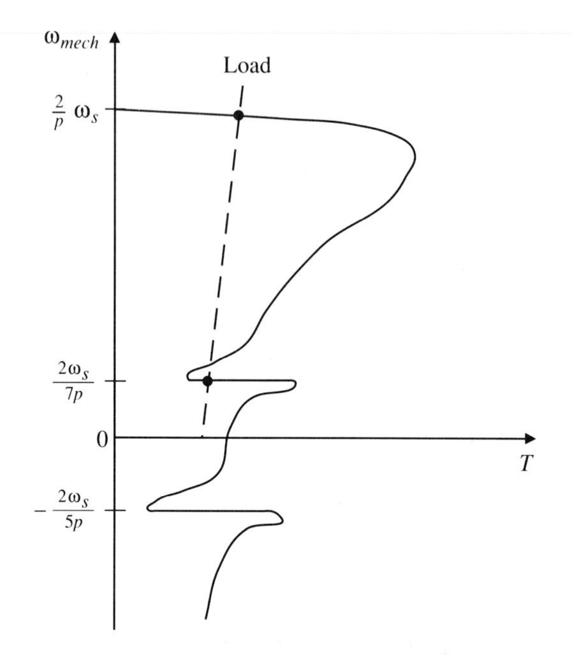

Figure 10.33 Speed-torque relation with large fifth and seventh space harmonic effects.

friction between wheels and track. Steeper slopes can be climbed even in wet or icy weather. Braking can be achieved with minimal wheel wear and can be made more reliable.

The major properties of a linear induction motor can be appreciated by evolution of the cylindrical machine shown in Fig. 10.34(a). The stator has sinusoidally distributed windings of which only the center conductor is shown. The rotor of this machine consists of an iron core surrounded by a sheet of aluminum or copper conductor. Suppose the cylindrical machine is cut along the dotted axis shown and unrolled. The part that was the stator becomes the two-pole primary of the linear induction motor of Fig. 10.34(b). This primary might typically be attached to the bottom of the vehicle. The rotor becomes a conductor-iron sheet, usually somewhat wider than the primary, which is typically attached to the road bed. For this arrangement, electric power must be conveyed to the vehicle through collectors on rails or overhead wires. An alternative arrangement for traction application provides a conductor-iron sheet on the underside of the vehicle with sections of primary located at intervals along the track.

If a three-phase supply with angular frequency ω_s is connected to the sinusoidally distributed primary windings, a sinusoidal flux-density wave will be produced across the air gap to the secondary conductor-iron sheet. If the sequence is *a-b-c*, this wave will travel to the left in Fig. 10.34(b) at a velocity

$$v_s = \frac{z\omega_s}{2\pi} = zf_s \qquad \text{m/s} \tag{10.136}$$

where z is the wavelength of a two-pole primary section and f_s is the primary frequency in hertz. This traveling wave of flux density will induce a sinusoidally distributed voltage in the secondary conductor sheet, the magnitude of which will be proportional to the difference between the synchronous wave velocity v_s and the relative velocity v of the primary with respect to the secondary. The resultant current density across the conducting sheet will interact with the flux density to produce a thrust to the left on the secondary and to the right on the primary. Thus, ideally, the thrust is expected to be proportional to $(v_s - v)$.

Conceptually, a linear induction motor can be represented by the same configuration of equivalent circuit as has been developed for a rotating induction machine, as in Fig. 10.29. By analogy, the thrust F is ideally expected to be given by

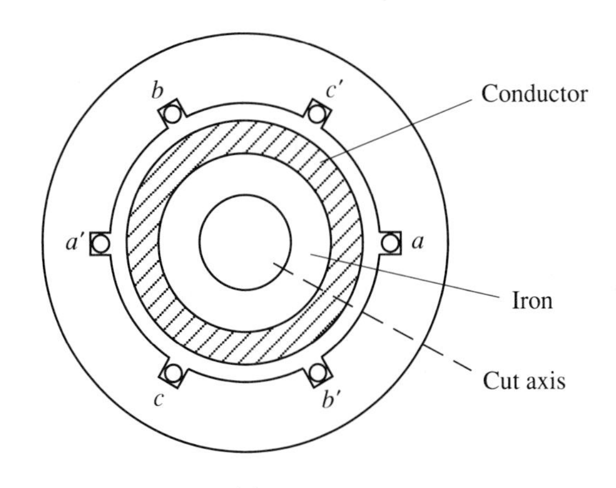

(a)

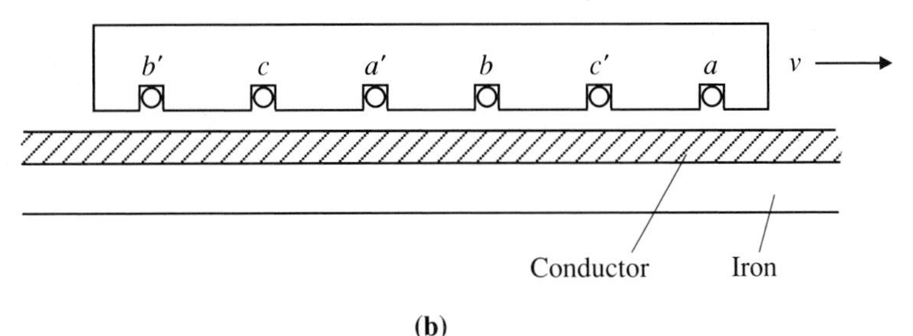

(b)

Figure 10.34 (a) Rotating induction machine. (b) Linear induction machine.

$$F = \frac{air\ gap\ power}{v_s} \quad \text{N} \tag{12.137}$$

Practically, a linear motor differs from a rotating motor in a number of ways. The number of poles in the primary need no longer be a multiple of two. Typically, primaries with five, six, seven, or eight poles are constructed; fractional poles are also possible.

The primary of the linear induction motor has new secondary conductor entering its leading edge and has secondary conductor leaving its lagging edge. By Lenz's law, the secondary current in the entry region acts to prevent the buildup of gap flux. As a result, the average gap flux density for the first one or two poles near the entry edge may be significantly lower than under later poles in the motor. Also, by the same Lenz's law, a current persists in the secondary conductor, tending to maintain the flux density near the exit edge. This current produces resistive losses without corresponding development of thrust. Therefore, these end effects cause the linear motor to have a lower thrust and a lower efficiency than would be expected from the ideal relation of Eq. (10.137). These effects are naturally more pronounced at high speed.

Large linear motors have been used in transportation, materials handling, extrusion processes, and the pumping of liquid metals. Small linear motors are used as sliding door closers and curtain pullers, for example. Typical speed-thrust curves are shown in Fig. 10.35. The maximum thrust normally occurs near standstill for small motors and is of the order of 5 kN/m^2.

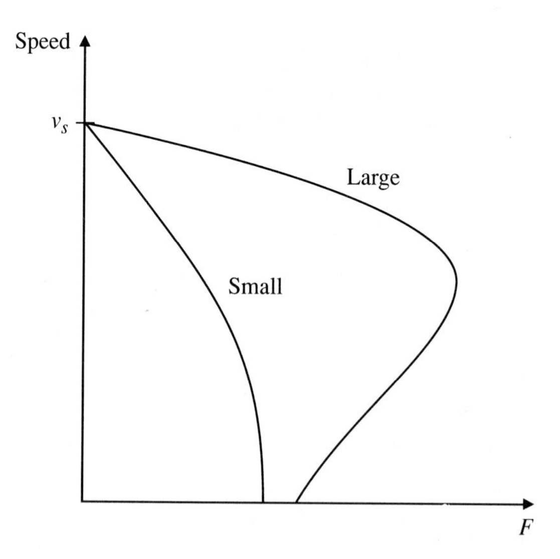

Figure 10.35 Typical speed-thrust relations for large and small linear induction motors.

PROBLEMS

10.1 A 460-V, 1770-r/min wound rotor induction motor has the following parameters, all referred to the stator number of turns:

$R_s = 0.225$	Ω	$L_{\ell s} = 1.89$	mH
$R_r = 0.45$	Ω	$L_{\ell r} = 1.89$	mH
$L_m = 73.7$	mH	$N_{se}/N_{re} = 1.88$	

(a) Develop a transient Γ-model equivalent circuit for the machine with the rotor short-circuited.

(b) Suppose a balanced three-phase resistor with 0.3 Ω/phase is connected to the rotor terminals. Develop an appropriate transient Γ model.

(Section 10.1.1)

10.2 For the machine of Problem 10.1, the rotor circuit is open circuited. A three-phase, 60-Hz sinusoidal source of 460 V rms (L-L) is applied to the stator terminals at the instant that the voltage on phase a is maximum positive.

(a) Derive an expression for the stator current space vector.

(b) Plot the three stator phase current waveforms for at least one cycle.

(Section 10.1.1)

10.3 The induction machine for which data are given in Problem 10.1 has its rotor windings short-circuited. The 460-V, 60-Hz supply is applied to the stator terminals at time zero. Estimate the peak transient torque produced by the machine and plot this torque for the first few cycles. The effect of the magnetizing inductance may be ignored.

(Section 10.1.2)

10.4 A three-phase, four-pole, 60-Hz wound-rotor induction machine is to be used as a phase shifter. The leakage reactance is 0.16 Ω/phase and the magnetizing reactance is 2.1 Ω/phase. The winding resistance may be ignored. The effective stator-to-rotor ratio (including the ratio γ) is 1:0.8. The stator is connected to a 230-V (L-L), three-phase supply. The rotor is displaced in the direction of field rotation through 15 mechanical degrees from alignment with the corresponding stator phase windings. Determine the internal voltage (L-N) phasor and the internal impedance/phase of a single-phase Thevenin equivalent circuit as seen from the rotor terminals.

(Section 10.1.3)

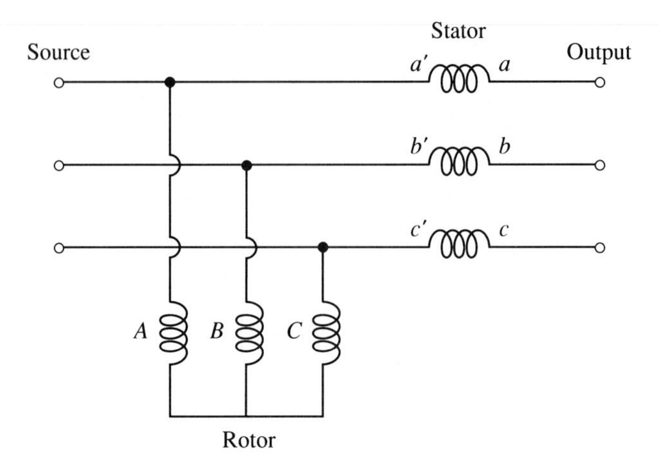

Figure 10.36 Induction regulator connection for Problem 10.5.

10.5 A three-phase wound-rotor induction machine can be operated as an induction regulator to provide a source of variable three-phase voltage at line frequency. The connection of the windings is as shown in Fig. 10.36. All six terminals of the three stator phase windings must be available for this connection. Consider a machine with the following parameters, referred to the stator turns: $L_L = 5.3$ mH, $L_M = 40$ mH, R_R and R_s negligible, $\gamma N_{se}/N_{re} = 1.2$. The rotor side (or left side) of the system in Fig. 10.36 is connected to a 220-V (L-L), three-phase, 60-Hz source.

(a) Determine the maximum and minimum values of the rms output voltage per phase on open circuit.

(b) Determine the internal impedance/phase of the regulator.

(Section 10.1.3)

10.6 A two-pole, 60-Hz wound-rotor motor has a leakage reactance of 0.2 Ω/phase and a magnetizing reactance of 2.0 Ω/phase. The resistances of the windings are negligible and the effective turns ratio $\gamma N_{se} : N_{re}$ is unity. The machine is employed as a variable three-phase inductive load by connecting its stationary windings as shown in Fig. 10.37.

(a) Derive an expression for the impedance per phase of the machine as a function of the rotor angle β.

(b) What should be the angle β if the load is to take 25 kVA from a three-phase, 115-V (L-L), 60-Hz source?

(Section 10.1.3)

10.7 A three-phase, six-pole, wound-rotor motor is to be employed as a variable-frequency source. For this purpose, the stator is connected to

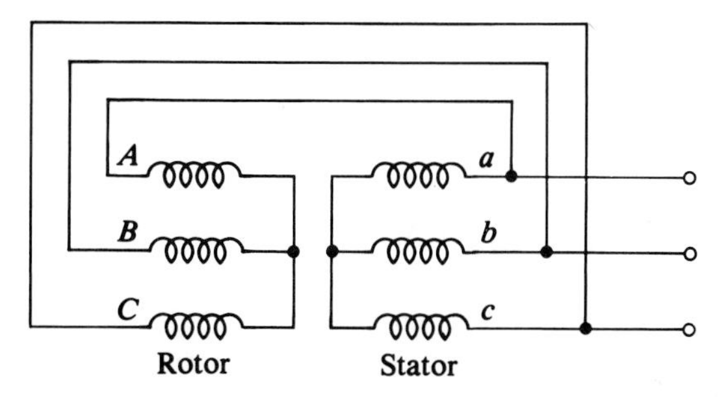

Figure 10.37 Connection of induction machine as a variable inductor for Problem 10.6.

a 60-Hz, 440-V (L-L), three-phase supply and the machine is driven at a variable speed by a prime mover. When the rotor is stationary, the open-circuit, line-to-line voltage at the rotor terminals is 220 V (L-L).

(a) Determine the speed range required to give a frequency range of 20 to 150 Hz.

(b) Determine the corresponding range of open-circuit output voltage (L-L).

(c) Assuming an ideal machine, determine the power supplied or absorbed at both of the frequency limits of (a) by (i) the stator source and (ii) the driving machine when the power delivered to the rotor circuit is 20 kW.

(Section 10.1.3)

10.8 A three-phase, six-pole, 440-V, 60-Hz induction motor has the following per-phase equivalent-circuit parameters all referred to the stator:

$$R_s = 0 \qquad\qquad L_L = 6.27 \quad \text{mH}$$

$$R_R = 0.318 \quad \Omega \qquad L_M = 103 \quad \text{mH}$$

Rotational losses are negligible, and the rotor resistance may be regarded as independent of frequency. When connected to a source of rated voltage and frequency, determine:

(a) the starting torque

(b) the maximum or breakdown torque

(c) the speed at which breakdown torque occurs

(d) the torque per unit of speed change for operation near synchronous speed

(e) the approximate speed at which the motor will drive a load requiring a torque of 2 N·m per rad/s of speed

(f) the rotor frequency for the condition of (e).

(Section 10.1.3)

10.9 A three-phase, six-pole, 220-V, 60-Hz wound-rotor induction motor has an effective stator-to-rotor ratio $\gamma N_{se} : N_{re}$ of 1.25. The machine may be regarded as ideal. A balanced wye-connected load of 3 Ω resistance in parallel with 2200 μF capacitance in each phase is connected to the rotor terminals. When the machine is rotating at 350 r/min, with rated stator voltage and frequency, determine:

(a) the effective impedance per phase as seen from the stator terminals

(b) the total power delivered by the supply

(c) the power delivered to the rotor circuit

(d) the mechanical power

(e) the shaft torque.

(Section 10.1.3)

10.10 A water-wheel turbine that develops a mechanical power of 1 MW at a speed between 600 and 625 r/min drives a three-phase, 12-pole induction generator. The stator windings of the generator are connected to a 4000-V (L-L), 60-Hz, three-phase distribution system. The magnetizing reactance of the induction machine is 33 Ω/phase, the leakage reactance is 3 Ω/phase, and the rotor resistance as seen from the stator is 0.5 Ω/phase. The resistance of the stator windings is negligibly small, and other losses in the generator may be ignored.

(a) Determine the speed of the turbine when the generator is delivering 1 MW to the electrical system.

(b) What value of capacitance per phase connected in wye at the stator terminals would be required to bring the terminal power factor to unity?

(Section 10.1.3)

10.11 For comparative purposes, the parameters of an induction machine are usually best expressed in per unit of the ratings of the machine. A 60-Hz, six-pole, 440-V (L-L), 12.1-A, 1170-r/min, 7.5-kW induction machine has a stator resistance of 0.15 Ω, a leakage reactance of 4.3 Ω, a magnetizing reactance of 48 Ω, and an effective rotor resistance of 0.4 Ω.

(a) Determine the base values of the voltage (L-L), phase current, impedance, frequency, synchronous speed, power per phase, total apparent power, and torque.

(b) Determine the per unit values of the equivalent circuit parameters.

(c) Determine the per unit values of the speed and the output torque under rated load conditions.

(d) What determines the ratio of the rated torque to the base torque?

(Section 10.1.3)

10.12 A variable-speed drive consists of a squirrel-cage induction motor supplied with three-phase, near-sinusoidal voltage of controllable magnitude at variable frequency from an electronic inverter. The motor has a rating of 460 V (L-L), 60 Hz, 100 kW, 1770 r/min, 140 A. The mechanical load requires a torque of 3 N·m per rad/s of speed. Stator resistance, leakage inductance, and losses may be ignored.

(a) To drive the load at a speed of 1200 r/min, what should be rms voltage (L-L) and the frequency of the inverter output?

(b) If the magnetizing inductance of the motor is 2.5 pu, what will be the source power factor for the condition of (a)?

(Section 10.2.1)

10.13 A three-phase, four-pole, 220-V, 60-Hz squirrel-cage induction motor has the following equivalent circuit parameters:

$$R_s = 0 \qquad X_L = 0.6 \quad \Omega$$

$$R_R = 0.3 \quad \Omega \qquad X_M = 15 \quad \Omega$$

The motor is supplied from an inverter that provides near-sinusoidal, variable-phase voltages with variable frequency. The ratio of source voltage to source frequency is kept constant at a value appropriate for the machine ratings.

(a) Sketch speed-torque curves for the system at inverter frequencies of 60, 20, and 3 Hz.

(b) Estimate the inverter frequency to drive a load requiring a constant torque of 150 N·m at a speed of 100 r/min.

(c) Determine the required source voltage for the operating condition of (b).

(Section 10.2.1)

10.14 An induction motor is rated at 2300 V (L-L), three-phase, 60 Hz, 1770 r/min, 150 A. Its no-load current is 50 A. Stator resistance, leakage inductance, and rotational losses may be ignored.

(a) Draw an equivalent circuit for the motor and give the values of its parameters.

(b) Suppose the motor is supplied from a simple voltage source inverter, a controlled rectifier, and a 2300-V (L-L), 60-Hz, three-phase supply.

Find the maximum no-load speed at which the motor can be operated with rated stator flux linkage.

(Section 10.2.1)

10.15 An induction motor has the following ratings: 4.6 kV (L-L), 60 Hz, six pole, 10 MW, 1192 r/min. It is to drive a compressor requiring a torque of 75 kN·m at a speed of 1000 r/min. A simple voltage-source inverter supplied from a three-phase, phase-controlled rectifier is to be used. The available utility supply is at 4.6 kV (L-L), three-phase, and 60 Hz. Making any reasonable assumptions, estimate:

(a) the required frequency of the inverter

(b) the required direct-link voltage

(c) the delay angle of the rectifier.

(Section 10.2.1)

10.16 A refrigerator compressor is to be driven at variable speed over a range of 500 to 3000 r/min. The mechanical power required is 2 kW at maximum speed. A 230-V, 60-Hz single-phase utility supply is available. It is proposed to use a controlled bridge rectifier and a simple voltage inverter to supply a three-phase, four-pole squirrel-cage induction motor. At rated load, the rotor frequency of the motor may be assumed to be 5% of the stator frequency. The stator resistance, leakage inductance, and rotational losses may be ignored.

(a) Specify an appropriate value of rated voltage for the motor.

(b) Estimate the rated phase current of the motor assuming the magnetizing current to be 0.5 pu.

(c) Determine the maximum frequency of the inverter.

(d) Assuming the compressor torque to be independent of speed, determine the delay angle of the rectifier for the minimum speed condition.

(Section 10.2.1)

10.17 A variable-speed drive consists of a 460-V (L-L), four-pole, 60-Hz, 50-kW, 1770-r/min induction motor supplied from a controlled rectifier and a simple voltage-source inverter. The electrical supply is at 460 V (L-L), three phase, and 60 Hz.

(a) The no-load current of the motor with rated stator voltage and frequency is 24 A. All losses except rotor losses can be ignored and the leakage inductance can be neglected. Estimate the fundamental stator current and power factor when this motor is delivering rated torque.

(b) The starting current for this motor is listed by the manufacturer as 600 A. Estimate the fifth and seventh harmonic stator currents when the motor is providing rated torque at 1200 r/min.

(c) By what factor should the motor torque be derated because of the fifth and seventh harmonics in the stator current?

(d) Find the sixth harmonic torque due to the interaction of rated fundamental flux linkage with the fifth harmonic current found in (b).

(Section 10.2.2)

10.18 A 460-V (L-L), 60-Hz, 1770-r/min induction motor has the following parameters for a T-form equivalent circuit, all referred to the stator turns:

$$R_s = 0.18 \quad \Omega \qquad L_{\ell s} = 1.89 \quad \text{mH}$$
$$R_r = 0.22 \quad \Omega \qquad L_{\ell r} = 1.89 \quad \text{mH}$$
$$L_m = 33.7 \quad \text{mH}$$

(a) Develop a transient equivalent circuit for the motor of the inverse-Γ form (as in Fig. 10.20).

(b) Suppose this motor is supplied from a current source inverter using the system shown in Fig. 10.22. If the stator current is 50 A and the rotor frequency is held at its value for rated operation of the motor, determine the torque produced.

(c) Describe why the torque in (b) is independent of the speed.

(Section 10.3)

10.19 A Class-D, four-pole, three-phase squirrel-cage induction motor is rated at 415 V (L-L), 50 Hz, 20 A. It is supplied from a three-phase, 415-V, 50-Hz source through a variable-voltage controller. The motor drives a fan for which the power is proportional to the cube of the speed. The motor parameters in per unit are $L_M = 2.5$ and $R_R = 0.1$. The parameters L_L and R_s as well as the rotational losses may be ignored.

(a) With rated voltage applied to the motor, the steady-state phase current is 15 A. Estimate the speed of the fan for this condition.

(b) Determine the rms value of the component of fundamental voltage (L-L) required to operate the fan at a speed of 1200 r/min.

(c) For the condition of (b), is the motor operating within its stator current rating?

(Section 10.4)

10.20 A fan requires a power of 20 kW at its rated speed of 1100 r/min. It is to be driven by a wound-rotor induction motor fitted for rotor-power recovery using the system shown in Fig. 10.25. The available utility supply is at 460 V (L-L), 60 Hz, three phase. The motor may be considered as ideal.

(a) Choose an appropriate number of poles for the motor.

(b) If the fan is to have a minimum speed of 750 r/min, determine the appropriate value for the effective turns ratio N_{se}/N_{re}. The firing angle of the inverter should not exceed 170°.

(c) To operate the fan at its rated speed, what should be the firing angle of the inverter?

(d) At rated speed, determine the value of the direct link current.

(e) If the magnetizing current of the motor is assumed to be 0.4 pu based on its rating, estimate the stator current of the motor when driving the fan at rated speed.

(Section 10.5)

10.21 A three-phase, 220-V (L-L), 400-Hz, two-pole induction machine has a leakage inductance of 8 mH and a rotor resistance of 4 Ω/phase. Its stator resistance and magnetizing current may be ignored. Suppose a single-phase, 220-V, 400-Hz supply is connected between two stator terminals leaving the third open circuited.

(a) Sketch an equivalent circuit showing the appropriate interconnection of the sequence networks.

(b) Derive an expression for the source current as a function of motor speed. Evaluate the current at standstill.

(c) Derive an expression for the motor torque as a function of speed and use it to evaluate the torque at half synchronous speed.

(Section 10.6)

10.22 A three-phase, six-pole machine stator has 36 slots. The coils each have ten turns. Each coil is short pitched by one slot.

(a) Determine the effective number of turns per phase.

(b) Evaluate the winding factor for the fifth space harmonic.

(Section 10.7)

10.23 A linear induction motor has a wavelength of 480 mm and has five poles. It can be modeled by an equivalent current with the following parameters:

$$R_s = \text{negligible} \qquad L_L = 1.2 \quad \text{mH}$$

$$R_R = 0.28 \quad \Omega \qquad L_M = 6 \quad \text{mH}$$

(a) Determine the synchronous velocity of the motor when operating on a 20-Hz supply.

(b) Determine the thrust, the mechanical power, the input current, and power factor when the motor is operated on a 300-V (L-L), 40-Hz supply and the speed is 45 km/h.

(Section 10.8)

GENERAL REFERENCES

Adkins, B. and Harley, R.G., *The General Theory of AC Machines*, John Wiley and Sons, New York, 1975.

Alger, P.L., *Induction Machines—Their Behavior and Uses*, 2nd ed., Gordon and Breach, New York, 1970.

Bose, B.K., *Power Electronics and AC Drives*, Prentice-Hall, Englewood Cliffs, New Jersey, 1989.

de Jong, H.C.J., *AC Motor Design with Conventional and Converter Supplies*, Clarenden, Oxford, 1976.

Dewan, S.B., Slemon, G.R., and Straughen, A., *Power Semiconductor Drives*, John Wiley and Sons, New York, 1984.

Dubey, B.K., *Power Electronics and AC Drives*, Prentice-Hall, Englewood Cliffs, New Jersey, 1986.

IEEE Standard Test Procedures for Polyphase Induction Motors and Generators, IEEE Standard 112, Institute of Electrical and Electronics Engineers, New York, 1984.

Kovacs, K.P., *Transient Phenomena in Electrical Machines*, Elseviers, Amsterdam, and Publishing House of the Hungarian Academy of Science, 1984.

Krause, P.C., *Analysis of Electric Machinery*, McGraw-Hill, New York, 1986.

Laithwaite, E.R., *Induction Machines for Special Purposes*, George Newness, London, 1966.

Levi, E., *Polyphase Motors*, John Wiley and Sons, New York, 1984.

Nasar, S.A. and Boldea, I., *Linear Motion Electric Machines*, Wiley Interscience, New York, 1976.

Vienott, C.G., *Theory and Design of Small Induction Motors*, McGraw-Hill, New York, 1957.

Yamamura, S., *AC Motors for High Performance Applications*, Marcel Dekker, New York, 1986.

Yamamura, S., *Theory of Linear Induction Motors*, 2nd ed., University of Tokyo Press, Tokyo, 1978.

CHAPTER

11

Synchronous Generators, Motors, and Drives

Synchronous machines are characterized by the direct link between their steady-state mechanical speed ω_{mech} and their supply frequency ω_s. For a p-pole machine, $\omega_{mech} = (2/p)\omega_s$ rad/s. Some basic concepts of synchronous machine operation have been introduced in Chapter 6. This chapter examines in more detail the modeling of a synchronous machine with a view to developing an ability to predict both its transient and steady-state properties. This will be followed by a discussion of specific operating properties that are of particular significance in important areas of application.

A major application area for synchronous machines is that of generators for supply of three-phase electric power to power transmission and distribution networks. The provision of adequate analytical models for these generators is of vital importance to power system engineers who are concerned with supplying a wide range of consumers with reliable, continuous, electric power of near-constant voltage and frequency, in spite of disturbances that continually occur on the system.

Synchronous generators for use in 50- or 60-Hz power systems may have ratings up to more than 1000 MVA. The larger ratings are generally more economical but it is usually not advantageous from a reliability standpoint to have more than about 10% of the total capacity of a system in one machine. Those generators that are driven by steam or gas turbines are usually of cylindrical-rotor construction with two poles operating at 3000 or 3600 r/min, although some four-pole, 1500- to 1800-r/min machines also are used. Generators driven by diesel engines are frequently used in remote locations and typically have four or six poles. Low-head waterwheel turbines usually have optimum speeds in the range 60 to 300 r/min. Synchronous generators for use with these turbines usually have salient-pole rotors with 20 to 120 poles. Impulse turbines are frequently used at high-head hydraulic sites, and these have intermediate to high speeds up to 3600 r/min. Special generators such as those providing 400-Hz supply for aircraft operate at speeds up to 24 000 r/min.

Another application area for synchronous machines is in electric drives, occasionally at constant speed but more frequently at variable speed. These drives vary in power rating from the very small to the very large. Typical application areas include robotics, machine tool drives, disk drives, ship propulsion, and pipeline pumps. All of the various synchronous motor types may be used in drive applications: wound field with either cylindrical or salient-pole rotors, permanent magnet, reluctance, or hysteresis. Each has advantages in particular applications.

Wound-field synchronous motors are larger, more complex, and more expensive than squirrel-cage motors of equal power rating. However, because they may be operated at leading or capacitive power factor they are attractive, particularly for large drives. Permanent-magnet motors are economical, efficient, and capable of good transient response. They are used extensively in low-power drives but the power range is being continually extended to larger power ratings. Reluctance machines have the advantage of simplicity of construction and low cost; they are being used in an increasing range of drive applications.

11.1

Modeling of Synchronous Machines

Whether a synchronous machine is used as a large generator or a small drive motor, prediction of its performance depends on having an adequate analytical model. An elementary equivalent circuit suitable for the steady state was developed in Section 6.2. Attention is now focused on the development of models that can be used for transient as well as steady-state conditions and that can represent the effects of magnetic nonlinearity and saliency.

11.1.1 • Cylindrical-Rotor Machines

Let us consider first a synchronous machine with a three-phase stator winding, and a cylindrical rotor with a single field winding as shown in Fig. 11.1. Our objective is to develop an equivalent circuit model and its accompanying equations from which the transient behavior of this machine can be predicted.

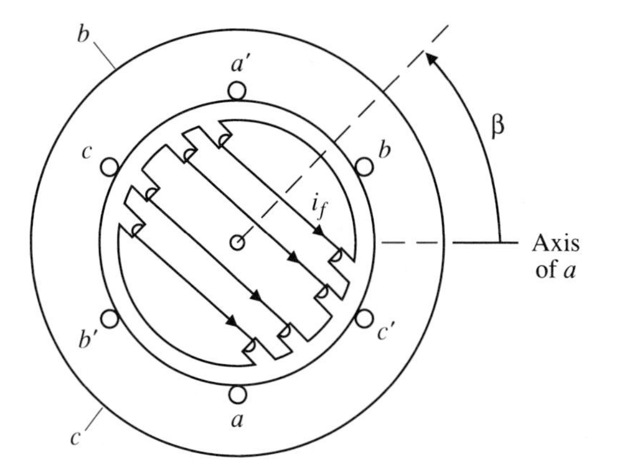

Figure 11.1 Winding relations for cylindrical-rotor, two-pole synchronous machine.

In Section 10.1, a transient model was developed for an induction machine. The structure of a cylindrical-rotor synchronous machine is similar to that of an induction machine with the exception that the synchronous machine has only a single distributed winding on its rotor. Therefore, most of the analysis of Section 10.1 is applicable. The instantaneous stator currents can be represented, from Eq. (10.10), by the space vector

$$\vec{i}_s = i_a + j\,\frac{i_b - i_c}{\sqrt{3}} \qquad \text{A} \tag{11.1}$$

in the stator reference frame.

The rotor of the synchronous machine contains a single winding with an equivalent sinusoidally distributed number of turns N_{fe} carrying a field current i_f. In the induction motor analysis, all space vector quantities were referred to the axis of stator phase a. In the case of the synchronous machine, it is necessary to maintain the identity of the axis of the field winding. On the other hand, our main interest is normally in the stator variables. Therefore, it is usually convenient to refer all rotor variables to the equivalent number of turns N_{se} of the stator. A field current i_f in the N_{fe} turns of a two-pole rotor produces an air-gap mmf of $N_{fe}i_f/2$. From Eq. (5.15) or Eq. (10.17), a vector current $\vec{i}$ in a three-phase winding of N_{se} turns per phase produces an air-gap mmf of $3\,N_{se}\vec{i}/4$. Thus, the effect of the field current in terms of stator turns but referred to the field axis can be represented by the current space vector

$$\vec{i}_F = \frac{2}{3}\,\frac{N_{fe}}{N_{se}}\,i_f\epsilon^{j0} \qquad \text{A} \tag{11.2}$$

If the rotor axis is to be used as a reference, all stator variables must now be referred to this frame of reference. As seen from the rotor axis , the axis of stator phase a is at an angle of $-\beta$. Thus, the stator flux linkage $\vec{\lambda}_s$ can be transformed into a rotor-reference space vector $\vec{\lambda}'_s$ by

$$\vec{\lambda}'_s = \lambda_s \epsilon^{-j\beta} \qquad \text{Wb} \tag{11.3}$$

and a similar expression transforms the stator current vector $\vec{i}_s$ of Eq. (11.1) to

$$\vec{i}'_s = \vec{i}_s \epsilon^{-j\beta} \qquad \text{A} \tag{11.4}$$

The machine may now be represented by the flux-linkage model of Fig. 11.2(a). The magnetizing inductance L_m is the same as was developed for the induction machine in Eqs. (5.50) and (10.20). The stator leakage inductance is $L_{\ell s}$, and $L_{\ell F}$ is the equivalent leakage inductance of the field winding referred to N_{se} turns.

To develop a voltage-current model, we start with the stator voltage relation in the stator frame of reference:

$$\vec{v}_s = R_s \vec{i}_s + p\vec{\lambda}_s \qquad \text{V} \tag{11.5}$$

Expressing this in terms of the transformed variables $\vec{v}'_s$ and $\vec{i}'_s$ gives

$$\begin{aligned} \vec{v}'_s &= \vec{v}_s \epsilon^{-j\beta} \\ &= R_s \vec{i}'_s \epsilon^{j\beta} \epsilon^{-j\beta} + (p\vec{\lambda}'_s \epsilon^{j\beta})\epsilon^{-j\beta} \\ &= R_s \vec{i}'_s + (p + j\omega_0)\vec{\lambda}'_s \qquad \text{V} \end{aligned} \tag{11.6}$$

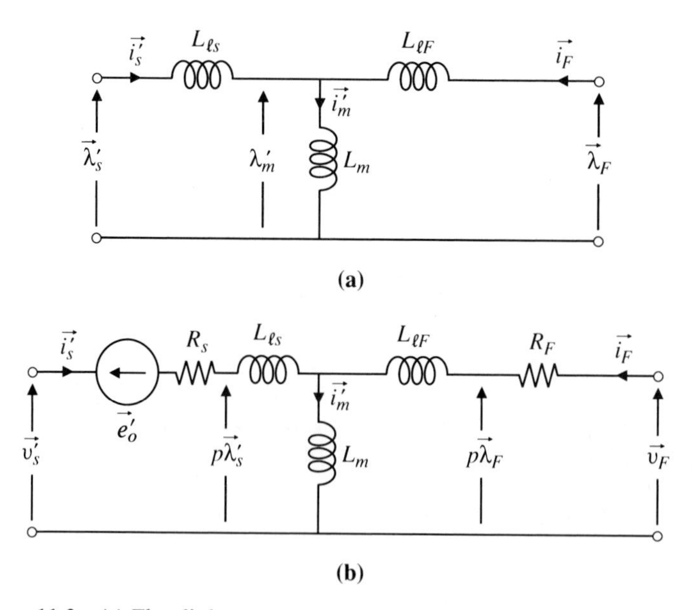

Figure 11.2 (a) Flux linkage-current equivalent circuit. (b) Voltage-current equivalent circuit.

because the speed for a two-pole machine is $p\beta = \omega_0$. The machine may now be represented by the transient equivalent circuit model of Fig. 11.2(b). In this circuit, the equivalent field voltage v_F is related to the actual field voltage v_f by the turns ratio,

$$v_F = \frac{N_{se}}{N_{fe}} v_f \qquad \text{V} \tag{11.7}$$

With constant field current and flux linkage, the field voltage v_f is related to the field resistance R_f by

$$v_f = R_f i_f \qquad \text{V} \tag{11.8}$$

Thus, the equivalent field resistance R_F is related to the resistance R_f by

$$R_F = \frac{v_F}{i_F} = \frac{3R_f}{2}\left(\frac{N_{se}}{N_{fe}}\right)^2 \qquad \Omega \tag{11.9}$$

The source voltage $\vec{e}_0'$ in Fig. 11.2(b) is given by

$$\vec{e}_0' = j\omega_0 \vec{\lambda}_s' \qquad \text{V} \tag{11.10}$$

i.e., it is the voltage generated by rotating through the stator flux linkage λ_s' at speed ω_0.

The instantaneous mechanical power of the machine by analogy with Eq. (10.43) is given by

$$p_0 = \frac{3}{2}\,\Re(\vec{e}_0'^{\,*}\vec{i}_s') \qquad \text{W} \tag{11.11}$$

Thus, the instantaneous torque is

$$\begin{aligned} T &= \frac{p_0}{\omega_{mech}} \\ &= \frac{3p}{4\omega_0}\,\Re(\vec{e}_0'^{\,*}\vec{i}_s') \\ &= \frac{3p}{4}\,\Im(\vec{\lambda}_s'^{\,*}\vec{i}_s') \qquad \text{N}\cdot\text{m} \end{aligned} \tag{11.12}$$

As an example of the use of this circuit in a transient analysis, consider determining the phase voltages of an open-circuited stator when a direct voltage V_f is applied to the field at time $t = 0$ with the machine running at constant speed ω_0. The rotor angle is

$$\beta = \beta_0 + \omega_0 t \qquad \text{rad} \tag{11.13}$$

From Eq. (11.10), the equivalent field voltage referred to the stator turns is

$$\vec{v}_F = \left(\frac{N_{se}}{N_{fe}}\right) V_f \epsilon^{j0} \qquad \text{V} \tag{11.14}$$

From the equivalent circuit of Fig. 11.2(b), with $\vec{i}_s' = 0$,

$$\begin{aligned}\vec{i}_F &= \frac{\vec{v}_F}{R_F + p(L_{\ell F} + L_m)} \\ &= \frac{V_F}{R_F}(1 - \epsilon^{-t/\tau_F})\epsilon^{j0} \qquad \text{A}\end{aligned} \tag{11.15}$$

where $\tau_F = (L_{\ell F} + L_m)/R_F$ is the field time constant with open-circuited stator, commonly referred to as the *open-circuit time constant*. The stator voltage vector referred to the rotor axis is, from Eq. (11.6),

$$\begin{aligned}\vec{v}_s' &= (p + j\omega_0)\vec{\lambda}_s' \\ &= (p + j\omega_0)L_m\vec{i}_F \\ &= L_m\frac{V_F}{R_F}\left[\frac{\epsilon^{-t/\tau_F}}{\tau_F} + j\omega_0(1 - \epsilon^{-t/\tau_F})\right] \qquad \text{V}\end{aligned} \tag{11.16}$$

The real part of this expression is the voltage induced in the stator by application of the field voltage source. The second quadrature term is the voltage generated by rotation. In most large synchronous machines at rated speed, $\omega_0 \gg 1/\tau_F$, and the real part of $\vec{v}_s'$ frequently can be ignored. The stator voltage space vector in its own reference frame is then given by

$$\vec{v}_s = \vec{v}_s'\epsilon^{j\beta} = \vec{v}_s'\epsilon^{j(\beta_0 + \omega_0 t)} \qquad \text{V} \tag{11.17}$$

The individual phase voltages can now be found by resolving $\vec{v}_s$ on to the a, b, and c axes. Figure 11.3 shows typical graphs of $\vec{\lambda}_s'$, $p\vec{\lambda}_s'$, and the phase voltage v_a on a time axis.

For steady-state operation, it is most convenient to have a model that is referred to the stator, preserving the actual stator voltage and current phasors. The space vector $\vec{i}_F$ of rotor current in Eq. (11.2) can be transformed to a vector $\vec{i}_f'$ in the stator reference frame by use of the expression

$$\vec{i}_f' = \vec{i}_F\epsilon^{j\beta} = \frac{2}{3}\frac{N_{fe}}{N_{se}}i_f\epsilon^{j(\omega_s t + \beta_0)} \tag{11.18}$$

because $\omega_0 = \omega_s$ for a two-pole machine in the steady state. This alternating current source now can be represented by the phasor

$$\vec{I}_f = \frac{\sqrt{2}}{3}\frac{N_{fe}}{N_{se}}i_f\epsilon^{j\beta_0} = n_f i_f\angle\beta_0 \tag{11.19}$$

A similar transformation through the angle $\epsilon^{j\beta}$ can be applied to all the machine variables, resulting in the steady-state equivalent circuit of Fig. 11.4(a).

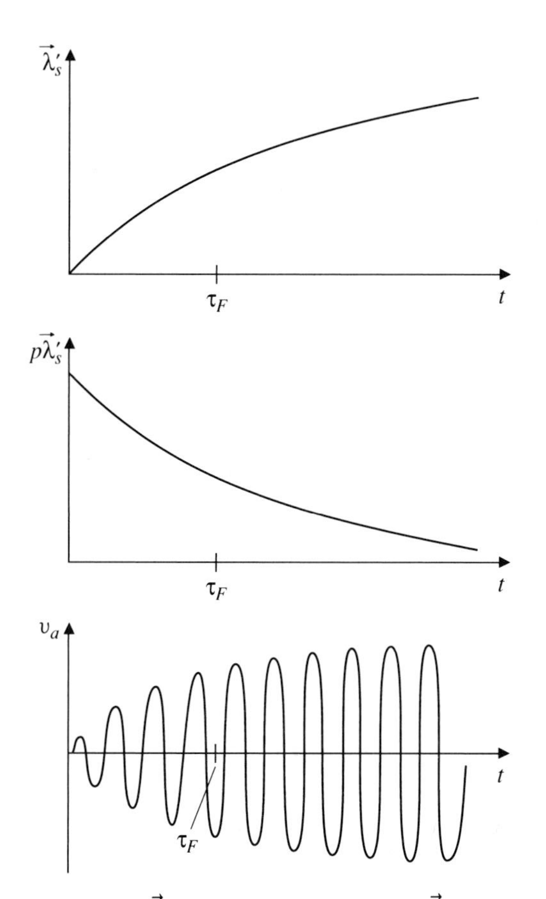

Figure 11.3 Flux linkage $\vec{\lambda}'_s$, its time rate of change $p\vec{\lambda}'_s$, and the phase-a voltage with step field voltage and open-circuited stator.

The parameters $L_{\ell F}$ and R_F of the rotor branch have been retained in this circuit, but because they are in series with a current source they have no effect on the terminal behavior of the machine.

The circuit of Fig. 11.4(a) can be further simplified by the use of Norton's theorem, as shown in Fig. 11.4(b), where

$$\begin{aligned}
\vec{I}_F &= \frac{L_m}{L_{\ell s} + L_m}\vec{I}_f \\
&= \frac{L_m}{L_s} n_f i_f \angle \beta_0 \qquad\qquad (11.20) \\
&= n i_f \angle \beta_0 \qquad \text{A}
\end{aligned}$$

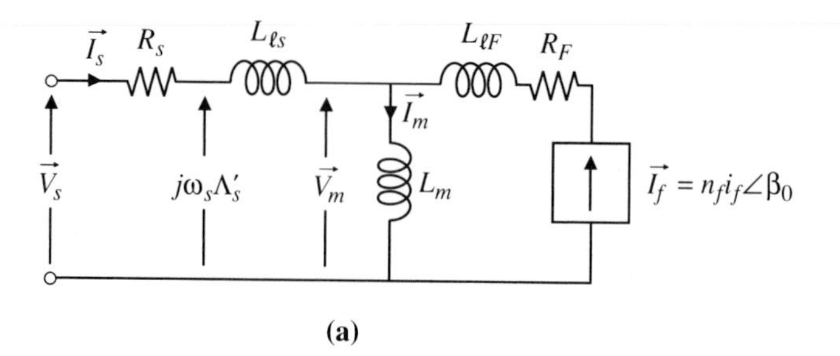

(a)

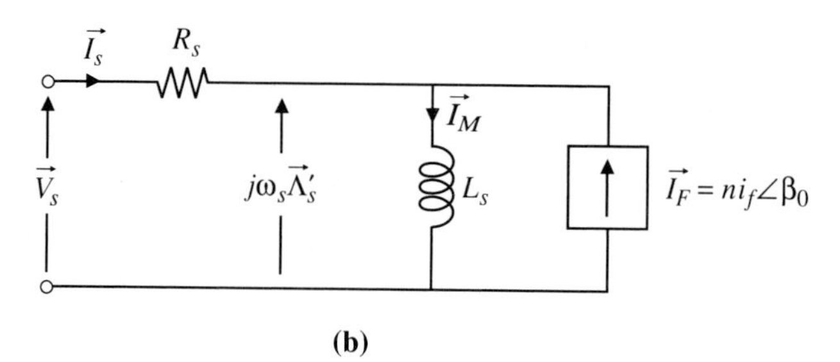

(b)

Figure 11.4 (a) Steady-state equivalent circuit. (b) Simplified form.

in which the effective current ratio is

$$n = \frac{L_m}{L_s} n_f$$

The quantity $L_s = L_{\ell s} + L_m$ is commonly known as the *synchronous inductance* and is the same parameter as the inductance L_M in Chapter 6. Phasor diagrams for a synchronous motor with a somewhat capacitive stator current are shown in Fig. 11.5 for both equivalent circuits of Fig. 11.4.

An expression for the steady-state torque may be obtained by use of Eq. (11.12). In the steady state, the stator voltage is frequently known. Therefore, it is usually convenient to use the stator voltage phasor $\vec{V}_s$ as a reference and assign to it an angle of zero. Ignoring the stator resistance, the stator flux linkage phasor is

$$\vec{\Lambda}_s = \frac{\vec{V}_s}{j\omega_s} \tag{11.21}$$

From the circuit of Fig. 11.4(b), the stator current is

$$\vec{I}_s = \vec{I}_M - \vec{I}_F = \frac{\vec{V}_s}{j\omega_s L_s} - \vec{I}_F \tag{11.22}$$

Thus, noting that the phasor quantities are rms values, from Eqs. (11.12) and (11.19), the torque is

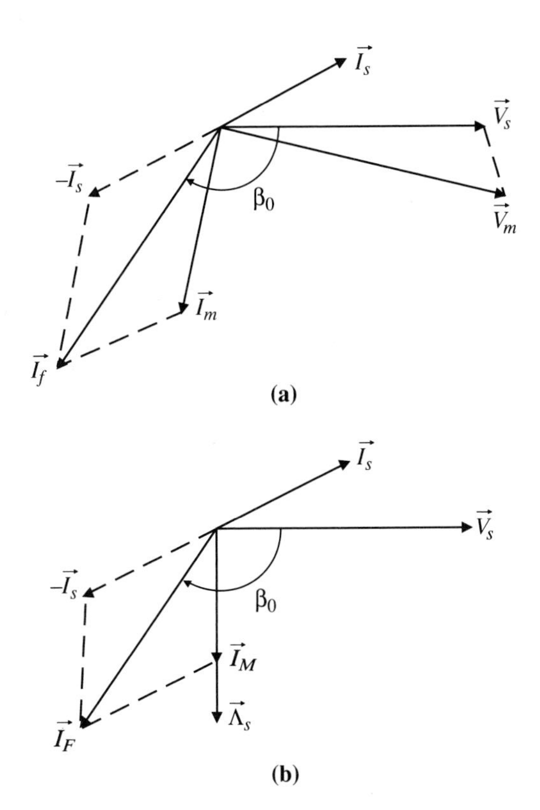

Figure 11.5 Phasor diagram for synchronous motor using (a) circuit of Fig. 11.4(a) and (b) circuit of Fig. 11.4(b).

$$
\begin{aligned}
T &= \frac{3p}{2}\,\mathcal{I}(\vec{\Lambda}_s^*\vec{I}_s) \\
&= -\frac{3p}{2}\,\frac{V_s}{\omega_s}\,I_F\cos\beta_0 \qquad\qquad (11.23) \\
&= -\frac{3p}{2}\,\Lambda_s n i_f\cos\beta_0 \qquad \text{N}\cdot\text{m}
\end{aligned}
$$

This torque expression emphasizes a central feature of a synchronous machine, i.e., that it acts in the steady state as an alternating current source. When connected to a supply of constant alternating voltage and constant frequency, zero torque occurs when the angle β_0 is $-\pi/2$, as shown in Fig. 6.3. In this condition, the field current phasor I_F is colinear with the stator flux linkage $\vec{\Lambda}_s$. Torque in the direction of rotation is produced as a motor when the angle β_0 of this alternating current source $\vec{I}_F$ is delayed. As a generator, the angle of this current source is advanced beyond $-\pi/2$, producing a torque opposite to the direction of rotation.

11.1.2 • Representation of Magnetic Saturation

Modeling of magnetic saturation effects in a cylindrical-rotor synchronous machine has many of the same features as developed for an induction machine in Section 10.1.4. As discussed there, the mmf required by the stator teeth and yoke is dependent on the stator flux linkage $\vec{\lambda}_s$ and can be represented by a nonlinear inductance L_{ty}, as shown in the flux linkage-current model of Fig. 11.6. Similarly, the mmf required by the rotor teeth and core is dependent on the rotor flux linkage $\vec{\lambda}_F$ and can be represented by the nonlinear inductance L_c in Fig. 11.6.

This rather complex model of the saturated machine can be simplified when most of the saturation occurs in either the rotor or the stator. For most operating conditions, the rotor flux linkage is greater than the stator flux linkage in a synchronous machine because the excitation is provided by the field current. Suppose the current component i_{ty} through inductance L_{ty} in Fig. 11.6 can be ignored. The remaining circuit can be simplified without loss of accuracy by use of the inverse-Γ transformation developed in Section 10.3. The result is shown incorporated into a transient voltage-current circuit in Fig. 11.7. By analogy with Eqs. (10.86) to (10.90), the relations involved in the transformation are

$$\gamma = \frac{L_m}{L_m + L_{\ell F}} \tag{11.24}$$

$$\vec{\lambda}'_F = \gamma \vec{\lambda}_F; \qquad \vec{i}'_F = \frac{\vec{i}_F}{\gamma} \tag{11.25}$$

$$L'_L = L_{\ell s} + \gamma L_{\ell F} \tag{11.26}$$

$$R'_F = \gamma^2 R_F \tag{11.27}$$

and

$$L'_M = \gamma L_m \parallel \gamma^2 L_c \tag{11.28}$$

where ∥ signifies "in parallel with." The corresponding steady-state equivalent circuit is shown in Fig. 11.7(b).

In situations where there is significant saturation in both stator and rotor, the equivalent circuits of Fig. 11.7 may still be used as an approximation,

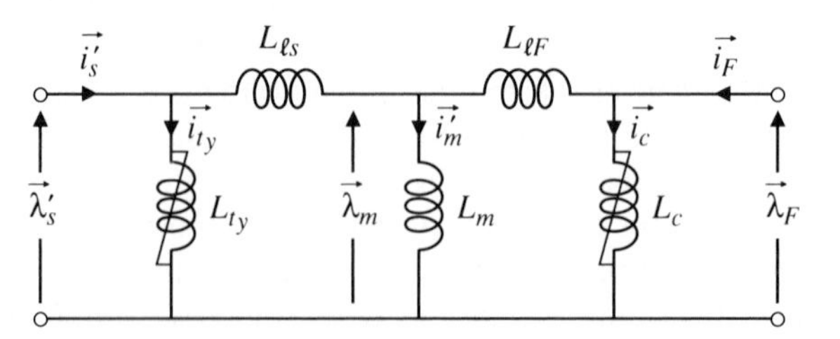

Figure 11.6 Flux linkage-current circuit including magnetic saturation.

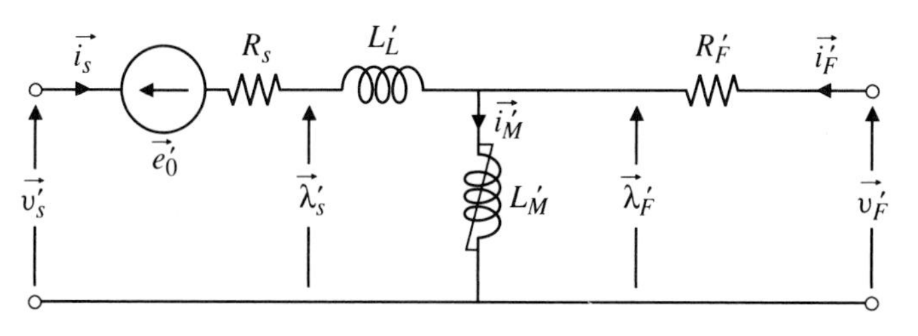

(a)

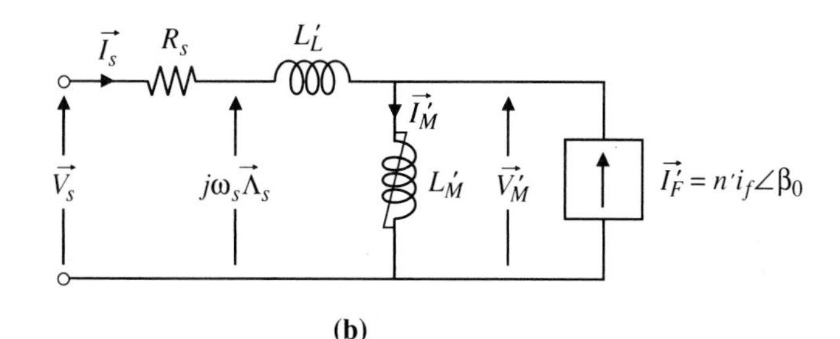

(b)

Figure 11.7 Equivalent circuits including rotor saturation, (a) transient and (b) steady state.

inserting a value of the leakage inductance L'_L somewhat less than given by Eq. (11.26). In the extreme case where all the saturation is in the stator, the appropriate value of leakage inductance approaches zero.

11.1.3 ▪ Modeling of Saliency

High-speed, turbine-driven generators generally have the structure shown in cross section in Fig. 11.1. This can usually be considered as a magnetically cylindrical rotor although the presence of the rotor slots produces a somewhat lower reluctance path for flux in the rotor winding axis than in the axis perpendicular to it. Thus, even this machine has some rotor saliency. Low-speed hydraulic-turbine-driven generators typically have many salient poles with concentrated field windings around each pole, as shown in Fig. 6.5.

To develop an analytical model for a salient-pole machine, consider the two-pole cross section of Fig. 11.8(a). The stator is cylindrical and carries phase windings that will be assumed to be sinusoidally distributed. When the field winding is energized, it produces a nonsinusoidal flux density distribution of the form shown in Fig. 11.8(b). Only the fundamental space component of this wave couples with the sinusoidally distributed stator winding. Therefore, there is a value of current vector $\vec{i}_F$ in the equivalent stator turns N_{se} which produces the same fundamental flux density as field current i_f in the N_f field turns. The effective current ratio for a transient space-vector model using peak values is

$$\vec{i}_F = \sqrt{2}n_f i_f \epsilon^{j0} \qquad \text{A} \tag{11.29}$$

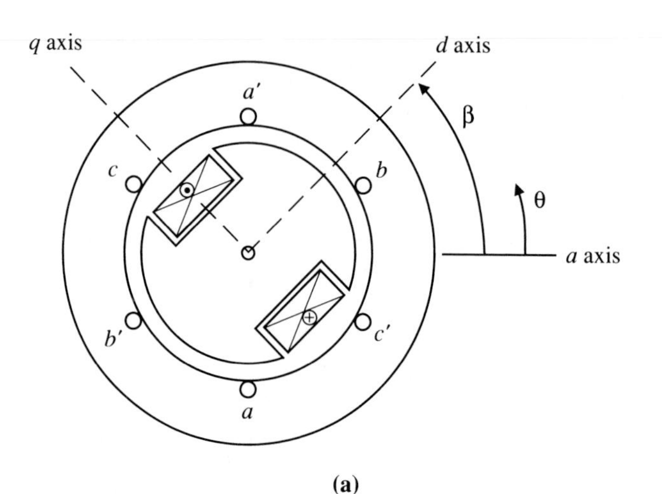

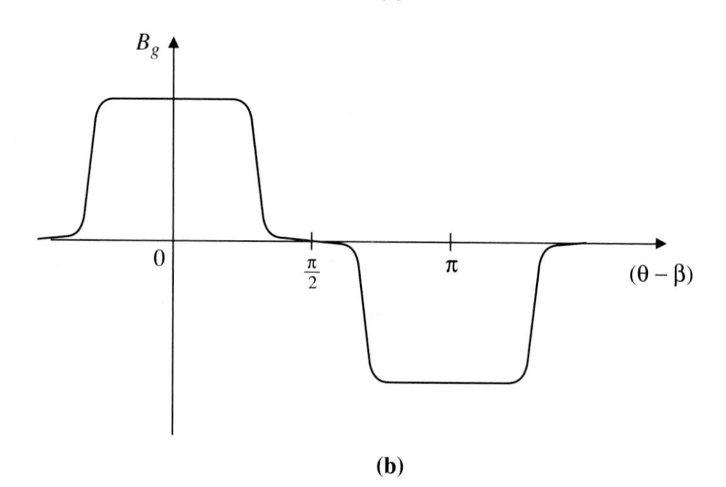

Figure 11.8 (a) Cross section of a salient-pole machine. (b) Air-gap flux density due to field current.

The natural reference axis for a salient-pole machine is the rotor axis, normally referred to as the *direct axis* (d). The rotor has another axis of symmetry at right angles to the direct axis and thus is known as the *quadrature axis* (q).

In the stator reference frame, the stator phase variables can be expressed as $\vec{v}_s$, $\vec{\lambda}_s$, and $\vec{i}_s$. For the salient-pole machine, these are transformed to the rotor reference frame by

$$\vec{\lambda}_s' = \vec{\lambda}_s \epsilon^{-j\beta} = \lambda_{sd} + j\lambda_{sq} \tag{11.30}$$

$$\vec{i}_s' = \vec{i}_s \epsilon^{-j\beta} = i_{sd} + j i_{sq} \tag{11.31}$$

and

$$\vec{v}_s' = \vec{v}_s \epsilon^{-j\beta} = v_{sd} + jv_{sq} \tag{11.32}$$

From Eq. (11.6), these transformed variables are related by

$$\vec{v}_s' = R_s \vec{i}_s' + (p + j\omega_0)\vec{\lambda}_s' \qquad \text{V} \tag{11.33}$$

Because of the orthogonality of the direct and quadrature axis, and assuming no magnetic saturation, currents in one axis produce no flux linkage in the other axis. The direct-axis flux linkages of the stator and the field (transformed to N_{se} turns) are related to the direct-axis stator current and the field current by the magnetizing inductance L_{md} in the direct axis and the leakage inductances $L_{\ell s}$ and $L_{\ell F}$. The quadrature-axis magnetizing inductance L_{mq} has a value smaller than L_{md} because of the large effective air gap in the quadrature axis. The stator leakage inductance in this axis is usually assumed to be the same as in the direct axis although sometimes a lower value may be appropriate because of the higher reluctance to circumferentially directed air-gap flux. The overall flux linkage-current relation can be expressed in matrix form as

$$\begin{vmatrix} \lambda_{sd} \\ \lambda_{sq} \\ \lambda_F \end{vmatrix} = \begin{vmatrix} L_{\ell s} + L_{md} & 0 & L_{md} \\ 0 & L_{\ell s} + L_{mq} & 0 \\ L_{md} & 0 & L_{\ell F} + L_{md} \end{vmatrix} \begin{vmatrix} i_{sd} \\ i_{sq} \\ i_F \end{vmatrix} \qquad \text{Wb} \tag{11.34}$$

or as in the pair of independent flux linkage circuits of Fig. 11.9.

The terminal voltage relations for the salient-pole machine can now be derived by inserting Eq. (11.34) into Eq. (10.33) to give

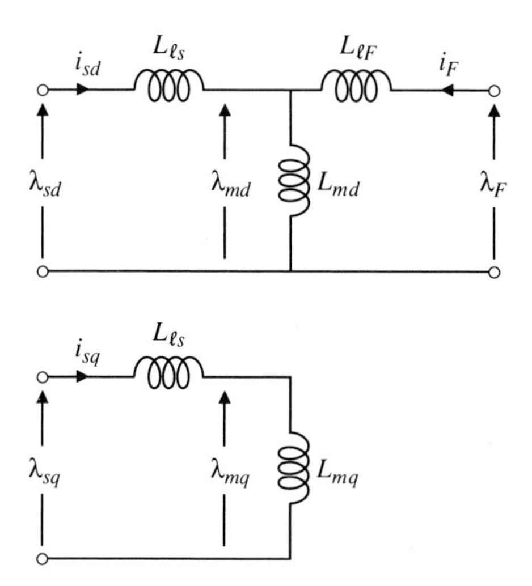

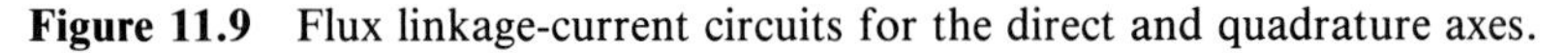

Figure 11.9 Flux linkage-current circuits for the direct and quadrature axes.

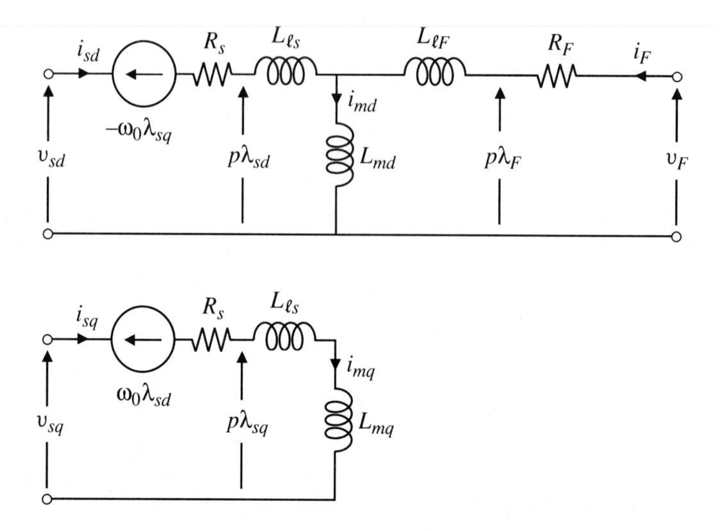

Figure 11.10 Direct and quadrature axis equivalent circuits for a synchronous machine.

$$\begin{vmatrix} v_{sd} \\ v_{sq} \\ v_F \end{vmatrix} = \begin{vmatrix} R_s + L_d p & -L_q \omega_0 & L_{md} p \\ L_d \omega_0 & R_s + L_q p & L_{md} \omega_0 \\ L_{md} p & 0 & R_F + L_F p \end{vmatrix} \begin{vmatrix} i_{sd} \\ i_{sq} \\ i_F \end{vmatrix} \quad \text{V} \qquad (11.35)$$

where $L_d = L_{\ell s} + L_{md}$, $L_q = L_{\ell s} + L_{mq}$, and $L_F = L_{\ell F} + L_{md}$. These relations can be represented in the pair of equivalent voltage-current circuits of Fig. 11.10. The only coupling between these two linear circuits occurs in the generated voltages $\omega_0 \lambda_{sd}$ and $-\omega_0 \lambda_{sq}$. These can be visualized as the effect of the stator winding, transformed to the rotor frame of reference, rotating at angular velocity ω_0 through the stationary flux linkages in the two axes.

Synchronous generators and motors are frequently fitted with shorted damper windings in slots on the pole faces, as shown in Fig. 11.11. In a gener-

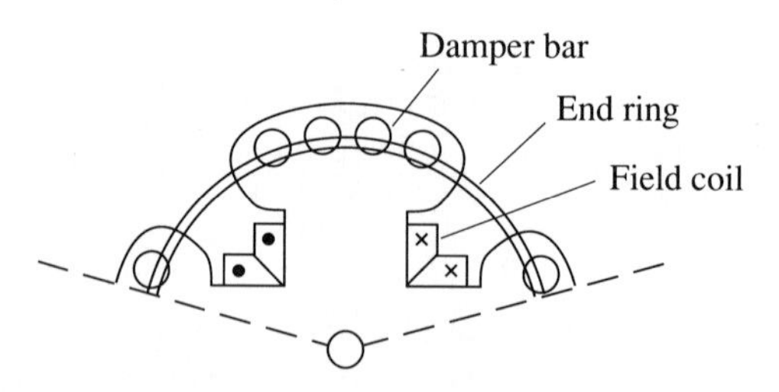

Figure 11.11 Section of a six-pole salient rotor with damper winding.

ator with an unlaminated cylindrical rotor, the conducting steel is equivalent to shorted damper windings. These damper windings act to attenuate mechanical oscillations about the synchronous speed, as will be discussed later in Section 11.3. In a salient-pole machine, the end connections of the damper windings are usually continuous around the machine. Thus, the damper windings act in both direct and quadrature axes but not equally. In the steady state, the flux linkages of these damper windings are constant and their currents are zero. However, in a transient condition, the damper flux linkages may vary inducing currents in the shorted windings; thus, the effect of dampers should be included in the transient voltage-current model of the machine.

Let us denote the damper windings in the two axes by the subscripts *kd* and *kq*. These windings will link with the fluxes produced by the stator and field currents and will also have leakage fluxes due to their own currents. They may be included in the transient model, as shown in Fig. 11.12, where $L_{\ell kd}$ and $L_{\ell kq}$ are the leakage inductances of the damper windings and R_{kd} and R_{kq} are their effective resistances, all referred to equivalent stator turns N_{se}. The field leakage inductance $L_{\ell F}$ now represents flux which links with the field but not with the direct-axis damper or the stator. The inductance L_{mkd} represents that component of flux that links with both the field and direct-axis damper but not with the stator. In some models, this last inductance is omitted.

In some large synchronous machines, the effective leakage inductances and resistances of the damper windings may be significantly affected by the frequency of currents induced in the dampers. This effect can be included in the equivalent circuit model in the same way as was developed for the double-cage or deep-bar rotor of an induction machine in Section 10.1.5.

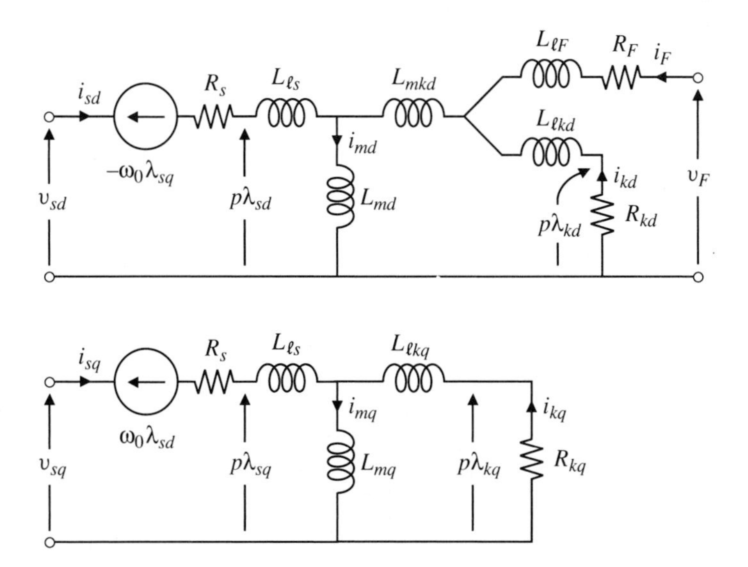

Figure 11.12 Equivalent circuits including the effects of damper windings.

The effects of magnetic saturation may be included in the circuit models of Figs. 11.10 and 11.12 in much the same manner as discussed for cylindrical-rotor machines in Section 11.1.2. It is frequently assumed that the quadrature axis has a linear flux linkage-current relation because of its large effective air gap. Saturation in the stator and field structure in the direct axis can then be represented by the insertion of appropriate nonlinear elements similar to those in Fig. 11.6. In some situations, the assumption of no quadrature axis saturation may not be sufficiently accurate because, under load conditions, saturation tends to be concentrated around one of the pole tips due to the net mmf of stator and rotor.

The transient torque of the salient-pole machine may be derived by expansion of Eq. (11.12) using Eqs. (11.30) and (11.31):

$$\begin{aligned} T &= \frac{3p}{4}\,\mathcal{I}(\vec{\lambda}_s^{\prime *}\vec{i}_s^{\,\prime}) \\ &= \frac{3p}{4}\,\mathcal{I}[(\lambda_{sd} - j\lambda_{sq})(i_{sd} + ji_{sq})] \qquad (11.36) \\ &= \frac{3p}{4}\,(\lambda_{sd} i_{sq} - \lambda_{sq} i_{sd}) \qquad \text{N}\cdot\text{m} \end{aligned}$$

In the steady state, all the variables in the equivalent circuits of Figs. 11.10 or 11.12 are constant and the voltages across the inductances are zero. In steady-state analysis, the stator voltage and current are frequently known and it is desired to find the required field current and the load angle β_0. Instead of being able to choose the reference angle arbitrarily as in the cylindrical-rotor machine, the reference for a salient-pole machine must be the direct or field axis. This can be found as follows.

For simplicity, let us use the simplified axis circuits of Fig. 11.13(a) and (b), where $\omega_0 = \omega_s$ and $i_f' = i_F L_{md}/L_d$. The stator resistance drop has been ignored for simplicity. The stator voltage space vector can be expressed as

$$\begin{aligned} \vec{v}_s^{\,\prime} &= v_{sd} + jv_{sq} \\ &= -\omega_s L_q i_{sq} + j\omega_s L_d i_{sd} + j\omega_s L_d i_f' \qquad (11.37) \\ &= j\omega_s L_q \vec{i}_s^{\,\prime} + j\omega_s (L_d - L_q) i_{sd} + j\omega_s L_d i_f' \qquad \text{V} \end{aligned}$$

Given $\vec{v}_s^{\,\prime}$ and $\vec{i}_s^{\,\prime}$ for a given load condition, the field current is given by

$$i_f' = \frac{\vec{v}_s^{\,\prime} - j\omega_s L_q \vec{i}_s^{\,\prime}}{j\omega_s L_d} - \left(\frac{L_d - L_q}{L_d}\right) i_{sd} \qquad \text{A} \qquad (11.38)$$

Examination of Eq. (11.38) shows that all terms must be real. Thus, the vector $\vec{v}_s^{\,\prime} - j\omega_s L_q \vec{i}_s^{\,\prime}$ must lie in the quadrature axis, as shown in the vector diagram of Fig. 11.13(c). Now that the direct and quadrature axes have been located, the current component i_{sd} can be found by projection of $\vec{i}_s^{\,\prime}$ and the field current evaluated.

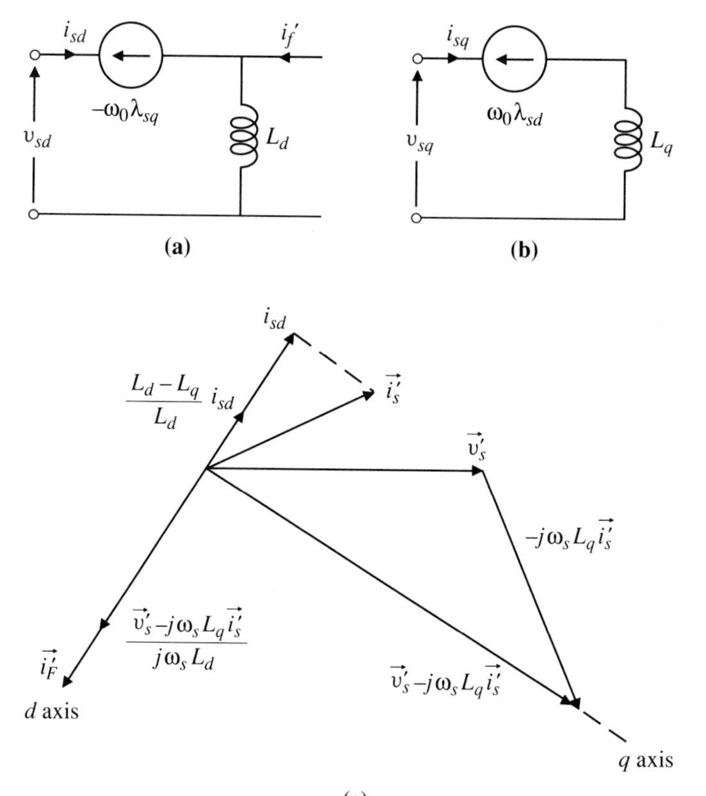

Figure 11.13 (a) Direct-axis model. (b) Quadrature axis model. (c) Space vector relation to determine axes given stator variables.

The torque of a cylindrical rotor machine in the steady state is proportional to the field current and to the cosine of the load angle as stated in Eq. (11.23). A similar term would be expected in the torque expression for the salient-pole machine. However, an additional term would be expected because of the tendency of the low-reluctance direct axis to align with the air-gap flux. To derive this term, let us consider a three-phase synchronous reluctance machine, i.e., a salient-pole machine without field windings. Suppose a voltage $\vec{V}_s = V_s\angle 0$ is applied to the stator windings when the direct axis of the rotor is at angle β_0 to the voltage phasor, as shown in Fig. 11.14. Ignoring stator resistance, the stator flux linkage Λ_s will lag the stator voltage by an angle $\pi/2$. This flux linkage can be considered as the sum of the two phasors, one Λ_{sd} along the direct axis and one Λ_{sq} along the quadrature axis. The component Λ_{sd} along the direct axis will require a component I_{sd} of magnetizing current, inversely proportional to the direct-axis inductance L_d, while the component Λ_{sq} will require a quad-

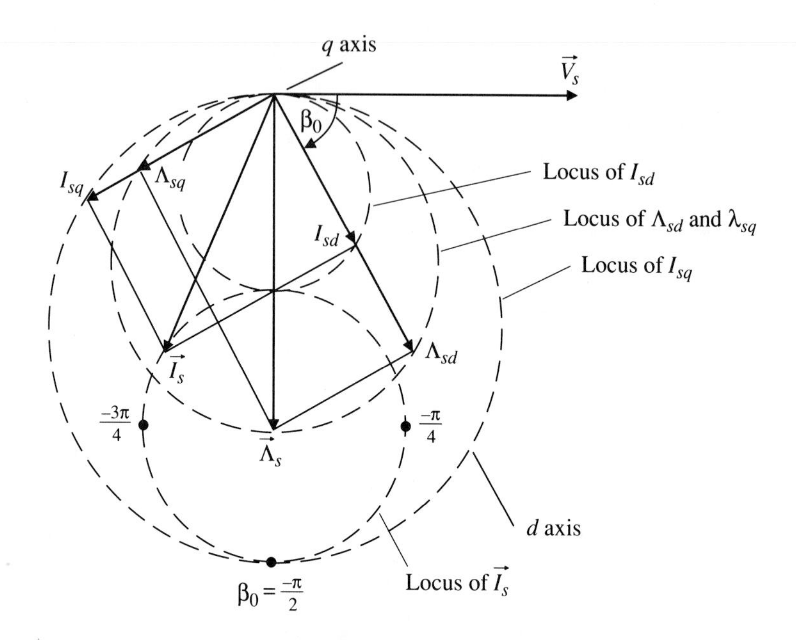

Figure 11.14 Phasor diagram for synchronous reluctance machine.

rature axis current $I_{sq} = \Lambda_{sq}/L_q$. The total stator current $\vec{I}_s$ will be the sum of these two current components.

These components are shown in the phasor diagram of Fig. 11.14 for a condition where the rotor direct axis leads the stator flux linkage, as is the case with generating action. It is noted that the stator current, directed into the stator, has a negative real component. As the angle β_0 is varied, all the phasor components follow the circular loci shown.

From Fig. 11.14, the stator flux linkage can be expressed as

$$\vec{\Lambda}_s = \Lambda_s \sin \beta_0 \angle \beta_0 - \Lambda_s \cos \beta_0 \angle (\beta_0 + \pi/2) \tag{11.39}$$

The stator current is then given by

$$\begin{aligned} \vec{I}_s &= \Lambda_s \left[\frac{-\sin \beta_0 (\cos \beta_0 + j \sin \beta_0)}{L_d} - \frac{j \cos \beta_0 (\cos \beta_0 + j \sin \beta_0)}{L_q} \right] \\ &= \Lambda_s \left[\frac{-j}{L_d} + \left(\frac{L_d - L_q}{L_d L_q} \right) (\sin \beta_0 \cos \beta_0 - j \cos^2 \beta_0) \right] \\ &= \Lambda_s \left[\frac{-j}{L_d} + \frac{L_d - L_q}{L_d L_q} \left(\frac{1}{\tan \beta_0 + j} \right) \right] \end{aligned} \tag{11.40}$$

Expressed in terms of the stator voltage V_s and the reactances $X_d = \omega_s L_d$ and $X_q = \omega_s L_q$, the stator current is

$$\vec{I}_s = \vec{V}_s\left[\frac{1}{jX_d} + \frac{X_d - X_q}{X_d X_q(\tan\beta_0 + j)}\right] \quad \text{A} \tag{11.41}$$

Therefore, the synchronous reluctance machine may be represented by the steady-state equivalent circuit of Fig. 11.15(a) which has a form similar to that

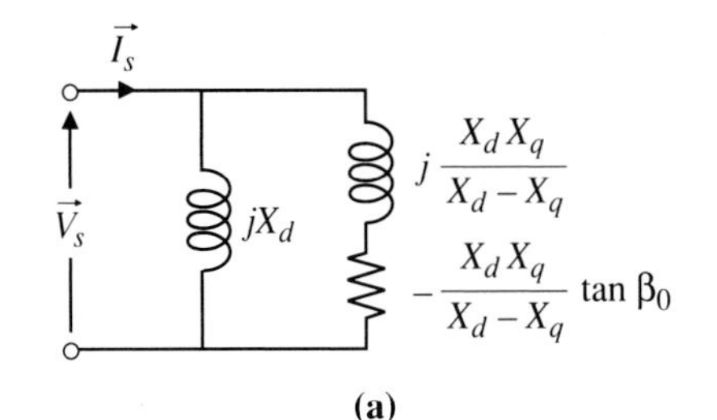

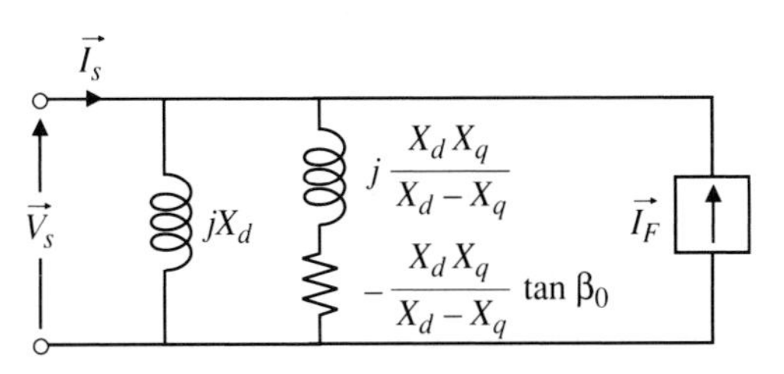

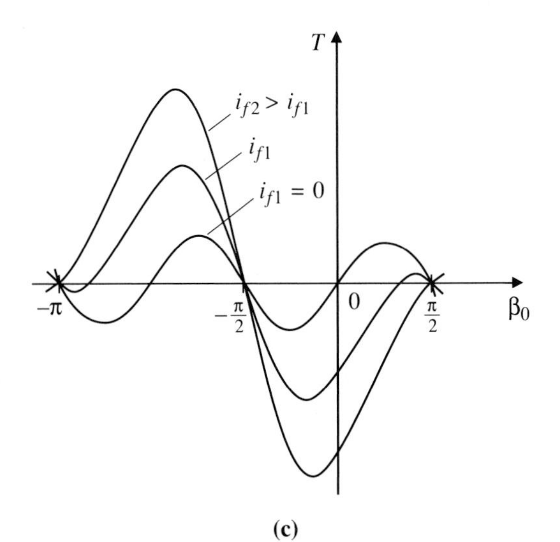

Figure 11.15 (a) Equivalent circuit for reluctance machine. (b) Equivalent circuit for salient-pole synchronous machine. (c) Torque-angle curves for salient-pole synchronous machine.

of an induction machine. When the angle β_0 is $-\pi/2$, i.e., in line with the stator flux, the resistive part of the impedance is infinite. When the angle β_0 is advanced to $-\pi/4$, the resistance has a negative value equal in magnitude to the reactance in series with it. This is the condition for maximum power output as a generator. A retardation of the angle β_0 to $-3\pi/4$ gives the condition for maximum power input as a motor.

The air-gap power per phase as a reluctance machine is

$$\begin{aligned} P_g &= \Re(\vec{V}_s^* \vec{I}_s) \\ &= \Re\left[\frac{X_d - X_q}{X_d X_q} \frac{V_s^2}{(\tan\beta_0 + j)}\right] \\ &= \frac{X_d - X_q}{2X_d X_q} V_s^2 \sin 2\beta_0 \qquad \text{W} \end{aligned} \tag{11.42}$$

The torque is therefore

$$\begin{aligned} T &= \frac{3P_g}{\omega_{mech}} \\ &= \frac{3p}{2\omega_s}\left(\frac{X_d - X_q}{2X_d X_q}\right) V_s^2 \sin 2\beta_0 \qquad \text{N}\cdot\text{m} \end{aligned} \tag{11.43}$$

For a salient-pole synchronous machine, the reluctance motor model of Fig. 11.15(a) now can be combined with the cylindrical-rotor model of Fig. 11.4(b) to produce the model of Fig. 11.15(b). The total torque of the salient-pole machine is a combination of Eqs. (11.23) and (11.43):

$$T = \frac{3p}{2\omega_s}\left[\frac{X_d - X_q}{2X_d X_q} V_s^2 \sin 2\beta_0 - V_s I_F' \cos\beta_0\right] \qquad \text{N}\cdot\text{m} \tag{11.44}$$

Torque-angle curves are shown for three values of field current in Fig. 11.15(c).

In many synchronous machines, the difference between the direct and quadrature axes inductances is not large. The torque term involving the field current is typically much larger than the reluctance term. For these, a steady-state analysis ignoring saliency is frequently of sufficient accuracy.

EXAMPLE 11.1 A six-pole, 60-Hz, three-phase reluctance motor is connected to a 220-V (L-L) supply. Its direct axis reactance is 15 Ω and its quadrature axis reactance is 3 Ω. Stator resistance and rotational losses may be ignored.

(a) Develop an equivalent circuit for the motor.

(b) Determine the maximum torque that the motor can develop before losing synchronism.

(c) Determine the maximum power factor at which the motor can operate and the mechanical power output for this condition.

Solution

(a) The parameters for the equivalent circuit of Fig. 11.15(a) are

$$X_d = 15 \quad \Omega; \qquad X_q = 3 \quad \Omega$$

$$\frac{X_d X_q}{X_d - X_q} = \frac{15 \times 3}{15 - 3} = 3.75 \quad \Omega$$

$$V_s = \frac{220}{\sqrt{3}} \angle 0 = 127 \angle 0$$

(b) From Eq. (11.42), the maximum torque occurs at $\beta_0 = -3\pi/4$ and is equal to

$$\hat{T} = \frac{3 \times 6 \times 127^2}{2\pi \times 60 \times 2 \times 3.75} = 102.7 \quad \text{N}\cdot\text{m}$$

(c) Maximum power factor occurs when the phasor I_s is tangential to its locus, as shown in Fig. 11.16. From Fig. 11.14, the dimensions of the right-angle triangle are

$$oa = \frac{V_s}{2}\left(\frac{1}{X_q} + \frac{1}{X_d}\right) = \frac{V_s}{2}\frac{X_d + X_q}{X_d X_q}$$

$$ab = \frac{V_s}{2}\left(\frac{1}{X_q} + \frac{1}{X_d}\right) = \frac{V_s}{2}\frac{X_d - X_q}{X_d X_q}$$

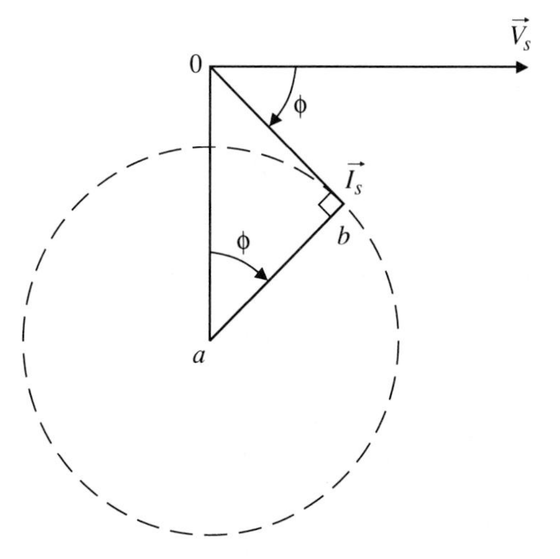

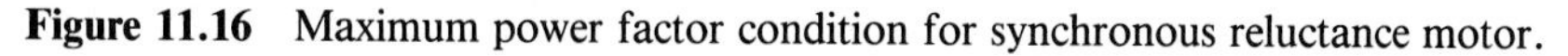
Figure 11.16 Maximum power factor condition for synchronous reluctance motor.

Thus, the maximum power factor is

$$\cos\phi = \frac{oa}{ob} = \frac{X_d - X_q}{X_d + X_q} = \frac{12}{18} = 0.67$$

For this condition, the stator current is

$$I_s = oa \sin\phi = \frac{127}{2}\,\frac{18}{15 \times 3}\,(1 - 0.67^2)^{1/2} = 18.93 \qquad \text{A}$$

and the mechanical power is

$$P_{mech} = 3V_sI_s\cos\phi = 3 \times 127 \times 18.93 \times 0.67 = 4.81 \qquad \text{kW}$$

EXAMPLE 11.2 A salient-pole synchronous generator is rated at 100 MVA, 13.8 kV, 60 Hz, and 180 r/min. Its direct-axis reactance is 2.38 Ω and its quadrature axis reactance is 1.44 Ω. A field current of 900 A is required to produce rated stator voltage on no load.

(a) The generator is to supply a power of 90 MW at 0.9 power factor inductive at rated voltage to a load. Determine the required field current and the load angle.

(b) With a field current of 1500 A and a stator voltage of 13.8 kV, determine the stator power and power factor when the load angle is −60°.

Solution

(a) The rated stator voltage is

$$V_s = \frac{13\,800}{\sqrt{3}}\angle 0 = 7970\angle 0$$

On no load, with $i_f = 900$ A, the field current source in the equivalent circuit of Fig. 11.15(b) is

$$I_F = \frac{V_s}{X_d} = \frac{7970}{2.38} = 3348 \qquad \text{A}$$

Thus, the effective field to stator current ratio is

$$n = \frac{I_F}{i_f} = \frac{3348}{900} = 3.72$$

For the given load, the stator current, directed into the machine, is

$$I_s = -\frac{90 \times 10^6}{3 \times 7970 \times 0.9}\angle\cos^{-1} 0.9 = 4184\angle 154.2°$$

To find the machine axes, from Eq. (11.38),

$$\vec{V}_s - jX_q\vec{I}_s = 7970\angle 0 - j1.33 \times 4184\angle 154.2 = 11\,560\angle 25.7$$

This establishes the quadrature axis at 25.7° and the direct axis at −64.3°. Then,

$$i_{sd} = 4184\cos(154.2 + 64.3) = -3274 \qquad \text{A}$$

The equivalent field current is, from Eq. (11.38),

$$I_F = \frac{11\,560}{2.38} - \left(\frac{2.38 - 1.33}{2.38}\right)(-3274) = 4857 + 1444 = 6301 \qquad \text{A}$$

Thus, the field current is

$$i_f = \frac{6301}{3.72} = 1694 \qquad \text{A}$$

The angle β_0 is −64.3°.

(b) From Eq. (11.41), the stator current is

$$\vec{I}_s = 7970\angle 0\left[\frac{1}{j2.38} + \frac{2.38 - 1.33}{2.38 \times 1.33(\tan -60° + j0)} - I_F\angle\beta_0\right]$$

$$= 4018\angle 168.2° \qquad \text{A}$$

The generator output power is

$$P_s = -3\Re(\vec{V}_s^*\vec{I}_s) = -3 \times 4018 \times 7970\cos 168.2 = 94 \qquad \text{MW}$$

The power factor is

$$pf = |\cos 168.2| = 0.98$$

11.1.4 ▪ Measurement of Parameters

An equivalent circuit model of a cylindrical-rotor synchronous machine is shown in Fig. 11.17(a). The parameters of this model, including the nonlinear inductance, can be derived from a set of measurements made on the machine. The resistances R_s and R_f of the windings can be measured with direct current. Suppose the machine is driven as a generator at speed $\omega_0 = \omega_s$ electrical rad/s. With the stator open circuited, the stator phase voltage can be measured with varying field current to obtain the open-circuit saturation curve shown in Fig. 11.17(b). Actually, line voltages will usually be easier to measure and may also be more appropriate. The phase voltage is equal to the line voltage divided by $\sqrt{3}$. Next, suppose the generator is provided with a balanced load that is as purely inductive as feasible. At various values of field current, this load is adjusted to take rated stator current I_{sb}, and the stator phase voltage is measured. The result is the zero-power-factor inductive load curve of Fig. 11.17(b).

A phasor diagram for the zero-power factor load condition is shown in Fig. 11.17(c). The voltage drop across the stator resistance can generally be ignored, partly because it is small and partly because it is in quadrature with the other voltage phasors. If the circuit model of Fig. 11.17(a) is accurate, the two

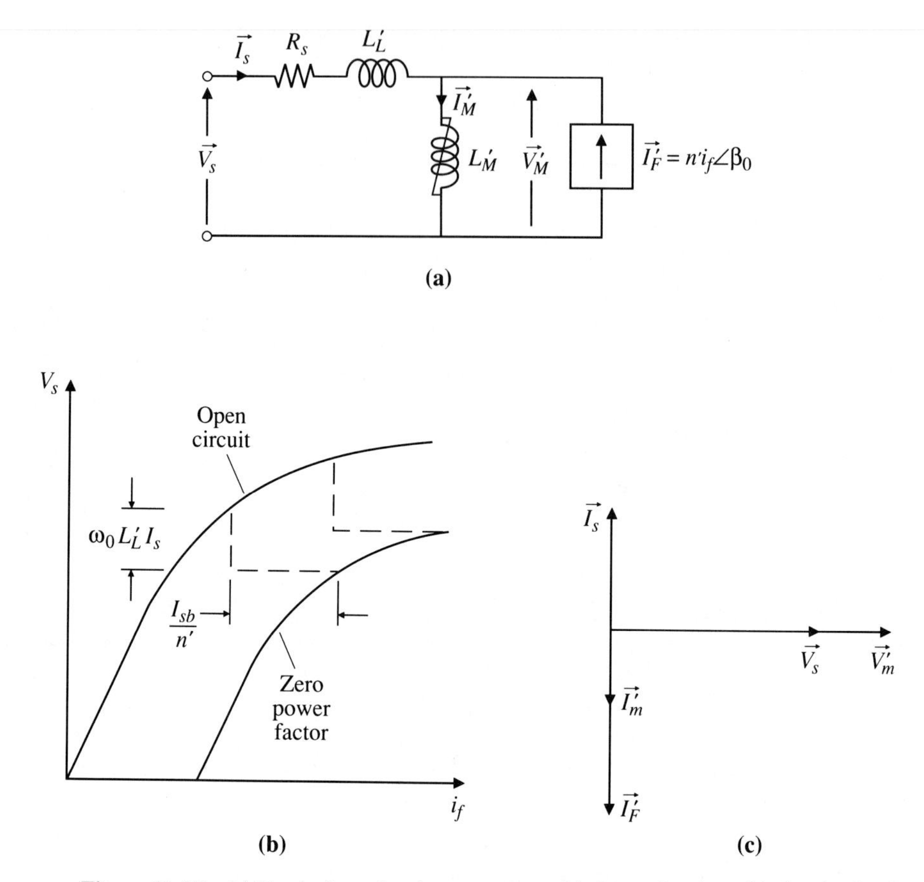

Figure 11.17 (a) Equivalent circuit parameters. (b) Open circuit and inductive load test curves. (c) Phasor diagram.

curves in Fig. 11.17(b) should have the same nonlinear shape because both experience the same nonlinearity due to the magnetizing branch. The open-circuit curve is a direct measure of the magnitude relation between $V_M' = V_s$ and the magnetizing current I_M' divided by the effective field to stator current ratio n'. With an inductive stator current of magnitude I_{sb},

$$V_M' = V_s + \omega_s L_L' I_{sb} \tag{11.45}$$

and

$$I_F' = I_M' + I_{sb} = n' i_f \tag{11.46}$$

Thus, the zero-power factor curve should be identical in shape to the open-circuit curve but shifted down by an amount $\omega_s L_L' I_{sb}$ and to the right by an amount I_{sb}/n' as shown. These displacements can best be determined by sketching a transparent copy of one of the curves and shifting it to obtain a best fit on the other. The displacement of the origin of the copied curve then gives the parameters L_L' and n' because ω_s and I_{sb} are known. The open-circuit curve may now be rescaled as $\Lambda_s = V_s/\omega_s$ and $I_M' = n'i_f$ to describe the nonlinear inductance L_M'.

If the two nonlinear curves do not fit perfectly, it is an indication of the imperfection of the equivalent circuit model of Fig. 11.17(a) due to the presence of significant stator saturation. However, a best fit will produce a compromise model of sufficient accuracy for most purposes.

These measurements provide all the parameters for the transient equivalent circuit of Fig. 11.7(a) except for the effective rotor resistance R_F'. From Eqs. (11.9), (11.19), and (11.27) and the measured value of R_f, this can be derived as

$$R_F' = \frac{R_f}{3(n')^2} \quad \Omega \tag{11.47}$$

For a salient-pole machine, the above measurements give the direct axis parameters. The unsaturated values of the quadrature-axis inductance L_q and the direct-axis inductance L_d can be measured if the machine is connected to a three-phase source with a small stator voltage V_s at frequency ω_s, and if the machine is driven at a speed ω_0 slightly below or above ω_s. The stator flux will sweep past the direct and quadrature axes in turn, and at a slow enough rate to produce negligible currents in the damper and field windings. The rms stator current will vary between $I_{s(\min)}$ and $I_{s(\max)}$ as the direct and quadrature axes align with the stator flux. Then,

$$L_q = \frac{V_s}{\omega_s I_{s(\min)}} \quad \text{and} \quad L_q = \frac{V_s}{\omega_s I_{s(\max)}} \tag{11.48}$$

11.2

Electrical Transient Performance

Adjustment of the field current of a synchronous generator or motor connected to a constant-voltage system provides for control of power factor. In a free-standing synchronous generator, it also provides control of stator voltage. Therefore, it is important to know how rapidly the machine responds to changes in its field-supply voltage. It is also important to know what stator currents flow after a sudden change at the stator terminals, such as a short circuit, so that protection systems can be properly designed. This section examines some of these electrical transient situations.

A linear transient equivalent circuit for a cylindrical-rotor synchronous machine has been developed in Section 11.1 and is shown in Fig. 11.2(b). That section contains a simple example showing the exponential rise of the envelope of open-circuit stator voltage following application of a step of field voltage, as shown in Fig. 11.3. The open-circuit field time constant τ_F associated with this transient is generally in the range of 3 to 10 s for large synchronous machines. It decreases as the size of the machine decreases and may be less than 0.1 s for small laboratory machines.

Let us consider connecting a direct voltage to the field when its stator is short circuited. While this is a somewhat unrealistic situation, it gives insight into further aspects of transient behavior. For simplicity, the stator resistance will be neglected. Before $t = 0$, the stator flux linkage $\vec{\lambda}_s$ is zero. After $t = 0$, this flux linkage continues to be zero because the stator voltage $\vec{v}_s$ is zero. When these variables are transformed to the rotor reference frame, their corresponding variables are also zero. The result is the equivalent circuit of Fig. 11.18(a) with $\vec{v}_s' = 0$ and $\vec{\lambda}_s' = 0$.

When a step of direct voltage V_f is applied to the field winding, the corresponding variable in the transient model of Fig. 11.18(a) is

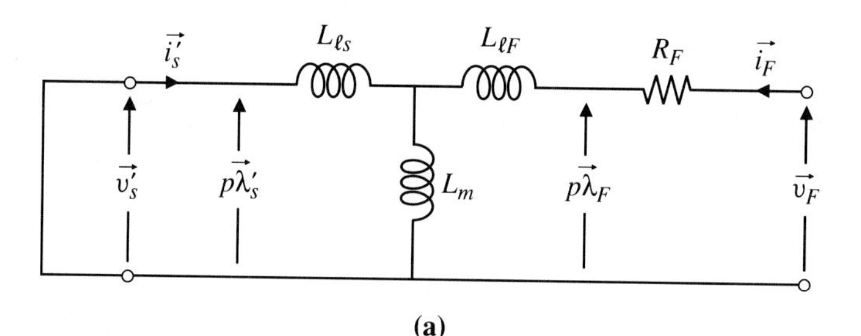

(a)

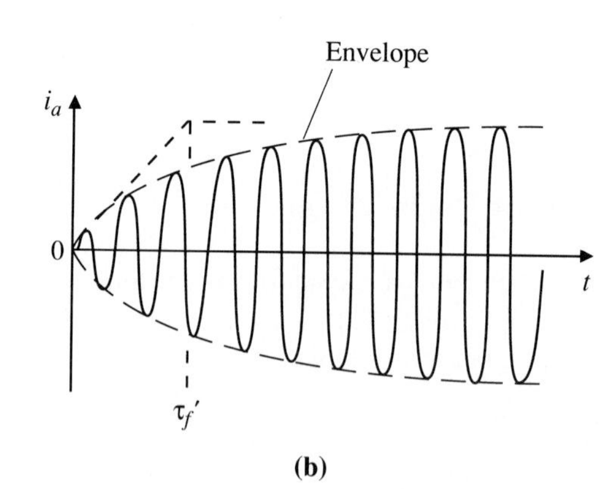

(b)

Figure 11.18 (a) Equivalent circuit for machine with stator short circuited. (b) Transient current in phase *a*.

$$\vec{v}_F = \left(\frac{N_{se}}{N_{fe}}\right) V_f \qquad \text{V} \tag{11.49}$$

The field current is then related to this voltage by

$$\vec{v}_F = R_F \vec{i}_F + L'_F p \vec{i}_F \qquad \text{V} \tag{11.50}$$

where

$$L'_F = L_{\ell F} + \frac{L_{\ell s} L_m}{L_{\ell s} + L_m} \tag{11.51}$$

The solution for the current from these expressions is

$$\vec{i}_F = \frac{\vec{v}_F}{R_F}(1 - \epsilon^{-t/\tau'_f}) \qquad \text{A} \tag{11.52}$$

where

$$\tau'_f = \frac{L'_F}{R_F} \qquad \text{s} \tag{11.53}$$

The field-circuit time constant τ'_f with shorted stator is known as the *transient time constant* of the machine in power system literature. Its value is generally in the range of 0.5 to 3 s for large synchronous machines.

The field current $\vec{i}_F$ divides between the inductances $L_{\ell s}$ and L_m in Fig. 11.18(a) in inverse proportion to their magnitudes. Thus,

$$\begin{aligned} \vec{i}'_s &= -\frac{L_m}{L_{\ell s} + L_m} \vec{i}_F \\ &= -\frac{L_m}{L_{\ell s} + L_m} \frac{\vec{v}_F}{R_F}(1 - \epsilon^{-t/\tau'_f}) \qquad \text{A} \end{aligned} \tag{11.54}$$

The actual stator current vector can now be derived as

$$\vec{i}_s = \vec{i}'_s \epsilon^{j\beta} \qquad \text{A} \tag{11.55}$$

where $\beta = \beta_0 = \omega_s t$. The individual phase currents are then found by resolving $\vec{i}_s$ on the phase axes. A phase current transient is shown in Fig. 11.18(b) for a relatively small synchronous machine with a small transient time constant. Typically, a 60-Hz machine would have 90 cycles of current within a time constant of 1.5 s.

Let us now consider a synchronous machine that is initially unloaded and excited with a field current $i_f = v_f/R_f$. Suppose a three-phase short circuit occurs at or near the stator terminals at time $t = 0$. A knowledge of the maximum value of the stator current is important in the mechanical bracing of the stator windings and also in the specifications of the circuit breakers in the power system to which the machine may be connected.

The flux linkage $\vec{\lambda}_F$ of the field cannot change instantaneously. With a shorted stator, its rate of change is governed by the time constant τ'_f which may have a value of several seconds. In a number of situations, we are interested in events that occur in a relatively short time after the short circuit. For example, circuit breakers typically operate in three to eight cycles of the supply frequency following a short circuit. For such short periods of time, it is reasonable to assume that the field flux linkage $\vec{\lambda}_F$ remains constant at the value that it had before $t = 0$.

Suppose all the relevant variables in the circuit of Fig. 11.18(a) are transformed to the stator reference frame. The result is the equivalent circuit of Fig. 11.19(a) in which the constant field flux linkage vector $\vec{\lambda}_F$ is considered to produce an alternating source voltage represented by the phasor

$$\vec{V}_F = j\omega_s\vec{\lambda}_F/\sqrt{2} \qquad \text{V} \tag{11.56}$$

at frequency ω_s. It is now convenient to simplify this circuit by use of Thevenin's theorem to give the circuit of Fig. 11.9(b) in which

$$\vec{V}' = \frac{L_m}{L_{\ell F} + L_m}\vec{V}_F \qquad \text{V} \tag{11.57}$$

and

$$L' = L_{\ell s} + \frac{L_{\ell F}L_m}{L_{\ell F} + L_m} \qquad \text{H} \tag{11.58}$$

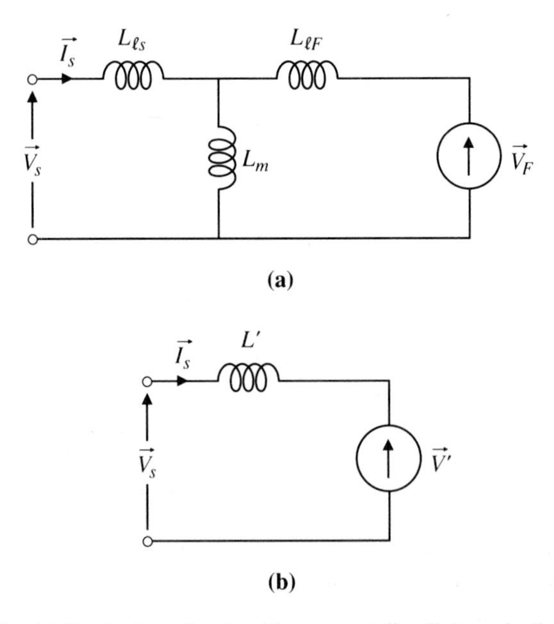

Figure 11.19 (a) Equivalent circuit with constant flux linkage in the field. (b) Simplified circuit.

The parameter L' is denoted as the transient inductance. For a large synchronous machine, its value is typically in the range 0.15 to 0.5 pu, based on the machine rating. The quantity $\vec{V}'$ is the voltage back of this transient inductance. Its value is equal to the voltage at an open-circuited stator before the short circuit is applied. Division of the transient reactance into this voltage gives the stator current phasor immediately after the short circuit:

$$\vec{I}_s = \frac{\vec{V}'}{j\omega_s L'} \quad \text{A} \tag{11.59}$$

The stator current does not remain at the value of Eq. (11.59) but decays at a rate governed by the transient time constant τ_f' of Eq. (11.53) toward its steady-state value:

$$\vec{I}_{ss} = \frac{L_m}{L_{\ell s} + L_m} \frac{\vec{v}_F}{\sqrt{2}R_F} \quad \text{A} \tag{11.60}$$

Thus, the alternating component of the stator phase current is

$$\vec{I}_s = \vec{I}_{ss} + \left(\frac{\vec{V}'}{j\omega_s L'} - \vec{I}_{ss}\right)\epsilon^{-t/\tau_f'} \quad \text{A} \tag{11.61}$$

There must be an additional transient component of stator current because the stator flux linkage cannot change instantaneously to zero from its value before the short circuit and the stator phase currents cannot immediately jump to new finite values. If the stator resistance were zero, the stator flux linkage would remain constant in each phase winding at its value just before $t = 0$. These unidirectional initial flux linkages of the stator windings decay exponentially toward zero at the stator or *armature time constant* τ_a. This time constant is equal to the inductance L' per phase as seen from the stator terminals with flux linkage $\vec{\lambda}_F$ constant, divided by the stator resistance per phase, i.e.,

$$\tau_a = \frac{L'}{R_s} \quad \text{s} \tag{11.62}$$

The value of this time constant is typically in the range 0.1 to 0.4 s for large synchronous generators.

A typical set of stator phase currents following a three-phase short circuit is shown in Fig. 11.20. All currents start at zero for this case of an initially open-circuited machine. The outer envelope of the alternating currents is governed by Eq. (11.61) and is symmetrical about the unidirectional transient with time constant τ_a. In the usual case where both τ_a and τ_f' are large with respect to a half-period of the alternating current, the peak value of the short-circuit current in one of the phases can approach the value

$$\hat{i}_s = 2\sqrt{2}\,\frac{V'}{\omega_s L'} \quad \text{A} \tag{11.63}$$

i.e., twice the peak value of the alternating component.

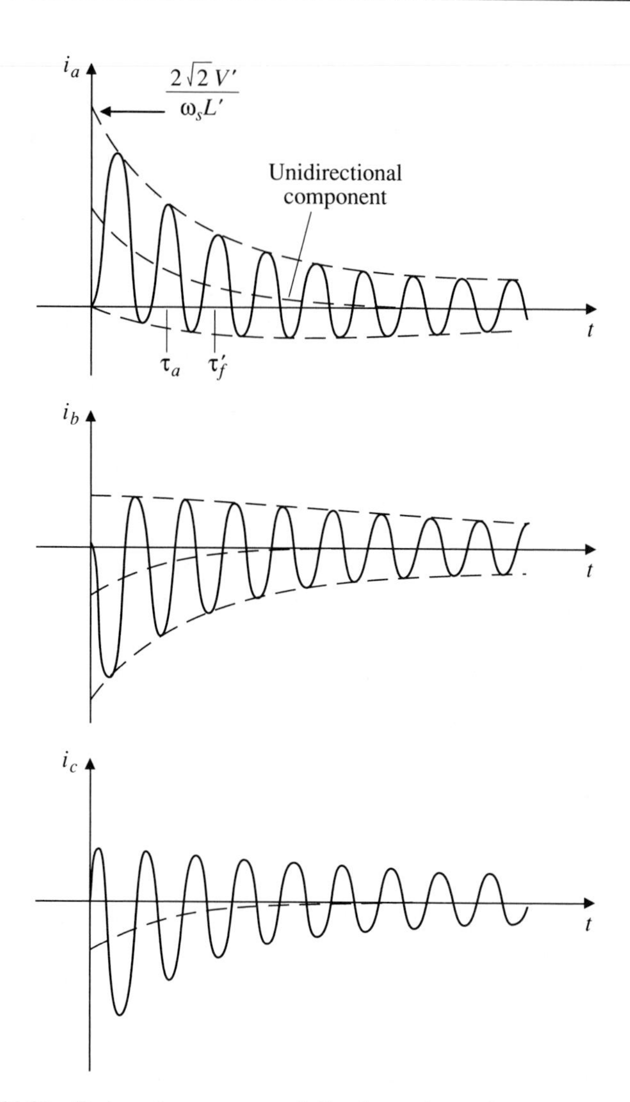

Figure 11.20 Stator phase currents following a three-phase short circuit.

When the unidirectional transient current of time constant τ_a in the stator windings is transformed to the rotor reference frame, the result is a decaying oscillating term of fundamental frequency ω_s in the transformed stator current $\vec{i}_s'$ and thus in the field current vector $\vec{i}_F$ as well. The field current i_f following a stator short circuit is shown as a function of time in Fig. 11.21. Its average value is the exponential term given in Eq. (11.52). The actual current begins at its initial value V_f/R_f and decays exponentially to the same value. Continuity of the current through time $t = 0$ is maintained by the oscillatory term of time constant τ_a and frequency ω_s.

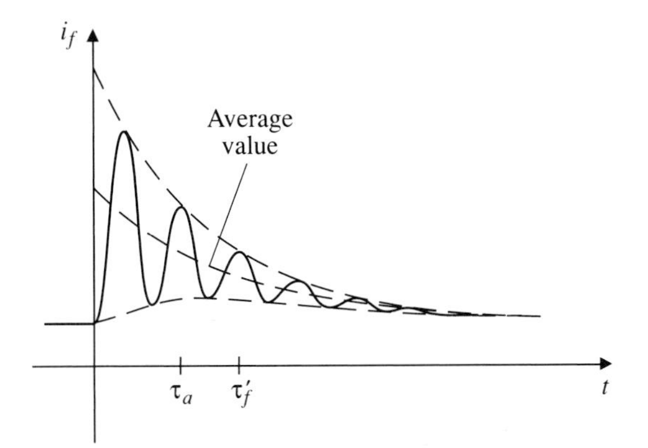

Figure 11.21 Field current following three-phase short circuit on the stator.

The equivalent circuits of Fig. 11.19 may be used, not only for short-circuit analysis, but also for study of a variety of events that occur shortly after a change in operating conditions such as the switching on of a load. The first step in analysis of these situations is the determination of the variables prior to the change. In particular, the initial value of the flux linkage $\vec{\lambda}_F$ must be evaluated. In precise analysis, it is only the component λ_{Fd} of this vector in the direct or field axis that is held constant through $t = 0$ by the action of the field winding current. In many analyses, it is sufficiently accurate to assume that the total vector $\vec{\lambda}_F$ remains constant as it would if the rotor contained a full set of three-phase field windings.

Thus far in this section, we have ignored the possible effects of damper windings. Immediately following any change in operating condition, the flux linkage of the damper windings must also remain constant. In the d and q axis equivalent circuits of Fig. 11.12, the flux linkages λ_{kd}, λ_{kq}, and λ_F would all be continuous through time $t = 0$. If any saliency is ignored, i.e., $L_{mq} = L_{md} = L_m$, $L_{\ell kq} = L_{\ell kd} = L_{\ell k}$ and $R_{kd} = R_{kq} = R_k$, the fundamental frequency condition immediately following a change can be represented by the circuit of Fig. 11.22(a). This can be simplified to the form of Fig. 11.22(b) in which

$$L'' = L_{\ell s} + L_m \| (L_{mk} + L_{\ell F} \| L_{\ell k}) \qquad \text{H} \qquad (11.64)$$

This quantity L'' is commonly known in power system literature as the *subtransient inductance* and $\vec{V}''$ is the phasor of voltage back of subtransient reactance. The value of L'' is somewhat less than the transient inductance L' and is normally in the range 0.1 to 0.4 pu for large synchronous generators.

During the first few cycles after $t = 0$, the short-circuit current may be somewhat greater than shown in Fig. 11.20 because of the effect of the damper windings preserving their flux linkages. This additional component of current, shown in Fig. 11.22(c), decays at a time constant τ'' which is equal to the sys-

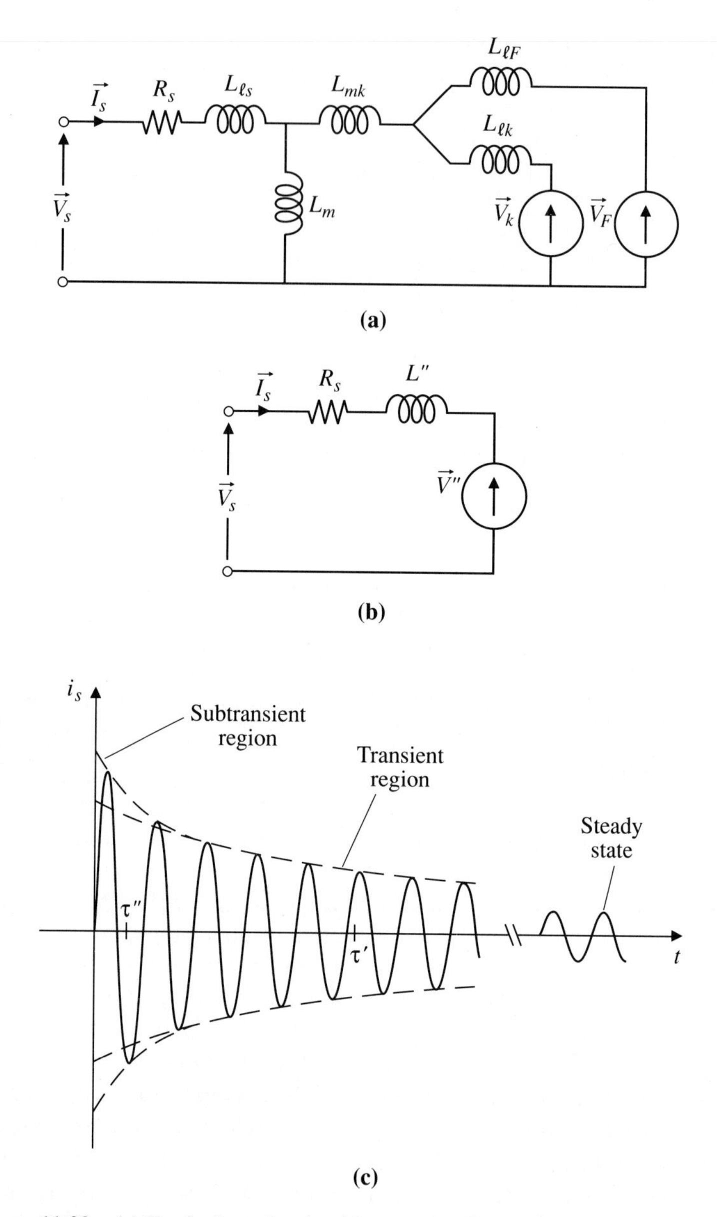

Figure 11.22 (a) Equivalent circuit with constant flux linkage in field and damper. (b) Simplified circuit. (c) Short-circuit current including subtransient component.

tem inductance as seen from the damper windings divided by the damper resistance. With no saliency,

$$\tau'' = \frac{1}{R_k}\left[L_{\ell k} + L_{\ell F} \| (L_{mk} + L_{\ell s} \| L_m)\right] \qquad \text{s} \tag{11.65}$$

This subtransient time constant is relatively small, typically 0.01 to 0.05 s for large generators. After a time of about 3 τ'', the damper currents can be ignored and the transient circuits of Fig. 11.19 can be used. However, the time constant τ'' is long enough to be significant in determining the maximum value of stator short-circuit current and in determining the current to be interrupted by fast-acting circuit breakers.

The transient and subtransient parameters of a synchronous machine are normally measured by applying a short circuit and recording the phase currents. Alternatively, the stator impedance may be measured as a function of frequency at standstill.

A situation of practical importance is the determination of the transient drop in the stator voltage of a generator when a large inductive load such as occurs in starting an induction motor is connected to the machine terminals. Suppose this load can be approximated by an inductive reactance X_e per phase. Neglecting subtransient effects and stator resistance, the stator voltage phasor immediately after the load is connected is

$$\vec{V}_s = \frac{X_e}{X_e + X'} \vec{V}' \quad \text{V} \tag{11.66}$$

where $X' = \omega_s L'$ and $\vec{V}'$ is the open-circuit stator voltage before the load is connected. If the field voltage were to remain constant, the stator voltage would decay at time constant τ to a steady-state value $\vec{V}_{ss}$ which, from Fig. 11.4(b) and Eq. (11.20), is

$$\vec{V}_{ss} = \frac{X_e X_s}{X_e + X_s} n \frac{V_f}{R_f} \quad \text{V} \tag{11.67}$$

The time constant τ is intermediate in value between τ' and τ_f, and can be shown to be

$$\tau = \frac{X_e + X'}{X_e + X_s} \tau_f \quad \text{s} \tag{11.68}$$

Figure 11.23 shows how the stator voltage magnitude might change with time after connection of the load. If the stator voltage is to be restored to its original value, some form of voltage regulator must be employed. The equivalent circuits of Fig. 11.2 and 11.7 are in convenient form for transient studies of voltage regulator action. With the simple inductive load just considered, the field current $\vec{i}_F$ is related to the field voltage $\vec{v}_F$ by

$$\vec{i}_F = \frac{\vec{v}_F}{R_F(1 + s\tau)} \quad \text{A} \tag{11.69}$$

using τ from Eq. (11.68). A rapid detection of the initial drop in stator voltage followed by a quick increase in field voltage provides a stator voltage response of the type shown in Fig. 11.23.

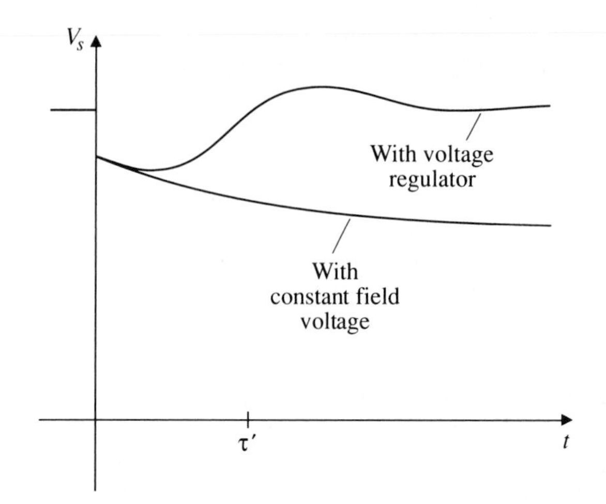

Figure 11.23 Stator voltage following the sudden application of an inductive load.

11.3

Electromechanical Transient Performance

An important class of synchronous machine problems is concerned with predicting the ability of the machine to remain in synchronism with the power system to which it is connected after a sudden change in operating conditions. For a motor, the disturbance may arise from a sudden change in its mechanical load or a loss of electrical supply for a few cycles. For a generator, the most severe disturbances arise from short circuits in the power system resulting in a loss of electrical power output from the machine.

The dynamic behavior of a rotating machine is reflected in the differential equation

$$J\frac{d\omega_{mech}}{dt} = T - T_{shaft} \qquad \text{N}\cdot\text{m} \tag{11.70}$$

where T is the torque generated by the machine in the direction of rotation and T_{shaft} is the torque applied to the shaft by the mechanical load for a motor or the prime mover for a generator.

If synchronism is to be maintained following a disturbance, the instantaneous speed $\omega_0 = \omega_{mech}p/2$ in electrical radians cannot vary appreciably from its synchronous value ω_s. Because of the high inertia of the synchronous machine and of its attached mechanical system, the changes in speed during a cycle of the electrical supply are relatively small. Thus, during a mechanical oscillation, the electrical part of the system may frequently be analyzed by use of phasor methods. From Section 11.2, we note that a synchronous machine connected to a constant alternating-voltage power system responds to changes in its electrical operating condition at the transient time constant τ_f' of Eq. (11.53). With large synchronous machines, this time constant may be several seconds in duration while mechanical oscillations are generally found to have a period in the range 0.25 to 1.5 s. A solution of sufficient accuracy often can be obtained by assuming the field flux linkage to remain constant during at least the first period of mechanical oscillation, thus allowing the use of the transient equivalent circuit model of Fig. 11.19. This approach allows us to examine the basic principles involved in electromechanical transient analysis. Where greater accuracy is required, the full set of system differential equations may be integrated to obtain a solution.

For a two-pole machine, the rotor position at any instant is denoted by the angle β where

$$\beta = \omega_s t + \beta_0 \qquad \text{rad} \tag{11.71}$$

It is usually convenient to use the phase angle of the stator voltage as a reference, making $\vec{V}_s = V_s \angle 0$. The equivalent field current $\vec{I}_f$ in Fig. 11.4(a) then has the phase angle β_0. The source voltage $\vec{V}_F$ in Fig. 11.19(a) arises from the constant flux linkage λ_F of the field winding. Strictly, it is only the component λ_{Fd} of the flux linkage vector λ_F along the direct or field axis which remains constant through the instant of a disturbance. In the stator reference frame, the flux linkage $\vec{\lambda}_{Fd}$ will have the angle β_0, and in the equivalent circuit of Fig. 11.19(a) its corresponding voltage will have the phase angle $\beta_0 + \pi/2$ because

$$\vec{V}_F = j\omega_s \vec{\lambda}_{Fd}/\sqrt{2} \qquad \text{V} \tag{11.72}$$

The source voltage $\vec{V}'$ in the simplified circuit of Fig. 11.19(b) now can be expressed as

$$\vec{V}' = V' \angle (\beta_0 + \pi/2) = V' \angle \delta \tag{11.73}$$

where, for convenience, δ is an angle 90° ahead of β_0.

The torque produced by the machine when in this transient mode can be expressed as

$$\begin{aligned} T &= \frac{3P_g}{\omega_{mech}} \\ &= \frac{3p}{2\omega_s} \Re[\vec{V}'^* \vec{I}_s] \\ &= \frac{3p}{2\omega_s} \Re\left[\vec{V}'^* \left(\frac{\vec{V}_s - \vec{V}'}{j\omega_s L'}\right)\right] \\ &= -\frac{3p}{2\omega_s} \frac{V'V_s}{\omega_s L'} \sin\delta \qquad \text{N·m} \end{aligned} \tag{11.74}$$

Suppose we consider ω_0 in Eq. (11.71) to be a constant and allow δ which is a measure of rotor angular position to vary with time. For a p-pole machine, the mechanical speed is then given by

$$\omega_{mech} = \frac{2}{p}\frac{d\beta}{dt} = \frac{2}{p}\omega_s + \frac{2}{p}\frac{d\delta}{dt} \qquad \text{rad/s} \tag{11.75}$$

The acceleration of Eq. (11.70) is then given by

$$\frac{d\omega_{mech}}{dt} = \frac{2}{p}\frac{d^2\delta}{dt^2} \qquad \text{rad/s}^2 \tag{11.76}$$

As a first example of an electromechanical transient, consider a synchronous motor connected to a supply with constant voltage $\vec{V}_s$. Suppose, at $t = 0$, the mechanical torque is suddenly increased from T_1 to T_2. Before $t = 0$, ignoring mechanical losses in the machine, the operating condition shown in Fig. 11.24 is $T = T_1$ and $\delta = \delta_1$. When the load is increased to T_2, the rotor decelerates toward $\delta = \delta_2$ where the torque T produced by the motor equals T_2. During this deceleration, kinetic energy is being removed from the inertia J. The change ΔW_k in this kinetic energy is equal to the net torque on the mechanical system integrated over the mechanical angle $\delta(2/p)$ traversed, i.e.,

$$\Delta W_k = \frac{2}{p}\int_{\delta_1}^{\delta_2} (T - T_2)\, d\delta \qquad \text{J} \tag{11.77}$$

At angle $\delta = \delta_2$, the speed of the machine will be somewhat less than synchronous speed; therefore, the angle δ will continue to decrease. However, the torque T produced by the motor now is greater than T_2. The net torque acts to accelerate the machine and restore its lost kinetic energy. The speed reaches its

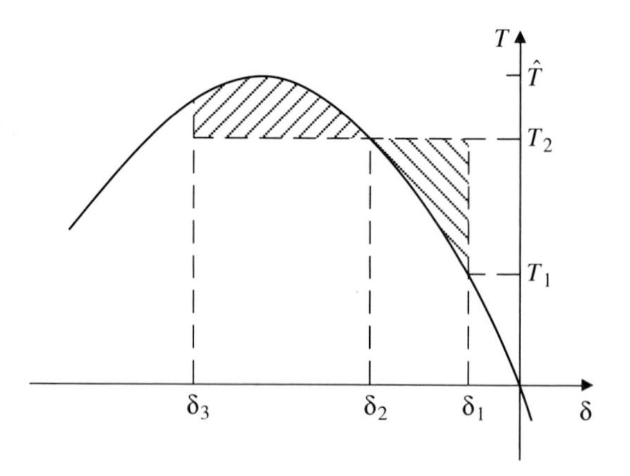

Figure 11.24 Equal-area method of determining the stability of a synchronous machine.

synchronous value again at $\delta = \delta_3$ for which the total change in kinetic energy since the start is zero, i.e.,

$$\Delta W_k = \frac{2}{p} \int_{\delta_1}^{\delta_3} (T - T_2)\, d\delta = 0 \tag{11.78}$$

which can be restated as

$$\int_{\delta_1}^{\delta_2} (T - T_2)\, d\delta + \int_{\delta_2}^{\delta_3} (T - T_2)\, d\delta = 0 \tag{11.79}$$

In Fig. 11.25, this condition exists when the two shaded areas, representing the two integrals of Eq. (11.79), are equal. When this condition can be met, the machine remains in synchronism.

At $\delta = \delta_3$, the machine is being accelerated. When the angle again reaches δ_2, the speed will be above synchronous value; thus, the angle will swing beyond δ_2. In the absence of any damping torque, the angle will again reach its initial value δ_1 and the oscillation will be repeated. In practice, the oscillation is eventually damped out by the damper windings, and the steady-state condition of $\delta = \delta_2$ is reached.

Suppose we consider a slightly different situation in which the initial torque T_1 is reduced but the final load torque T_2 is unchanged. As shown in Fig. 11.25, the energy represented by the area A between the load torque line T_2 and the T-δ curve cannot be restored by the time the acceleration ceases at $\delta = \pi - \delta_2$ because the available area B is less than the area A. Thus, the angle δ will continue to decrease and the machine can be expected to come to a standstill.

This graphical or equal-area approach is useful for obtaining an understanding of simple electromechanical transients. However, it does not give the

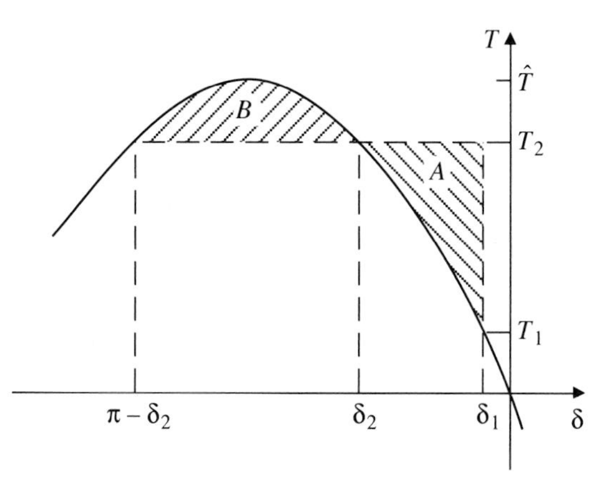

Figure 11.25 Situation where synchronism is lost. Area $B <$ Area A.

rotor angle δ as a function of time and is therefore not adequate for situations involving a sequence of system changes that occur at specific times. As an example of such a sequence, consider a generator connected to a power system containing a number of other generators and loads. Suppose a short circuit occurs in the power system. Prior to the short circuit, the generator is delivering power to the system and has a shaft torque $T_{shaft} = T$ from its prime mover. The short circuit has the effect of reducing the stator voltage, possibly to zero for a three-phase short near its terminals. From Eq. (11.74), this will reduce the torque produced by the machine. The prime mover torque, however, continues to act, and the machine will be accelerated. Within a short time, generally from 0.05 to 0.15 s, the relays of the power system will detect and locate the short circuit and appropriate breakers will disconnect the faulty section of the system.

During this sequence of events, the angle δ can be evaluated by solution of the equation

$$J \frac{2}{p} \frac{d^2\delta}{dt^2} = T - T_{shaft} \qquad \text{N}\cdot\text{m} \tag{11.80}$$

where both torque terms are normally negative. As the torque T is a nonlinear function of the angle δ, the solution is normally obtained by numerical integration. At each time step in the integration, the torque T is determined by use of an appropriate equivalent circuit model of the synchronous machine and its connected electrical system. The integration is carried on until it is determined whether the angle δ will be restored to a steady-state value or whether it will continue to be accelerated by the prime mover.

Figure 11.26 shows typical curves of δ as a function of time for a generator following a short circuit in the connected power system. It is noted that the

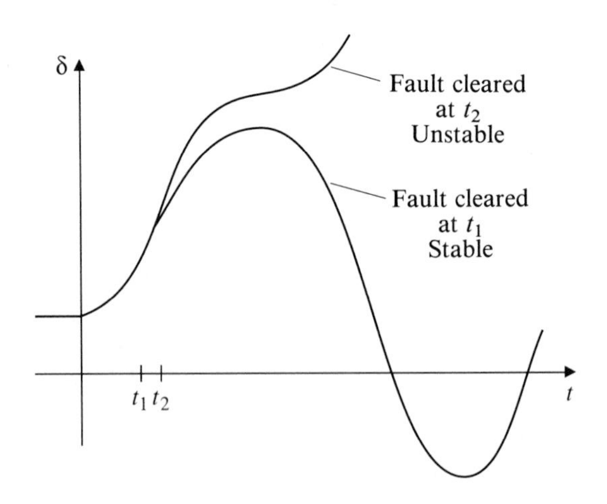

Figure 11.26 Variation of rotor angle δ of synchronous generator after short circuit on power system.

ability of the generator to remain in synchronism depends on the time required for the circuit breakers to operate.

If the disturbance is small, the change in angle δ is also small. The torque-angle relation may then be linearized for small excursions in δ away from a value δ_1 by use of the approximation

$$\sin\delta = \sin\delta_1 + \cos\delta_1(\delta - \delta_1) \tag{11.81}$$

as shown in Fig. 11.27(a). Insertion of this approximation into Eq. (11.74) converts Eq. (11.80) to the form

$$J\frac{2}{p}\frac{d^2\delta}{dt^2} = -\hat{T}\sin\delta_1 - \hat{T}\cos\delta_1(\delta - \delta_1) - T_{shaft} \tag{11.82}$$

Suppose we consider a motor for which the shaft torque is changed by a small increment from T_1 to T_2 at $t = 0$, as shown in Fig. 11.27(b). After $t = 0$, the system is described by the linear differential equation

$$J\frac{2}{p}\frac{d^2\delta}{dt^2} + (\hat{T}\cos\delta_1)\delta = (T_1 - T_2) + (\hat{T}\cos\delta_1)\delta_1 \tag{11.83}$$

The solution for the equation is

$$\delta = \delta_1 + \frac{T_1 - T_2}{\hat{T}\cos\delta_1}(1 - \cos\omega_n t) \quad \text{rad} \tag{11.84}$$

The angle δ oscillates about an average value δ_2, as shown in Fig. 11.27(c). Its angular velocity of oscillation is

$$\omega_n = \left[\frac{\hat{T}\cos\delta_1}{2J/p}\right]^{1/2} \quad \text{rad/s} \tag{11.85}$$

Practically, damping action will cause this oscillation to decay as shown in Fig. 11.27(d). While some damping action may come from speed-dependent torque in the load, the most important damping action arises from currents generated in the damper windings of the machine rotor as the speed departs from its synchronous value, i.e.,

$$\omega_{mech} = \frac{2}{p}(\omega_s + \omega_n) \quad \text{rad/s} \tag{11.86}$$

The angular velocity ω_n is the negative of the angular frequency ω_r of rotor currents as discussed in relation to induction machines. For a rotor of effective resistance R_k and stator flux linkage Λ_s, the damping torque due to the oscillation can be expressed from Eq. (10.66) as

$$T_d \approx \frac{3p}{2}\frac{\Lambda_s^2\omega_r}{R_k} = \hat{T}_d\omega_r = -\hat{T}_d\omega_n = -\hat{T}_d\frac{d\delta}{dt} \quad \text{N}\cdot\text{m} \tag{11.87}$$

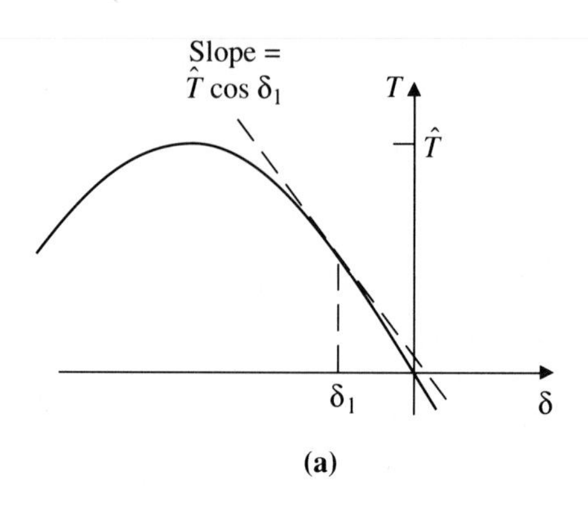

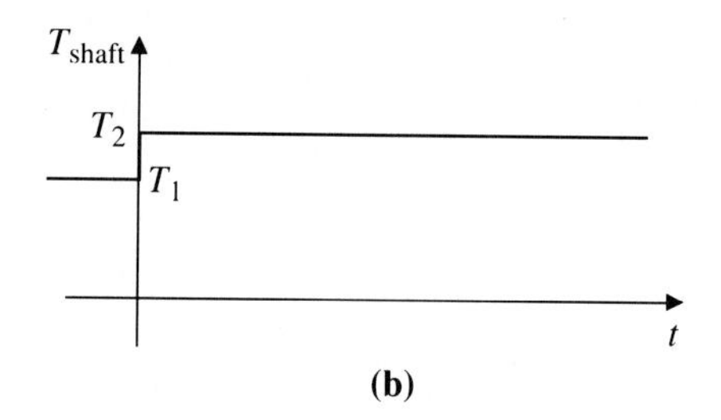

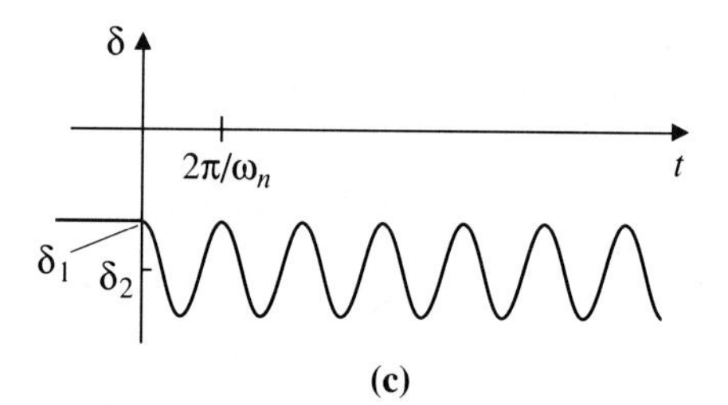

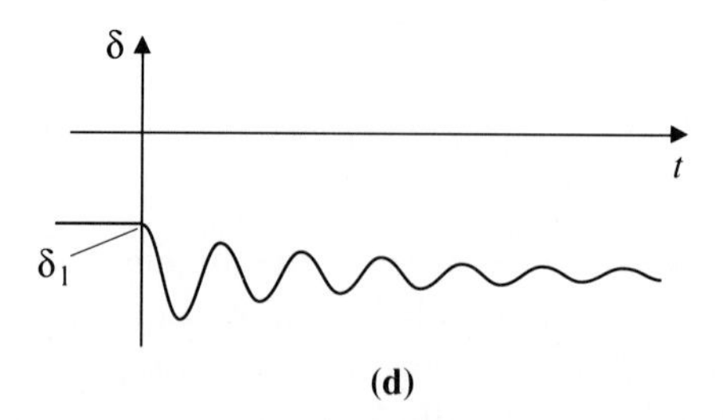

Figure 11.27 (a) Linearization of torque-angle relation about $\delta = \delta_1$. (b) Step change of load torque on synchronous motor. (c) Undamped oscillation of rotor angle δ. (d) Damped oscillation of rotor angle δ.

If this torque is included, the dynamic behavior of the synchronous machine is represented by the differential equation

$$J\frac{2}{p}\frac{d^2\delta}{dt^2} + \hat{T}_d\frac{d\delta}{dt} = T - T_{shaft} \qquad \text{N}\cdot\text{m} \tag{11.88}$$

Sustained mechanical oscillations of considerable magnitude can occur in a synchronous machine that has a periodically varying shaft torque. An example would be a generator driven by a multicylinder reciprocating engine, particularly if one cylinder is misfiring. A further example would be a motor driving a reciprocating compressor. Serious mechanical oscillation may occur if the frequency of a perturbing torque is approximately equal to the natural frequency ω_n of the machine.

11.4

Permanent-Magnet Synchronous Motors

Permanent-magnet motors supplied from inverters have become increasingly attractive for application in a wide range of speed applications, particularly following the introduction of neodymium-iron-boron and samarium-cobalt magnet materials. Most applications are for low- and medium-power levels but the range is continually being extended. Permanent-magnet motors can produce more steady-state and transient torques than can an induction machine of the same frame size; they also can have greater efficiency.

Permanent-magnet machines were introduced briefly in Section 6.6. This section will include a fuller description of some common types of permanent-magnet machines and their transient and steady-state properties.

11.4.1 • Machines with Surface-Mounted Magnets

A common type of permanent-magnet machine is shown in cross section in Fig. 11.28. Small motors typically have this two-pole construction. For larger ratings, four or more poles are more common. The stator contains a three-phase distributed winding, similar to that of an induction machine. The rotor has radially directed permanent magnets mounted on a steel core. The air gap between rotor and stator is made as small as mechanical consideration will conveniently allow.

Permanent-magnet materials for use in most motors have essentially linear demagnetization characteristics of the type shown in Fig. 1.18. As long as the magnet flux density B_m is greater than a critical value B_D, the characteristic may be expressed by the relation

$$B_m = B_r + \mu_r\mu_0 H_m \qquad \text{T} \tag{11.89}$$

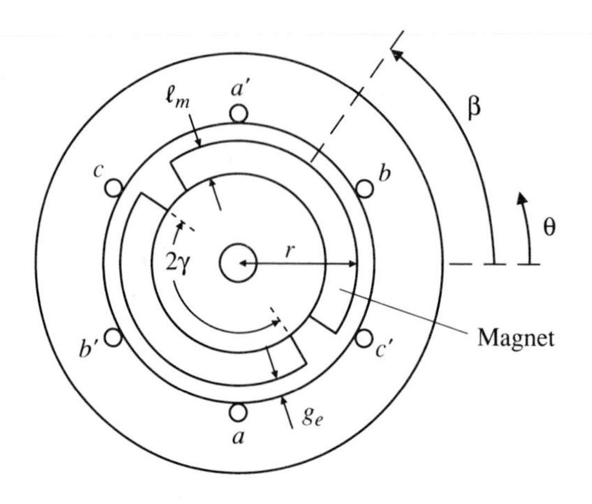

Figure 11.28 Cross section of permanent-magnet machine.

where B_r is the residual flux density (of about 1.1 T for neodymium magnets and about 0.35 T for ferrite magnets) and μ_r is the relative recoil permeability which is usually about 1.05, i.e., little greater than for air. Application of the Circuital Law in a path crossing the air gaps of Fig. 11.28 and assuming ideal iron in stator and rotor gives

$$\ell_m H_m + g_e \frac{B_g}{\mu_0} = 0 \tag{11.90}$$

where ℓ_m is the radial length of the magnet and g_e is the effective length of the air gap. Flux continuity requires that the magnet flux density B_m be equal to the air-gap flux density B_g. A combination of Eqs. (11.89) and (11.90) gives the relation

$$B_g = \frac{\ell_m}{\ell_m + \mu_r g_e} B_r \quad \text{T} \tag{11.91}$$

The steel lamination used in stators have a maximum useful flux density of about 1.8 T. If teeth and slots are about equal in width, the average flux density in the air gap then can be about 0.9 T. With neodymium magnets, this can be achieved with magnets of radial length about 5 mm with an effective air gap of about 1 mm. With samarium cobalt magnets for which $B_r \approx 0.85$ T, the stator teeth would be made somewhat narrower than the stator slots. With ferrite materials, the air-gap flux density is generally less than 0.3 T, and very narrow stator teeth are used.

Let us now develop an equivalent circuit model for the permanent-magnet machine. A rotor magnet of length ℓ_m can be considered to be equivalent to a coil located around the magnet periphery with a constant mmf $\ell_m B_r/\mu_r \mu_0$. This mmf acts over the angular width 2γ of the magnet. As shown in Fig. 6.7, the peak value of the fundamental space component of this mmf is

$$\hat{\mathfrak{F}} = \frac{4}{\pi} \frac{\ell_m B_r}{\mu_r \mu_0} \sin \gamma \qquad \text{A} \tag{11.92}$$

directed along the angular axis β of the rotor. Thus, using Eq. (6.7), the current space vector representing the magnet as seen from the stator is

$$\vec{i}_F = \frac{4\mathfrak{F}}{3N_{se}} = \frac{16}{3\pi} \frac{\ell_m B_r \sin \gamma}{\mu_r \mu_0 N_{se}} \epsilon^{j\beta} \qquad \text{A} \tag{11.93}$$

where $\beta = \beta_0 + \omega_s t$.

The magnet mmf acts on a path consisting of the actual air gap between the magnet and the stator teeth plus the length of the magnet. For practical purposes, this can be considered as equivalent to a steel-to-steel gap of $(g_e + \ell_m)$ because the relative permeability of the magnet material is nearly unity. From Eq. (5.22), the magnetizing inductance as seen from the stator is

$$L_m = \frac{3\pi}{8} N_{se}^2 \frac{\mu_0 \ell r}{g_e + \ell_m} \qquad \text{H} \tag{11.94}$$

where ℓ is the axial length of the rotor and r is its outer radius. Because of the large steel-to-steel gap, the magnetizing inductance of a permanent-magnet machine is relatively small in value, being typically in the range of 0.25 to 0.4 pu based on rating, in contrast to the range of 2 to 5 pu for induction machines.

Because of the near-unity relative permeability of the magnet materials, surface-mounted permanent-magnet machines exhibit almost no saliency. The stator leakage inductance is essentially the same as for an induction machine, i.e., typically in the range of 0.1 to 0.15 pu. The magnet material has a fairly high resistivity and the rotor provides little damper effect. Thus, the machine can be adequately represented by the transient equivalent circuit of Fig. 11.29(a) in the stator reference frame. In contrast with a synchronous machine with wound rotor, the permanent-magnet machine model contains a rotor current source $\vec{i}_r$ of constant magnitude even during transients.

Magnetic saturation in this type of machine is frequently of negligible effect. The mmf required for the large steel-to-steel gap is much larger than that required for the stator and rotor steel.

Because both the inductances $L_{\ell s}$ and L_m are constants, the circuit of Fig. 11.29(a) can be simplified using Norton's theorem to give that of Fig. 11.29(b) where

$$L_M = L_{\ell s} + L_m \tag{11.95}$$

$$\vec{i}_f' = \frac{L_m}{L_M} \vec{i}_F \tag{11.96}$$

The equivalent circuits of Fig. 11.29 can be used for steady-state analysis at frequency $\omega_s = \omega_0$. The phasor variables $\vec{V}_s$, $\vec{I}_s$, $\vec{\Lambda}_s$, $\vec{\Lambda}_m$, $\vec{I}_F$, and $\vec{I}_f'$ are substituted for their corresponding space vectors.

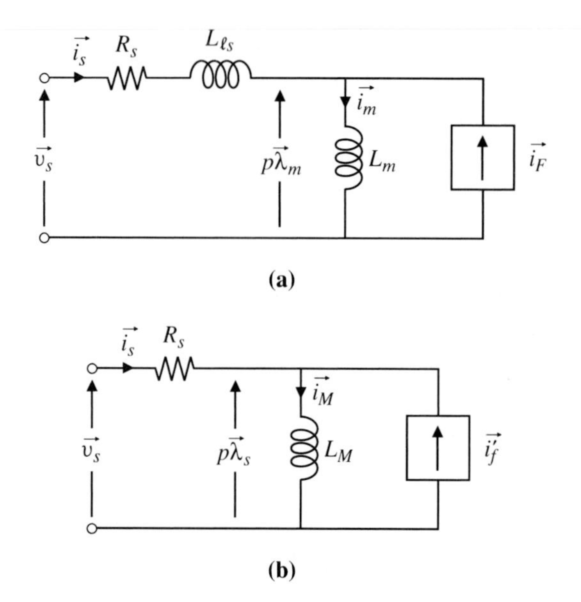

Figure 11.29 (a) Transient equivalent circuit for surface-mounted permanent-magnet machine. (b) Simplified form.

To protect the permanent-magnet material from demagnetization, the flux density at no point in the magnet material should be reduced below the critical value B_D identified in the *B-H* characteristics of Fig. 1.18. This flux density B_D may be either positive or negative for neodymium magnets, the value increasing with increase in temperature. For ferrites, the value decreases with temperature. For samarium cobalt, the value of B_D is highly negative. The air-gap flux density produced by the stator current is given, by adaptation of Eq. (5.17), as

$$\hat{B}_s = \frac{3N_{se}\mu_r\mu_0}{4(\ell_m + \mu_r g_e)}\,\vec{i}_s \tag{11.97}$$

The peak value of flux density produced by the magnet has the value B_g given in Eq. (11.91). The criterion for magnet protection is that

$$B_D < B_g - \hat{B}_s \tag{11.98}$$

The maximum tolerable value of instantaneous stator current is therefore

$$\begin{aligned}\hat{i}_s &= \frac{4(\ell_m + \mu_r g_e)}{3N_{se}\mu_r\mu_0}\left[\frac{\ell_m B_r}{\ell_m + \mu_r g_e} - B_D\right] \\ &= \frac{4}{3N_{se}\mu_r\mu_0}\left[\ell_m B_r - (\ell_m + \mu_r g_e)B_D\right] \quad \text{A}\end{aligned} \tag{11.99}$$

The major danger to the magnet arises from a three-phase short circuit at the stator terminals. Suppose the effective rotor source current at the instant of short circuit is expressed as

$$\vec{i}_F = \hat{i}_F \epsilon^{j\omega_s t} \tag{11.100}$$

The short-circuit stator current is then given by

$$\begin{aligned} \vec{i}_s &= \frac{-j\omega_s L_m}{R_s + j\omega_s(L_{\ell s} + L_m)} \hat{i}_F(\epsilon^{j\omega_s t} - \epsilon^{-t/\tau_s}) \\ &\approx -\frac{L_m}{L_{\ell s} + L_m} \hat{i}_F(\epsilon^{j\omega_s t} - \epsilon^{-t/\tau_s}) \end{aligned} \tag{11.101}$$

where $\tau_s = (L_{\ell s} + L_m)/R_s$. The peak magnitude of this stator current transient will occur at approximately $\omega_s t = \pi$ where the rotating space vector of current is aligned with the fixed-current vector decaying at the stator time constant τ_s. In small machines, this time constant is frequently small with respect to a half period of the fundamental frequency. In large machines, it may be equal to several periods. In any case, the machine must be designed with sufficient stator leakage inductance to restrict the peak stator current on short circuit to the value given in Eq. (11.99).

When operating a permanent-magnet motor from an inverter supply, it is desirable to have the stator current more or less in phase with the stator voltage to make best use of inverter capacity. A phasor diagram for such an operating condition is shown in Fig. 11.30. Because of the relatively high value of magnetizing current in a permanent-magnet machine, the magnitude of the angle β_0 between the stator voltage and the equivalent rotor current for unity power factor is just greater than 90°. The phase angle of the stator current is quite sensitive to both the angle β_0 and to the stator voltage magnitude.

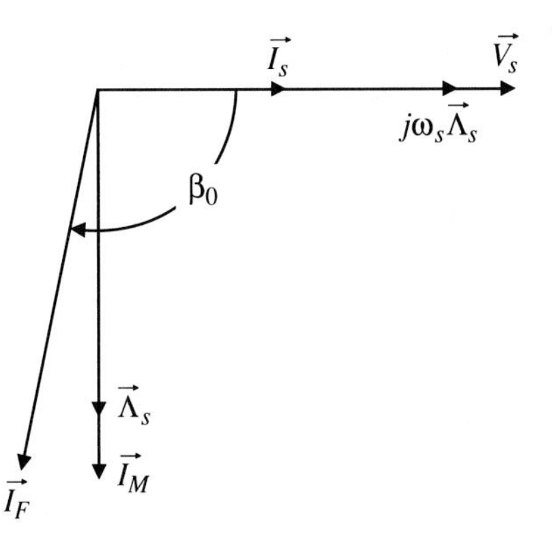

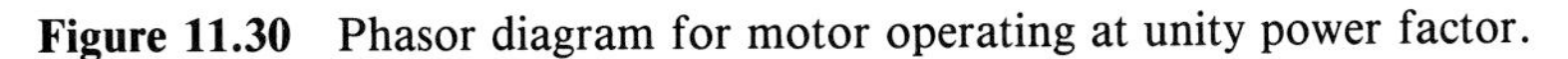

Figure 11.30 Phasor diagram for motor operating at unity power factor.

One of the concerns in designing a permanent-magnet motor is the reduction of the cogging torque that arises from the interaction of the rotor magnets with the stator teeth. Careful choice of magnet width and shape is required to keep the reluctance as seen by the magnet as constant as possible during rotation. In some designs, the stator slots are skewed. An alternative design has a stator with no teeth. A set of continuous stator windings is fitted in the gap between the cylindrical yoke and the air gap. This type of motor is virtually free of torque ripple.

11.4.2 ▪ Machines with Inset Magnets

An alternative form of rotor construction is the inset magnet arrangement shown in Fig. 11.31. In this, the magnets are set into the steel rotor so that steel sections project outward to the air gap. The action of the magnets in this machine is essentially the same as for the surface-mounted magnet machine of Section 11.4.1, but the magnetic system is salient. The relative permeability of the magnet material is near unity, and the steel-to-steel air gap in the direct or magnet axis is much larger than in the quadrature axis.

For transient analysis, this machine may be represented by the pair of equivalent circuits of Fig. 11.32 which are adapted from Fig. 11.10. The equivalent magnet current i_F referred to the stator turns is as derived for the surface-mounted magnet in Eq. (11.93). The quadrature axis magnetizing inductance L_{mq} will be comparable in magnitude to that of an induction machine because of the small effective air gap over much of the quadrature axis arc. The direct-axis magnetizing inductance L_{md} will be substantially less than L_{mq} and approximately of the same value as given by Eq. (11.94). Thus, this machine has a saliency that is opposite in nature to that encountered in salient-pole synchronous machines in Section 11.1.3.

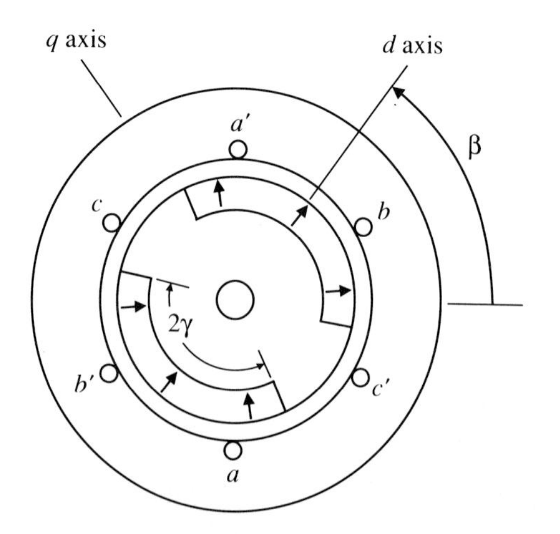

Figure 11.31 Machine with inset magnets.

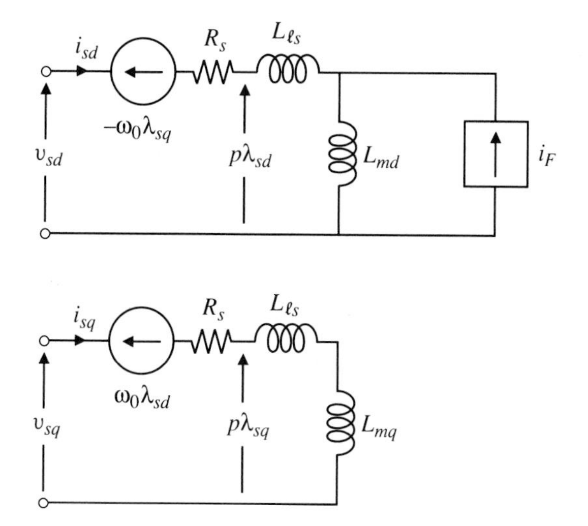

Figure 11.32 Transient equivalent circuits for a machine with inset magnets.

For steady-state analysis, the equivalent circuit of Fig. 11.15(b) is conceptually useful. Because, in this case, $X_d < X_q$, the series reactance in this circuit is negative. The corresponding relation between torque and rotor angle is shown in Fig. 11.33. For the same magnet arc 2γ, the peak torque of a motor with inset magnets can be somewhat larger than that of a surface-mounted type. An additional advantage is that the net flux of the inset magnet machine can be

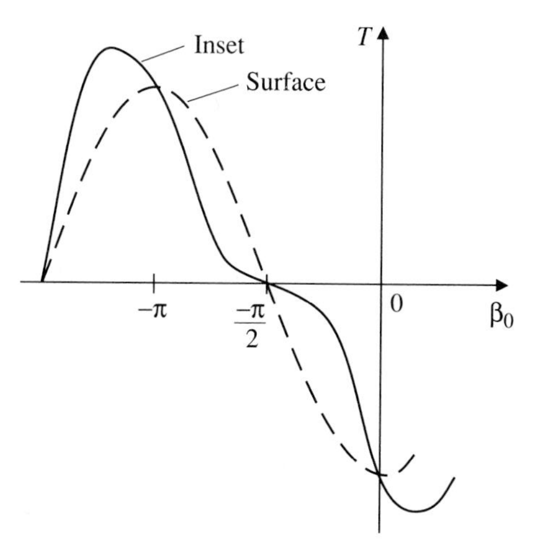

Figure 11.33 Torque-angle relation for inset-magnet machine with surface-mounted magnet machine for comparison.

weakened for operation at high speed with an inverter of limited voltage. This weakening is accomplished by use of a direct-axis component of stator current producing reversed flux density on each side of the magnet.

11.4.3 • Machines with Circumferential Magnets

Ferrite magnets are preferred for many permanent-magnet machines because of their low cost. When ferrite magnets are used in the radially directed surface mounting as shown in Fig. 11.28, the air-gap flux density is relatively low. The gap density can be substantially increased by arranging the ferrite magnets in a circumferential direction sandwiched between steel pole arcs, as shown in Fig. 11.34.

The relation between gap flux density B_g and magnet flux density B_m can be derived by noting that the air-gap flux for one pole is equal to the flux from two adjacent magnets. Thus,

$$B_g\left[\frac{2\pi r - p\ell_m}{p}\right] = 2w_m B_m \tag{11.102}$$

The maximum possible width w_m of the magnet occurs when the radius r_c of the nonmagnetic core is such that the magnets just touch, i.e., where $r_c = p\ell_m/2\pi$. Thus,

$$w_m = r - r_c = r - p\ell_m/2\pi \qquad \text{m} \tag{11.103}$$

Combining Eqs. (11.102) and (11.103) gives

$$B_g = \frac{p}{\pi} B_m \qquad \text{T} \tag{11.104}$$

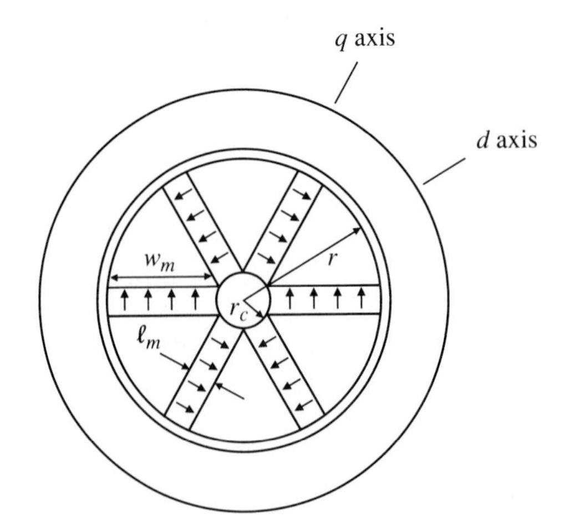

Figure 11.34 Machine with circumferential magnets.

With ferrite magnets with a working flux density of about 0.25 to 0.3 T, the number of poles is usually designed to be eight or greater to achieve the desired gap flux density.

In this type of machine, the direct axis is centered on a steel pole while the quadrature axis is centered on the edge of a magnet. Direct-axis flux encounters the long effective gaps ℓ_m through the magnets in passing to adjacent poles, while quadrature axis flux can complete its rotor path entirely in the steel of the pole. Therefore, this machine is magnetically salient with $L_{md} < L_{mq}$. The transient equivalent circuit of Fig. 11.32 is appropriate for analysis. The torque-angle relation is similar to that of Fig. 11.33.

Permanent-magnet motors that require a capability to start on a constant-frequency supply are frequently made with a rotor structure similar to that of Fig. 11.34 except that a squirrel cage is built into slots or holes around the rotor surface to provide induction motor action when the speed ω_0 is less than the synchronous speed ω_s. When the machine pulls into synchronism, the cage continues to act as a damper.

11.4.4 ▪ Hysteresis Motors

A special form of permanent-magnet motor is the hysteresis motor in which the rotor typically consists of a ring of permanent-magnet material mounted over a central core. When this central core is made of high-permeability steel, the magnet material has a radially directed flux density. If the central core is non-magnetic, the flux from the air gap is directed circumferentially around the magnet arc.

In contrast to the permanent-magnet machines discussed earlier in this section, the magnet material for this machine is chosen so that its operating point can be cycled around its *B-H* characteristic by the stator mmf. A typical *B-H* loop is shown in Fig. 11.35(a). As the stator mmf $\vec{\mathfrak{F}}_s$ rotates around the machine, the air-gap flux density has the space wave B_θ shown in Fig. 11.35(b). Therefore, the fundamental component of this flux density $\vec{B}_{1g}$ and its corresponding stator flux linkage $\vec{\lambda}_s$ lag behind the stator current vector $\vec{i}_s$, as shown in Fig. 11.35(c). The machine produces a torque

$$T = \frac{3p}{4}\, \mathcal{I}(\vec{\lambda}_s^* \vec{i}_s) \qquad \text{N·m} \tag{11.105}$$

If it is assumed that the shape of the *B-H* loop is independent of the rotor frequency $\omega_r = \omega_s - \omega_0$, the space waves of Fig. 11.35(b) apply at any frequency and the torque will be independent of the rotor speed ω_0 as long as $\omega_0 < \omega_s$. The materials used in most hysteresis motors are alloys of iron, nickel, cobalt, vanadium, and other metals and have enough conductivity to provide some additional induction motor torque. When synchronous speed is reached, the torque depends on the angle β_0 as in other synchronous machines. A typical speed-torque relation is shown in Fig. 11.35(d).

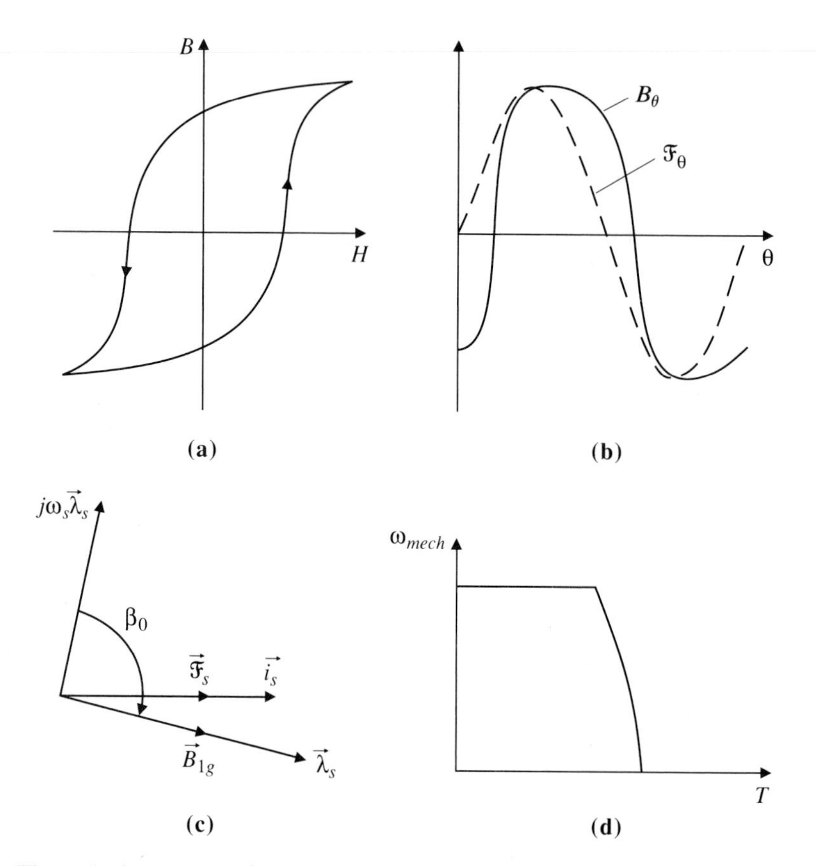

Figure 11.35 Hysteresis motor: (a) *B-H* loop. (b) Space waves of stator mmf and gap flux density. (c) Vector diagram. (d) Speed-torque relation.

Hysteresis motors provide very smooth, ripple-free torque as is required, for example, for tape drives and phonograph motors. The stator current is nearly constant for the whole speed range up to synchronous speed. Their low efficiency confines their use to low power applications. They are frequently used in timing motors.

11.5

Variable-Frequency Synchronous Motor Drives

Speed control can be obtained using a synchronous motor supplied from any of the variable-frequency inverters or cycloconverters described in Sections 8.4

and 8.5. In many respects, the principles involved are similar to those discussed in relation to induction motor drives in Sections 10.2 and 10.3.

Because of the unique relation between stator frequency and mechanical speed for a synchronous machine, accurate speed control can be achieved. Also, because of the ability to adjust field current in wound-field machines, greater control of operating properties such as power factor is possible than with induction machines. On the other hand, synchronous machines are subject to loss of synchronism under transient conditions unless special control measures are taken.

There are two basic modes of operation for synchronous-motor drives: (a) open loop, where the stator frequency, voltage, and/or current are directly controlled, and (b) closed loop, where the frequency is governed by information obtained usually from a rotor position sensor.

11.5.1 ▪ Open-Loop Drive Systems

A block diagram for a typical open-loop speed control system is shown in Fig. 11.36. A controlled rectifier supplies variable direct voltage through a link to an inverter which in turn supplies a synchronous motor. In its simplest form, both the frequency of the inverter and the direct-link voltage are made proportional to the demand speed ω^*_{mech}. In the steady state, the mechanical speed for a p-pole machine is

$$\omega_{mech} = \frac{2}{p}\,\omega_s \qquad \text{rad/s} \tag{11.106}$$

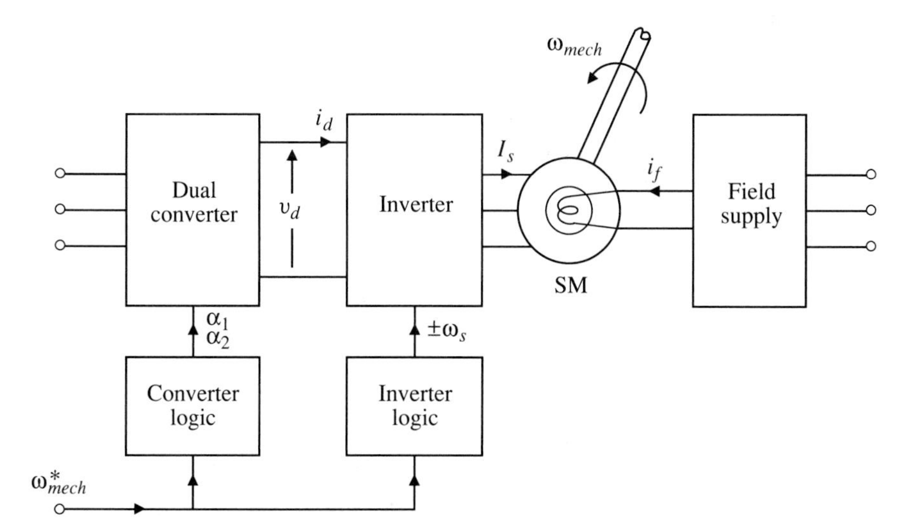

Figure 11.36 Synchronous motor drive operating in open loop.

Because the frequency ω_s of the inverter can be very accurately controlled by a crystal oscillator, a correspondingly accurate control of steady-state motor speed can be achieved.

For a wound-rotor synchronous machine with field current i_f, the steady-state torque is, from Eq. (11.23),

$$T = -\frac{3p}{2} \Lambda_s n i_f \cos \beta_0 \qquad \text{N}\cdot\text{m} \tag{11.107}$$

where Λ_s is the rms value of stator flux linkage, n is the effective current ratio, and β_0 is the angle between the stator-induced voltage and the field axis. The machine may be represented in the steady state by the equivalent circuit of Fig. 11.4(b) for example, from which the stator voltage is seen to be

$$\vec{V}_s = R_s \vec{I}_s + j\omega_s \vec{\Lambda}_s \qquad \text{V} \tag{11.108}$$

The steady-state operating range of the drive system is dependent on several ratings. For the motor, these will include the rated values of both the stator current and the field current, each being dependent on the heat dissipation properties of the cooling system. There will also be a rated value of stator flux linkage designated as Λ_{sb}, which is limited by the magnetic properties of the steel laminations. For the controlled-rectifier, voltage-source-inverter system of Fig. 11.36, there will be a maximum value of the link voltage $\hat{v}_d$ and thus a maximum value for the rms value of the fundamental stator voltage designated as V_{sb}. Thus, ignoring the stator resistance drop, the drive can operate at full flux linkage and therefore full torque up to a speed

$$\omega_{sb} = \frac{V_{sb}}{\Lambda_{sb}} \qquad \text{rad/s} \tag{11.109}$$

For higher-speed operation with constant stator voltage, the flux linkage Λ_s of the wound-field machine can be reduced in inverse proportion to the speed by reduction of the field current. With rated stator current, the available steady-stator torque will be proportional to the flux linkage and thus will be inversely proportional to the speed. The available mechanical power will then be constant up to the maximum speed of the drive. If the direct-voltage link is supplied from a dual converter, the drive will be capable of operating in all four quadrants as indicated in Fig. 11.37. Reversed speed is achieved by reversal of phase sequence.

Over much of the speed range below ω_{sb}, it is satisfactory to make the stator voltage directly proportional to frequency or speed. For operation at very low speed, an additional compensation for the stator resistance drop is required.

If a simple voltage-source inverter is used, as described in Section 8.5, there will be substantial harmonic components in the stator current. These will cause 6th, 12th, and higher harmonic torques similar to those described for induction motor drives in Section 10.2.2. For smooth operation at low speeds, it is generally necessary to use a pulse-width-modulated inverter with a more nearly sinusoidal voltage waveform.

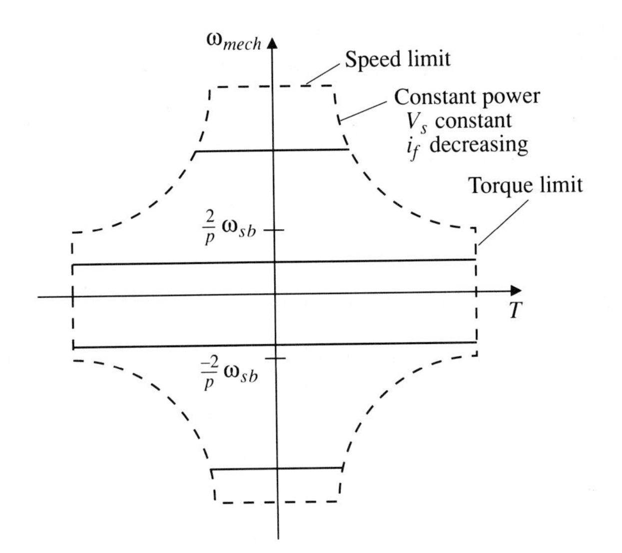

Figure 11.37 Steady-state operating range for a synchronous motor drive.

A typical phasor diagram for a synchronous motor is shown in Fig. 11.38. If the stator current is made somewhat leading or capacitive by adjustment of the field current, the switches in the inverter will be naturally commutated. This feature is frequently exploited in large power drives because of the limitation in turn-off capability of large semiconductor switches at high frequency.

In an open-loop system, the rotor angle β_0 adjusts itself to changes in the load torque. If the magnetizing current of the motor is significantly smaller than the stator current, it is seen from Fig. 11.38 that the angle β_0 has a large negative value with a cosine approaching the value -1. Thus, when operating at full torque and a constant value of field current, there may be very little margin of steady-state torque available if load is increased, i.e., there is danger of losing

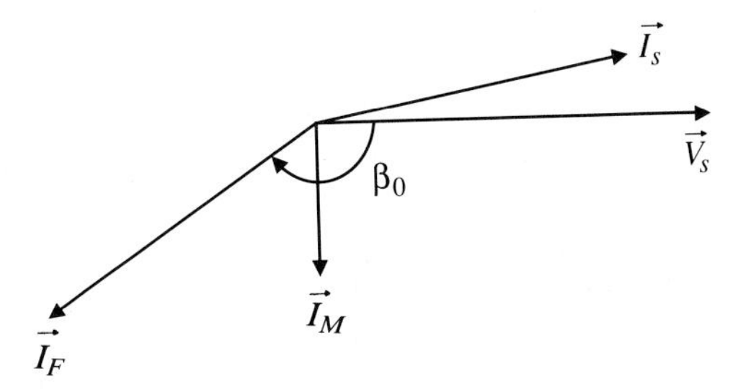

Figure 11.38 Phasor diagram for synchronous motor with large magnetizing inductance.

synchronism. Because of this, open-loop drives usually use synchronous motors with smaller magnetizing inductances than for induction machines, i.e., usually less than 1 pu.

Open-loop drives are employed only with loads that provide a steady predictable torque, free of significant transient disturbances that might otherwise produce instability and loss of synchronism. They are also restricted to situations where the speed and frequency are changed very slowly.

11.5.2 ▪ Self-Controlled Drive Systems

The major shortcoming of an open-loop synchronous motor drive is its liability to lose synchronism following a transient in load torque or a too-rapid change of inverter frequency. This limitation can be overcome by using the position of the motor shaft to determine the inverter switching operations. This self-controlled system, shown in block diagram form in Fig. 11.39, has an angular position sensor attached to its rotor. Signals from this sensor are processed in the control logic and used to operate the switches in the inverter. If a simple voltage source inverter is used, the six switching operations per cycle then occur at predetermined values of the rotor position. This allows the angle β_0 and the quantity cos β_0 in the torque expression of Eq. (11.107) to be held constant independent of the instantaneous speed.

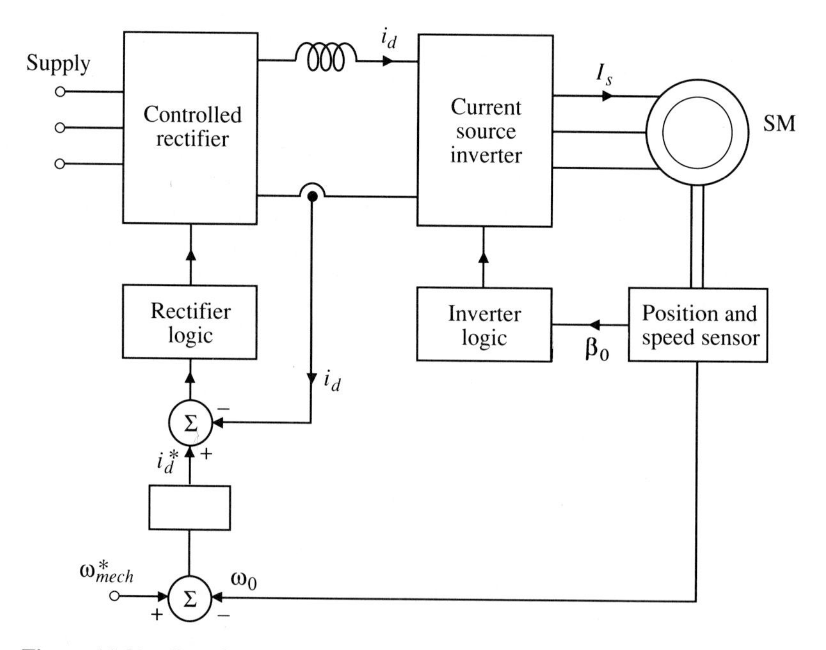

Figure 11.39 Speed control system using synchronous motor with shaft position sensor.

In a self-controlled drive system, the speed is no longer determined independently. If, for example, the stator voltage magnitude V_s were held constant, as well as the field current i_f, the stator flux linkage Λ_s would be inversely proportional to speed or frequency. From the torque expression of Eq. (11.107), it is seen that the result would be a constant power characteristic where

$$T\omega_{mech} = -3V_s n i_f \cos\beta_0 \qquad \text{W} \tag{11.110}$$

Increase in load torque would result in a decrease in speed and therefore in frequency which would result in an increase in stator flux linkage, increasing the torque produced by the motor. The operation would then be somewhat similar to that of a series commutator machine.

To produce a speed control system, the speed derived from the position sensor or from a tachometer is used to control the input to the inverter. Either a voltage-source or current-source inverter may be used; the latter is shown in the system of Fig. 11.39. This system controls the magnitude I_s of the stator current while the position sensor and the logic control the angle of this current. A typical phasor diagram for a synchronous motor with stator-current control is shown in Fig. 11.40. From the equivalent circuit of Fig. 11.4(b),

$$\vec{\Lambda}_s = L_s(\vec{I}_s + \vec{I}_F) \qquad \text{Wb} \tag{11.111}$$

The torque can now be derived as

$$\begin{aligned} T &= -\frac{3p}{2}\,\mathcal{I}(\vec{\Lambda}_s^*\vec{I}_s) \\ &= -\frac{3p}{2}\,L_M I_s n i_f \sin\delta \qquad \text{N}\cdot\text{m} \end{aligned} \tag{11.112}$$

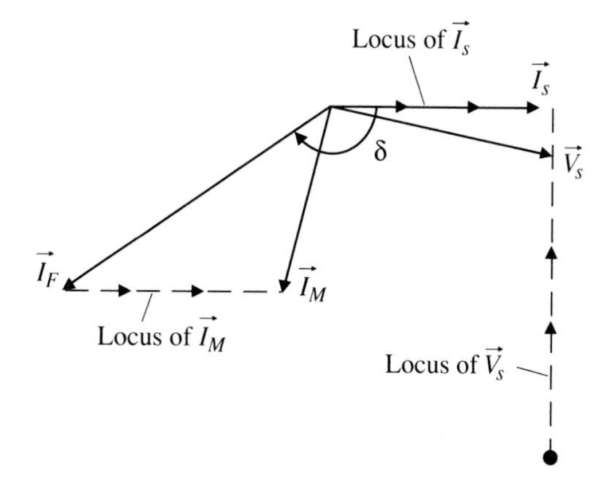

Figure 11.40 Phasor diagram for current-driven synchronous motor with constant angle δ showing loci as stator current is increased.

where δ is the angle of $\vec{I}_F$ with respect to $\vec{I}_s$. Thus, with constant field current i_f and constant angle δ, the torque is directly proportional to the stator current and thus also to the link current i_d. As the magnitude of the stator current is varied at constant angle δ, the net magnetizing current I_M and the stator voltage V_s follow the loci shown in Fig. 11.40. As the stator current decreases, the magnetizing current and the stator voltage increase until at $I_s = 0$ all of the field current is used for magnetization.

Shaft position sensing can be achieved in a number of ways. For systems where six switching operations are required per cycle, the sensor system illustrated in Fig. 11.41 is applicable for a two-pole motor. The three proximity detectors are spaced $\pi/3$ rad from one another. The three signals that indicate the presence or absence of the rotor-mounted shutter then uniquely define six 60° intervals in a revolution. For a p-pole machine, the detectors are spaced $\pi/3$ electrical rad from one another and the shutter has $p/2$ large-diameter sectors. Signals from the detectors may be used to operate the inverter switches. It may be noted that operation in both directions of rotation is possible because the sequence of the detector signals reverses with reversed speed. Either magnetic or optical proximity detectors may be used.

For more precise position measurement, the most frequently employed sensor consists of a set of concentric optical rings each having alternate opaque and clear sections. With optical detectors, each ring gives one bit in the angular position information, i.e., ten rings, for example, give a position indication each 0.35°. Using a shaft position sensor, a synchronous motor can be used as an accurate position control. The equivalent circuit model developed in Section 11.1 can be used for transient analysis of such a system.

At low power levels, the dominant type of drive is the electronically switched permanent-magnet motor introduced in Section 6.7. The motors for these drives differ from those considered earlier in this chapter in that the stator windings have uniform conductor distribution, as shown in Fig. 6.8(a), in-

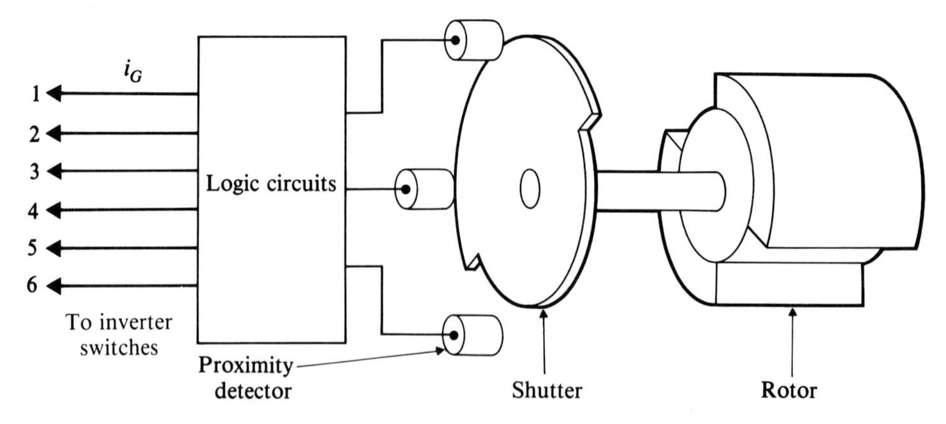

Figure 11.41 Rotor shaft-position encoder.

stead of a near-sinusoidal distribution. Torque is produced by the interaction of a rectangular flux density from the magnets with a rectangular stator current distribution. A typical position control system using such a switched permanent-magnet motor is shown in Fig. 11.42. A position sensor on the shaft produces a signal representing the rotor angle β. This is compared with the demand angle β^*, and the difference is used to produce a demand value of current i_d^*. This in turn is compared with the measured value of the link current i_d from the direct-voltage supply, and the difference is processed in the switching logic to activate the switches in such a way as to maintain the desired link current within a small-magnitude band. In addition, the position sensor signal is used to select the appropriate pair of operating switches so that the link current is carried through the two stator windings in series corresponding to the rotor angle arc.

Expressions for torque as a function of link current, generated voltage as a function of speed, and speed as a function of torque for this type of machine are given in Eqs. (6.26), (6.27), and (6.28). The operating properties of these drives are similar to those of commutator machines.

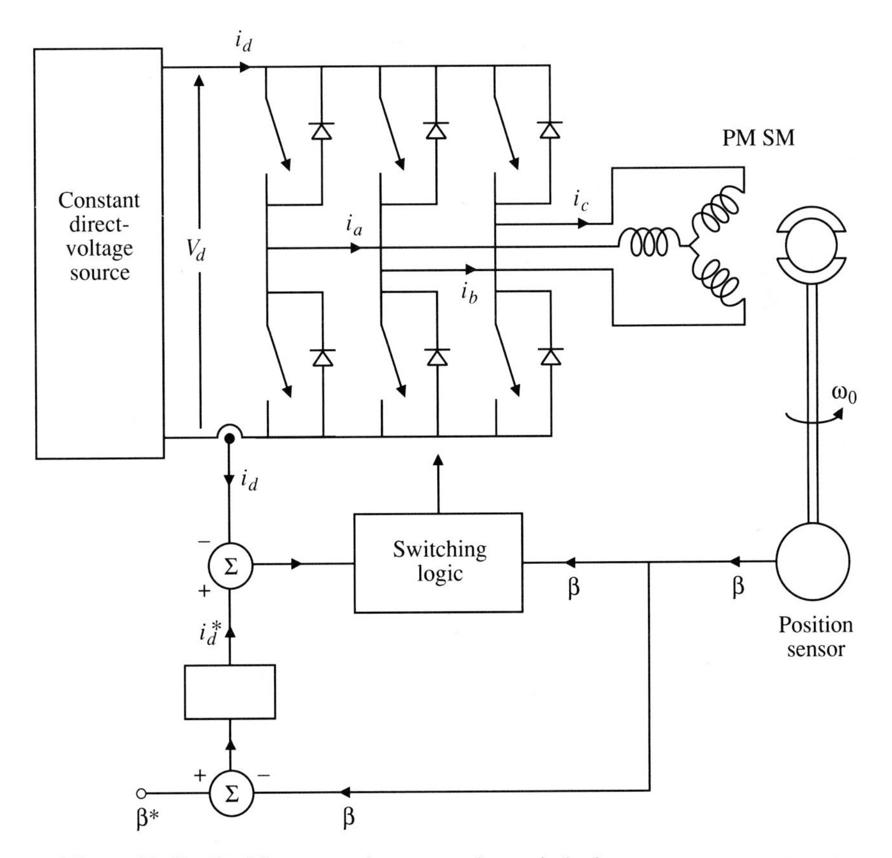

Figure 11.42 Position control system using switched permanent-magnet motor.

11.6

Switched Reluctance Motor Drives

The combination of a reluctance motor with an appropriately controlled power converter provides a drive that has a number of attractive features: simplicity of both motor and converter, ruggedness, wide speed range, and low rotor inertia. A cross section of a typical reluctance machine is shown in Fig. 11.43(a). The stator has six salient poles, each with a concentrated coil. The rotor has

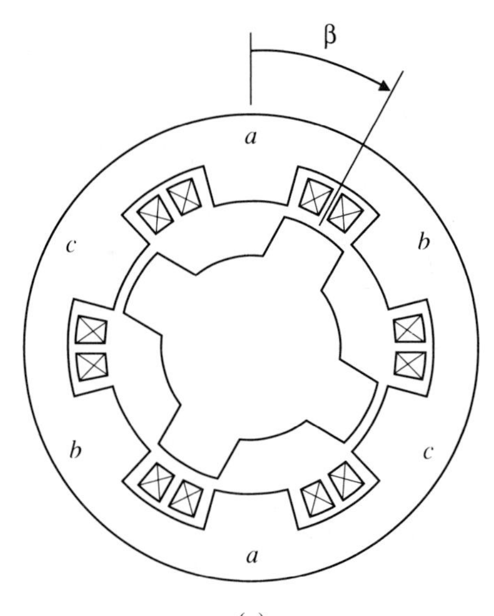

(a)

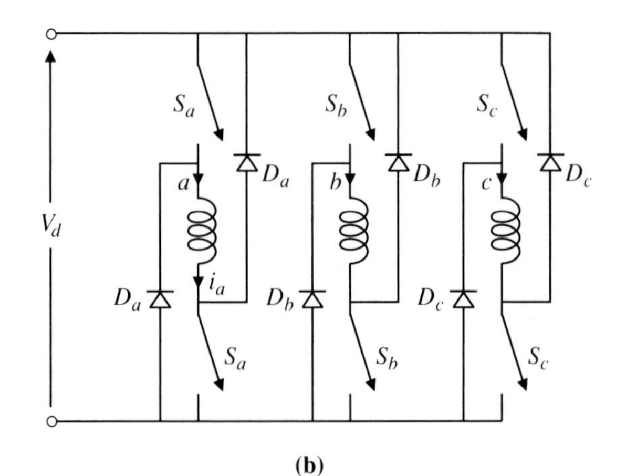

(b)

Figure 11.43 (a) Reluctance motor. (b) Power converter.

four salient poles with no windings. The stator coils on opposite poles are connected in series. When a current i_a is passed through the phase-a windings in the machine of Fig. 11.43(a), two of the rotor poles tend to rotate to align with the energized pair of stator poles. If now the current is switched to the windings of phase b, thére will be a torque tending to continue the counterclockwise rotation.

The three stator-winding phases can be supplied from the power converter shown in Fig. 11.43(b). Ideally, only one phase carries current at any instant. The direction of the current is not significant for torque production in a reluctance machine.

The introduction to reluctance motors in Section 3.7 was based on the simplifying assumption of ideal stator and rotor iron, providing a torque that was proportional to the square of the phase current and proportional to the rate of change of inductance with angle (Eq. [3.45]).

If a reluctance motor is operated in the magnetically linear range, its available torque is relatively small. In practice, motors are operated so that the energized poles are driven significantly into saturation in order to increase the torque. Therefore, prediction of the motor-operating properties requires use of the concepts introduced in Section 7.8. It was shown there by analogy with Eq. (7.71) that the torque with constant phase current could be derived as the rate of change of the coenergy of the magnetic field with respect to the rotor angle β:

$$T = \left.\frac{\partial W'}{\partial \beta}\right|_{i=constant} \qquad \text{N}\cdot\text{m} \tag{11.113}$$

Suppose that only the coils a in Fig. 11.43(a) are conducting a current i_a. The flux linkage λ_a of this phase circuit is shown in Fig. 11.44(a) as a function of i_a for several angular positions of the rotor. When two of the rotor poles are located along a line between phases a and b as shown for $\beta = -\pi/6$ in Fig. 11.43(a), the flux linkage per stator ampere is low and the λ_a-i_a relationship is nearly linear. When the rotor poles are aligned with the phase-a poles at $\beta = 0$, they are highly saturated and the λ_a-i_a relation is nonlinear. From Eq. (7.69), the coenergy $W'_{a\beta}$ at any rotor position β is defined as

$$W'_{a\beta} = \int_0^{\hat{i}_a} \lambda_a \, di_a \qquad \text{J} \tag{11.114}$$

i.e., equal to the area between the λ_a-i_a curve and the i_a axis up to the current i_a for the appropriate value of β in Fig. 11.44(a). Therefore, the torque can be found by evaluating the coenergy of Eq. (11.114) for a number of angular positions β and using the differences between these values in Eq. (11.113) to obtain the average torque over each of the increments in β. A typical torque-angle relation with constant phase current i_a is shown in Fig. 11.44(b).

For the machine in Fig. 11.44(a), the torque from phase a is responsible for revolving the rotor from $\beta = -\pi/6$ to $\beta = 0$. If phase b then is energized, the rotor will rotate toward $\beta = \pi/6$. Thus, energizing the phase coils in the sequence a-b-c produces rotation in the positive direction over the range $\beta = -\pi/6$

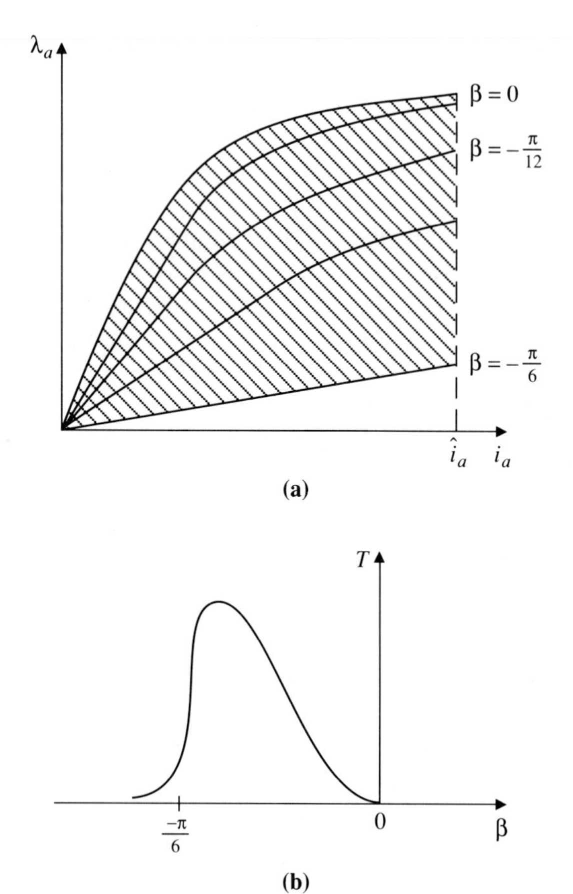

Figure 11.44 (a) Flux linkage-current curves for several rotor positions. (b) Typical torque-angle relation with constant phase current.

to $\beta = \pi/3$, i.e., over one-quarter of a revolution. Each stator phase is energized four times as the four rotor poles complete one revolution. Reversed rotation is achieved by reversing the phase sequence to *a-c-b*.

A number of combinations of stator poles p_s and rotor poles p_r can be used effectively in a switched reluctance motor. The machine in Fig. 11.43(a) is usually designated as a 6/4 machine. In order to ensure that the rotor is never in a position where the torque due to the current in any phase is zero, the ratio p_s/p_r or its inverse should not be an integer. Each of the $p_s/2$ stator phases is energized p_r times per revolution. If the current in a phase could be maintained constant over the angular displacement π/p_s, the mechanical work done over this interval would be, from Eq. (11.113),

$$W_{mech} = W'_{(\beta=0)} - W'_{(\beta=-\pi/p_s)} \quad \text{J} \tag{11.115}$$

It would be represented by the total hatched area in Fig. 11.44(a). The average torque would be

$$\bar{T} = \frac{W_{mech}}{\Delta\beta} = \frac{W_{mech}p_s}{\pi} \quad \text{N·m} \tag{11.116}$$

In designing a reluctance motor, the choice of the pole widths is of critical importance. The wider the poles, the greater is the maximum flux linkage that can be achieved and the greater is the value of coenergy at $\beta = 0$ in Fig. 11.44(a). However, the narrower the poles, the lower will be the flux linkage when the rotor is at the position $\beta = \pm\pi/p_s$. There is an optimum width to maximize the difference in the coenergy terms of Eq. (11.114). It also may be noted that the use of the minimum air gap consistent with mechanical limitations produces the maximum slope of the unsaturated part of the λ_a-i_a curve in Fig. 11.44(a), thus contributing to maximizing the coenergy at this angular position.

The power converter of Fig. 11.43(b) provides for individual control of the current in each of the stator phases. A constant-link voltage V_d is provided from a rectifier or from another direct-voltage source. A set of proximity sensors provides information of the position of the rotor poles relative to the stator. A typical waveform of an achievable pulse of stator current is shown in Fig. 11.45. The two switches S_a for phase a are turned on at rotor angle $\beta = -\pi/p_s$ by a sensor detecting a pole edge. At this angle, the inductance L of phase a is low and reasonably linear. The current rises at a time constant $\tau = L/R$ toward a value V_d/R where R is the effective resistance of the circuit, including the source. When the phase current, which is also the supply current, reaches the desired value, the switches S_a are turned off. The phase current continues to flow through the diodes D_a. The phase winding is now connected to a supply of voltage $-V_d$. Thus, the current decays at essentially the same time constant toward a value $-V_d/R$. The switches S_a are then operated in such a manner as to limit the current i_a to a small band about the desired

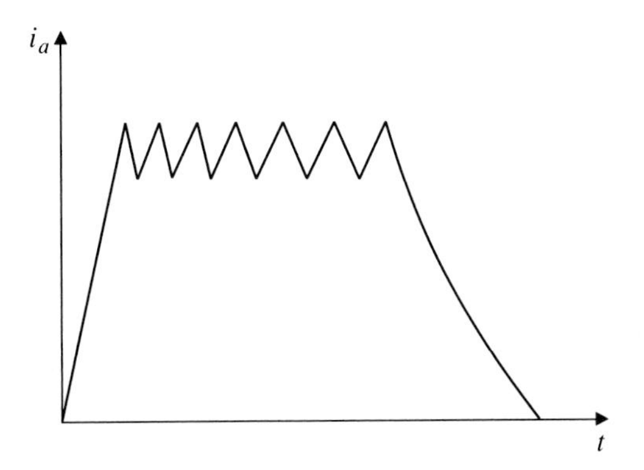

Figure 11.45 Phase current waveform.

value. As the rotor moves toward $\beta = 0$, the effective inductance of the phase increases and also becomes nonlinear. When the switches S_a are turned off at near $\beta = 0$, the phase current decays at a relatively long time constant because this is the position of maximum inductance. The energy stored in the magnetic field is returned to the link supply or is dissipated in the resistance.

One of the undesirable features of a switched reluctance motor drive is the ripple in its torque at a basic frequency of $p_s p_r/2$ multiplied by the rotor speed in revolutions per second. This ripple is caused by a number of factors: the non-uniformity in the rate of change of coenergy with angular rotation as shown in Fig. 11.44(a), the delay in establishing the current in the phase, and the relatively long decay in the current at the end of the pulse as shown in Fig. 11.45. This last factor can produce a reversal of instantaneous torque as the rotor pole pulls away from the stator pole while still carrying flux. Minimization of this ripple torque requires detailed consideration of the nonlinear dynamic relations for the complete drive system. One measure that can be taken to reduce torque ripple is to turn on the current somewhat earlier and to begin the final current decay somewhat earlier. Also, the current profile can be shaped so as to make the torque more nearly constant during each phase period.

An advantage of the switched-reluctance motor over the permanent-magnet motor is that, if a stator short circuit occurs, the stator current decays quickly. In the permanent-magnet motor, the short-circuit current persists as long as the machine continues to rotate.

11.7

Stepper Motor Drives

A stepper motor is one that is designed to rotate through a specific angle in response to each pulse of supply current applied to it. Its applications are dominantly in positioning devices such as recording heads in disk drives, numerically controlled machine tools, and paper drives in printers and typewriters.

While there are many configurations of stepper motors, they may be grouped into two basic types, one having a reluctance rotor and the other having a permanent-magnet rotor. The reluctance machine shown in Fig. 11.43(a) is an elementary example of the former type. Current pulses applied sequentially to the phases a, b, and c each produce a rotation of $\pi/12$ or 30°. More frequently, step sizes down to less than 1° are desired. An example of a typical stepper motor is shown in cross section in Fig. 11.46. This machine has 32 rotor teeth or poles. Each of the six stator poles has three teeth. As it is shown in Fig. 11.46, the rotor teeth have been aligned with the phase a stator teeth by a current in stator phase a. The number of teeth on the rotor has been chosen so that the teeth of the phase-b poles lead the adjacent rotor teeth. For equal tooth and slot

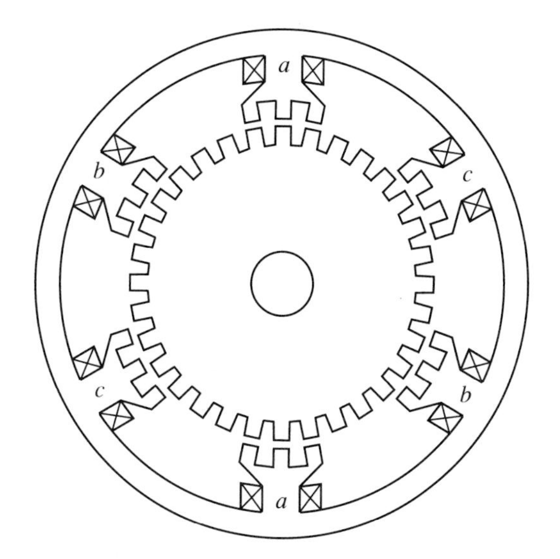

Figure 11.46 Reluctance-type stepper motor.

widths, the lead is two-thirds of a tooth width. Application of a current pulse to phase *b* following removal of the current in phase *a* produces a positive or counterclockwise rotation of angle $\Delta\beta$, where

$$\Delta\beta = \frac{360}{2 \times 32} \times \frac{2}{3} = 3.75^\circ \tag{11.117}$$

In general, the number of steps per revolution for this type of machine is $p_s p_r/2$, which, for $p_s = 6$ and $p_r = 32$ in the example, is 96.

Another form of a reluctance-type stepper motor has a multistack configuration, one stack of which is shown for a typical machine in Fig. 11.47. In each stack, the rotor and stator have equal numbers of teeth. Adjacent stator teeth have equal and opposite magnetomotive forces produced by windings carrying the phase current. On the same shaft, there are n similar stacks. Each of these has either its rotor or its stator displaced sequentially from its neighbor by a tooth or pole pitch divided by n, and each is excited by a phase current. Using this simple construction, the number of steps per revolution is made equal to np_r. While a three-stack configuration is common, larger numbers of stacks can be used to obtain smaller step sizes.

A simple example of a permanent-magnet stepper motor is shown in Fig. 11.48. The cylindrical-magnet rotor tends to align with the stator phase that is excited. In contrast with the reluctance-type stepper motor, the direction of the phase current is significant in this machine. Energization of the phases *a*, *b*, and *c* in sequence produces 60° steps of forward rotation. An advantage of

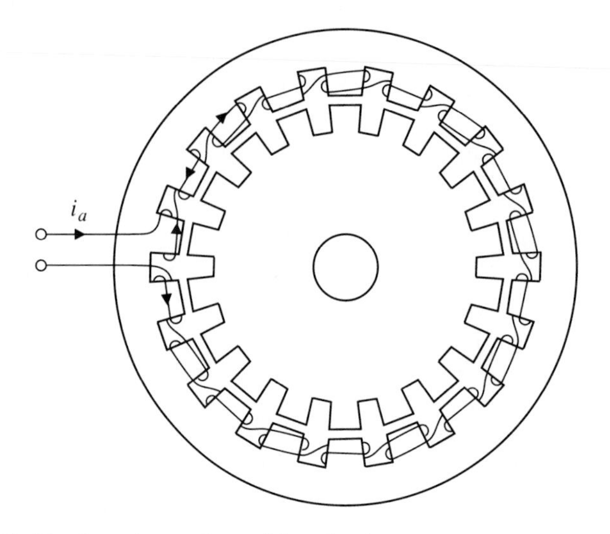

Figure 11.47 One phase of a multistack reluctance-type stepper motor.

this type of machine is that the rotor tends to remain in its last position when phase current is removed. For a reluctance machine to hold its position, a continuous phase current is required.

Stepper motor drives are particularly suitable for use in digital control systems. Usually, a mechanical load can be positioned to a desired angle simply by counting the number of applied pulses. In contrast with the position control sys-

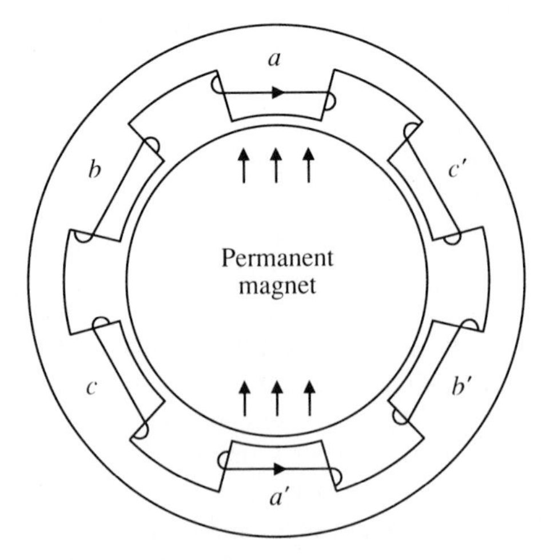

Figure 11.48 Permanent-magnet stepper motor.

tems discussed in earlier chapters, no rotor position sensor is required in most applications. The maximum rate at which the angular position of a stepper motor can be changed is governed by the ability of its rotor to remain in synchronism with the stator in much the same manner as described for a synchronous machine in Section 11.3.

PROBLEMS

11.1 The parameters for a three-phase, 440-V (L-L), 1200-r/min, 60-Hz, 10-kVA wound-rotor machine all referred to the stator turns are as follows:

$$R_s = R_r = 0.4 \quad \Omega \qquad L_m = 127 \quad \text{mH}$$

$$L_{\ell s} = L_{\ell r} = 7.43 \quad \text{mH} \qquad N_{se}/N_{re} = 0.65$$

This machine is to be used as a synchronous generator. Direct-field current is to be supplied to the rotor circuit, entering at phase A and leaving at phase terminals B and C connected together.

(a) Sketch a steady-state equivalent circuit of the T form, as in Fig. 11.4(a), inserting the values of all reactances, resistances, and the current ratio.

(b) Determine the field current required to produce rated terminal voltage on open circuit when the machine is driven at rated speed.

(c) Suppose the machine is to supply a power of 10 kW to a balanced resistive load at rated terminal voltage and frequency. Determine the required field current and the rotor angle.

(Section 11.1.1)

11.2 Steady-state analysis of the machine in Problem 11.1 can be somewhat simplified by use of the Γ form of equivalent circuit, as was shown for the induction machine in Fig. 10.8.

(a) Derive the values of the parameters for this equivalent circuit.

(b) Repeat (b) and (c) of Problem 11.1 using this model and satisfy yourself that the results are the same as those obtained using the T model.

(Section 11.1.1)

11.3 Another alternative equivalent circuit for the synchronous machine is the inverse-Γ form as shown in Fig. 11.7(b). Using the parameters given for the wound-rotor machine in Problem 11.1,

(a) Determine the value of the parameters for a steady-state inverse-Γ model.

(b) Repeat (b) and (c) of Problem 11.1 using the model and again satisfy yourself that the results are the same as obtained in Problems 11.1 and 11.2.

(Section 11.1.2)

11.4 Suppose the machine for which the parameters are given in Problem 11.1 is used as a synchronous motor. Its stator terminals are connected to a 440-V (L-L), 60-Hz, three-phase supply. The field circuit is connected to a 12-V battery that has negligible internal resistance. Stator resistance may be ignored.

(a) Select the most appropriate form of steady-state equivalent circuit from Problems 11.1, 11.2, and 11.3 and use it to derive the value of the torque produced by this motor as a function of the rotor angle β_0.

(b) Suppose the load torque is gradually increased until it has just exceeded the maximum value found in (a). Noting that the battery appears as a short circuit to all frequencies except zero, at approximately what speed will the load now be driven with the machine operating as an induction machine?

(Section 11.1.1)

11.5 Parameters and field connections of a 440-V, 10-kVA, 60-Hz wound-rotor machine have been given in Problem 11.1.

(a) Sketch a transient equivalent circuit of the *T* form shown in Fig. 11.2(b) and derive the value of its instantaneous current ratio n_f.

(b) Suppose the machine is driven as a generator at rated speed with its stator terminals open circuited. A direct-voltage source with negligible internal resistance is connected at time zero to the rotor field circuit. At what time constant will the field current rise toward its steady-state value?

(c) Repeat (b) assuming the direct-voltage source has an internal resistance of 0.5 Ω.

(d) Suppose the generator is loaded by a three-phase inductor that requires 10 kVA at essentially zero power factor with rated voltage and frequency. If the field source in (b) is connected at time zero, approximately what time will be required for the stator terminal voltage to reach 95% of its final rms value? The effects of stator resistance may be ignored.

(e) Sketch the waveform of the stator current space vector as a function of time for the condition of (d).

(f) Sketch waveforms of the three stator phase currents as a function of time for about one time constant for the same conditions as in (e).

(Section 11.1.1)

11.6 Repeat Problem 11.5 using a transient equivalent circuit of the inverse-Γ form shown in Fig. 11.7(b). Assure yourself that the results are the same as obtained in Problem 11.5.

(Section 11.1.2)

11.7 A 4.6-kV (L-L), three-phase, 60-Hz, 3-MVA synchronous generator, driven at rated speed with its stator terminals open circuited, gave the following stator voltage V_s (L-L)-field current i_f data:

i_f (A)	0	20	30	40	60	80	100	120
V_s (V)	50	1450	2200	2850	4080	4900	5400	5850

Let us assume that all of the significant saturation in this machine is in the rotor. The equivalent circuit model of Fig. 11.7 is therefore appropriate. In this circuit, the inductance L'_L is 2.45 mH/phase, the ratio n' relating field current to rms stator current is 15.1. The resistance measured between two stator terminals is 0.07 Ω. Supposing that generator is to supply a load of 2.5 MW at rated voltage and 0.9 power factor inductive or lagging, determine the required field current i_f and the load angle β_0.

(Section 11.1.2)

11.8 A salient-pole synchronous motor is rated at 10 MW, 13.8 kV (L-L), 60 Hz, and 720 r/min. Its direct-axis reactance is 24 Ω and its quadrature-axis reactance is 13 Ω. The stator resistance and the rotational losses may be ignored. A field current of 210 A is required to produce rated terminal voltage on no load.

(a) Determine the torque when the field current is 340 A and the angle β_0 is $-115°$.

(b) Estimate the maximum steady-state torque that the motor can supply with a field current of 340 A before synchronism is lost.

(Section 11.1.3)

11.9 A four-pole, 550-V (L-L), 60-Hz, three-phase reluctance motor has a direct-axis reactance of 8.0 Ω and a quadrature-axis reactance of 3.0 Ω as seen from its stator terminals. Stator resistance and rotational losses may be ignored.

(a) Determine the maximum shaft torque that this motor can produce when connected to a supply of rated voltage and frequency.

(b) Because the power factor of this machine is relatively low, it is desirable to operate it at its maximum power factor condition. Determine the value of this maximum power factor and the power output for this condition.

(Section 11.1.3)

11.10 A salient-pole synchronous generator has a direct-axis reactance of 1.0 pu and a quadrature-axis reactance of 0.6 pu. The generator is to be operated with a terminal voltage of 1.0 pu supplying a current of 1.0 pu to a load with a lagging power factor of 0.8.

(a) Determine the angle β_0 of the machine for this condition.

(b) Determine the field current required for this condition.

(Section 11.1.3)

11.11 Consider the use of a salient-pole synchronous machine as a synchronous reactor to act as an adjustable capacitive or inductive load on a power system. Suppose the machine has direct and quadrature-axis reactances X_d and X_q and has negligible losses and negligible mechanical load. It is connected to a power system with constant voltage V_s per phase.

(a) Suppose the field current is reduced to zero. Show that the complex power per phase is given by

$$\vec{S} = -j\frac{V_s^2}{X_d} \qquad \text{VA}$$

(b) As the field current is increased from zero, the inductive load is decreased. Show that it becomes zero at the value of field current

$$i_f = \frac{V_s}{nX_d} \qquad \text{A}$$

(c) A large inductive load is sometimes required on power systems with long transmission lines in order to control the system voltage. To increase the inductive load beyond the value obtained in (a), consider reversing the direction of the field current. The rotor will remain in synchronism as long as a deviation of rotor angle β_0 from $-\pi/2$ produces a restoring torque. Show that the machine can remain in synchronism until the field current is reduced to

$$i_f = -\frac{X_d - X_q}{X_d X_q}\frac{V_s}{n} \qquad \text{A}$$

(d) For the condition of (c), show that the complex power per phase is

$$S = -\frac{jV_s^2}{X_q} \qquad \text{VA}$$

(Section 11.1.3)

11.12 A 4.6 kV (L-L), three-phase, 3-MVA cylindrical-rotor synchronous machine is tested with variable 60-Hz voltage and with no mechanical load to determine its parameters. Tests were made with (i) near-zero stator current and (ii) rated stator current capacitive or overexcited. The results were as follows:

Field current (A)	Stator (L-L) voltage (V)	
	(i)	(ii)
20	1450	
30	2200	0
40	2850	750
60	4080	2140
80	4900	3350
100	5400	4200
120	5850	4800

(a) Determine the leakage inductance L'_L and the current ratio n' for an inverse-Γ type equivalent circuit for this machine.

(b) Draw a graph relating the rms magnetizing flux linkage Λ'_M to the rms magnetizing current I'_M.

(c) Determine the field current required to operate the machine as a motor with rated load at 0.8 power factor leading.

(Section 11.1.4)

11.13 Suppose the machine described in Problem 11.12 is to be used as a synchronous reactor. It has no mechanical load and its losses are negligible. It is connected to a three-phase, 50-Hz power system with a voltage of 4.0 kV (L-L). What field current is required when the machine is acting as a 2.5-MVA capacitive load on the system?

(Section 11.1.4)

11.14 The wound-rotor machine for which parameters are given in Problem 11.1 is to be used as a synchronous generator. Its field circuit is connected with the field current entering at phase A and leaving at phase terminals B and C connected together.

(a) Sketch a transient equivalent circuit of the inverse-Γ form, inserting the value of all parameters.

(b) Determine the open-circuit field time constant τ_f, i.e., with the stator open circuited.

(c) Determine the transient time constant τ', i.e., the field time constant with stator short circuited and R_s ignored.

(d) Determine the armature time constant τ_a.

(e) Suppose the generator is initially supplying rated load at unity power factor. A three-phase short circuit occurs at time zero. Derive the stator current space vector following the short circuit. Estimate the maximum short-circuit current that will flow in any phase.

(Section 11.2)

11.15 A salient-pole, 60-Hz synchronous generator has the following parameters expressed in per unit of rating:

$$X_d = 1.25, \quad X_q = 0.7, \quad X_d' = 0.3, \quad X_d'' = X_q'' = 0.17$$

Its time constants are:

$$\tau_f = 6.25 \text{ s}, \quad \tau' = 1.5 \text{ s}, \quad \tau'' = 0.045 \text{ s}, \quad \tau_a = 0.15 \text{ s}$$

Develop direct and quadrature-axis equivalent circuits of the Γ form for this generator, including the per unit values of all resistances and inductances, and note any assumptions made.

(Section 11.2)

11.16 A synchronous motor has a synchronous reactance of 0.9 pu and a transient reactance of 0.35 pu. Its stator resistance may be neglected, and it may be considered to be magnetically linear. The motor is initially synchronized to a balanced supply of 1.0 pu voltage by adjusting its field current to the value that produces an open-circuit voltage of 1.0 pu.

(a) With the field current left unchanged, load torque is gradually applied. Find the maximum per unit torque that can be applied without losing synchronism.

(b) Let us assume that the transient time constant is large. Estimate the maximum torque that can be applied suddenly after initial synchronization without losing synchronism. It can be assumed that this torque will be applied for a period that is much less than the transient time constant.

(c) Suppose the load torque is gradually increased to 1.0 pu and the field current adjusted to give an input power factor of unity. Approximately how much more additional torque can be added suddenly without losing synchronism?

(Section 11.3)

11.17 A 4.5 MW, 1000 r/min synchronous motor is supplied from a 13.2 kV (L-L), three-phase, 50-Hz distribution line. The motor has a transient reactance of 14 Ω, and the transient time constant may be considered to be long. Its stator resistance may be ignored. The combined polar moment of inertia of the motor and its mechanical load is 5000 kg·m^2. Initially, the motor is supplying a mechanical power of 5000 kW to its load, and the field current is adjusted to obtain unity power factor at its stator terminals. When lightning causes a temporary flashover on the distribution line, the circuit breakers connecting the motor to the line open for a period of 0.3 s to allow the fault to clear and then reclose.

(a) Derive the torque angle relationship for the motor prior to the fault.

(b) Determine the initial rotor angle.

(c) Determine the rotor angle after operating for 0.3 s without electric supply.

(d) Assuming that the load torque remains unchanged, will the motor remain in synchronism after the reclosure?

(Section 11.3)

11.18 A two-pole, 1000-MVA, 50-Hz, 16-kV (L-L) synchronous generator is brought up to rated speed, its field current is adjusted to give rated terminal voltage, and it is synchronized to a 16-kV power system that can be assumed to have negligible internal impedance. The transient reactance of the generator is 0.4 pu based on its rating. Losses in the generator may be ignored. The polar moment of inertia of the generator and its driving turbine is 10^4 kg·m^2.

(a) Determine the angular velocity of the speed oscillation that follows application of a small-step increment of turbine torque.

(b) When this generator is connected to the power system and operated with a shorted field circuit, it operates as an induction generator, producing an output power of 15 MW at a speed of 0.990 pu of synchronous speed. For the condition of (a), determine the angular velocity and the relative damping factor of the speed oscillation following an incremental torque change.

(c) Suppose that, because of improper design of the control gear on the turbine valves, the turbine torque T contains an oscillating term in addition to its average torque, i.e.,

$$T = 5 + 3 \sin t \qquad \text{kN·m}$$

Determine the peak-to-peak value of oscillation in the rotor angle β_0. Include the damping effect of (b).

(Section 11.3)

11.19 A two-pole aircraft motor has a three-phase stator winding with 24 sinusoidally distributed turns per phase. The rotor has a pair of permanent magnets arranged as shown in Fig. 11.28. The magnet material is neodymium-iron-boron having a demagnetization curve of the form shown in Fig. 1.18 with a residual flux density of 1.15 T and a relative recoil permeability of 1.05. The rotor radius is 40 mm and its length is 60 mm. The air-gap length is 0.5 mm and the radial thickness of the magnet is 5 mm. The magnets cover two-thirds of the rotor surface.

(a) Determine the average air-gap flux density above each magnet.

(b) Determine the rms value of the fundamental space component of air-gap flux density.

(c) Derive the equivalent current source and the magnetizing inductance for a steady-state equivalent circuit for this motor.

(d) If this motor is rotated at 24 000 r/min, what is its open-circuit terminal voltage per phase?

(e) Suppose the magnet material must always have a flux density greater than zero to prevent demagnetization. What is the largest rms value of stator current that can be safely supplied to this machine?

(f) With the stator current of (e) and a supply frequency of 400 Hz, what is the mechanical power and torque produced by the machine? Losses may be ignored.

(Section 11.4.1)

11.20 An eight-pole, permanent-magnet motor is constructed with circumferential magnets arranged as shown in Fig. 11.34. The permanent-magnet material is a ferrite with a residual flux density of 0.35 T and an incremental relative permeability of 1.06. The thickness of the magnets is 8 mm. The rotor radius is 25 mm and the effective air-gap length is 0.5 mm. The magnets have the maximum possible width for their thickness.

(a) Determine the average flux density in the air gaps above the iron poles.

(b) Determine the rms value of the fundamental component of the air-gap flux density.

(Section 11.4.3)

11.21 A method of assessing the torque-producing capability of a hysteresis motor is to determine the power dissipated as hysteresis loss in the rotor when operating at a speed below synchronism. From this loss power, the air-gap power and thus the torque may be estimated. The estimate will be higher than can be practically achieved because of loss factors that have been ignored. Consider a two-pole, 400-Hz, three-phase motor with a rotor consisting of a cylindrical ring of permanent-magnet material of radius 20 mm, length 40 mm, and thickness 5 mm mounted on a solid iron core. The permanent-magnet material has a hysteresis loop that may be approximated by an idealized model of the form shown in Fig. 1.17 with $B_r = 1.2$ T and $H_c = 10$ kA/m. Suppose the stator voltage and current are sufficient to cause the whole volume of permanent-magnet material to make a complete excursion around the loop for each revolution of the magnetic field relative to the rotor. Eddy current losses may be ignored.

(a) Determine the hysteresis loss in the rotor at standstill.

(b) Determine the torque due to this hysteresis effect.

(c) Show that the torque is independent of the rotor speed below its synchronous value.

(d) Determine the maximum mechanical power that the motor can produce when operated at rated frequency.

(Section 11.4.4)

11.22 A 375-kW, 720-r/min, 2300-V, 60-Hz, three-phase synchronous motor has a magnetizing inductance of 30 mH. A field current of 15 A produces a rated terminal voltage when driven at rated speed at no load. The motor is supplied from an inverter with controllable, near-sinusoidal voltage and variable frequency. The mechanical load is a pump that requires a torque of

$$T_L = 0.75\omega_{mech}^2 \qquad \text{N}\cdot\text{m}$$

Losses in the motor may be ignored.

(a) When driving the load at a speed of 720 r/min, the inverter is adjusted to give rated flux linkage and the field current is adjusted to give unity power factor. Determine the values of the stator flux linkage (rms), the stator current, and the field current.

(b) Suppose the speed is to be reduced to 240 r/min, keeping the stator flux linkage and the field current at the same values as found in (a). Determine the stator voltage, the stator current, and the power factor.

(c) To what value should the field current in (b) be reduced to achieve unity power factor?

(Section 11.4.1)

11.23 A synchronous motor has the following ratings: 4.16 kV (L-L), three phase, six pole, 60 Hz, 10 MW, $L_M = 12.75$ mH. All losses may be ignored. When operated as an open-circuited generator at rated speed, a field current of 50 A produces rated terminal voltage. The motor is to drive a compressor requiring a torque of 75 kN·m at a speed of 1000 r/min. The motor is to be supplied from a simple voltage-source inverter which in turn is supplied from a controlled rectifier. To provide natural commutation of the inverter, the fundamental power factor of the motor is to be 0.95 leading. Determine the required values of the direct-link voltage, the frequency, and the field current.

(Section 11.5.1)

11.24 A drive consists of a synchronous motor supplied by a simple voltage-source inverter which in turn is supplied from a controlled rectifier, as shown in the system of Fig. 11.36. The motor rating is 75 kW, 1200 r/min, 440-V (L-L), 60 Hz, three phase. Its magnetizing reactance at rated frequency is 2.2 Ω, and its losses are negligible. The field current ratio n is 12. The available utility supply is at 460 V (L-L), 60 Hz, and three phase.

(a) Determine the maximum value of motor speed up to which this drive can provide its rated value of torque.

(b) Assuming unity power factor operation, find the required field current for the condition of (a).

(c) Suppose the speed is to be increased to 2700 r/min. If the rated stator current of the motor is not to be exceeded, what is the maximum available torque and mechanical power at this speed?

(d) What should be the field current for the condition of (c)?

(e) What should be the firing angle of the rectifier to provide rated torque at a speed of 300 r/min?

(Section 11.5.1)

11.25 A switched reluctance motor of the form shown in Fig. 11.43(a) has four rotor poles and six stator poles. All poles are 30° in arc width. The flux linkage-current relations for a typical stator phase are shown in idealized form for the maximum- and minimum-reluctance rotor position in Fig. 11.49.

(a) Assuming ideal switching action in a power converter of the form of Fig. 11.43(b), find the average torque that can be produced by this motor with a supply current of 4 A.

(b) Ignoring losses, what supply voltage would be required to provide the torque of (a) at a speed of 300 rad/s?

(Section 11.6)

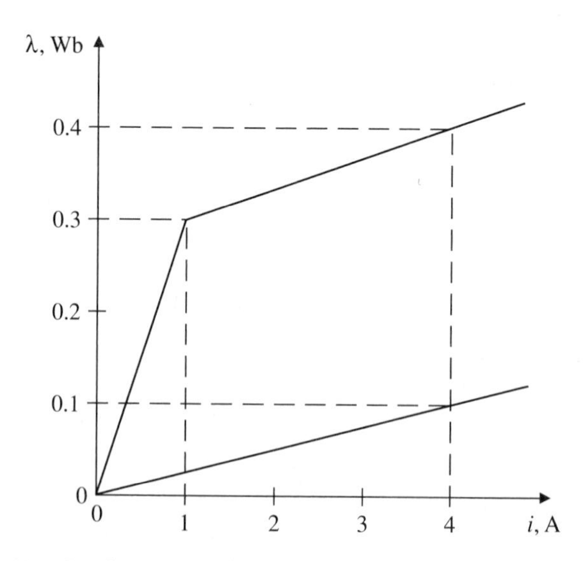

Figure 11.49 Flux linkage-current relations for switched reluctance motor of Problem 11.25.

11.26 A switched reluctance motor drive has a motor of the configuration shown in Fig. 11.43(a) and a drive circuit as shown in Fig. 11.43(b). Each phase winding, consisting of two coils in series, has a resistance of 10 Ω. When the rotor poles are aligned with the stator poles, the phase inductance is 20 mH. When the rotor poles are between phases, the inductance is 3 mH. The direct-supply voltage is 48 V.

(a) Estimate the time required for the phase current to reach 95% of its final value after the phase is switched on.

(b) Assuming a reasonably low rotor speed, determine the energy stored in each phase at the instant when the poles are aligned for that phase.

(c) If the phase switches are turned off at the instant of alignment, what time is required for the phase current to decay to zero?

(d) Determine the proportion of the stored energy in the phase that is returned to the supply.

(Section 11.6)

11.27 A stepper motor is required with a step size of 2°. Consider the stepper motor structure illustrated in Fig. 11.46 and Fig. 11.47 and determine several appropriate combinations on numbers of stator and rotor poles and numbers of stacks to achieve this step size.

(Section 11.7)

GENERAL REFERENCES

Acarnley, P.P., *Stepping Motors: A Guide to Modern Theory and Practice*, 2nd ed., Peregrinus, London, 1982.

Dewan, S.B., Slemon, G.R., and Straughen, A., *Power Semiconductor Drives*, John Wiley and Sons, New York, 1984.

Gross, C.A., *Power System Analysis*, John Wiley and Sons, New York, 1979.

Kenjo, T., *Stepping Motors and Their Microprocessor Controls*, Clarendon Press, Oxford, 1984.

Krause, P.C., *Analysis of Electric Machinery*, McGraw-Hill, New York, 1986.

Levi, E., *Polyphase Motors*, John Wiley and Sons, New York, 1984.

Miller, T.J.E., *Brushless Permanent-Magnet and Reluctance Motor Drives*, Clarendon Press, Oxford, 1989.

Stevenson, W.D., *Elements of Power System Analysis*, 3rd ed., McGraw-Hill, New York, 1977.

Walker, J.H., *Large Synchronous Machines*, Clarendon Press, Oxford, 1981.

APPENDIX A

SI Unit Equivalents

Property	*SI Unit*	*Equivalents*
Length	1 meter (m)	3.281 feet (ft) 39.37 inch (in.)
Angle	1 radian (rad)	57.30 degrees
Mass	1 kilogram (kg)	2.205 pounds (lb) 35.27 ounces (oz) 0.0685 slugs
Force	1 newton (N)	0.2248 pounds (lbf) 7.233 poundals 10^5 dynes 102 grams force
Torque	1 newton-meter (N·m)	0.7376 pound-feet (lbf·ft)
Moment of inertia	1 kilogram-meter^2 (kg·m^2)	23.7 pound-feet^2 (lb·ft^2) 5.46×10^4 ounce-inches^2 0.738 slug-feet^2 10^7 gram-centimeter^2 (g·cm^2)
Energy	1 joule (J)	1 watt-second 0.7376 foot-pounds (ft·lb) 2.778×10^{-7} kilowatt-hours 0.2388 calorie (cal) 9.48×10^{-4} British thermal unit (BTU) 10^7 ergs
Power	1 watt (W)	1 joule/second 0.7376 foot-pounds/second 1.341×10^{-3} horsepower (hp) 3.412 BTU/hour
Resistivity	1 ohm-meter (Ω·m)	10^8 micro-ohm/centimeter 6.015×10^8 ohm-circular mil/ft
Magnetic flux	1 weber (Wb)	10^8 maxwells or lines 10^5 kilolines

continued

Property	*SI Unit*	*Equivalents*
Magnetic flux density	1 tesla (T)	1 weber/meter^2 10^4 gauss 64.52 kilolines/inch^2
Magnetomotive force	1 ampere(turn) (A)	1.257 gilberts
Magnetic field intensity	1 ampere/meter (A/m)	2.54×10^{-2} ampere/inch 1.257×10^{-2} oersted
Frequency	1 hertz (Hz)	1 cycle/second

APPENDIX B

Physical Constants

Quantity	*Symbol*	*Value*	*Unit*
Magnetic constant	μ_0	$4\pi \times 10^{-7}$	N/A^2 or H/m
Electric constant	ϵ_0	8.854×10^{-12}	$C^2/N \cdot m^2$ or F/m
Acceleration of gravity	g	9.807	m/s^2
Electron mass	m_e	9.1×10^{-31}	kg
Proton mass	m_p	1.67×10^{-27}	kg
Quantum of charge	Q_e	1.603×10^{-19}	C
Bohr magneton	p_B	9.27×10^{-24}	$A \cdot m^2$

APPENDIX C

Resistivity and Temperature Coefficient of Resistivity of Some Conductive Materials

Material	ρ_{20} *(ohm-meter)*	α_{20} $(°C)^{-1}$
Copper (annealed)	1.72×10^{-8}	3.93×10^{-3}
Copper (hard drawn)	1.78×10^{-8}	3.82×10^{-3}
Aluminum	2.7×10^{-8}	3.9×10^{-3}
Sodium	4.65×10^{-8}	5.4×10^{-3}
Nickel	7.8×10^{-8}	5.4×10^{-3}
Lead	2.2×10^{-7}	4.0×10^{-3}
Tungsten	5.5×10^{-8}	4.5×10^{-3}
Iron	9.8×10^{-8}	6.5×10^{-3}
Mercury	9.58×10^{-7}	8.9×10^{-4}

APPENDIX D

Wire Table

Gauge No. (B & S)	Diameter	
	Millimeters	Inches
0	8.252	0.3249
1	7.348	0.2893
2	6.543	0.2576
3	5.827	0.2294
4	5.189	0.2043
5	4.620	0.1819
6	4.115	0.1620
7	3.665	0.1443
8	3.264	0.1285
9	2.906	0.1144
10	2.588	0.1019
11	2.304	0.0907
12	2.052	0.0808
13	1.829	0.0720
14	1.628	0.0641
15	1.450	0.0571
16	1.290	0.0508
17	1.151	0.0453
18	1.024	0.0403
19	0.912	0.0359
20	0.813	0.0320

Gauge No. (B & S)	Diameter	
	Millimeters	Inches
21	0.724	0.0285
22	0.643	0.0253
23	0.574	0.0226
24	0.511	0.0201
25	0.455	0.0179
26	0.404	0.0159
27	0.361	0.0142
28	0.320	0.0126
29	0.287	0.0113
30	0.255	0.01003
31	0.227	0.00893
32	0.202	0.00795
33	0.180	0.00708
34	0.160	0.00630
35	0.142	0.00559
36	0.127	0.00500
37	0.114	0.00449
38	0.102	0.00402
39	0.089	0.00350
40	0.079	0.00311

ANSWERS TO PROBLEMS

Chapter 1

1.1 a) 80 μT, b) 320 μT. **1.2** a) 6630, 3980 A/m, b) 2.554 μWb, c) 6.385, 6.25 mT. **1.3** 255 V. **1.4** a) 4.89×10^8 A/Wb, b) 38.71 Ω. **1.5** a) 1620, b) 2.37 mm, 2.17 Ω. **1.6** a) 15.54 J/m^3, b) 1.6 mJ. **1.7** a) 0.329 H, b) 0.0533 J. **1.8** a) 0.64 T, b) 1.41 T. **1.9** a) 1.51 T, b) 60 kA/m. **1.10** a) 1.56 A, b) 3180, 1.64 H. **1.11** a) 43.5 V, b) 67.5 mA, c) 43.6 V. **1.12** 8 V. **1.13** a) 40.3 J/m^3, b) 0.22 J, c) 4.4 W, d) 193. **1.14** a) 379 W, b) 251 W. **1.15** a) 2.93 ms, b) 0.55 A. **1.16** a) 500 V, b) 0.095, 0.0713, 0.0596 A. **1.17** 0.0191, 0.0064, 0.0038 A. **1.18** 122 Ω, 53 Ω, 224 V. **1.19** 1.74 A. **1.20** 0.95 mWb. **1.21** a) 1.42 mm, b) 0.33 A. **1.22** 48.9 mm, 4.2 mm. **1.23** a) 0.34×10^7, 11.9×10^7 A/m, b) 23.2 mH, c) 23.6 mH, 1.4%. **1.24** a) 0.212, 0.292, 0.159 MA/m, b) 0.102, 0.0716 Wb.

Chapter 2

2.1 b) 3. **2.2** a) 575 V, 17.39 A, 43.48 A, b) 5.29 Ω. **2.3** 2.88∠25.84 Ω. **2.4** b) 3.6 H, 540 Ω, 4.17 μF. **2.5** a) 7.5, b) 177, 23.6 V, 4.69, 35.2 A. **2.6** 11.18. **2.7** a) 13.75, 2.2 Ω, c) 110 V, d) 0.098. **2.8** 0.075 H. **2.9** a) 0.932 Ω, 0.78 H, b) 64.72 Ω, 0.49 H. **2.10** a) 5.5 A, b) 177 W, c) 39.8 Ω, 0.127 H, 226 Ω. **2.11** 0.44 m^2. **2.12** 337 kVA. **2.13** a) 2300 V, 87 A, 26.45 Ω, 70.2 mH, b) 0.0117, 0.068, 48.28, c) 0.0207, 0.0197. **2.14** 0.865 Ω, 1337 Ω, 1936 Ω, 2.27 Ω. **2.15** a) B, b) $3212. **2.16** a) 0.59, 0.967, b) 0.962. **2.17** 0.0387. **2.18** a) 0.976, b) 0.0389. **2.19** a) 0.0895, b) 11.17.

Chapter 3

3.1 51 μA. **3.2** 853 N out. **3.3** 419 N/m. **3.5** 780 N. **3.6** b) No, c) Yes, d) 55.9 N. **3.7** 14.26 N. **3.8** a) 0.38 N·m, b) 1.19 V, c) 47.8 W. **3.9** 18.9 mV. **3.10** a) 0.96 V, b) 0.16 MN/m^2 or MPa, c) 1.06 V, d) 2120 W. **3.11** 1.41 MW. **3.13** a) 11.88 V, b) 12.56 kA, c) 149.1 kW. **3.14** Yes. **3.15** a) 50 Hz, b) 123.1 V. **3.16** c) 0.795 A. **3.17** a) 0.329 H, b) 0.18 N, c) 5.33 N. **3.18** b) 6.28 N, c) 6280 Pa or N/m^2. **3.19** b) 853 N, c) 64.1 N. **3.20** a) 503 N, b) 758 N, c) 537 N, 732 N. **3.21** a) 4.19 mm, b) 5730 N, c) 581 kg, d) 9.3 m/s^2. **3.22** 11.6 N. **3.23** b) 3.53 mm, c) 8.9 J, d) 1.91 J. **3.24** b) 1.43 rad, 0.085 V. **3.25** a) 4.77 A, b) 0.915 N·m, c) 1.44 J. **3.26** b) 0.597 A, c) 3.75 N·m/A. **3.27** a) 0.0275 cos 377t, b) ±377 rad/s, c) 14.35 N·m, d) 5.41 kW. **3.28** a) 314, 157 rad/s, b) 0.375, 0.5 N·m, c) 117.8, 78.5 W.

Chapter 4

4.1 7.85 mWb/A. **4.2** 4.2 mm. **4.3** a) 13.6 mWb, b) 12.6 mWb and 1.11 A.
4.4 a) 141.3, b) 469 V, c) 149.2 N·m. **4.5** a) 0.114 Wb, b) 47.9 kN·m.
4.6 82.6 rad/s. **4.7** a) 450 V, b) 859 N·m, c) 90 kW.
4.8 a) 60 Ω, 34.7 H, 0.6 Ω, 0.72 N·m/A^2. **4.9** 61.8 N·m. **4.10** 154.8 V.
4.11 57.1 rad/s. **4.12** a) 117.3 A, b) 168 N·m. **4.13** 279.9 rad/s. **4.14** a) 2.7, b) 166.7 A, c) 1.35. **4.15** a) 287.5 V, 6.25 kW, b) 460 V, 5 kW.
4.16 0.23, 0.11 Ω. **4.17** a) 2.01, b) 3.51 Ω. **4.18** a) 273.5 N·m, b) 148.3 rad/s.
4.19 10%. **4.20** 207.1 W. **4.21** 2.2, 3.1 A. **4.22** 91 V.

Chapter 5

5.1 a) 230.7 V, b) 1.31 H, c) 0.56 A. **5.2** 14 N·m. **5.3** a) 3.66 mm, b) 2.43 Ω, c) 18.14 rad/s, d) 3.55 kW. **5.4** b) 181.9 rad/s, c) 23.5 A.
5.5 a) 39.9 N·m, b) 39.9 N·m, c) 76.1 rad/s. **5.6** 68.23 A, 0.94.
5.7 1.47 A, 0.83, 3.71 N·m, 436.7 W. **5.8** 0.87, 2.6 A.
5.9 a) 0.66 Ω, 84 mH, 4.7 mH, 0.384 Ω, b) 290 N·m, c) 147.5 rad/s.
5.10 9.98 N·m, 476 rad/s. **5.11** a) 156.4 V, 21.3 Hz, b) 136.9 V, 18.67 Hz, c) 3.52 kW. **5.12** 310 V, 40.68 Hz. **5.13** a) 747.7 W, b) 0.872, c) 4.54 N·m.
5.14 0.22. **5.15** a) 76.3 N·m, 110 rad/s, b) 41.92, 47.06 N·m.
5.16 110, 100.5, 43.9 N·m.

Chapter 6

6.1 6, 40, 1200 r/min; 4, 26, 1800 r/min. **6.2** a) 0.135 H, 0.833, b) 2.62.
6.3 a) 16.85 A, b) 9.04 A, 0, c) 85.1 N·m, d) 14.9 A. **6.4** a) 16.85 A, b) 85.1 N·m. **6.5** a) 3.485 Ω, 3.092, b) 707 A, c) 3092 A.
6.6 a) 15.6 A, −106°, b) 17.91 A, −101.1°. **6.7** 20.6°. **6.8** a) 32.8 A, −110°, b) 11.52 kN·m, 869 kW. **6.9** 335 kVA. **6.10** a) 6.14 kN·m, b) 8.28 kN·m.
6.11 a) 104.9 V, b) 1.21. **6.12** a) 3.04 Ω, 11.93, b) 0.79, c) 8.07 mH, 0.79.
6.13 a) 9.0 A, b) 17.2 kVA, 9.71 A. **6.14** 117.8 V (L-N). **6.15** 82.5 V, 1257 rad/s.
6.16 a) 0.092 N·m, b) 0.96 A, c) 107. **6.17** 316.8 rad/s.

Chapter 7

7.1 c) 4.52 H, e) 8.4 V. **7.2** b) 830 Ω, c) 64.4 Ω. **7.3** b) 63.3 V, 494 A, c) 15.65 kW, d) No. **7.4** 9.6 Ω, 21.7 mH, 72.4 mH, 16 Ω. **7.5** b) 420 W.
7.6 b) 211 μF. **7.7** a) 230, b) 4.2 mm, c) 430. **7.9** 0.206 mH.
7.10 k^2, k^2, k^4, k^3, k, k^3, k^{-1}, k^3, k, k^{-1}, k, k, k, k^{-1}, k. **7.11** a) 34.9 mm, b) 3.99 A/mm^2, c) 3645, d) 1.02 mm. **7.12** a) 90, 392, b) 1.035 kVA, c) 26.2 W, d) 30 W, e) 0.946. **7.13** 65.9 mm, 2706, 150. **7.14** 21 mm, 1.05×10^{-4} m^3.
7.15 a) 0.0909, b) 15.2 Hz, 14.6 kHz. **7.16** 8.86 A, 5.21.
7.17 Δ/Y: 220, 190.5, 15.74, 18.18, 1.15, 0.87; Y/Δ: 381, 110, 9.09, 10.5, 3.46, 0.29.
7.18 1328 V, 6.02 A, 69.6 A. **7.19** a) 15.8, 424.4 kV, 2110, 78.5 A, b) 17.33 kV.

7.20 17.32 MVA. **7.21** b) $1.058\angle 19.1°$. **7.22** c) No. **7.23** c) Yes.
7.24 a) 125, 83.3 A, b) 200 kVA. **7.25** 98, 108 mm. **7.26** a) 160, b) 32.2 Ω,
c) 0.187%, d) 0.124°, e) High. **7.27** 2429 N. **7.28** a) 16 J, b) 10.2 N·m,
c) 1.7 N·m. **7.29** 18.4 N·m.

Chapter 8

8.1 a) 17.15°, b) 0.86. **8.2** a) 90, 31.27°, b) 0, 0.82, c) 5.88 kVA. **8.3** 31.8°.
8.4 3.47 mH. **8.5** 158.3°. **8.6** a) 51.8, 48.9°, b) 12.3 A, c) 0.627. **8.7** 1.02 mH.
8.8 a) 0.1 ms, b) 570 V, c) 7.5, 150 A, d) 6.0 A. **8.9** 0.27 ms. **8.10** a) −1.65 V,
b) 0.86 ms. **8.11** a) 428.8, 247.6 V, b) 449.1, 259.3 V. **8.12** 130. **8.13** a) 659.2 V,
b) 61.8 V, c) 883 V.

Chapter 9

9.1 a) 7.5, b) 4.82 kg·m^2, 0.89 N·m·s, c) 87.7 kW.
9.2 a) 4905 N, 33.3 rad/s, 400 rad/s, 122.6 N·m, 49.05 kW,
b) 9255, 33.3, 400, 231.4, 92.55, c) −6924, −13.33, −160, −173.1, 27.7.
9.3 24.82 kN, 1.42 kN·m. **9.4** a) 2.01 N·m/A, b) 0.091 N·m·s, c) 122 V,
d) 3.22 Ω. **9.5** a) 2.47, b) 61.8 N·m, c) 4.49 N·m/A^2. **9.6** a) 3.38 s, b) 144 Ω.
9.7 a) 9.0 F, b) 0.602 Ω, c) 0.376 s. **9.8** a) 138 rad/s, b) 115.6 rad/s.
9.9 a) 0.114 s, b) $17.24(1 - \epsilon^{-t/0.114})$, $233\epsilon^{-t/0.114}$, c) 0.0754, 0.0193 s.
9.10 a) 47.2 V, b) 61.1°, c) 82.2. **9.11** 18.9 s. **9.12** 8.07 s.
9.13 72.1 s^{-1}, 137 rad/s. **9.14** b) 114, c) 0.15 s. **9.15** a) 174 N·m, b) 0.21,
c) 0.934. **9.16** a) 0.62, b) 0.53. **9.17** b) 600 V, 327 A, 196.2 kW, c) 218, 109 A.
9.18 3.16 mH, 4720 μF. **9.19** a) 5.3 N·m, b) 85.9 rad/s. **9.20** b) 17.22 kW, 28.7 A,
c) 21.42, d) 2.54, 1.27. **9.21** a) 25.04°, b) 0.82, c) 56.4 rad/s, d) 13.8 A.
9.22 a) 53.6°, b) 101.3°, c) 0.57, d) No. **9.23** a) 61.1, b) 310 V, c) 200 A,
d) 175 A. **9.24** a) 12.87 Ω, b) 29.4 Ω. **9.25** 118.9 rad/s, 33.6 A.
9.26 a) 244.4 rad/s, b) 0.76, c) 334.4 rad/s.

Chapter 10

10.1 a) 3.93 mH, 75.7 mH, 0.473 Ω, b) 1.12 Ω.
10.2 b) $13.18\epsilon^{-j1.56}(\epsilon^{j377t} - \epsilon^{-t/0.34})$ A. **10.3** 750 N·m.
10.4 $101.6\angle -30°$, $j0.102$ Ω. **10.5** a) 0, 262 V, b) $j1.76$ Ω.
10.6 b) 30.6°. **10.7** a) 800, −1800 r/min, b) 73.3, 550 V, c) 60, 40 kW, 8, 12 kW.
10.8 a) 86.1 N·m, b) 325.8 N·m, c) 1038 r/min, d) 4843, e) 119.6 rad/s,
f) 2.89 Hz. **10.9** a) $3.27\angle -60.43°$, b) 7.32 kW, c) 5.18 kW, d) 2.14 kW,
e) 58.3 N·m. **10.10** a) 618.8 r/min, b) 110.7 μF.
10.11 a) 254 V, 12.1 A, 21 Ω, 60 Hz, 125.7 rad/s, 3073 VA, 9221 VA, 73.36 N·m,
b) 0.007, 0.205, 2.29, 0.019, c) 0.975, 0.835. **10.12** a) 310 V, 40.68 Hz, b) 0.84.
10.13 b) 14–16 Hz, c) 30–34 V. **10.14** a) 0.07 H, 0.156 Ω, b) 198.5 rad/s.
10.15 a) 50.36 Hz, b) 4955 V, c) 37.1°. **10.16** a) 161.5 V, b) 8.7 A,
c) 105.3 Hz, d) 78°. **10.17** a) 68.23 A, 0.94, b) 24.1, 12, 3 A, c) 0.92,

d) 101.8 N·m. **10.18** b) 239 N·m. **10.19** a) 1405 r/min, b) 199 V, c) Yes. **10.20** a) 6, b) 0.381, c) 102.6°, d) 48.6 A. **10.21** b) 5.39 A, c) 0.06 N·m. **10.22** a) 142.6, b) 0.067. **10.23** a) 9.6 m/s, b) 5170 N, 54.6 kW, 266 A, 0.713.

Chapter 11

11.1 b) 4.87 A, c) 13.7 A, −21.3°. **11.2** a) 50.7 Ω, 1.03.
11.3 a) 5.45 Ω, 45.3 Ω, 1.15. **11.4** a) −52.7 cos β_0, b) 123.7 rad/s. **11.5** a) 1.089, b) 0.336 s, c) 0.25 s, d) 0.357 s. **11.7** 75.3 A, −67.5°. **11.8** a) 106 kN·m, b) 189 N·m. **11.9** a) 167 N·m, b) 0.455, 28.1 kW. **11.10** a) −70.56°, b) 1.77.
11.12 a) 2.44 mH, 15.1, c) 105.5 A. **11.13** 122.9 A.
11.14 a) 0.4 Ω, 14.44 mH, 120.1 mH, 0.357 Ω, b) 0.338 s, c) 0.0361, d) 0.036 s, e) 117 A. **11.15** d: 0.003, 1.25, 0.395, 0.0007, 0.392, 0.026; q: 0.003, 0.7, 0.225, 0.015. **11.16** a) 1.11, b) 2.05, c) 1.4.
11.17 b) −0.312 rad, c) −1.17 rad, d) Yes. **11.18** a) 8.14 rad/s, b) 8.12 rad/s, 0.065, c) 0.01 rad. **11.19** a) 1.041 T, b) 0.81, c) 188.8 A, 372 μH, d) 176.5 V, e) 171.2 A, f) 90.65 kW, 36 N·m.
11.20 a) 0.667 T, b) 0.48 T. **11.21** a) 482.6 W, b) 0.192 N·m, d) 482.6 W.
11.22 a) 3.522 Wb, 80.64 A, 18.2, b) 442.5 V, 26.29 A, 0.34, c) 15.04 A.
11.23 4440 V, 314 rad/s, 160.5 A. **11.24** a) 138.2 rad/s, b) 12.64 A, c) 291.8 N·m, 82.5 kW, d) 9.46 A, e) 76.86°. **11.25** a) 1.91 N·m, b) 143 V.
11.26 a) 0.6 ms, b) 230 mJ, c) 1.4 ms, d) 60%.
11.27 For 11.46: 6 & 60, 8 & 45, 4 & 90; For 11.47: 2 & 90, 3 & 60, 4 & 45.

Index